Limit State Design
of Reinforced Concrete Structures

Limit State Design
of Reinforced Concrete Structures

P. Dayaratnam

Formerly Professor and Head of Civil Engineering, IIT Kanpur
Formerly Vice Chancellor of Jawaharlal Nehru Technological University, Hyderabad
Chairman, Southern Regional Committee, AICTE, Chennai

Oxford & IBH Publishing Co. Pvt. Ltd.
New Delhi

CBS

(A Unit of CBS Publishers & Distributors Pvt Ltd)

New Delhi • Bengaluru • Chennai • Kochi • Kolkata • Mumbai
Hyderabad • Jharkhand • Nagpur • Patna • Pune • Uttarakhand

CBS Publishers & Distributors Pvt Ltd
204 FIE, Patparganj Industrial Area, Delhi-110 092
E-mail: delhi@cbspd.com, cbspubs@airtelmail.in

© 2004. P. Dayaratnam
Last Reprint   2017
ISBN 978-81-204-1597-3

Printed at Chaman Enterprises, New Delhi.

4-Y17-11

Preface

Reinforced concrete has been dominating the construction industry for more than eight decades. Most engineers in construction and design of civil engineering have to have basic knowledge in concrete. The reinforced concrete is a compulsory subject to all civil engineering students. Codes of practice and specification on materials, design and construction are updated continuously. Research and developmental work on cement, concrete, reinforcement and construction is a challenge in developed and developing countries. Limit State Design was evolved during late sixties of the last millennium from the experience of working stress and ultimate strength design methods and now most popular method during the last two decades. Indian Code of Practice on Plain and reinforced Concrete (IS: 456-1978) introduced the method in 1978, while continuing the use of Working Stress Design. The fourth revision of the code in 2000 introduced a number of changes on specifications and emphasized durability considerations. The working stress design method is now moved to annexure of the code. This book aims at enlighten the students and practicing engineers on Limit State Design of Concrete Structures with specific emphasis on the design considerations on durability of concrete structures. The design is an art, depends on experience and innovation of the engineer. The author having been in the field of teaching, research and consultancy for more than forty years has evolved the book to suit the needs of the student and the practicing engineer. The book gives in depth understanding of the concepts in design, detailing and practicability of the design; and enables the designer to see the economics and optimality of the design. The continuous changes of the modern concrete practice is kept in mind in developing the theory and powerful illustrative examples.

The first chapter introduces the concepts in structural design. The student may not be able to appreciate the chapter in the first reading as he is yet to get the familiarity of the design but gets a depth into the art of design if read after the end of the course. The second chapter deals with concrete technology in nutshell. The concrete technology is a vast subject so an attempt is made to bring in most relevant topics for student and practicing engineer for quick reference for better design understanding. The chapter includes objective questions to enable the reader to get a good perspective of the subject. Something may look right but not quite, better options available to the engineer are made more apparent through objective questions. The third and fourth chapters deal with the strength design of beams of rectangular and non-rectangular sections. An introduction to the design of prestressed concrete structures is presented in chapter five, as civil engineer needs to have some fundamental knowledge in that aspect. Chapter six deals with the serviceability design of beams as emphasized by the code. The seventh chapter presents design of staircases, curved beams and ties that are often used in buildings. Chapter eight deals with the design of variety of concrete slabs and chapter nine presents an introduction to the yield line theory of concrete slabs. Yield line theory of slabs even though not practiced now, but the concept of the yield lines give a depth in the safety margins in slabs. It is also hoped that the method becomes operational soon.

Columns are very important members in any structure. The design of a column subjected to combined compression and bending involves long iterative equations of equilibrium. The student and the practicing engineer shouldn't be burdened with the long tedious calculations. The design of columns with reliable aids is presented in chapters ten and eleven thus avoiding the tedium of long computations. Concrete foundations such as footings, rafts, piles and pile caps are important members, so chapters twelve and thirteen present design of foundations. Integration of members in a system is important to ensure the overall soundness of a structure. Chapters fourteen and fifteen illustrate the design of industrial and multistory building frames. The importance of detailing at the joints is highlighted. Chapter sixteen presents design of retaining walls that are common in road and other constructions. Design procedure follows simple principles of statics and stability; but the overall safety and durability of a structure depends on the art of structural detailing of members and the interconnectivity. Many examples with illustrative sketches of reinforcement detailing are presented in all chapters. The author's earlier book on the design of reinforced concrete structures has been popular for many years. The working stress design of reinforced concrete is almost outdated so this is a natural evolution of the earlier book for modern method of design.

Very often, the reinforcement detailing is left to bar bender or mason. Life of a structure is curtailed by wrong detailing practices of bar bender and poor supervision. The book makes it a point to impress on the importance of reinforcement details. The market is flooded by variety of structural analysis computer software. Engineering colleges and the training centers provide training in structural analysis software. Software prints out results based on the input and idealization of the structural skeleton. Sometimes, engineers are misled by long output that may or may not be relevant to the actual structure. It is important that the user of the software is an experienced structural engineer who can model the structure to suit the software for reliable results.

—*P. Dayaratnam*

Contents

Concepts in Structural Design

1.1 INTRODUCTION

A structural member configuration is arranged based on the functional and architectural aspects of the building, site and environmental conditions. It is also influenced by the overall economy, construction time and sequence, available equipment and facilities, the background and training of the manpower. The structural skeleton includes column, floor slabs, beams, domes, staircases; the interconnecting joints, supports, lift wells and similar facilities that carry the loads. Once the skeleton and the construction materials are selected, the structural engineer makes preliminary estimate of the size of members. For example the thickness of a slab is controlled by safety and serviceability, constructability, cables and ducts embedded in the slab, acoustics and thermal conditions, and water or moisture impermeability.

Structural Analysis gives the forces and deformations of the members. The method of structural analysis depends on the codes of practice and available facilities. Most commonly applied method is the elastic and linear analysis. One must also realise that there is no uniqueness in sizes and details of the members even for a given skeleton and loads. Reasonable margins and variations are possible in any one problem. For a given size of the member, it may be possible to design the reinforcement to meet the safety and serviceability limits. The beam may be designed as under-reinforced or doubly reinforced within certain limits of the size of the beam. Any arbitrary size of member can't be manipulated with the reinforcement details, but often-reasonable margins do exist in most cases. In such cases, the cost of the element depends on the relative cost of the materials. Cost of steel in India is much more when compared with that of concrete.

The experience of architect or structural engineer helps the choice of the preliminary sizes of the members. The most optimal choice is not necessarily unique, and there is always a room for improvement in the design. The design is often referred as an art, and is true to a limited extent. An optimal or near optimal design solution is arrived by the experience and the intuitive nature of the designer. What is optimal at one location need not necessarily be optimal in all locations, what is optimal at one time may not be optimal at all times. Economy of a design solution depends on many factors, time and space coordinates and strategies. Some parameters are difficult quantify in money, and further the specifications on the durability of the structure are not fixed in time and space. Design, analysis and redesign are iterative to a certain level. The size and reinforcement details of a simply supported beam can be obtained directly from the bending moment and shear force without actually assuming the size, however the weight of the beam has to be assumed to include in the total bending moment acting on the beam. In general, the structural analysis demands member properties such as area of cross section, moment of inertii, torsional constants and the location of the centre of gravity, and the reference of the principal axes. Most engineers can acquire the art of structural design by integrating the design principles, economic considerations of material and methods of construction.

Computer software has taken over the analysis and even design of structures. However the modeling of the structure and the economic considerations are within the imaginative thinking of engineer. The so-called super-duper software may promise every thing but one shouldn't be carried away by such promises. The present software has no direct logical thinking. The results of any software depend on the input. The input includes the structural properties, material behaviour, load computations and their combinations and the method of analysis and design. A poor or improper modeling of the structure cannot be compensated by the computing power of the software however powerful it may be. The method of structural design has to go through well-established principles and they are:

1. Structural configuration model should suit the architectural configuration of the building,
2. Establish Structural skeleton, that is the location of columns, beams, slabs etc., and their connectivity,
3. Assume preliminary sizes of the members based on practical limitations,
4. Select basic loads based on the function and location of the building. The basic loads include dead weights, live loads, wind load wind coefficients, earthquake load coefficients etc.
5. Analyze the structure for critical forces in the members. The method of analysis and design depend on the acceptable code of practice. A number of computer software is available for structural analysis and design,
6. The member size and shape, the reinforcement details etc. are obtained by considering the forces and the code specifications,
7. If the member sizes converge to the preliminarily selected ones, the detailing of the reinforcement is carried out. Otherwise revise the member sizes and redo the structural analysis,
8. Quantity estimates.

Figure 1.1 indicates the milestones in structural design process.

1.2 STABILITY OF STRUCTURES

The stability and soundness of the structure are the primary aims of any method of design. Structure or any part of the structure shouldn't get damaged or even de-figured if the design loads act on the structure. Further, the structure should perform satisfactorily under the design loads.

A structure is said to be stable if it doesn't undergo large or unlimited deformations under designed or less than the designed loads. A structure may become unstable because of poor support conditions or poor structural configurations and strengths. A simply supported beam resting on roller supports on both sides is an example of the static instability of the structure. A force component in the direction of the rollers can't be restrained by the supports. Consequently the beam will move rigid bodily even for a very small load. This is called *static unstability*. This type of support condition can't satisfy the static equilibrium. The algebraic sum of the forces in three mutually perpendicular directions must be zero. And the algebraic sum of the moments about any three mutually perpendicular directions must also be zero. This static equilibrium condition can be expressed as:

$$\sum f_{ij} = 0, \text{ for } i = 1 \text{ to } n \text{ and } j = x, y, \text{ and } z \tag{1.1}$$

$$\sum M_{kj} = 0, \text{ for } k = 1 \text{ to } m \text{ and } j = x, y, \text{ and } z \tag{1.2}$$

Where
f_{ij} = i^{th} force vector acting in the jth direction, (j corresponds to the x or y or z directions respectively),

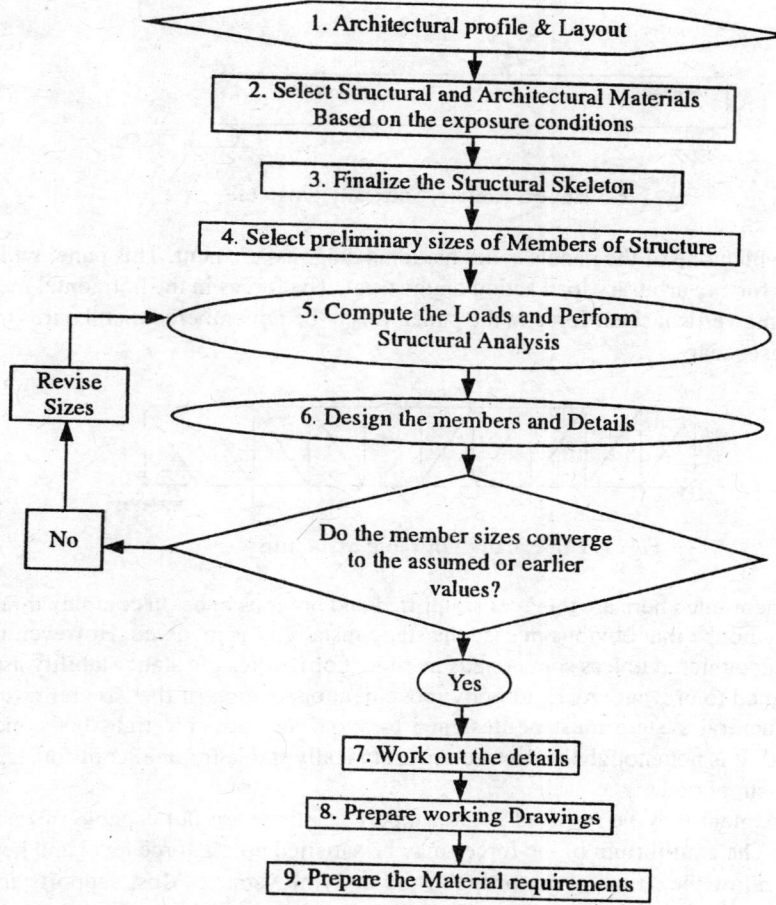

1. Architectural profile & Layout

2. Select Structural and Architectural Materials Based on the exposure conditions

3. Finalize the Structural Skeleton

4. Select preliminary sizes of Members of Structure

5. Compute the Loads and Perform Structural Analysis

Revise Sizes

6. Design the members and Details

No

Do the member sizes converge to the assumed or earlier values?

Yes

7. Work out the details

8. Prepare working Drawings

9. Prepare the Material requirements

Fig. 1.1 Milestones in Structural Design Process

$M_{kj} = k^{th}$ moment vector acting about the jth direction, (j corresponds to the x or y or z axes respectively),
n = total number of the forces in the system,
m = total number of the moment vectors in the system,
The Eq. 1.1 can be expanded for each of the principal directions as:

$$\Sigma f_{ix} = 0; \quad \Sigma f_{iy} = 0; \quad \Sigma f_{iz} = 0; \quad for\ i = 1\ to\ n \tag{1.3}$$

The moment equilibrium of Eq. 1.2 can be expanded similarly.

Static Stability: The static equilibrium conditions are the most powerful requirements to be satisfied for stability of a structure. The equilibrium in any three mutually perpendicular directions leads to equilibrium in any direction.

Figure 1.2 illustrates a beam on rollers that can be moved in the horizontal direction with any small force having a component in the horizontal direction. This is an over simplified example that illustrates the static instability of a structure. Another simple example of statically unstable structure is illustrated

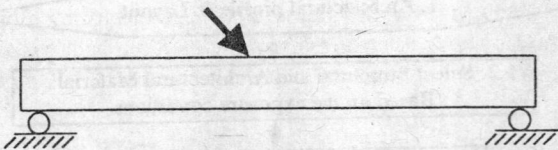

Fig. 1.2 Beam on Rollers, Statically Unstable

in Fig. 1.3 by truss in which one of the panels is not having a diagonal element. This panel will undergo undefined deformation for any arbitrary load acting on the truss. The forces in the horizontal members of the truss can't balance the vertical shear force in the panel. It is to be remembered that the truss members are not designed to resist shear.

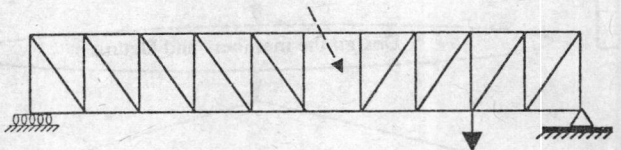

Fig. 1.3 Internally Unstable Structure

The two examples mentioned here are the over simplified and obvious ones. In complex three-dimensional structures, it may not be that obvious in case the static instability is involved. However, this is not a common instability encountered unless some one is negligent or ignores the static stability aspect. The supports must be designed to prevent any rigid body movement or rotation of the structure; further, the connectivity in the structural system must be designed to avoid any possible rigid body movements under any arbitrary load. It is not enough that a structure is statically stable for one set of forces, but must be suitably linked and supported.

The rigid body movement may be initiated if the support reactions are not capable of resisting the forces on the structure. The equilibrium of the forces may be satisfied upto a force level and beyond that the supports yield and allow the structure go through rigid body movement. Most supports are considered rigid when compared to the stiffness of the structure. Such an assumption is not always true.

Some structures such as dams, retaining walls, have continuum support of the ground. The force may be due to earthquake, wind, Hydraulic thrust or drag or a combination of the said set. Consider the case of a dam built across a river as shown in Fig. 1.4. The hydrostatic force causes overturning of the dam

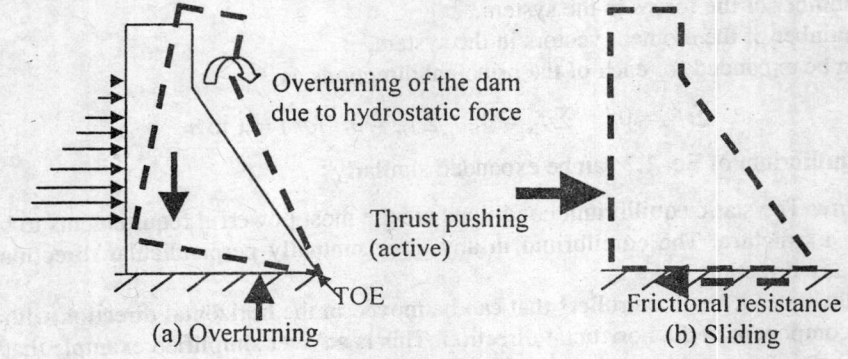

Fig. 1.4 Stability of Dam: Overturning and Sliding

while the weight of the dam provides the restoring couple. If the overturning moment by the active forces is greater than the stabilising (also called restoring) moment, then the dam will over turn about the toe of the dam. Similarly the pushing force is greater than the frictional resisting and other passive forces, the dam will tend to slide. The elastic deformations are very small when compared to the rigid body movements caused by the instability problem. This type of stability doesn't depend on the strength of the materials but on the mass and other static forces of the system. Adequate safety must be provided to prevent instability problem. The restoring moment must be greater than the overturning moment by a factor greater than one. The ratio of the restoring moment to the overturning moment is called factor of safety and it is taken in the range of 1.2 to 2.0 depending on the importance of the structure. In ordinary retaining walls the factor can be 1.5 in ordinary load conditions. A reduction factor of 0.9 is applied to dead loads that provide restoring effect to be on the safer side. A similar combination of factors of safety can be give to the stability against sliding.

The hydrostatic force on the dam is $= F_h = \gamma h^2/2$ and it acts at $h/3$ from base

Where h = head of water and γ = density of water
The overturning moment of the hydrostatic force is:

$$M_o = F_h\, h/3 = \gamma h^3/6$$

The restoring moment due to the weight of the dam is:

$$M_r = Wa$$

Where W = weight of the dam acting at a distance of a from the toe of the dam, then the static stability criterion is given by:

$$M_r > M_o$$

The effect of the soil pressure either active or passive is neglected in the assessment for first order accuracy and for simplicity of explanation. The assessment of the stability criterion is expressed by factor of safety.

$$Factor\ of\ Safety\ against\ overturing = \frac{M_r}{M_o}$$

Similarly the factor of safety against sliding of the dam is given by:

$$Factor\ of\ Safety\ against\ sliding = \frac{\mu W}{F_h}$$

In which μW = is the frictional resistance of against sliding offered by the ground,
$\qquad \mu$ = Coefficient of friction between the ground and the dam.
Most simple expressions are given here to illustrate the static stability concept, one must.consider all the forces acting in the system in computing the factor of safety.

Geometric Stability

The geometry of a structure may be such that the equilibrium of forces is not satisfied in the original configuration of the system but may be met in the deformed state of the structure.

A finite deformation is produced to satisfy the equilibrium of the forces. Take the case of a two-link chain connected to rigid supports as shown in the Fig. 1.5. A concentrated load acts at the middle hinge.

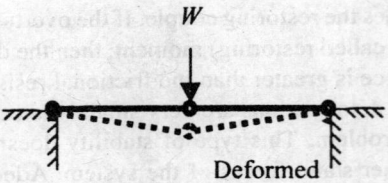

Deformed

Fig. 1.5 Geometric Instability

The central load placed at the middle joint can be resisted by vertical components of the two links on either side of the load. The equilibrium is not satisfied if the link members on either side of the joint are horizontal. The load forces the joint to move down through a finite distance until the connecting links are able to carry the load through their vertical components. The connecting links get reoriented in the inclined directions providing equilibrium of forces at the joint. From now onwards the structure is stable but not at the original configuration. This type of structure is called geometrically unstable structure, and such structures may be used only for temporary purposes.

Structural Stability

A structure deform in a unique shape upto a point, as the loads increase there comes a load when the equilibrium of the structure is satisfied for an undefined magnitude of the deformation. At such a *critical load,* there is no defined magnitude of the deformation. *Leonhard Euler (1707 – 1783)* is the first one who identified this multiple positions of equilibrium state of simple column and called the phenomenon as *elastic buckling* of a column. Buckling of column is a phenomenon at which a perfectly straight column that is under axial load can have an equilibrium state in a laterally deformed configuration also. The buckling is feasible in structures that are subjected to direct or bending compression. A beam bending can buckle laterally and such load is called *lateral buckling load.* Portal frames, shell structures, arches etc have buckling loads and they must be designed to be safe against buckling. Fig. 1.6 illustrates the buckling phenomenon of a simply supported column loaded with axial load. The mode of buckling shown is called the first mode and there are infinite possible modes of the buckling of the column. For every mode shape, there is a unique buckling load.

Fig. 1.7 illustrates a possible buckling mode of a portal frame. Several possible buckling modes of the frame exists depending on the relative stiffness of the members. The buckling mode indicated in the

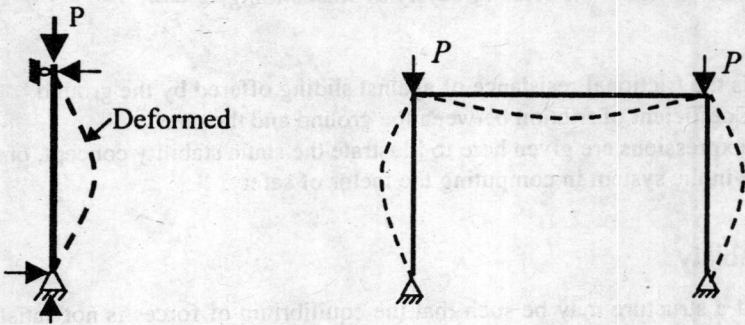

Fig. 1.6 Column Buckling   **Fig. 1.7 Frame Buckling**

figure is probably the lowest mode corresponding to the lowest buckling load. Depending on the relative stiffness of the beam and columns, three more alternatives exist and they are shown in Fig. 1.8. The lateral buckling of the beam introduces another type of buckling leading to twisting of the frame. All structures and elements of the structure must be designed to be structurally stable.

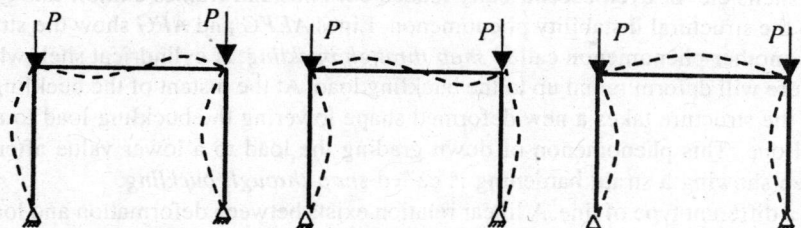

Fig. 1.8 Other Three Possible Buckled Modes of Frame

Strength Stability

A structure that is statically and geometrically stable resists the loads and deforms gradually in a unique deformed shape having a one to one correspondence with the load. The load deformation relation can be linear or even non-linear depending on the properties of the material and the structure. Figure 1.9 illustrates different stability or instability problems associated with a structure. The horizontal axis indicates the deformation and the load axis is vertical.

Line AB is a horizontal line starting from the origin or from zero loads and it indicates the static instability of a structure. Line AC is almost flat starting from zero level indicating a large deformation to start with, and then line CD is like strain-hardening phenomenon. It is load-deformation in the deformed position of the structure. Line ACD illustrates the geometric instability process. Geometrically unstable structures are not suitable as permanent structures, however they are often used for temporary purpose. Line $AEFG$ indicates increase of load without any deformation and then sudden deformation without

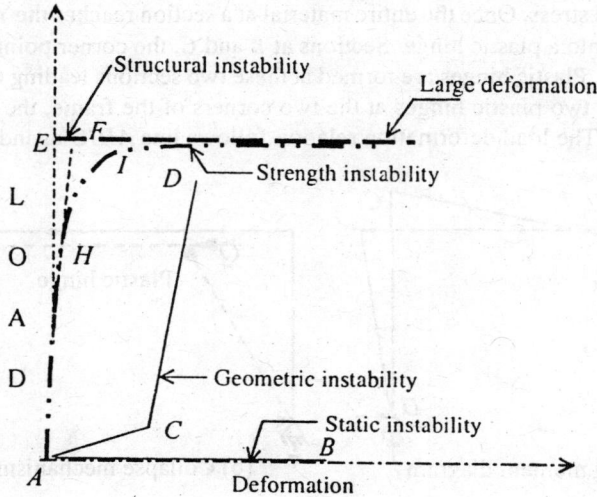

Fig. 1.9 Illustration of Types of Instabilities

further increase in the load. This type of behaviour is the column buckling and the point E is called the *bifurcation point*. The structure all of a sudden goes into large unbounded deformation and this indicates structural instability of some structures. There are some structures, which deform gradually upto a point and then undergo unbounded deformation indicating the instability and this is shown by line AFG. Arches, frames, shells etc. or even eccentrically loaded columns and frames exhibit this type of behaviour. This is also the structural instability phenomenon. Lines $AEFG$ and AFG show the structural instabilities. There is another phenomenon called *snap through buckling*. A cylindrical shell when subjected to external pressure will deform radial up to the buckling load. At the instant of the buckling of the shell, the geometry of the structure takes a new deformed shape lowering the buckling load to a value lower than the original one. This phenomenon of down grading the load to a lower value after initiation of buckling and then showing a strain hardening is called *snap through buckling*.

Line $AHIG$ is a different type of line. A linear relation exists between deformation and load up to point H. The material in the structure is elastic and linear upto this point. One or more locations start yielding, then non-linearity in the deformation sets in. The deformation increases faster than the load. More and more points in the structure yield, some sections also yield, and then from here onwards rapid rate of increase in the deformation takes place, finally the deformation line becoming asymptotic with the horizontal line. This line is the one that shows the strength Limit State and may also called as *strength instability*. Even though large deformation takes place as the load increases beyond the yielding of a section, the deflection is uniquely defined. However, the structure generates into a mechanism leading to large deformation and collapse. A mechanism formed out of plastification of one or more sections in a structure is called *Plastic mechanism*. The deformation is said to be large when its magnitude is large when compared to the elastic deformations and in fact it is comparable to the geometry of the structure. Elastic deformations are usually small when compared to the member sizes.

Fig. 1.10(a) illustrates a portal frame having two hinged supports and subjected to a horizontal load at top. It also indicates the bending moment diagram. As the load, Q increases from zero level, the frame undergoes elastic deformation upto a point. If the material at the critical section starts yielding, a non-linearity in the deformation sets in. Linearity of a material implies that the load deflection relation is linear and the principle of superposition holds good. A section is plastified if all the material in that section is subjected to yield stress. Once the entire material at a section reaches the Yield State, the section is said to be have formed into a plastic hinge. Sections at B and C, the corner points are the critical ones that reach the yielding first. Plastic hinges are formed at these two sections leading to unstable state. With two hinges at the base and two plastic hinges at the two corners of the frame, the structure tends into a mechanism and collapses. The load deformation relation follows line $AHIG$ as indicated in Fig. 1.9.

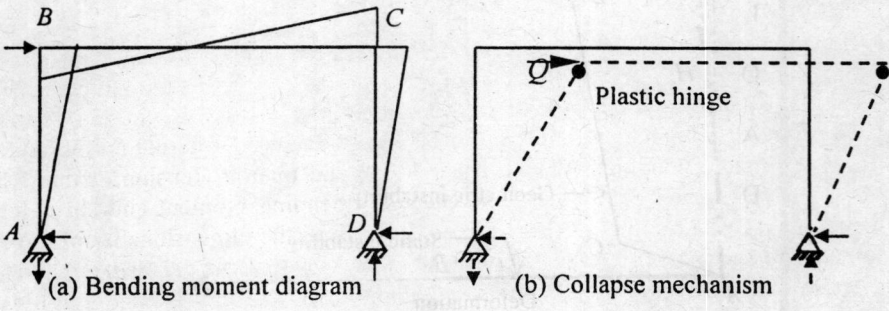

(a) Bending moment diagram   (b) Collapse mechanism

Fig. 1.10 Collapse Mechanism of Portal, Strength Limit

Dynamic Instability

The mass and stiffness of a structure combined in a certain form gives a characteristic performance. A structure when given slight excitation, vibrates with amplitudes of vibration. This amplitude is measurable and depends on the mass and stiffness of the structure. The number of oscillations it takes per second is called the *frequency* of vibration. The amplitude and the frequency are a measure of the vibration. The excitation may be due to wind, earthquake, machine vibrations or vehicular movements. If the external excitation is stopped, the vibration of the structure damps out and the structure returns to its normal position. However, for a set of unique excitations, the structure continues to vibrate even after the removal of the external excitation. The frequency at which the structure vibrates with undefined magnitude of amplitude is called the *natural frequency*. The lowest natural frequency is the critical of all and occurs first. The structure is in an unstable situation during the period of natural vibration. The instability associated with the vibration of the structure is called *dynamic instability*. The designer must ensure that the natural excitations such as movement of vehicles and machines, wind, blast and other don't come close to the natural frequency of the structure. This problem is severe in towers, bridges, machine foundations, cables etc. The structural stability and vibrations are very similar problems and are more dominant in slender elements and structures.

1.3. METHODS AND PHILOSOPHY IN STRUCTURAL DESIGN

For many centuries, the master builder was considered to be the designer. Some rules of design existed in the pre-modern period. Regulated engineering practice and education was in operation for about four centuries, but nationally accepted codes of practice came in existence by the beginning of the twentieth century. Code of practice was developed based on the experience, research and available technology. Three methods of design are popular now in the world and they are:

1. Limit State Design (LSD)
2. Load and Resistance Factor Design (LRFD)
3. Working Stress Design (WSD)

The order of listing of the methods is arbitrary and the usage of the method changes with country. In addition to these, other methods were practiced to limited extent and these are:

1. Ultimate Strength Design (USD)
2. Plastic Design (PD)

A brief philosophy of the methods is discussed in this section.

Working Stress Design (WSD)

The stresses developed in the material of the elements of structure due to the action of the external forces should be less than the allowable set of values. Similarly, the deflections in the structure should be limited to a set of allowable values. The actual stresses are computed from the internal forces on the members using the principles of mechanics of solids. The possible types of internal forces on the members are tension, compression; bending moment, shear force, and torsion. Tension, compression and torsion are about the longitudinal axis of the element, while bending moment and shear forces are transverse to the main axis. The number of these forces depends on the dimensionality of the element. One bending moment and one shear force exist on plane element (two dimensional element) and there are two moments and two shear forces for spatial element. Compressive force creates a possibility of buckling. The elastic structural analysis is used in the working stress design. The structural analysis gives deformations and member forces. From the member forces, one can obtain the maximum stresses from simple principles of mechanics of solids.

The allowable stress in a material for a given state of stress or stress combination is specified by the code of practice. The allowable stress is a fraction of characteristic strength which is either yield or proof or ultimate or buckling stress. Yield or proof stress is the basis in determining the allowable stresses in ductile materials. Ultimate stress is the basis in case of brittle materials. Buckling stress is the reference limit value in case of compression members. The characteristic strength is evolved with experience, experimental data and magnitude of acceptable risk. The governing condition of the working stress design consideration is:

$$\sigma_{ij} \leq \sigma_{aij} \tag{1.4}$$

in which σ_{ij} = actual stress in the ith type of element in the jth load condition and

σ_{aij} = allowable stress in the ith type of element in the jth load condition

For example, the allowable tensile stress in high yield deformed reinforcement bars with proof stress of 415 N/mm^2 is 230 N/mm^2 under the normal working load condition. Whereas the allowable stress in the gust wind load condition is increased by 1.33 times that is allowed in the normal load condition. Similarly, the allowable bending compressive stress in M20 concrete is 7 MPa. This is with reference to crushing strength of 150mm concrete cube. On the other hand, the allowable direct compressive stress is 5 MPa. These stresses can be increased in the wind, earthquake or handling load conditions. Such loads act only for a short duration when compared to the normal working load.

Similarly, the governing condition for deflection limitation can be expressed as:

$$v_i \leq v_{ai} \tag{1.5}$$

in which v_i is the actual deflection and v_{ai} is the allowable value.

Working stress design is probably the earliest one to come in the form of code of practice. This method was popular till late fifties then out of practice in many countries. It is an admissible method by the Indian code of practice however it is less practiced in the design of reinforced concrete structures at present. The working stress method is referred in the annex of the code IS: 456-2000. The working stress design is consistent with the elastic structural analysis.

The structural safety is based on *factor of safety* concept. The factor of safety is *roughly* defined as:

$$Factor\ of\ safety = yield\ or\ ultimate\ strength/allowable\ stress \tag{1.6}$$

This concept of factor of safety was in use many decades, but now the allowable stresses are listed by the code without reference to the factor of safety. One must take the definition as a rough guideline but not in strict sense. The value of factor of safety may vary from 1.2 to 12 depending on several factors. The lowest value of 1.2 is applicable in case of steel members in bending when subjected to gust load condition. This is with reference to the yield or proof strength of steel. The ultimate strength of steel may be 1.4 times the yield strength. The duration of the gust wind is only about two to five minutes and its occurrence may be once in fifty years. Maximum stress occurs at the extreme fibre of a bending member. A redistribution of stresses takes place in the other fibres of the section after the extreme fibre has yielded. Therefore, consideration of all these factors are reflected in the in the lower value of the factor of safety. On the other end of the spectrum, the factor of safety listed as 12 is applicable to brick or stone masonry with reference to the ultimate strength of the brick or stone. Masonry construction consists of stone or brick, mortar, and type of joints and workmanship. The statistical variation of the ultimate strength is much wider than that of steel. The strength of the masonry may be only one third of the brick strength or even less. There may be openings for door or window in the masonry walls and further the application of the load transfer on to the masonry is not well defined. One has to normalize many of the uncertainty factors. The factors are associated with basic quantities of structural material, type of element, type of stress distribution in the element, interconnectivity of the elements, method and quality of fabrication or construction, and type of workmanship and supervision. Further the type and rate of occurrence of the load, and the importance and design life of the structure should also be considered in arriving at the allowable stresses. Most of the recommendations on the allowable stresses or factor of safety have been arrived based on the experience, experimentation, theoretical knowledge, and intuition and engineering judgement. There is always a room for improvement and refinement.

The main question that is often asked is what is the relation between the factor of safety and the actual collapse or failure of the structure? This question can be answered in simple elements such as ties and simply supported beams made of well-defined material such as steel. Many uncertainties do exist in loads, materials and safety of the structure. One must be able to take a calculated risk in providing the safety margins. The stresses are not directly measurable and have no one to one correspondence with the collapse of the structure. Even though the method is satisfactory at the working load level but can't define the margin of safety with reference to the failure of the structure. At the same time, it is difficult to establish relative safety levels between different types of structures and materials. The spirit of the working stress design is still maintained in other methods to some extent even though the method is considered outdated. For example, limiting stresses and deflections in the modern methods are still valid.

Ultimate Strength Design (USD)

The ultimate strength design is chosen as the second in the order of presentation, because it got evolved after working stress design. The strength of a member is a measurable quantity. Tensile strength of a tie member, bending capacity of a beam or the buckling or strength of a compression member can be computed theoretically and verified by experiments. A good agreement between the theoretical and experimental results of ultimate strengths exists. Since the working loads are also measurable, it appeared reasonable to link the loads on the structure and the ultimate strength of the members. The safety margin of the structure is therefore applied to the loads and the strengths of the members. The working loads are multiplied by factors referred as *load factors* and the effects caused by such loads on the members are matched with the strength of the members. The governing criteria of the ultimate strength design method may be expressed as:

The strength of the member must be more than the ultimate force on the member

$$R \geq F_u \tag{1.7}$$

In which R is the resistance of the member and F_u is the ultimate force on the member.

F_u = critical of the forces caused by the ultimate loads. The ultimate load is expressed as:

$$W_{ui} = \Sigma \gamma_{ij} W_j \text{ for } j = 1 \text{ to } k \text{ and } i = 1 \text{ to } n$$

Where W_{ui} is defined as the ultimate load in the ith combination of the working loads,

W_j is the jth working load in the ith combination of loads; j^{th} load may refer to dead load, live load, wind load, or earthquake load etc.,

γ_{ij} is the jth load factor in the ith load combination.

There are a number of loads that act on a structure. All possible loads may not act at the same time and further the safety margin in different load combination can be different. For example the normal live load condition can be written as:

$$W_{u1} = \gamma_{1d} W_d + \gamma_{1l} W_l \tag{1.8}$$

Where W_{u1} = the ultimate load in normal live load condition,

$\quad W_d$ = dead load

$\quad W_1$ = normal live load

$\quad \gamma_{1d}$ = load factor applied to the dead load in normal load condition, it is usually = 1.2 to 1.5

$\quad \gamma_{1l}$ = load factor applied to live load in normal live load condition, it is usually = 1.5 to 1.8

Similarly the ultimate load in the wind load condition is given by:

$$W_{uw} = \gamma_{2d}W_d + \gamma_{2l}W_l + \gamma_{2w}W_w \qquad (1.9)$$

Where W_{uw} = the ultimate load in wind load condition,

$\quad W_w$ = wind load

$\quad \gamma_{2l}$ = load factor applied to live load in wind load condition, it is usually = 0.5 to 1.0

$\quad \gamma_{1w}$ = load factor applied to wind load in wind load condition, it is usually = 1.5

The ultimate load is an imaginary load if such a load comes on the structure, the structure is likely to fail or collapse. In the limit state design, one is working with upward factored load and actual strength of the member.

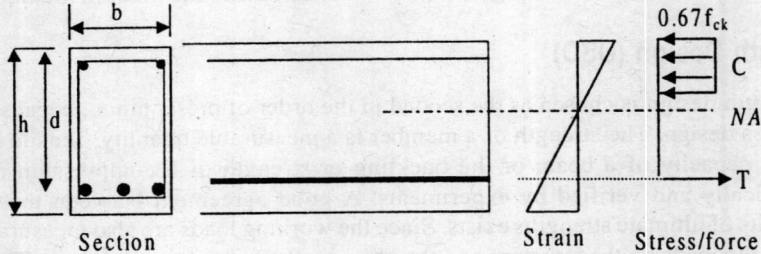

Fig. 1.11 USD Strain-stress Idealization – Rectangular Block

The strength of the member is calculated using the limit strength of the member and total material in the section reaching the appropriate strength. The limit strength is equal to the yield stress or ultimate stress multiplied by a factor called *strength reduction factor*. In case of steels, the yield/proof stress is multiplied by a reduction factor of the order 0.85. However in the case of concrete, the ultimate stress of the prism sample is taken instead of the cube. The prism strength is about 0.67 times the cube strength. Figure 1.11 illustrates the concept of the ultimate strength design. The plane section remains plane even at the time of crushing of the concrete is indicated in the figure. The idealized rectangular stress distribution is also indicated. The values C and T are the total compression from concrete and T is the total tension from the reinforcement. The structural safety of a member is ensured and it is taken for granted that the safety of the structure is thus provided by the safety of each of the members. The stress-strain curve of concrete is idealized as a rectangular stress block having a maximum strength of 0.67 times that of the cube. Similarly, the stress-strain relation of the steel is idealized as rigid plastic having design strength equal to the yield or proof strength. The moment capacity of a reinforced concrete rectangular section is given by:

$$M_{rb} = Kbd^2f_{ck} \qquad (1.10)$$

In which

$\quad M_{rb}$ is the balanced moment capacity of a rectangular section, it is the value in which the yield stress in steel and crushing of concrete occur simultaneously,

b and d are width and effective depth of the section, f_{ck} strength of the concrete
$K = 0.225$ where the material reduction coefficient set equal to 0.87.

The moment capacity of an under-reinforced concrete section is given by:

$$M_r = A_{st} f_y (d - 0.425 x_u) = 0.8 A_{st} f_y d \qquad (1.11)$$

In which M_r is the resisting moment capacity, A_{st} is the area of cross section of the reinforcement, f_y is the proof strength of the reinforcement, x_u is the neutral axis distance from the top compression fibre.

The elastic structural analysis is applied in the ultimate strength design using the pseudo-ultimate loads. There is some dichotomy in the method. At the instant of the member reaching its strength, it is beyond the linear or elastic state but yet the analysis is linear. As this is a statically admissible analysis, the safety of the structure is ensured, in fact more margin of safety is provided by this analysis. In plastic or limit analysis, redistribution of internal moments or forces takes place. The error is on the safer side by not admitting the redistribution of forces in the elastic analysis.

The ultimate strength design is a simple method aimed at a desired margin of safety. The expressions and the governing conditions were also simple, so the method became popular in fifties and sixties. One of the disadvantages of the method was that the performance of the structure at actual working loads was not given a consideration. The method has given way to limit state design in late sixties.

Limit State Design (LSD)

Limit state design is the method that is currently popular and evolved during late sixties of the last millennium. The evolution is from the combined merits of working stress and ultimate strength design methods. The strengths of the earlier methods were incorporated in the limit state design method. The two states for which design is aimed are:

Strength limit state and Serviceability limit state: Detailed discussion on the method is given as the book deals with the subject itself.

Load and Resistance Factor Design (LRFD)

The load resistance factored design is very similar to the limit state design. The limit state design is practiced in Europe and Asia, while the load resistance factored design is practiced in the American continent. The loads are either known or measurable and similarly the resistance of a section can be calculated or measured through experiment to a reasonable accuracy. These two quantities are multiplied by factors to ensure safety of the structure.

Plastic Design (PD)

A structure when subjected to external loads undergoes elastic deformation to start with. As the load increases, some points of the material reach yield state and, non-linearity sets in the deformation behaviour of the structure. The load deformation curve will indicate flattening as more material undergoes yielding. The material is assumed ductile like steel. The rate of flattening of the curve depends on the stress strain relation of the material. Plastic design considers the total ductility of a structure. The structure is analyzed by plastic analysis that permits total collapse of a member or structure. The method is applied to steel structures, and not to concrete structures.

1.4 LOADS

The loads that are likely to act on a structure are specified by the codes of practice or by the user. The main code on the forces or loads for buildings in India is Structural Safety Standards, Load on buildings IS: 875. The earthquake loads are specified by another code IS: 1893. Similarly, the Indian Road Congress specifies the loads on bridges. The specified loads may or may not act on a structure in its lifetime but likely to act, therefore the structure is designed to resist the loads. The loads may be classified in the following groups:

❖ *Dead Loads (DL)*, also called as fixed loads. The loads fixed in position and in magnitude with reference to a structure are called dead loads. Weights of the members, floors, walls, foundations etc. are considered as dead loads. Unit weight of some of the commonly used structural and other materials in civil engineering are given in table 1.1 for ready reference.

Table 1.1 Commonly Used Material in Civil Engineering

Material	Unit weight (kN/m^3)	Material	(kN/m^2)
Sand	17-18	Fire insulation boards	0.03-0.05
Stone	16-19	Fibre boards	0.02-0.05
Brick	10-14	Particle boards	0.09-0.12
Slag	7-8	Gypsum board	0.07-0.15
Heavy duty Brick	24-26	Plywood	0.07-0.12
Brick Masonry	20	Expanded metal	0.04-0.06
Cement	13-14	5mm glass	0.13
Concretes:		10mm mastic asphalt	0.21
Foam	8-10	20mm cement plaster	0.41
Light-weight	10-14	1.6mm CGI sheet	0.13
Lean	22	6mm ACC sheet	0.13
Plain	24	40mm PCC	0.85
Reinforced	25	Terrazzo floor	0.85
Prestressed/ HPC	25-26	Roof tile in mortar	0.85-1.5
Stone masonry	20-26		
Timber	7-11		
Grain	7-8		
Flour sacks	3-6		
Textile bundles	8-17		

❖ *Live Loads (LL)* also referred as imposed loads. The loads that are movable and moving without any accelerations or impacts are called live loads. People, furniture or any other movable objects in buildings that are not fixed in position are the typical examples. These loads act primarily on floors and roofs of buildings, bridge footpath etc. These loads are specified by the codes of practice.

Table 1.2 lists some selective live loads on floor slabs. The load is treated as uniformly distributed over the floor area. A concentrated load is indicated in another column and it is to be placed at location that causes maximum bending and shear forces. A reduction of five- percent live load is given for every 50 m^2 subject to a maximum of twenty five percent. However no reduction is allowed for machine, impact, water loads etc. Reduction of loads has to be applied to columns, walls and foundation in multistory buildings. Zero percent reduction for top floor, that is roof or terrace. 10 percent for the 2nd from top, 20 percent for the 3rd from top, 30 percent for the fourth from top. Forty percent reduction in load from 5th to 10th floors from top, and fifty percent beyond the tenth floor till ground floor. The live load on flat roofs having access is 1.5 kN/m^2 and it is 0.75 kN/m^2 where no access is provided. This is further subjected to a minimum of 4.5 kN and 1.9 kN per span respectively. In case of

Table 1.2 Recommended Live Loads on Floors

Type of floor	UDL (kN/m²)	Concentrated (kN)	Type of floor	UDL (kN/m²)	Concentrated (kN)
Dwelling houses					
Corridors, stairs,	3	4.5	Rooms	2	1.8
Institutional Buildings					
Class, dining	3	2.7	Bed, toilets	2	1.8
Corridor, Stair	4	4.5	Plant	4-5	4.5
Office Buildings					
Rooms (no store)	2.5	2.7	Rooms + store	4	4.5
Banking	3	2.7	Corridors	4	4.5
Hostels, Hotels					
Kitchen, Stair	3	4.5	Living, bed, bath	2	1.4
Restaurants	4	2.7	Plants	5	4.5
Garage (light)	2.5	9	Garage (heavy)	5	9
Commercial Buildings					
Office	2.5	2.7	Retail shop	4	4.5/2.7
Corridor	4	3.6	Plants	5	6.7
Assembly Halls					
Fixed seat area	4		Free Seating	5	3.6
Restaurants etc	4	4.5	Stages	5	4.5
Storage Buildings					
Work houses	2.4/ m ht.	7	Cold Storage	5/m ht.	9
Corridors	4	4.5	Plant rooms	7.5	4.5
Industrial Buildings					
No Machinery	2.5	4.5	Light machine	5	4.5
Medium machines	7	4.5	Heavy machine	10	4.5
Corridors, stairs	5	4.5	Kitchens	3	4.5

sloping roofs, the design live load is $(0.75 - 0.2(\theta - 10^0))$ kN/m² subject to a minimum of 0.45 kN/m², where θ is the slope of the roof.

❖ *Wind Loads (WL).* Because of either obstruction or constraint against passage of wind, pressure windward side and suction on the leeward side are developed. The wind can also set vibrations in the structure that adds accelerating forces of the mass of the structure. The wind force often controls the design of tall towers, industrial structures, ropeways, long span bridges and slender elements. The country is divided into a number of zones of basic wind speeds. The design wind speed is given by:

$$V_z = k_1 k_2 k_3\, V_b \tag{1.11}$$

Where

V_z = design wind speed, V_b = basic wind speed in metres per second,

k_1 = probability factor or also called risk coefficient,

k_2 = terrain, height, and structure size factor,

k_3 = topography factor (ground contour factor)

The wind pressure p_z in Newtons per square metre is given by:

$$p_z = 0.6\, V_z^2 \tag{1.12}$$

The wind load on a structure depends on obstruction area, configuration, shape, size and few other factors and details of the structure. The wind load is usually given by the basic wind pressure multiplied by the obstruction area and coefficients associated with the structure.

❖ *Earthquake Load (EL)*. Earthquake load is also referred as seismic load. The earthquake causes random ground motion that introduces vibrations in a zone of influence. The response of the structure to the earthquake is a function of the type of soil, foundation, and mass and mass distribution of the structure, stiffness of the structure, and the duration and the intensity of the ground motion. The dynamic forces introduced by the response of the structure, cause internal forces and vibrations in the structure. The entire country is divided into a number of basic seismic zones based on the observed tectonic conditions. The earthquake force on a structure is computed from the basic earthquake zone and factors associated with the foundation, structure and the importance. The allowable bearing pressure in earthquake condition can be increased by 25 to 50 percent of that under normal live load condition. The permissible stresses in materials can be increased by 33 percent of those under normal live load conditions. Alternatively, the partial load factors applied to loads in the limit state design can be reduced.

❖ *Moving Loads (ML)*. Vehicles as they move on a bridge or on any structure, sudden changes in the motion or operation of the vehicle may take place. Such sudden changes produce impact in addition to the gravitational load. The acceleration or deceleration of a vehicle also causes drag or frictional loads on the structure. The impact and the drag forces are functions of the response and configuration of the structure. A crane operating on a gantry girder introduces the impact, lateral and drag forces on the girder. The impact force is damped out as it is transmitted to the columns, and foundations. Only a part of the impact force is transmitted to the supporting columns and foundations. The impact coefficients as functions of the type of structure and the type of the vehicle are recommended normally by the appropriate codes.

❖ *Temperature Load*. This is referred as thermal load. Daily or annual variation in the temperature cause changes in the sizes of the elements of a structure. Some process within the structure may also cause the temperature changes. For example, a chimney, which transmits hot gases and smoke, is subjected to variable temperature along the thickness and along the height of the wall. The shortening or elongation of the members introduces differential strains and consequential internal forces in the structure. In a statically determinate structure, a uniform change in temperature in members causes strains but not stresses. Differential temperature between the top and bottom faces of a bridge cause differential strains and consequently stresses. The variation of temperature across the thickness of a member produces stresses. The free elongation or shortening (ΔL) caused by the temperature t is:

$$\Delta L = \alpha L t \tag{1.13}$$

in which α is the coefficient of linear expansion of the material of the element, L is the length of the member.

A redistribution of strains takes place to maintain the compatibility of the structure. Therefore, stresses are caused in statically indeterminate and continuum structures. Fire in a building raises the temperature of the members. *Fire loading* is not treated like temperature load. The damage due to fire is altogether different and each building must be designed to withstand a rated fire. The fire rating of the buildings is the duration of fire that the building can withstand. Fire protection must be given for the designed fire rating.

❖ *Snow and Ice Loads*. Some regions in the country are subjected to snow. The snow as it accumulates on the roofs on structures; its weight becomes a snow load. The heights of snow to be considered in the calculation of the weight are listed by the code for different regions. An expression to determine the snow load is based on the height, type of the structure and the slope of the roof are available. The snow when freezes, becomes ice load. The thickness of ice formation is also recommended by the codes.

❖ *Impact Load.* Sudden drop of weight, sudden changes in the loads or their motions cause impact loads. Movement of vehicles on bridges, travelling cranes on gantry girders, drop hammers on anvil, hammering in buildings etc. induce impact loads. The impact load is directly connected with the weight and sudden changes or drops. Impact factor when multiplied with the load that is causing the impact results in to the impact load that is equivalent of gravitational load. In most cases, the impact loads are treated as gravitational loads for the purpose of analysis except in the case of drop hammers.

❖ *Blast and Explosion Loads.* An explosion releases sudden increased volume of gases and consequently blast of air. The sudden increase in the volume of the gases and air generates waves that travel in all directions. The explosive may also contain some solid particles. The mass of the air or gases along with the wave motion of the explosion or blast is converted into dynamic load on the structure. The wave front and the vibrations generate response from the structure. Only explosion and blast prone structures are designed to withstand the blast loads.

❖ *Erection and Fabrication Loads.* Some structures or elements are prefabricated or pre-cast and erected at site. Erection loads are short duration loads and therefore, lower load factors are assigned to the erection loads.

❖ *Repetitive or Fatigue Loads.* Frequent repetitions and reversal of loads on material cause fatigue. The effect of the fatigue decreases the capacity of the material. Trainloads moving on rail bridges are classical example of repeated loads on civil engineering structures. Live loads on buildings, normal moving loads on roads and even the wind or earthquake loads are not classified under the repeated loads as the frequency of repetitions is low so as to cause any fatigue effect. The allowable stresses under repeated loads are reduced to account for loss of strength due to fatigue.

❖ *Earth Pressure Load.* The exposed earth mass has a tendency to slide and occupy an equilibrium slope. The tendency of sliding of the mass of the earth puts a pressure on the structure that retains the soil. Retaining walls, abutments, underground structures are subjected to earth pressures. The earth pressure is a function of head of earth and nature of the soil. The sandy soils have a greater tendency to slide when compared with the clayey types. The pressure generated by the tendency of sliding of soil is called *active earth pressure*. If a structure tries to push the earthen mass, the resistance of the earth against the tendency of the structure movement is called the *passive earth pressure*. The earth pressure along with moisture or earthquake has added loading effects. The force increases with height of filling and is linear along the depth and it is called hydrostatic type of force variation. Fig. 1.12 illustrates the hydrostatic type of force variation.

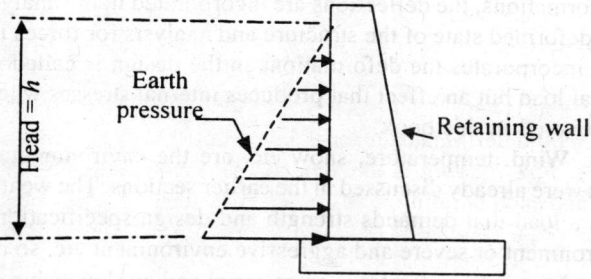

Fig. 1.12 Hydrostatic Type of Force on Retaining Wall

❖ *Hydrostatic Load.* The horizontal force of water on the supporting surface is called hydrostatic pressure. At any point in water, the intensity of pressure is same in all directions. This type of equal pressure in all directions is called hydrostatic state of pressure. Water tanks, dams, harbor and river structures, and similar structures are subjected to hydraulic force or pressure. Fig. 1.12 illustrates the hydrostatic force variation along the depth of water.

❖ *Dynamic Force.* A force that is caused by the mass and acceleration is called dynamic force. Accelerating or decelerating bodies generate the dynamic forces. The structure responds to the dynamic loads and then its behavior is modified. Wind load, earthquake load, blast load, and moving vehicles with acceleration etc. dynamic loads. Some structures are sensitive to the dynamic loads and therefore such ones to be analyzed dynamic performance.

❖ *Foundation Settlement.* The soil that supports a structure settles down due to the load. Uniform settlement of the soil doesn't cause any internal forces in the structure. However, unequal or differential settlements of the foundations introduce forces on to most of the structures. The settlement is not by itself a load but as the supports of the structure settle, deformations take place in the structure. Such deformations in the structure modify the loads transmitted to the soil and then modify the settlement of the soil. This communication between the soil and the structure is called *soil structure interaction*. Many sensitive structures optimize their performance through soil-structure interaction.

❖ *Creep and Shrinkage.* Concrete shrinks as it gets hardened and also with decrease in the moisture content. In statically determinate structures, the shrinkage may induce cracking. In indeterminate structures, the changes in the member length interfere with the compatibility of deformation. Internal forces are developed with differential shrinkage. The free shrinkage deformations are modified by the compatible deformations. Creep is the deformation under constant load. Creep deformation depends on the quality of concrete, and level of stress in the material. The creep or shrinkage by itself is not a load but cause internal forces in the structure due to differential deformation caused by them, therefore considered as loads.

❖ *Prestress.* Some members or structures are prestressed by either external jacking or internal prestretching of some components. Steel cables are suitably placed in position and stressed to a desirable level by external jacks, and then the force is transferred to the concrete. Sometimes the concrete is cast over stressed wire and the force is transferred to the concrete as when the concrete hardens. Prestressed concrete beams, tanks and bridges are often used in construction. Many methods of prestressing are practiced in construction.

❖ *Secondary Forces.* The deformations caused by the loads are small when compared to the actual lengths of the members. The locations of loads are considered with respect to the undeformed configuration of the structure and the analysis is performed with such load locations. However, in some situations of large deformations, the deflections are incorporated in the analysis. The structural analysis that considers the deformed state of the structure and analysis for forces is called nonlinear analysis. The analysis that incorporates the deformations in the design is called secondary analysis. This again is not an external load but an effect that produces internal stresses when incorporated. Material non-linearity is not covered in this book.

❖ *Environmental Loads.* Wind, temperature, snow etc. are the environmental forces that act on any structure. These loads were already discussed in the earlier sections. The weather effect, environmental pollution also acts as a load that demands strength and design specifications. The structure may be exposed to mild environment or severe and aggressive environment etc. so must be designed considering the exposure conditions. Specifications on material and workmanship must be made depending on the exposure conditions.

OBJECTIVE QUESTIONS

Four statements for each objective question are listed. One that is most appropriative is to be selected. It is possible that some of the statements may be suitable for limited applications or in few situations.

1.01 Structural configuration of a building is dominated by:
  a) Functional and architectural aspect of the building.
  b) Front architectural appearance of the building.
  c) Direction of wind and light effects of the site.
  d) Constructional feasibility of the building.

1.02 Structural skeleton of a framed building consists of the following:
  a) Beams and Columns.
  b) Beams, Columns, and Slabs.
  c) Beams, Columns, Slabs and Foundations.
  d) Walls, Beams, Columns.

1.03 Primary Super-structure of a building includes:
  a) Beams, Columns, and Staircases.
  b) Beams, Columns, and Slabs.
  c) Beams, Columns, Slabs and Foundations.
  d) Walls, Beams, Columns.

1.04 Most commonly used structural analysis in design of concrete structures is:
  a) Limit state analysis.
  b) Plastic Analysis.
  c) Load factor analysis
  d) Elastic and linear analysis

1.05 The member sizes in most structures can be determined through structural analysis & design:
  a) Uniquely.
  b) Complete flexibility exists in choice.
  c) Can be selected with in reasonable bounds.
  d) Completely controlled by the analysis.

1.06 The amount of reinforcement in most members of a structure can be determined through structural analysis & design:
  a) Uniquely.
  b) Complete flexibility exists in choice.
  c) Selected with in reasonable bounds.
  d) Controlled completely by the analysis.

1.07 A structure must be stable under the following situations:
  a) Under dead and live load conditions.
  b) Under wind or earthquake load conditions.
  c) Under moving or dynamic load conditions.
  d) Under any load the structure is likely be subjected.

1.08 Structure is said to be unstable under the following conditions:
  a) Undergoes large deformation under the designed loads.
  b) Collapses when subjected cyclonic or earthquake loads.
  c) Undergoes large deformation or rigid body movement even under small load.
  d) Collapses when subjected to limit load.

1.09 The following are the different types of instability of a structure:
  a) Static, geometric and structural instability.
  b) Sliding and overturning instability.
  c) Buckling and strength instability.
  d) Dynamic and static instability.

1.10 Static instability of a structure is indicated:
  a) Collapse of the structure when subjected to designed static loads.
  b) Undergoes large rigid body displacement under loads even less than the design load.
  c) Overturning or sliding of the structure for loads equal or less than the designed loads.
  d) Undergoes into a mechanism when subjected to designed loads.

1.11 Static instability of a structure is indicated:
  a) Equilibrium of forces is not satisfied under arbitrary small load.

b) Undergoes finite displacement under loads even less than the design load.

c) Equilibrium of forces of the structure is satisfied in the deformed shape but not in the original.

d) Undergoes into a mechanism when subjected to designed loads.

1.12 Geometric instability of a structure is indicated:

a) Equilibrium of forces is not satisfied under arbitrary small load.

b) Undergoes finite displacement under loads even less than the design load.

c) Equilibrium of forces of the structure is satisfied in the deformed shape but not in the original.

d) Undergoes into a mechanism when subjected to designed loads.

1.13 Geometric instability of a structure is indicated by:

a) Collapse of the structure when subjected to designed static loads.

b) Undergoes large rigid body displacement under loads even less than the design load.

c) Undergoes large displacement even under a small load but regains stability in the new location.

d) Undergoes into a mechanism when subjected to designed loads.

1.14 Structural instability of a structure is indicated by:

a) Collapse of the structure when subjected to designed static loads.

b) Undergoes large rigid body displacement under loads even less than the design load.

c) Undergoes into a mechanism when subjected to designed loads.

d) Deforms into an undefined magnitude of deformation at a set of loads.

1.15 Structural instability of a structure is indicated by:

a) Buckling of a member or the structure under a critical load.

b) Collapse or crushing of the structure.

c) Undergoes into a mechanism when subjected to designed loads.

d) Bad configuration not suitable for wind or earthquake loads.

1.16 Strength instability of a structure is indicated by:

a) Buckling of a member or the structure under a critical load.

b) Collapse of structure through a mechanism.

c) Undergoes large deformation for loads even less than designed loads.

d) The material of the structure yields with time.

1.17 Methods of structural design that are popular in present practice of concrete structures are:

a) Limit design and safe stress design.

b) Factor of safety design and limit design.

c) Limit states design and working stress design.

d) Ultimate strength design, Limit state design and working design.

1.18 Most general purpose computer software for structural analysis has the following facility:

a) Idealize the structure, compute the loads and analyse for the member forces.

b) Generate members and joints in the structure and establish member connectivity.

c) Members and joints must be read in either directly or indirectly as an input data.

d) Joint loads are generated once the joint coordinates are input.

1.19 Most general purpose computer software of structural analysis has the following facility:

a) Analyse the member forces and joint displacements.

b) Establish member connectivity once the coordinates of the joints and member sizes are input.

c) Loads on members are generated once the member properties are input data.

d) Joint loads are generated once the joint coordinates are input.

1.20 The governing principle of working stress design is:

a) Actual stress or deflection under service loads must be less than the permissible values.

b) Actual loads acting on the structure should not exceed the allowable loads.

c) The working stress under the critical load must be a fraction of the yield stress.

d) The working stress must be equal to Permissible stress.

1.21 The allowable stress in working stress design is equal to:

a) Yield or ultimate strength of material divided by load factor.

b) Yield or ultimate strength of material divided by partial safety factor.

c) Yield or ultimate strength of material divided factor of safety.

d) As specified by the code of practice.

1.22 Working stress design uses the following method of structural analysis:

a) Working stress analysis.

b) Elastic linear analysis.

c) Elastic analysis.

d) Factor of safety analysis.

1.23 The allowable stress in working stress design is based on:
a) Yield or ultimate strength of material.
b) Yield or buckling strength of material.
c) Elasticity of the material.
d) Working loads on the structure.

1.24 The governing criterion of Ultimate Strength Design is:
a) The ultimate strength of the material should be more than the ultimate load.
b) The strength of any section must be more than the design ultimate force on the section.
c) The ultimate strength design must account for ultimate strength of the material.
d) Ultimate load is equal to product of load factors and the working loads.

1.25 Ultimate design uses the following method of structural analysis:
a) Non-linear analysis.
b) Elastic linear analysis.
c) Ultimate load analysis.
d) Load factor analysis.

1.26 The ultimate load in Ultimate Strength Design is equal to:
a) Service load multiplied by the corresponding factor of safety.
b) A load corresponding to the collapse of the structure.
c) A load corresponding to an imaginary load leading to crushing of the material.
d) A load equal to product of load factors and the corresponding load factors.

1.27 The governing criterion of Limit State Design is:
a) The limit load must always be less than the ultimate load.
b) The limit state of a structure must always be more than the safe acceptable load level.
c) The structure must have adequate safety against collapse and perform satisfactorily in service.
d) The limit performance of the structure must be satisfactory under all loads.

1.28 Limit State Design uses the following method of structural analysis:
a) Non-linear analysis.
b) Limit analysis.
c) Limit state analysis
d) Elastic linear analysis.

1.29 The Limit load in limit State Design is equal to:
a) Service load multiplied by the corresponding factor of safety.
b) A load corresponding to the collapse of the structure.
c) A load corresponding to an imaginary load leading to crushing of the material.
d) A load equal to product of partial safety factors and the corresponding specified loads.

Key solutions to the questions:

	1	2	3	4	5	6	7	8	9	10
0+	a	c	b	d	c	c	d	c	a	b
10+	a	c	c	d	a	b	c		a	a
20+	d	b	a	b	b	d	c	d	a	d

Brief Review of Materials Associated with Reinforced Concrete

2.1 INTRODUCTION

Concrete is a solid mass formed from stone chips or pebbles as filler and bonded with cement mortar. This is not a technical definition but suggested to a beginner or laymen. It is a matrix of aggregate, cement and water. The cement when mixed with water results into complex compounds. Cement-paste is mixed with aggregate particles results into a solid mass. Concrete has been in use since ancient times. Egyptians, Greeks, Indians, and Romans have used lime as a cementing material in concrete foundations. The limestone (calcium carbonate) when heated, disintegrates into calcium oxide and carbon dioxide. Calcium oxide is a reactive compound and in the presence of water, it results into a binding material. Lime combined with active silica forms far stronger bonding compounds. Powdered brick has active silica and the combination of lime and brick powder was used as a cementing material for centuries. Brick powder is called *soorkee,* in India. Lime-soorkee concrete was in use for many centuries until cement-concrete. Cement, normally referred as Portland cement came into existence in the nineteenth century. So the cement-concrete came into practice in the later part of the 19th century. Three basic materials, namely aggregate, cement and water, mixed in desired proportion result into concrete of certain strength. Some admixtures are added to the concrete mix to obtain certain desired properties. This chapter is devoted to a review of the important characteristics of concrete. It is assumed that the student has some exposure to concrete technology and therefore the review is made brief.

2.2 AGGREGATES IN CONCRETE

Broken stone material, gravel and sand are called *aggregates* for production of concrete. Even small boulders are also called as coarse aggregate. The aggregates are broadly divided into two categories base on size, namely, fine and coarse aggregates. Aggregate, whether fine or coarse, must be hard, durable, clean and free from coal, mica, iron pyrites, shale, clay, seashells, alkalis, and other organic materials. Unless otherwise specified in this book, aggregate refers to that from stone and used in concrete constructions. *Fine aggregate* is the basic stone material having a particle size less than 5 mm, and 90 to 100 per cent of which pass through 4.75 mm sieve. Natural sand and finely crushed stone are treated as fine aggregate. *Coarse aggregate* is the stone material most of which is retained on 4.75 mm sieve. Gravel or crushed stone are the usual coarse aggregate. There are other types of aggregates, like *broken brick-aggregate, cinder-aggregate, light-weight, slag and heavy-weight aggregate*, etc. that are used in special concrete constructions. Coarse and fine aggregates together are used in concrete whereas only fine aggregate with either cement or lime is used in mortar for plastering and jointing. Particles larger than 4.75 mm are grouped into coarse aggregate, particles between 4.75 mm and 60 microns (mm= micro millimeter) are considered as fine aggregate, natural particles smaller than 60 microns and larger than 2 microns are grouped under *silt* and still smaller particle material is called *clay*.

Coarse aggregate is classified into two main groups: (i) *single-size aggregate* and (ii) *graded aggregate*. Single-size aggregate is based on a nominal size specification and contains about 85 to 100 per cent of the material that passes through the specified size of the sieve and zero to 25% of it is retained in the next lower sieve. About 90 to 100 percent of the aggregate passes through the sieve whose size is same as that of the specified and further 30 to 70 per cent pass through the next lower size sieve. Another type called *gap graded* aggregate is also used in concrete at times.

The fine aggregate is further classified into four sub-groups, as per the Indian Standards. Grade I is the coarsest and grade IV is the finest of the fine aggregate. For grade one, about 90 to 100 per cent of the material must pass through 4.75 mm sieve and about 60 to 80 percent must pass through the next lower standard sieve, namely 2.36 mm. The first three graded zones are usually acceptable for reinforced and prestressed concrete construction. However, the grade four is too fine and not recommended for concrete construction. It can be used in plastering.

All-in-aggregate: Aggregate containing both coarse and fine aggregates is called all-in-aggregate. All-in-aggregate is not normally used in structural concrete. It is desirable to mix appropriate quantities of fine and coarse aggregates to develop good compaction and required strength. Important properties of the aggregate are discussed briefly in this chapter.

Shape of aggregate: Aggregate shape can be either round, angular or a combination of the both. But it should not be flaky or oblong. It should not contain deleterious materials such as clay lumps, coal particles, soft material and too fine material that pass through 75-micron sieve. The maximum limit of combined deleterious materials in an aggregate is five percent. The angular aggregate having about the same intersecting angular faces and is neither oblong nor flaky is considered the best to develop good interlocking mechanism.

Size of a coarse aggregate is specified by a sieve size. A 40 mm aggregate normally means that 100 percent of the aggregate should pass through 63 mm IS sieve (that is next higher of 40 mm IS sieve), and 85 to 100 percent pass through 40 mm sieve. And about 0 to 30 percent may pass through the next lower size, i.e. 20 mm sieve. Commonly used size terminology of the aggregate is:

Very large aggregate 80 mm to 150 mm size,
Large aggregate is 40 mm to 80 mm,
Medium size aggregate is 20 mm to 40 mm, and
Small size coarse aggregate is 4.75 mm to 20 mm.

Very large and large aggregates are used in mass concrete. Large aggregate may be used in lean concrete, medium and small aggregate are used in plain, reinforced and high strength concretes. Each country has its own standard sieves to size the aggregate. The Indian Standard (IS) sieves are designated by 63, 40, 20, 16, 12.5, 10, 4.75, 2.38, 1.18, (all in mm), 600, 300, 150, and 75 microns. The size of the sieve is specified by the nominal size of the aperture either in millimeters (mm) or in micrometers (microns).

The important characteristics that one has to look for in aggregate are:
(1) Shape, size and surface texture,
(2) Specific gravity,
(3) Void ratio,

     (4) Bulk density,
     (5) Moisture content,
     (6) Porosity and absorption value,
     (7) Aggregate crushing strength,
     (8) Abrasion value,
     (9) Flakiness index,
    (10) Elongation index,
    (11) Presence of deleterious materials,
    (12) Soundness,
    (13) Alkali Aggregate reaction, and
    (14) Fineness Modulus.

Shape, size and surface texture of aggregate is either acquired from natural erosion or through crushing process. The roundness is indicated by the degree of wear of the faces of the particles. The compactness of the aggregate depends on the shape of the aggregate. Similarly the void in the aggregate depends on the shape and size distribution. The angularity of an aggregate is defined by an angularity factor. *Angularity factor* is the ratio of the volume of un-compacted aggregate to that of the glass spheres of specified grading. Angular aggregate requires more water when compared to the rounded one for the same level of compaction of the concrete. Equal dimensions in most faces of the aggregate are desirable for good consolidation and interlocking. The angularity number is equal to 67 minus the percentage of solid volume of the aggregate measured in a specific manner. The angularity number of the spherical aggregate is zero, which means that the percentage of the solid volume of rounded aggregate is 67. The angularity number indicates the percentage of voids in excess of the rounded aggregate. The surface texture is classified into rough, smooth and polished. The bonding of cement paste also depends on the surface texture of the aggregate. Rough texture is preferred for better bonding of the cement paste. The tensile strength of the mortar or concrete depends on the shape and texture of the aggregate.

Fineness Modulus (FM): The aggregate consists of particles of different sizes in a range. The aggregate is analyzed by sieving through a set of standard sieves. The *fineness modulus* is defined as the sum of the cumulative percentages of aggregate retained on a set of standard sieves numbered 63, 40, 20, 10, 4.75, 2.36, 1.18 mm, and 600, 300, and 150 micron. The aggregate is sieved through the standard sieves and the material retained on each of the sieves is weighed, the cumulative percentage of the material retained on each sieve is computed. Table 2.1 indicates the range of fineness moduli of the aggregate. Smaller fineness modulus means finer aggregate. The range of *FM* of fine aggregate is 1.0 to 3.5. The fineness modulus of 1.0 indicates the finest sand that can be used for plastering and not for mortar or for concrete. The fineness modulus of coarse aggregate is in the range of 5 to 7.5. Fineness modulus of aggregate is an important parameter in the design of the concrete mix.

Table 2.1 Ranges of Fineness Modulus of Aggregates

Fine Aggregate		Coarse aggregate			Mixed aggregate		
Very fine	Fine	20 mm	25 mm	40 mm	20 mm	25 mm	40 mm
1.0-2.0	2.0-3.5	5.8-6.4	6.0-7.0	6.5-7.5	4.7-5.2	5.0-5.6	5.3-6.0

Specific gravity of a material is the ratio of its weight to the weight of the equivalent volume of distilled water. Stone may contain pores or capillary voids. The absolute specific gravity is the one that excludes the pores in the solid mass. *Absolute specific gravity* is the ratio of the mass (weight) of the solid in vacuum to the mass of equal volume of distilled water. The absolute specific gravity is of academic interest. *Apparent specific gravity* of the material is the ratio of the oven dried mass at 100 to 110 degrees Celsius for 24 hours to the weight of the equal volume distilled water. The volume referred here includes the pores in the aggregate. The apparent specific gravity is the most commonly used term or quantity in the engineering design. The specific gravity of a material in any form either in large or small aggregate is same as that of the base stone from which it is derived from. It varies from 2.6 to 2.85 for most stone based aggregate. Specific weight is the product of the specific gravity and the unit weight of water.

The *bulk density* of aggregate is the weight of unit bulk volume of material. The bulk density is measured in two ways, one loose bulk density and the other compact bulk density. Aggregate is pored in a standard container and leveled to the top in a specified manner. The ratio of weight of the loose aggregate measured from the container to the volume of the container is called as *loose bulk density*. In case of compact bulk density, the container is filled in three layers, and each layer is compacted by tamping with a 16 mm standard rod to a specified number of times. The ratio of the weight of the compacted aggregate to the volume of the container is called *compact bulk density*. The compact bulk density is commonly referred as the bulk density. The bulk density depends on the void ratio and it varies from 0.54 to 0.65 times the specific weight of the material. The range of bulk density is 1500 kg/cum to 1900 kg/cum. High-density aggregate may have a bulk density in the range of 1800 kg/cum to 2100 kg/cum.

Bulk aggregate contains solid particles and void spaces in between the particles. *Void ratio* is the ratio of the volume of the voids to that of the solid particles. The void ratio is determined by the following expression:

$$\text{Void ratio} = 1 - \text{bulk density/specific weight} \qquad (2.1)$$

Void ratio of single-size aggregate is around 0.42 to 0.46 and in graded aggregates it is around 0.38 to 0.44. Less is the void ratio better is the grading of the aggregate. The void ratio can be determined from the amount of water that is needed to fill the voids of a measured volume of the dry aggregate.

Porosity, Moisture content and Absorption value: Aggregate contains pores some visible on the surface and others inside but not visible. If the pores are too small or the capillaries are very fine, the viscosity of water may delay the absorption of the water. *Saturated aggregate* is the one in which the pores are fully saturated with water. The saturation must be free from the surface wetness. In other words the percentage of saturation is measured in surface dry condition. The saturated aggregate that is allowed to surface dry in room temperature is called *air dry aggregate*. Aggregate dried in oven for a prolonged period is called *oven dry or bone-dry aggregate*. Aggregate when exposed to rain and atmosphere collects moisture on its surface and absorbs certain portion of it through the pores in the particles. The total water content in the aggregate is divided into two parts. One is the surface moisture and it is the difference in weight of the moist and the surface dry aggregate and is called moisture content. The other portion is the absorbed water and it is the amount absorbed in the pores of the aggregate and is equal to the difference in weight of the air dry and bone dry aggregates. Moisture content increases the bulk volume of fine aggregate whereas the absorbed water content does not change the volume of the aggregate. The increase in volume of the sand due to surface moisture film pushing the particles apart is called *bulking of sand*. The volume of sand increases with increase in the moisture content upto a certain limit, but further increase in the moisture content reduces the bulk volume of the sand. The maximum bulking of the sand depends on the type of sand and it is about 30 percent with normal occurrence around 8 percent of the moisture content. The ratio of the volume of moist sand and that of dry sand is called *bulking factor*. The bulking factor varies from 1.0 to 1.35 depending on the moisture content and the type of sand.

Strength of aggregate: The strength of an aggregate depends on the nature of the basic rock from which it is derived, size distribution, shape and texture of the aggregate and the impurities present.

Aggregate *crushing, value and aggregate impact value* are the usual indices used to specify the strength quality of an aggregate. The aggregate crushing value is determined by a simple test. As per the Indian standards, a dry sample of aggregate passing through 12.5 mm sieve and retained on 10 mm sieve is subjected to compression test in a specified manner under 400 kN load. The crushed material that passes through 2.36 mm sieve is measured and the percentage of that material with respect to the original sample weight is called the *aggregate crushing value*. A simple procedure with a simple apparatus is used for the determination of the crushing strength of aggregate. Higher the crushing value, lower is the strength of the aggregate. This is a misleading notation but was adopted for unknown reasons. It would have been better to measure the strength by the percentage of the particles retained rather than passing through the 2.36 mm sieve. A crushing value more than 40 reflects a weak aggregate. Crushing value less than 30 mean a strong aggregate. The following maximum limits of crushing value are recommended:

> *For lean and plain concrete: 45,*
> *for Plain concrete in dams and hydraulic structures: 40,*
> *for reinforced concrete in buildings: 40,*
> *for concrete roads and high strength concrete or High Performance Concrete: 30.*

The toughness of the aggregate is measured by the impact test. An impact value of upto 30 indicates good quality aggregate; however, a value upto 45 is acceptable for plain concrete. Aggregate crushing value or impact value for concrete should not be more than 45 for ordinary concrete works, and for wearing surfaces such as roadways and runways it should not be more than 30.

The strength of aggregate is also determined by Los Angeles test. A specified sample of aggregate is loaded into a rotating drum machine in which steel balls are also loaded. The mixture is subjected to a specified number of revolutions and the percentage loss of the weight of the aggregate indicates the weakness or the strength of the aggregate. The Los Angeles test has some good characteristics that indicate the strength of the aggregate under wear and tear conditions. But this and impact tests make more noise compared to the compression test. A minimum of 80 MPa as core crushing strength of rock is normally acceptable for good concrete aggregate. This test is of academic interest. In general either aggregate crushing or impact or Los Angeles tests are preferred.

Abrasion value of aggregate: Concrete roads and runways are subjected to constant wear. In addition to good compressive strength, the aggregate must have good wear resistance. There are special abrasion apparatus such as Los Angeles; Deval machines designed to test the wear of an aggregate. The percentage of material lost in wear in the abrasion test that is performed on a sample of aggregate in a specified manner is considered as the abrasion value of the aggregate. More material lost, more is the abrasion value, and therefore the wear resistance of the aggregate is inversely proportional to the abrasion value.

Flakiness Index: Thin and flaky pieces aggregate may result if the rock is not crushed properly in the aggregate crusher or if the basic rock has thin layers. Sieving the aggregate through a set of sieves having oblong openings indicates the order of magnitude of the flakiness of an aggregate. *Flakiness index* is the total weight of aggregate passing through a set of special sieves designed exclusively to measure the property, expressed as a percentage of the total weight of the sample. Flaky aggregates shouldn't be used for reinforced concrete.

Elongation Index: Oblong aggregate may result because of poor aggregate crushing mechanism or in sieve separation. Concrete produced with oblong aggregate results in poor strength of concrete. The oblong character of the aggregate is indicated by an index called elongation index. The nomenclature of the *elongation index* is not appropriate term to indicate the oblong property. However it is measured by sieving a sample of the aggregate through a set of sieves having oblong openings, specially designed to test the property. It is the total weight of the material retained on various length gauges, expressed as percentage of the sample.

Bond of aggregate: The cement is the bonding agent in concrete, and the shape, size, grading and texture of the aggregate influence the level of bond. The interlocking of the aggregate is an important property that will result in better bonding and better strength of concrete. Interlocking of the aggregate is inversely proportional to the flakiness index or the elongation index. As already mentioned earlier, the bond of the aggregate is directly proportional to the regular angularity and rough texture of the aggregate.

Soundness of an aggregate: Some aggregates disintegrate into smaller particles when exposed to weather and changing temperature even without load. Soundness is the property of the aggregate to withstand the weathering conditions. It depends on the chemical and physical properties of the aggregate. The soundness of an aggregate is measured by subjecting it to alternate wetting in saturated solution of sodium or magnesium sulphate and drying it in an oven through a set of cycles. The reduction in the particle size of the aggregate, which is obtained through a sieve analysis of the tested sample, indicates the soundness or unsoundness of the aggregate. The percentage absorption of moisture is a good indicator of soundness of the aggregate. Unsoundness is partly indicated by higher absorption of moisture. The soundness is more or less inversely proportional to the absorption.

Aggregate and alkali reaction: Sodium and potassium oxides are called as alkalis (Na_2O and K_2O). Small amounts of alkalis are normally present in cement. The alkalis in cement or in other materials of the concrete react with silica of the aggregate in the presence of moisture resulting into a gel. The *silica alkali gel* formation in the pores or in the planes of weakness aggregate increases the volume thus resulting in cracking of concrete. The gel also destroys the bond between the aggregate and the cement paste. The gel formation is a slow phenomenon and takes years before it is manifested through cracking. The rate of gel formation increases with increase in temperature and moisture content. It is also proportional to the porosity of the aggregate. The absence of moisture or high humidity avoids the gel formation thus the alkali reaction. Hydraulic structures must be designed with sound aggregate and low alkali cements to avoid the alkali reaction. Alkali aggregate reaction is measured by mechanical and chemical tests. Mortar bar with pre-assigned alkali content, that is stored over water at about 38° C and then tested for expansion in time.

2.3 PORTLAND CEMENT

Commonly used binding building material in pre-modern period was lime. Volcanic ash and powdered brick was also used along with lime as a binding material. Alumina and silica contained in the burnt clay powder and volcanic ash have the active solidification properties, and when they are combined with lime gave a good binding material better than lime. Joseph Aspdin, a mason from England, obtained material by grinding limestone and clay and then burning the mixture at clinkering temperature. The clinker was then ground along with small amount of gypsum into a fine powder. This powdery material was named as *Portland cement*. He is the first person to patent the material in 1824. May be there were others who used the technique to a limited extent even before 1824. It was named as Portland cement as it resembled a stone in Portland. The Portland cement is manufactured in very large scale all round the world. More than 3.6 billion tonnes of Portland cement was produced in the world during 2000. The installed capacity of Portland cement in India is about 120 million tonnes in year 2000. The basic principle of making cement even today is about the same as that of 19th century except that the quality and efficiency have improved considerably. The calcareous material such as limestone or chalk and argillaceous material such as clay or shale containing silica are mixed thoroughly and burnt at a clinkering temperature of about 1350° C to 1500° C. The clinker is then powdered with small amount of gypsum (about 3.5 percent of clinker). The two basic methods of manufacturing of Portland cement are dry process and Wet process. The principles of the methods of manufacturing are mentioned here. Fig. 2.1 illustrates the basic principle and components in the manufacture of cement. Water is the medium by which the cement develops combining mechanism, therefore these cement are also referred as *hydraulic cements*.

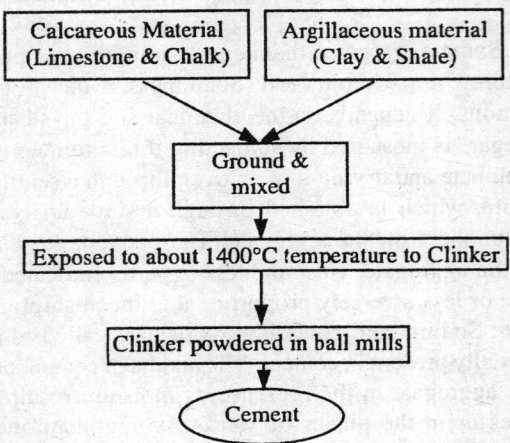

Fig. 2.1 Main Components in Cement Manufacture

Dry process: Crushed raw materials in appropriate proportions are ground into fine powders in a grinding mill. The dry powders are blended into a fine mixture in a blending silo with the help of compressed air. The dry mixture is sieved and fed into rotating granulators along with appropriate quantity of water and then air-dried to form into granules. The granules also called pellets are baked by hot air and fed into a rotating kiln where the material undergoes chemical changes at temperature in the range of 1350° C to 1500° C. The clinker formed in the kiln is cooled and ground with gypsum in ball mills. The powder ground to a satisfactory level (about 10^{12} particles per Newton) is separated as cement.

Wet process: Broken raw materials are mixed with water in separate wash mills to form fine mixtures. The mixtures are screened, mixed in appropriate proportions and the resulting slurry is pumped into storage tanks. The slurry is kept under constant stirring against any sedimentation. The slurry is fed into a rotary kiln from one end and pulverized coal is fed from the other end of the kiln and burnt. The slurry as it moves down in the kiln gets burnt and finally forms into clinker. The clinker is then let through cooling chambers and afterwards pulverized in ball mills along with gypsum. The powder ground to a satisfactory level is separated as cement.

The manufacture of cement is energy intensive, the wet process consumes more energy when compared to the dry process. More than fifty percent of the expenditure is the energy. Further one kilogram of cement production results in producing one kilogram of carbon dioxide as pollution into the air. The dust pollution from cement industry is controlled to a great extent. The cement is packed in 50-kilogram bags and marketed in most countries. In advanced countries, cement is also packed in drums or even transported in bulk. India is yet to pick up bulk packaging and transportation.

2.4 PRINCIPAL COMPOUNDS OF PORTLAND CEMENT

The burning up of calcareous and argillaceous material at high temperature results into a clinker made up of calcium oxide, silicate, aluminate and aluminoferrite. The main chemical compounds in the cement are Tricalcium silicate, Dicalcium silicate, and Tricalcium aluminate and Tetracalcium aluminoferrite. First tricalcium aluminate and Tetracalcium aluminoferrite are formed in the kiln, and then dicalcium silicate is formed, and after its saturation, tricalcium silicate is formed. Tricalcium silicate is mostly responsible in giving early strength of cement whereas the dicalcium silicate contributes to the strength at later stage. The tricalcium aluminate contributes towards the strength developed in the first 48 hours. The principal chemical compounds and the range of percentage of the quantity of them in ordinary Portland Cement are listed in Table 2.2.

The basic chemical compounds in Portland cement are CaO, SiO_2, Al_2O_3, Fe_2O_3, MgO, K_2O, and Na_2O. And they are combined to form the principal compound listed in 2.2.

Table 2.2 Chemical Compounds in Portland Cement

No.	Compound	Composition	Notation	%Range
1.	Tricalcium Silicate	$3CaO.SiO_2$	C_3S	45-58
2.	Dicalcium Silicate	$2CaO.SiO_2$	C_2S	15-32
3.	Tricalcium Aluminate	$2CaO.Al_2O_3$	C_3A	6-13
4.	Tetra Calcium Aluminoferrite	$4CaO.Al_2O_3Fe_2O_3$	C_4AF	6-12
5.	Minor Compounds	MgO, K_2O, Na_2O		3-6

Some of the chemical requirements of ordinary, rapid hardening and Portland pozzolana cements are given in table 2.3.

2.5   PHYSICAL PROPERTIES OF PORTLAND CEMENT

Important physical properties that are of interest to construction and quality control engineers are briefly described in this section. The specifications as per the Indian standards are given in a table 2.4. The important properties are:

❖ Fineness of cement,
❖ Normal consistency of cement,
❖ Initial setting time of cement,
❖ Final setting time of cement,
❖ Strength of cement,
❖ Soundness of cement,
❖ Heat of hydration of cement.

Table 2.3 Chemical Composition Requirements in OPC, Rapid Hardening and Pozzolana Cement

No.	Substance	Percentage limits
1.	Ratio of Lime to Silica, Alumina & Iron Oxide	$0.66 < p < 1.02$
2.	Ratio of Alumina to that of Iron Oxide	Not less than 0.66%
3.	Insoluble residue	Not more than 2%
4.	Magnesia	Not more than 6%
5.	Total sulpher content, sulphuric anhydride (SO_3)	Not more than 2.75%
6.	Total loss on ignition	Not more than 5%

Fineness of cement: The cement is ground to fine powder and impossible to grind it to a single size particle. The particle size is usually in a range and an average of the size of the particle is of interest. The surface area of particles per unit weight of cement indicates its fineness. There are experiments that compute the surface area of unit weight of a powder. The rate of development of strength of the cement depends on the fineness of the cement. The compound formation starts on the surface of the cement particles. Similarly the heat of hydration developed by the chemical reaction of cement with water depends on the surface area. The technology of grinding the cement powder is improving every day. The percentage residue on 90-micron sieve should not exceed 10 percent for ordinary Portland cement. American specification list 45-micron sieve instead of 90-micron sieve. The surface area of cement in square metres per one kilogram of cement is called *specific surface* of cement. The surface area should be in the range of 300 square metres per kilogram of cement. The minimum surface area that is required for ordinary Portland cement is 225 m^2 per kg, however this is on the lower side. The present day cements have a specific surface more than 300 square meters per kilogram. Some of the advantages of higher surface area of cement are:

❖ Faster reaction of cement with water, consequent faster development of strength,
❖ Marginal improved workability in cement paste,
❖ Less bleeding of cement paste (the term bleeding is explained later)

Some of the disadvantages of the fineness of cement beyond point are:

❖ More heat of hydration for finer cements,
❖ More shrinkage with higher fineness,
❖ More chances of cracking due to more early high heat of hydration,
❖ Faster reaction with alkali-aggregate.

The particle size of powder can be determined by measuring the free fall of the particles in an inert medium using the principle of fluid mechanics of free falling bodies. The concentration of turbidity of the cement powder in kerosene can be measured by passing light through the solution and the amount of light passed through gives an indication of the number of particles in the solution. *Wagner turbidimeter* is the common instrument that is used in the determination of the specific surface of cement. This method not only indicates the size but also the shape of the particles. *Air permeability* test is another commonly used test for the determination of specific surface. Air is allowed pass through a bed of the particle at constant velocity under constant pressure. A manometer measures the pressure drop of the air that passes through a bed of the powder. The surface area is then computed using the specific gravity of cement, porosity of the bed, velocity of flow, and the pressure drop through the bed. There is another method called *Blaine air permeability* test that uses slightly different principle of the permeability. A known volume of air is allowed pass through at an average pressure through a bed of known porosity. The time taken for the air to pass through is measured from which the specific surface is computed. Specific surface of very fine powder such as fine flyash and silica fume cannot be determined by the permeability method. The gas absorption techniques are used in the determination of the specific surface of very fine powders. The silica fume is about ten to fifty times finer than the cement. There could be a variation in the result of specific surface measured by different methods. The codes of practice will indicate the method and the minimum specific surface requirement.

Normal (standard) Consistency of Cement

Cement reacts with moisture and forms complex compounds. This chemical reaction is termed as hydration and heat that is generated during the reaction of cement with water is called *heat of hydration*. The rate of hydration, the chemical kinetics and the physical behavior of the cement paste depend on the amount of water added to the cement. The percentage of water required to produce a reasonable workable cement paste under prescribed testing is called *the normal consistency* or *standard consistency* of cement. An instrument called *Vicat* apparatus is used in the determination of standard consistency. A standard plunger of 10 mm diameter is allowed to penetrate partly through a cement paste. The percentage of water that allows the plunger to penetrate a specified distance through a cement-water paste is taken as the *standard consistency* of the cement. The test is conducted under specified humidity and temperature conditions in a specified duration of time after mixing the cement and water. The range of acceptable normal consistency of Portland cement is 26 to 33 percent of the mass of the cement.

Initial setting time of cement: Cement paste changes from semi-fluid state to stiffened state, then to hard state and finally solid mass state. The process of transforming the cement paste from semi-fluid state to stiffened or hard state is called setting of cement. The transformation of state is measured through a prescribed test procedure using Vicat apparatus. The time needed for the standard consistency cement paste to become stiffened in permitting 1.13 mm needle under specified weight, to penetrate through a specified distance in a standard mould is called *initial setting* of cement. The time that is required for the initial setting is called *initial setting time*. The cement mortar or concrete should give reasonable time for mixing, transporting, placing and finishing of the product. Concrete or mortar must be workable without developing bonding for about 30 minutes after the water is added to the Portland cement. The initial setting time is delayed by addition of some compounds in case the placement of cement takes more than 30 minutes.

False Setting: The premature stiffening or hardening of the cement paste is called false set. The false set of cement takes place much before the initial setting. In normal setting, heat of hydration takes place and no such heat is generated in the false set. The process of hardening of cement paste caused by the reaction of pure calcium aluminate may be the reason for false set. Or it may be due to the non-uniform distribution of gypsum in the cement. Gypsum delays fast setting of the cement. False set doesn't generate good bonding mechanism and hence the concrete can be re-mixed and finished in position. False set is avoided by addition of gypsum to the cement.

Final setting time: The time taken for initiation of hardening of cement paste from the time it is mixed with water is called final setting time. A standard procedure by Vicat apparatus measures the so-called initiation of hardening of the cement paste. The paste should become hard enough to withstand light pressure. A circular cutting edge needle fitted to the Vicat apparatus is gently lowered on to the surface of the stiffened mortar. The time that is taken to ensure that no indentation is made on the mortar specimen is called the final setting time. The concrete can resist minor disturbances or small pressures immediately after the final setting. However, it should not be disturbed till reasonable curing takes place. The chemical reactions between cement and water are very fast in the initial stages, considerable heat of hydration is generated during the first few hours of mixing of cement with water. As the reactions continue, the cement paste becomes harder and harder and the process may continue for a long period. The water that is added to the cement absorbs the heat of hydration in the initial stages may be upto twelve to twenty four hours depending on the water cement ratio. There after water has to be sprayed over the surface of the concrete to absorb the heat of hydration. The concrete or mortar can resist the loads only after it is cured as specified by the engineer. The final setting time of most Portland cements is about 10 hours.

Initial setting of cement indicates the beginning of noticeable stiffening of the cement paste and the final set indicates the beginning of the hardening of the paste. The normal consistency, initial and final setting times are determined by Vicat needle or by Gillmore needle tests. The limits of initial and final setting times are indicated in Table 2.4.

Strength of Cement

Strength of cement is determined by compression test on hardened cement-sand mortar specimen. Cement mortar cube specimens are made in the ratio of 1:3 cement and standard sand, and are cast in cubes of 50 cm^2 face. The specimens are cured in water for 28 days and tested under controlled conditions. The amount of water used in making the mortar is selected from the normal consistency of cement. The amount of water is equal to $(0.25p + 3)$ percent of the combined weight of cement and sand, in which p is the percentage of the normal consistency of cement. Three cement mortar cubes are tested after one day, three days, seven days and 28 days of curing. The average strength of three cubes tested at a time is taken as a basis for acceptability criteria. The codes of practice specify the minimum average strength required after one day, three days, seven days and twenty-eight days. The ordinary Portland cements of grade 33, 43 and 53 must give a minimum average strength of 33, 43, and 53 N/mm^2 respectively after 28 days of curing. The curing of the specimens is done in water at 20° C. Standard sand is a siliceous, round and natural sand. The sand particles are graded between 80-micron to 1.6 mm. In India, sand available at *Ennore* in Tamil Nadu is considered to be the standard sand. The absolute minimum strength of mortar cubes of any type of cement must give 33 N/mm^2.

Table 2.4 Physical Properties Required of Different Cements.

No.	Type of test	Ordinary grade 33	Rapid Hardening	Low Heat	Pozzolana
1	*Fineness of cement*				
	a) Residue by weight shouldn't exceed (percent) after sieving the cement through 90-micron IS sieve,	10	5	–	5
	Or				
	b) Specific surface (m^2/kg) by air permeability test method shouldn't be less than	225	325	320	300
2	*Setting times*				
	Initial setting time shouldn't be less than	30	30	60	30
	Final setting time shouldn't be more than	600	600	600	600
3	*Minimum average compressive strength of*	–		–	
	standard mortar cube	–		–	
	a) 24 ± 0.5 hours	–	16		
	b) 72 ± 1 hours	16	27.5	10	
	c) 7 days ± 2 hours	22	33	16	22
	d) 28 days ± 4 hours (14 days ± 4 hours for pozzolana)	33	–	35	33
4	*Soundness (maximum)*				
	a) Expansion (in mm) unaerated cement by Le Chatlier method	10	10	10	10
	b) Expansion (in mm) of aerated sample by being spread out to a depth of 75 mm at relative humidity of 50 to 80% for seven days	5	5	5	5
	c) Expansion unaerated cement by autoclave test (in percent)	0.8	0.8	0.8	0.8
5	*Heat of Hydration in calories per gram*				
	a) After 7 days	–	–	65	–
	b) After 28 days	–	–	75	–

* Cement grades 43 & 53 are to give minimum average strength of 43 MPa & 53 MPa respectively on the 28 days curing. The 7 days strength is about 67 percent of 28 days strength.

Soundness of Cement

The word soundness here refers to the tendency of expansion of hardened cement mortar with time. Most of the chemical reactions must take place during the 28 days of curing of the cement or concrete. Some compounds like free lime or magnesia present in the clinker are slow in reactions and take several months to hydrate. The volume of the hydrated free lime or magnesia is more than the original volume. Increase in volume or the expansion of the hardened cement will cause cracking of the mortar or concrete. The micro-cracking of concrete leads to deterioration and failure of concrete. That is why, expansion of cement paste beyond a point is said to be unsound. Le Chatelier apparatus tests the expansion of cement caused by the free lime. A cylindrical tube split along the vertical line of the cylinder and fitted with long thin handles. The cylinder is filled with cement paste with water content corresponding to the normal consistency of the cement. The assembly as a whole along with the cement paste is placed at 200° C and at 98% relative humidity for a prescribed period. The assembly is placed in water and the water is raised to boiling point in thirty minutes. The increase in the opening of the pointers is measured to determine the soundness of the cement. The admissible expansion of the pointers is specified by the codes of practice. This opening of the pointers is caused by the increase in the volume of the cement cylinder. The presence of magnesia can't be determined by the Le Chatelier test. An autoclave test has to be performed to identify the excess presence of magnesia as well as free lime. A 25.4 mm side square bar of 254 mm length (one inch by 10 inches bar) is cured for 24 hours in humid air. The cured bar is placed in autoclave, the temperature of the water is raised to 216° C at a pressure of 2 MPa, in 60 minutes and maintained for three hours. The expansion of the bar after cooling back to room temperature should be less than 0.8 mm.

The rise of temperature in the Le Chatlier or the autoclave test accelerates the hydration process and simulates the long term effects. Fine grinding of cement maximizes the free lime mixing with the other compounds and resulting in earlier hydration thus minimizing the unsoundness of the cement.

Heat of Hydration

The basic silicates and aluminates of cement combine with water resulting into insoluble solid compounds. Calcium silicate and tricalcium aluminate hydrates are the main resulting compounds. Calcium hydroxide is one of the other products in the hydration process. The heat developed during the chemical changes is called the heat of hydration. Typical chemical reactions are listed here.

$$3CaO.Al_2O_3 + 6H_2O \rightarrow 3CaO.Al_2O_3.6H_2O \tag{a}$$

$$3CaO.Al_2O_3 + 3(CaSO_4.2H_2O) + 25H_2O \rightarrow 3CaO.Al_2O_3\ 3CaSO_4.31H_2O \tag{b}$$

$$2(3CaO.SiO_2) + 6H_2O \rightarrow 3CaO.2SiO_2.3H_2O + 3Ca(OH)_2 \tag{c}$$

$$3(2CaO.SiO_2) + 6H_2O \rightarrow 3CaO.2SiO_2.3H_2O + 3Ca(OH)_2 \tag{d}$$

$$4CaO.Al_2O_3.Fe_2O_3 + 2Ca(OH)_2 + 10H_2O \rightarrow 3CaO.Al_2O_3.6H_2O + 3CaO.Fe_2O_3.6H_2O \tag{e}$$

The reaction of calcium aluminate is almost instantaneous and it may lead to false setting of the cement. The gypsum reacts with aluminate resulting calcium sulfoaluminate. Similarly the gypsum reacts with calcium aluminoferrite resulting calcium sulfoferrite. Most of the exothermic reaction of the cement with water takes place in about two hours. The cement paste becomes stiff during the hydration and then hardens into a solid mass. The calcium hydroxide that is generated from the hydrolysis of the calcium silicates provides an alkaline environment to the concrete.

Specific gravity of ordinary Portland cement varies from 3.12 to 3.16, and for all practical purposes, it is taken as 3.15. The percentage of voids in cement is around 40 and bulk density is about 18 to 20 kN/m^3. Usually cement is supplied in bags of 50 kg.

2.6 TYPES OF PORTLAND CEMENTS

Portland cement is produced in different grades and in different types in different countries. The classification of Portland cements varies with the country. The Indian code of practice classifies the Portland cement into the following classes.

❖ Ordinary Portland Cement (grades 33, 43 and 53)
❖ Rapid Hardening Portland Cement,
❖ Portland Slag Cement,
❖ Portland Pozzolana Cement (flyash based),
❖ Portland Pozzolana Cement (Calcined clay based),
❖ Low Heat Portland Cement,
❖ Sulphate Resisting Portland Cement,
❖ Hydrophobic Cement.

There are other types of cements available in the global market and also in the Indian Market. The number is rather too large, only very commonly used types are mentioned here.

❖ Portland Blastfurnace Cement,
❖ White Cement,
❖ Very Rapid Hardening Cement,
❖ Supersulphated Cement.

A brief description of commonly used cements is given.

Ordinary Portland Cement (OPC) Grades 33, 43 and 53

Ordinary Portland Cement is known as just cement or simply ordinary cement for laymen. OPC is the most commonly used cement in building construction. A grade notation refers the strength of the cement. Cement whose standard sand mortar cube made with normal consistency water level should give a minimum strength of 33 N/mm^2 at 28 days of curing is assigned grade 33. Similarly grades 43 and 53 are expected to give a minimum strengths of cement mortar at 43 MPa and 53 MPa respectively. The strength refers to an average strength of three mortar cubes of face area 50 cm^2. The strength of cement on 7 days curing is about two-thirds of that at 28 days strength. The cubes are made in a specific manner and tested on 28 days of curing in water. The expected strength after 2 days curing of grade 33 and 43 cement mortar cubes are 10 MPa (mega Pascal = one million Newton per square metre) and 20 MPa respectively. At the moment, the most commonly used grades of cement are 43 and 53 primarily because of marketing strategies of the cement companies. The cost difference between the different grades of cement is marginal, therefore, there is a tendency to select higher grade of cement even though it may not be needed. Cement grade 63 is also available in the global market. The fineness of the cement increases with the increase in the grade of the cement and so the rate of heat of hydration increases with the increase of the grade of the cement. Some countries specify the maximum strength of the cement at 28 days to avoid wrong use of high strength cements.

Rapid Hardening Cement

Special cement called rapid hardening cement, also called as early strength cement is produced to meet the demand of quick setting. The strength developed by rapid hardening cements in about 3 days is equal to that of the ordinary cement developed in 28 days. The factors that influence the early strength are chemical composition, degree of chemical combination of the raw materials, blending, grinding and burning of the raw materials and the fineness of cement. The initial setting time of the rapid hardening cement is still about 30 minutes. The specific surface of the cement is in the range of 450 to 600 square metres per kilogram of cement. The rapid hardening cement is used when formwork is to be removed at an early age or the construction is under water. The rate of heat development in this cement is also higher when compared with the ordinary cement. Very rapid hardening cement or ultra early strength cements are also produced with specific surface more than 700 square metres per kilogram. The percentage of tricalcium silicate in such cements is high in the range of 60 percent. Cement called *jet cement* that sets too fast is also produced in the advanced countries. This may develop more than five N/mm^2 within an hour of mixing with water.

Portland Pozzolana Cement (PPC)

Pozzolana is a material that contains active silica that reacts with lime in the presence of moisture, producing calcium silicate. Pozzolana by itself is not an active binding material, but in the presence of calcium oxide and moisture, develops the binding character. Flyash that is fine and rich in silica is the common pozzolanas. Amorphous porous material like silica fume has very high specific surfaces as high as ten times or even more than that of cement. There are natural pozzolanas such as diatomaceous earth. Cement produced by grinding Portland cement clinker and pozzolana with addition of gypsum is called Portland Pozzolana cement. A uniform blending of the Portland cement with fine pozzolana can also obtain it. The percentage of pozzolana in such cements varies from 10 to 30 percent. There is also some cement with much higher percentage of pozzolanic material. Portland pozzolana cement produces less heat of hydration and offers greater resistance to the attack of aggressive waters. The initial setting time of the cement is same as the ordinary cement. It develops the strength at a slower rate when compared to the ordinary Portland cement. However, the 28 days strength of PPC should be same as that of the OPC. At the moment it is reported that the consumption of Portland pozzolana cement in the world is higher than the ordinary Portland cement. India is yet to catch up with the world in the use of the pozzolana cement. It is particularly useful in marine and hydraulic structures and also for mass concrete construction.

Portland (Blast furnace) Slag Cement

Hot slag from blast furnace is granulated with cold water and then mixed with Portland cement clinker in different proportions. The grinding of the mixture with gypsum results in Portland Slag cement. Blending of OPC with ground granulated slag can also produce it. The mixture along with some gypsum is ground to the same fineness as that of Portland cement. This cement yields about the same strength as that of the ordinary Portland cement and can be used in place of ordinary cement. Slag is considered to be a waste (better call it a byproduct) product in the production of pig iron. It is a mixture of silica, alumina and lime. The percentage of lime, silica and alumina in the slag are about 45, 35 and 10 respectively. However the chemical and physical quality of the slag differs with the process and the rate of cooling. About 300 kg of slag is produced for one tonne of pig iron. As the slag contains lime and silica, the ground-granulated slag (*ggbs*) can also be used as a cementitious product to a limited extent. The percentage of slag in the slag cement may vary from 35 percent to 95 percent. The heat of hydration in the slag cement is less than that of the Ordinary Portland cement as the tricalcium and dicalcium content is less. This may be considered as low heat cement. The slag cements are further classified into two to three groups depending on the percentage of the slag. The seven days strength can be as low as forty percent of that at 28 days but this shouldn't discourage use of this cement for masonry and similar constructions.

Low Heat Cement

Low heat cement produces lower heat of hydration when compared to the ordinary Portland cement. The percentages of tricalcium and dicalcium silicates are less in the low heat cement. Such cements may also contain pozzolana or ground granulated slag to control the heat of hydration. These cements are useful in mass concreting in which it is difficult to cure the interiors of the concrete. The net calcium oxide content is also less in such cements.

High Alumina Cement

Bauxite and limestone are ground in appropriate proportions together and the mixture is subjected to high temperature to form clinker. The kiln used in the Portland cement is different. The main compounds are calcium aluminate, hydrates and hydrated colloidal alumina. The silica content is small and in the range of 5 to 10 percent. Bauxite is relatively more expensive when compared with clay so the cement is more expensive. The alumina hydrates very fast and attains strength at an early stage. About eighty percent of the strength is developed in the very first one day of curing. The 72 hours curing of the cement mortar gives about the same as that of Portland cement at 28 days curing. Most of the heat of hydration takes place in about first 24 hours, results in high heat of hydration. This cement can be used for underwater construction. Its use in mass concrete and concreting in dry weather is not desirable as it is difficult to control heat of hydration in mass concreting,. The cement also resists sulphate and even acidic attacks. It also has refractory characteristics and withstands high temperatures.

White Cement

White cement is made out of chalk or limestone free from impurities and having very little of oxides of iron. Oil or gas is used as a fuel in place of coal ash in the kiln to produce the clinker. The clinkering temperature is taken to about 1650°C. Grinding of the clinker is done in special mill so as to avoid contamination by iron oxide. The specific gravity of white cement is slightly less than that of the *OPC*. White cement is used for architectural purposes, in mosaic floors, in fixing glazed tiles, etc. The colour of the white cement may fade away with time so some colour pigments are often mixed to give a more stable colour.

Expansive (or Expanding) Cement

Expansive agents such as bauxite, calcite, etc. are burnt to form clinker. The clinker is then ground with Portland cement clinker to form cement. The tendency of shrinking of cement during drying is reduced in expansive cements. In the presence of water, calcium aluminate and calcium react to form calcium sulpho-aluminate hydrate resulting in expansion of the paste. A stabilizing agent such as Blastfurnace slag reacts with excess calcium sulphate and stabilizes the expansion. This cement usually contains 5 to 14 percent of expansive agent, 10 percent of stabilizer and the rest being Portland cement clinker.

Natural Cement

Cement rock containing clayey limestone upto 25 per cent is calcified and grounded to obtain natural cement. It is easier to manufacture natural cement and but it is somewhat inferior to Portland cement.

Masonry Cement

Masonry cement is obtained by inter grinding of mixture of Portland cement clinker with inert materials such as limestone, conglomerates, and dolomite, and gypsum and air entraining agent in suitable proportions. This has slow hardening and high workability and water retention properties. It is suitable for masonry construction.

Oil Well Cement

Hydraulic cement that is suitable for resisting high pressure and temperature in sealing water and gas pockets and setting castings during the drilling and repair of oil wells is called oil well cement. This cement contains reduced content of tricalcium aluminate and a retarding additive to suit the requirements.

Hydrophobic Cement

Ordinary Portland cement clinker when ground with an additive that imparts a water repelling property is called *hydrophobic cement*. The water repelling property is eliminated when the cement is mixed with water. This cement can be stored for long periods in wet and highly humid climates.

2.7 POZZOLANAS AND ADMIXTURES

Admixtures are the materials that improve some properties of cement such as setting time, workability, colour, and impermeability and resistance aggressive agents when added to the cement. Some of the commonly used admixtures are:

❖ Pozzolana,
❖ Flyash (powdered fuel ash),
❖ Rice husk ash,
❖ Silica fume,
❖ Metakoline,
❖ Ground granulated blast furnace slag.

Pozzolana

Pozzolana is a natural or artificial material containing large percentage of reactive silica. It is a fine powder of siliceous and aluminous material that reacts with calcium hydroxide in the presence of water to form cementing compounds. The silica in amorphous form reacts with calcium hydroxide to form cementing compound. The volcanic ash, Diatomaceous earth and burnt clay are the common natural pozzolanic material. *Flyash* and *rice husk* burnt at regulated temperature of about 600°C are main artificial pozzolanic materials. One has to be selective in fly ash to ensure sound pozzolanic action. Coal is pulverized and fed into boilers for generation of thermal power. The ash is collected either mechanically or electrostatics in the chimneys. This ash is called flyash. The ash when pulverized is called *pulverized fuel ash*. The specific surface of the flyash suitable for Pozzolana can vary from 300 to 600 square metres per kilogram. The flyash also contains small amount of unburned coal particles. The quality of flyash depends on the type of coal that is used in the burning. Flyash derived from Lignite is not considered to be very suitable unless it is processed. It is desirable to process the flyash before using it in the cement. In some sense, the flyash is not really an admixture since more than 30 percent of flyash can be added to the cement.

Burning kaolinitic clay at a temperature of about 700°C and grinding it to a fine powder of specific surface in the range of 700 square metres obtains the Metakaolin. The Pozzolana is blended with cement either at the grinding stage or at powdered stage in desired proportions. The Pozzolana blended cement is called Pozzolana cement or Pozzolana blended cement. The percentage of Pozzolana in blended cements varies with the desired property.

Following are some of the properties of pozzolanic cements when compared with ordinary Portland cements:

(a) Total curing time required is likely to be more may be of the order of 20 per cent,
(b) Heat of hydration developed is less than that of the ordinary Portland cement,
(c) Gains strength slower than that of Portland in the first 15 days but by 28 days it is about the same,
(d) Reacts less with salts and sulphates,
(e) Reduced permeability because of the fine particles of the pozzolana,
(f) Improved workability to some extent,
(g) Reduce bleeding.

Silica Fume

Silica fume is an important pozzolanic material used in high performance concrete. It is obtained as a by-product in ferrosilicon alloy furnaces. It is also called as *microsilica or condensed silica fume*. It has very high active silica content and having a low bulk density of the order 200 to 300 kg/m³. The specific gravity of the silica fume is in the range of 2.2 and particle size in the range of atleast ten times finer than the cement particles. The specific surface can be in the order of 3000 to 10,000 square meters per kilogram. Because of its fineness and amorphous nature, it reacts well and fills up the fine voids in the cement paste. Silica fume is used in high performance concrete in strengths of 60 to 120 MPa. The finest particles reduce the permeability of the concrete. Silica fume is supplied in compressed pallets and packed well so as to not to get powdered in transportation. Silica fume pallets are fed while mixing with cement in the concrete mixer. Silica fume is mixed in about five to ten percent by weight of cement. It is not really an admixture as it is added as an essential ingredient to achieve very high strengths and low permeability. For lack of an alternate terminology, it is grouped under admixtures.

Admixtures are really additives to cement in small quantities to improve specific character of concrete or mortar. Most of them are artificial and the main purposes of admixtures are:

(a) Improve workability,
(b) Accelerate setting time,
(c) Reduce setting time,
(d) Aid in curing,
(e) Decrease permeability,
(f) Improve wear resistance,
(g) Improve durability,
(h) Reduce shrinkage,

(i) Reduce weight,
(j) Reduce bleeding,
(k) Reduce heat of hydration,
(l) Impart colour.

The admixtures are usually finely ground powders having certain chemical or physical reactions with cement thus producing the desired action. Some admixtures are obtained in liquid form. The admixtures are added to concrete at the time of actual mixing or sometimes they are premixed with cement just before use. The quantity of admixture is usually limited to one to three percent in most cases. Excessive use of admixture can have secondary effects that are not desirable.

Workability admixtures: Powdered hydrated lime, diatomaceous earth, bentonite and fly ash can be used as admixtures to improve workability. However, use beyond certain proportion and the strength of concrete decreases, and shrinkage increases. A number of patented artificial admixtures are available in the market and they are usually referred as *plasticizers and super-plasticizers*. Calcium chloride, stearate, some oily compounds which function as wetting or dispersing agents are used as workability admixtures. Minute bubbles of air in concrete improve the workability. Foaming agent or a chemical that produces gas bubbles on reaction with cement is used as an *air entrained agent*. Natural resins, sulphonated soaps and oils are the common basic materials for this purpose. Aluminum and zinc powders, hydrogen peroxide are some of the elements used to produce gas in concrete to increase workability. Neutralized vinsol resin when added to cement disperses the cement particles and entraps air. Air entrained concrete is usually lighter and may be less strong when compared with ordinary concrete. Chloride in concrete accelerates corrosion of steel so it is not recommended for reinforced concrete construction. Ready mixed concrete is often pumped through pipeline to reach distances and heights. The slump of such concrete has to be more than 200 mm. Super-plasticizers are the most commonly used workability agents in pumped concrete.

Accelerators: Admixtures that accelerate the setting and hardening are called accelerators. Powdered calcium chloride is one of the commonly used accelerators. Calcium chloride of one to two per cent of cement content is likely to reduce the initial setting time by 10 to 15 minutes. Some soluble carbonates and silicates also help in reducing setting time. The heat of hydration is likely to be more in the first two days. Too much of accelerator can cause too early setting thus resulting in poor workability and consequently low ultimate strength. Chloride admixtures shouldn't be used in the reinforced concrete construction as the chlorine ion will cause corrosion of steel reinforcement. Some of the accelerators such as sodium silicate may result in poor strength and durability.

Retarders: Admixtures that prolong the setting and hardening of cement concrete are called retarders. Retarding agents are used when placement of concrete requires more time because of distance or transportation or other mechanical problems. Gypsum is one of the commonly used retarders. Indiscriminate use of retarders, can, however, affect the ultimate strength. Ready mixed concrete that has to be transported to long distances make use of retarders in combination with plasticizers.

Some of the trade names of admixtures marketed in India are listed for quick reference. This set is not complete and the actual product properties have to be obtained from the manufacturer. The information given here is taken from the information booklets available in the market.

Emcplast BV, MC-Plast AEA, MC-Mischoel LP, Conplast P211, P509, MAPEMIX R64, & N90, MAPEPLAST N30, R14 (Plasticizer or air entraining agents for concrete),

MC-Mischoel AEA, Cemix, MC-Sunplast, Cebex 112 (mortar plasticizers),

Zentrament Super, Zentrament F BV, Centriplast, Conplast SP337, Conplast SP430, MAPFLUID N200, R104, N100 (super plasticizers)

MAPEFLUID PZ504, PZ500 (Hyperplasticizers)

MC-Plast RP, Zentrament T5 BV, MC-Erstarrungsbremse K33, Conplast RP264, (Retarding Plasticizers)

MC-Retarder 060 (Retarder)

MC-Schnell SDS, MC-Schnell OC, Conplast NC (accelerating agents for concrete)

Impero, Sunseal DM, Dreiseal 330,MC-Special DM, Conplast X421Ic, Conplast WP90 (water proofing compounds),

Surfacit, Primex 150, 250,Roofex 2000, (concrete surface hardening liquid*), Hardcrete* (cement water proofing and hardening liquid), and *Colorum* (decorative waterproof cement paint).

Snowsol (stabilizing solution).

There are many more products available in the market.

The performance of admixture depends on the following aspects:

- Type of cement and its characteristics,
- Type of aggregate and its properties,
- Admixture property and the quantity,
- Sequencing of addition of the cement, water and admixture etc.,
- Interacting properties of the admixtures, and
- Methods of curing and temperature conditions.

2.8 WATER QUALITY IN CEMENT CONCRETE CONSTRUCTION

Water is one of the most important ingredients of concrete. Water used in mixing and curing of concrete should be free from solids, acids, alkalis, organic materials and salts. Potable and clean bathing water is normally acceptable for concrete mix but not necessarily always. The actual solids or salts present in drinking or bathing water may be higher than that required for durable concrete. The permissible limits of solids in water in making concrete are given in table 2.5. The water that is used in curing should also be of similar standard.

Table 2.5 Permissible Limits of Solids in Water (*pH* value not less than 6).

Material	Maximum limit
1. Suspended	2000 mg/litre
2. Organic	200 mg/litre
3. Inorganic	3000 mg/litre
4. Sulfates as SO_3	400 mg/litre
5. Chlorides	2000 mg/litre for plain concrete
	500 mg/litre for reinforced concrete

2.9 STEEL REINFORCEMENT

Reinforcement in concrete normally refers to steel bars or wires. Of late, other types of fibres are used to improve the tensile capacity of the concrete. Natural and artificial fibres are being used to a limited extent and such reinforcement is not covered in this book. The following steel reinforcement is in practice.

(a) Mild and medium tensile steel bars (*MS*).

(b) High Yield Strength Deformed bars (*HYSD* or *HYD*).

(c) Cold Twisted Deformed bars (*CTD*).

(d) Rolled structural steel in composite construction.

The reinforcing bars must be free from dust, rust, oil, paint, loose mill scales, wrinkles etc. before placing them in position. Usually 6 mm to 16 mm bars are used in slab and shell constructions, 10 mm to 28 mm bars are commonly used in beam and column constructions. Bars upto 40 mm are used in large beams, columns and foundations. Some important properties and other minimum requirements for reinforced bars are listed in table 2.6. The mild steel has a clear yield point that is considered as the reference strength of the steel. However, cold twisted deformed bars and high yield strength deformed bars don't have such a clear yield point. In such cases a stress called proof stress is taken as the reference value. *Proof stress* is the stress that corresponds to a residual strain of pre-assigned value. 0.2 percent residual strain is normally accepted as a pre-assigned value in most cases. The proof stress can be obtained from stress-strain curve of uniaxial tension test specimen.

Table 2.6 Minimum Requirements of Properties of Reinforcement Bars (stress in N/mm^2).

(The quality is associated with steels produced in India)

No	Property	MS-I	MS-II	MTS	HYSD/CTD	Hard drawn
1.	*Ultimate tensile*	410	370	570	490	560
2.	*Yield/proof stress*					
	Diameter: 6 to 40 mm	250	230	350	415	500
	Diameter > 40 mm	230	210	320	415	450
3.	*Percentage elongation*					
	Diameter < 10 mm	20	20	17	14.5	7.5
	Diameter > 10 mm	23	23	20	14.5	7.5
	(on Gauge length = 8D)					
4.	*Limiting tolerances*					
	In length: Length not specified	−25 mm; +75 mm				
	Maximum: Length specified	+75 mm				
	Maximum: Length specified	−50 mm				
	In weight					
	Diameter less or equal to 8 mm	±4%				
	Diameter > 8 mm	±2.5%				
	In diameter upto 25 mm	±0.5%				

The notations used in the table are: D = diameter
MS-I = mild steel grade 1, *MS-II* = mild steel grade 2, *MTS* = medium tension steel
HYSD = high yield strength deformed bars of grade 1,
CTD = cold twisted deformed bars; Hard drawn, cold drawn wires/bars

Mild steel bars are usually plain bars and have lower bond strength. The bars are likely to fail by bond even before reaching the yield strength unless proper hooks or anchorages are provided at the ends. Usage of plain bars except 6 mm bars is extinct in the present day reinforced concrete construction. At times small diameter bars such as 6 mm bars (wires) are used as ties or as secondary reinforcement. The deformed bars or cold twisted bars are used invariably in present day practice. The bond strength of the deformed bars is about 60 percent higher than the plain bars.

2.10 INTRODUCTION TO CONCRETE

Concrete is a hardened composite matrix consisting of cement, sand and course aggregate mixed with water and cured under moist condition. The composite wet mix ready for placement or placed in position but not hardened is called *green concrete*. Spray of water or supply of moisture to the setting concrete to gain strength and absorb heat of hydration is called *curing of concrete*. The word curing should not be misunderstood in the medical terminology as if the concrete is sick and needs curing. Civil engineers and builders should be interested in the properties of the green as well as hardened concrete. Concrete technology is a vast subject so only relevant properties of the concrete are discussed briefly in this chapter.

Major types of concretes in practice are:

(a) Lean Cement Concrete (*LCC*) may be called as Lean Concrete (*LC*), (*LC* may also refer to lime concrete)
(b) Plain Cement Concrete (*PCC* or *PC*),
(c) Reinforced Cement Concrete (*RCC* or simply *RC*),
(d) Prestressed Concrete (*PSC*), sometimes referred as high strength concrete (*HSC*),
(e) High Performance Concrete (*HPC*),
(f) Light Weight Concrete (*LWC*), and
(g) Heavy Weight Concrete (*HWC*).

The Indian Standard code (abbreviated as IS code) of practice on plain and reinforced concrete gives a classification of concretes as: *ordinary, Standard and High strength concrete*. The ordinary concrete has strength in the range of 10 N/mm^2 to 20 N/mm^2 (called $M10$ to $M20$ grades). The range of grades of standard concrete is $M25$ to $M55$. The grade of high strength concrete is in the range of $M60$ to $M80$. All grades are at an incremental value of 5 N/mm^2. This classification appears to be somewhat arbitrary and not aimed at the common usage.

Concrete is generally identified by its strength and referred as grade. A notation in specifying the grade of concrete by the Indian code of practice is prefixed by letter M. Grade $M10$ refers to a concrete having a characteristic strength of 10 N/mm^2. Similarly $M15$, $M20$, $M25$, $M30$, $M35$, $M40$, $M45$ etc are referred at an incremental value of 5 N/mm^2. The mother code IS:456-2000 lists grads upto $M60$, however, concretes upto grade $M80$ are produced in India at the moment. Concrete grade $M130$ is talked about in advanced countries. The days are not very far off when concrete of grade $M150$ will be used in construction. The notation of identifying the concrete may vary with a country. The grade identifies strength and this strength is referred as characteristic strength. The *characteristic strength is the strength of a material below which not more than five percent of the test results are expected to fall*. The strength data of most material follow normal distribution. If f_k is the characteristic strength of a material, then the probability of not more than a specified sample fall below it can be expressed as:

The probability that f (strength) should not fall below f_k is = 5 percent (2.1)

$$P(f < f_k) = P_f \qquad (2.2)$$

Where:

P = probability distribution function,
f_m = mean strength of the sample,
p_f = is the accepted probability,
f = sample strength, and

The above expression can be rewritten for normal distribution as:

$$\phi\left(\frac{f_k - f_m}{s}\right) = P_f \qquad (2.3)$$

Where

ϕ = Normally distributed probability function,

s = standard deviation of the sample strength.

The inverse of the probability equation can be expressed as:

$$\frac{f_k - f_m}{s} = -k \tag{2.4}$$

Where k is the accepted probability of failure,

A negative sign is assigned to k as the quantity is at the left tail end of the distribution.

For 5 percent acceptability of failure, the value of k can be obtained from normal distribution table as 1.65. Therefore the above equation reduces to:

$$f_m = f_k + ks \tag{2.5}$$

The significance of the above expression is to aim the mean strength of the samples higher than the characteristic strength by about the product of the acceptable probability and the standard deviation.

The characteristic strength of concrete (f_{ck}) can be written as:

$$f_{ck} = f_m - 1.65s \tag{2.6}$$

Where

f_{ck} = is the characteristic strength of concrete

Lean Cement Concrete

Lean cement concrete (LCC) also called lean concrete (LC) is made of very small proportion of cement. Lean concrete is not a structural concrete. It is used as a base-course in flooring and foundations, and at times it is also used as a filler material. The lean concrete is used to prepare a level surface on which structural concrete can be placed. Lean concrete can level the irregularities in excavation or the ground unevenness. It also helps in preventing the cement grout of structural concrete to seep into the soil. The strength of the lean concrete varies in the range of 5 MPa to 10 MPa. MPa refers to Mega Pascal, meaning million Pascals. A Pascal is equal to one Newton per square metre. Nominal mix proportions of 1:5:10 is supposed to give 5 MPa. This probably the leanest mix that is workable. The next higher lean mix is 1:4:8, likely to give strength of 7.5 MPa, commonly used for leveling course. The thickness of the leveling course of the lean concrete varies 75 mm to 150 mm depending on the ground conditions and the strength of the structural concrete that is placed over it. 120 mm thick layer is the most common one. The course aggregate used in the lean mixes is in the range of 40 mm to 63 mm size and depends on the thickness of the concrete. The size of the aggregate should be one third or less than the thickness of the concrete layer. Nominal mixes used in the lean concretes with mix proportions in volume are listed in table 2.7. The lean concrete can be broken easily by pickaxe.

Table 2.7 Nominal Mix Proportions by Volume.

Grade	(approximate)	Nominal mix
Lean concrete	M5	1:5:10
Lean concrete	M7.5	1:4:8
Plain concrete	M10	1:3:6
	M15	1:2:4
	M20	1:1.5:3

Phases of Concrete

Concrete goes through six phases of transformation. The approximate duration of the periods of these phases for ordinary Portland cement are:

- Phase 1: *Initial hydration*, first 15 minutes after mixing with water,
- Phase 2: *Induction period*, from 15 minutes to about 4 hours,
- Phase 3: *Setting process*, 4 hours to 8 hours,
- Phase 4: *Hardening period*, 8 hours to 24 about hours,
- Phase 5: *Curing period*, 24 hours to upto 28 days and
- Phase 6: *Service period:* about 28 days of curing on wards.

   In the first phase, the nucleation of cement hydration products is formed. The hygroscopic particles of cement react with water and chemical kinetics set in at a very fast rate. The surface of the cement is coated with the hydration products. Lot of ionic reactions also takes place in the short duration. Some admixtures may interface with the chemical kinetics of cement even at this stage. Reaction of the Tricalcium aluminate with water results into hydrates of calcium aluminates during the induction period. In case of inadequate supply of gypsum, flash set of the cement takes place. In case of excess of gypsum, false set of cement takes place during this period. Nucleation of gypsum crystals takes place during the period. The concrete can be re-mixed in flash or false set of cement. The setting of the other compounds of cement also introduced during this period.
   The third phase is dominated by the setting of cement. The cement compounds combine with water and result into hydrates of calcium silicates etc. The final setting time for most types of cements is 10 hours. Retarders that are often used in ready mixed concrete delay the initial set. Too much of the retarder can even effect the final setting time of the cement, and consequently the setting of the concrete. Too much delay in the final setting also affects the strength of the concrete. An admixture that reacts with the cement compound can effect the crystallization of the hydrates. The fourth phase is the hardening of concrete. Setting of cement even though leads to some hardening of concrete, the real combination of aggregate with hydrates and concrete solid development takes place during the fourth stage. The period is indicated as 24 hours but the plasticizer and the retarder again influence it. Incompatible two admixtures when used can delay the hardening for longer period. Plasticizer or super-plasticizer along with retarder is added to the ready mixed concrete. Sometimes it takes four or even more hours to transport and pump the ready mixed concrete (*RMC*) because of mechanical, power or formwork problems. That is why more than one dose of retarder is added to *RMC*.
   Concrete curing is the fifth phase. This phase is very important yet often neglected. Builder is interested in laying good concrete but doesn't pay same attention to the curing. The final state is the service state. It is often said that the concrete needs no maintenance. This is an over simplification. Concrete needs care and maintenance to sustain it for the designed life. A scheme of maintenance, quick repairs and protection against aggressive agents is needed for durable concrete.

Plain Concrete (Plain Cement Concrete)

The plain concrete is structural concrete used in foundations, retaining walls, roads, dams and similar applications. Weight of the structure plays an important role in many structures such as retaining walls, hydraulic dams. In such structures the tensile stresses developed are small when compared with the compressive stresses. In roads, some foundations and in floors where the plain concrete is used, load is mostly bearing type and doesn't cause much bending moment. $M10$ to $M30$ are the commonly used grades of plain concrete. There is no restriction to limit the plain concrete to grade $M30$ however the plain concrete constructions beyond $M35$ are rare. The size of the aggregate used in the plain concrete varies widely. The size of the aggregate can be as much as 150 mm, in mass concrete structures like in hydraulic dams. While in road construction, it can be 40 mm down to 20 mm, and similar size is used in the foundations. The size of the aggregate should be less than one sixth of the thickness of the structure or element. Curing of mass concrete is not as easy as the curing in thin and slender elements. Use of Portland pozzolana cement in mass concrete is better as the rate of heat of hydration is less.

Reinforced Concrete (Reinforced Cement Concrete, RC or RCC)

Most structural elements are subjected to bending in some form or other. The bending moment causes tensile and compressive stresses. The tensile strength of concrete is about one tenth of that in the compression. Reinforcement is provided to resist tension in concrete. Steel is much stronger than concrete in tension and even in compression, so steel reinforcement is used to resist tension force, in some cases even the compression. Concrete in which reinforcement is embedded is called reinforced concrete. The reinforcement bars should not slip from the concrete surface when subjected to forces. Further, the reinforcement must be protected against corrosion or rusting. Exposure of steel reinforcement to moisture and oxygen causes corrosion. Therefore, the concrete must be impermeable. Porous and honey combed concrete is not suitable to reinforced concrete or in fact any type of structural work. $M20$ to $M40$ grades of the concrete are normally used in the reinforced concrete. However, grade $M20$ concrete is not acceptable in many advanced countries even though it is permitted by the Indian code of practice for mild exposure condition. Even though there is no upper limit to the strength of concrete to be used in reinforced work, $M40$ is invariably considered as an upper bound either because of economics or because of other considerations such as deflections etc. The commonly used reinforcement is high yield deformed bars or cold twisted bars. The proof strength of these bars is in the range of 415, 500 and 550 N/mm^2. 20 mm size is the commonly used aggregate. The normal weight of the reinforced concrete is 24 kN/m^3 to 25 kN/m^3.

Prestressed Concrete

Concrete in which stresses are induced even before the external loads act on the structure is called *Prestressed Concrete*. Prestressed concrete structures are commonly in bridges, towers, water tanks, shell structures, folded plates, nuclear reactors, and long span girders. Tensioned high-tension steel wires and cables and anchored to the concrete apply prestressing force on the concrete. The grades of concrete used in the construction start from $M35$ onwards. Till recently, concrete grade $M60$ was about the upper limit in many constructions. However, with the introduction of high performance concrete in the range of $M80$ to $M120$ are coming into practice. 20 mm graded aggregate is used in the concrete. The water cement ratio is limited to the least value and is normally less than 0.45. The water-cement ratio as low as 0.28, along with super-plasticizers is used in high performance concrete. The tensile proof strength of the high tensile steel wires is of the order 1500 to 2000 N/mm^2. The shrinkage strain of concrete is comparable with the allowable strain in the mild and medium grade steels, so only high tensile steels are used in this construction. Prestressed concrete is the main solution for long span structures. The normal weight of the prestressed concrete is about 25 to 26 kN/m^3.

High Performance Concrete (HPC)

Very high strength concrete that has the desired strength and workability is called *high performance concrete*. The high performance concrete has come into practice since 1980's. Strong well-graded aggregate, *OPC* with silica fume and small amount of water are used in high performance concrete. Silica fume also called as micro-silica is a very fine powder (a fume). It is amorphous active silica, reacts with lime forming calcium silica hydrates in the microform. It fills up the finest voids in the concrete, making the concrete impermeable and very strong. Silica fume is more expensive when compared to cement. It is byproduct of Ferro-silica alloys and at the moment it is being imported in to India. The quantity of silica fume used in high performance concrete is about five to ten percent of cement. Use of super-plasticizer is a must to obtain good workability of the concrete. The present range of high performance concrete is *M*60 to *M*120. Sooner than expected, the strength of the concrete will go beyond 120 N/mm^2.

Light Weight Concrete

Light weight concrete uses lightweight aggregate and is lighter than the normal concrete. A porous aggregate results in to lightweight material. Further the filling up the voids in the coarse aggregate by fine aggregate is minimized and air content is increased. The aim is to obtain more voids in the concrete and at the same time make it homogeneous. The range of the density of the lightweight concrete is quite wide. The weight of the lightweight concrete varies from 400 kg/m^3 to 1900 kg/m^3. Some lightweight concrete is lighter than water and is used for thermal and sound insulation. Very light concrete is also called as *cellular concrete* and coarse aggregate is not used in such concrete. Further large volume of air bubbles is introduced into the concrete with lightest aggregate. Finally there is the structural lightweight concrete having a unit weight of 1400 to 1900 kg/m^3 and having a minimum compressive strength of 15 N/mm^2. Pumice stone, slag, cinder etc are the common lightweight aggregates. Artificial lightweight aggregate is manufactured using industrial byproducts.

Heavy Weight Concrete

Heavy weight concrete is the one that is made of heavy-density aggregate. Heavy-density aggregate such as hematite, an iron ore material is extensively used in heavy weight concrete. This concrete is used in industrial floors where considerable wear and tear is expected. It is also used in nuclear containment vessel constructions. The unit weight of heavy weight concrete is more than 2600 kg/m^3.

2.11 IMPORTANT PROPERTIES OF CONCRETE

A number of characteristics and properties of concrete are of importance to the structural engineer and builder. Many properties are interdependent and should be studied together. Some properties influence in manufacturing the concrete and some in resisting the loads and environmental forces. The important properties relevant to structural and construction engineer are:

❖ Workability of concrete,
❖ Segregation and Bleeding,
❖ Curing of concrete
❖ Shrinkage and Creep,
❖ Strength, and
❖ Durability of concrete.

The main factors that affect the characteristics of the concrete are:
❖ Quality of ingredients, such as aggregate, sand, cement, water and admixtures,
❖ Relative proportions of the ingredients,
❖ Making of green concrete, mixing time, transportation, laying etc.,

❖ Methods in making, laying, finishing of concrete,
❖ Protection of green concrete and curing,
❖ Temperature and weather conditions during laying and curing,
❖ Exposure conditions and maintenance.

Different properties of the concrete are briefly discussed in the following sections.

2.12 WORKABILITY OF CONCRETE

Workability of concrete is the property that indicates the ease in mixing, placing, compacting and finishing with least segregation of the particles. So the workability involves the following aspects:

❖ Ease in mixing the ingredients,
❖ Ease in placing the concrete in position,
❖ Ease in compacting,
❖ Achieving the homogeneity and no segregation, and
❖ Finishing the surface of the concrete.

Layman thinks that the concrete is more workability with more water. The laborer who handles the mixing of concrete is happy to add more water to make the concrete easily workable. More water in concrete mix may result in faster production, quick laying and finishing. But water beyond a point weakens the strength and durability of the concrete. The workability requirement depends on the size of the element, amount of reinforcement, location of placing the concrete and the weather conditions at the time of concreting. One needs more workable concrete if the percentage of the reinforcement is higher. The influences of different factors on the workability of concrete are discussed briefly here.

(a) *Water-cement ratio:* The workability increases with increase in the water-cement ratio, however the strength and durability of the concrete decrease if the ratio is beyound a point. It is therefore desirable to limit the water-cement ratio to the minimum possible based on the strength criterion. For high strength and high performance concrete, the water-cement ratio is to be limited to 0.45 or even down to 0.30. Even the water applied during curing of the concrete assists in the formation of the hydrates. Low water-cement ratio doesn't give workable concrete unless a super-plasticizer is added in making the concrete. It is advisable to lower the water-cement ratio and use a water reducing agent or super-plasticizer for better quality.

(b) *Size of aggregate:* Coarse aggregate improves the workability of the concrete. Larger the size of the aggregate, better is the workability. Higher the size of the aggregate, higher is the strength of the concrete. However the size of member, percentage of reinforcement, cover and surface finish specifications also control the size of the aggregate. 20 mm or 25 mm size of aggregate is recommended for reinforced concrete. The desired size of the aggregate depends not so much on the workability but on the type of structural elements, and cover to the reinforcement. Fine aggregate decreases the workability of the concrete.

(c) *Grading of aggregate:* A well-graded aggregate will result in better compaction and minimum segregation. The specification on size of the aggregate normally refers to the largest value used. But it is desirable to use at least two different sizes of the coarse aggregate for good concrete compaction and strength. For example one can use 20 mm and 10 mm combination in the coarse aggregate selection. Better workability and strength are obtained by the well-graded aggregate.

(d) *Shapes of aggregate:* The rounded aggregate demands least amount of water when compared with that for the other shapes. Angular aggregate with almost equal size faces is good for interlocking and strength. The angles between the faces of the aggregate should not be acute or obtuse. Angular aggregate is preferred for strength consideration aspect. Crushed aggregate is usually an angular shaped one.

(e) *Flakiness of the aggregate:* Flaky or oblong aggregate decreases the workability and also the strength of the concrete. Therefore such aggregate should not be used in making of concrete.

(f) *Aggregate-cement ratio:* More cement means more water for the same water-cement ratio. Consequently the increase in the aggregate content with respect to the cement decreases the ratio of the water to the total mass of the concrete. Normally the aggregate-cement ratio is chosen for a given water-cement ratio to obtain a desirable strength of the concrete. Higher aggregate-cement ratio decreases the workability as well as the strength of the concrete.

(g) *Air content in concrete mix:* Air in green concrete acts as a lubricant and increases the fluidity and consequently the workability. Sometimes air entraining agents are added to concrete to increase the workability at a lower water-cement ratio. However the entrapped air must be driven out by proper vibration for good consolidation of concrete.

(h) *Time of transit:* A minimum mixing time is needed for a given water-cement ratio to obtain good workability of concrete. The transit time of concrete from the mixer to the location of placement needs to be as minimum as possible. Green concrete gets stiffened during transit even before the initial setting of the cement. Five minutes of transit time is desirable, at the most ten minutes can be considered as a last resort. A retardant must be added if the transit time is more than ten minutes. Ready mixed concrete needs more transit time. It can be anywhere from 30 minutes to four hours. Addition of retardant in the transit mix is very common. The retardant may have to be added in periodically depending on the transit time.

(i) *Admixture to concrete:* Workability admixtures such as air entraining agent or plasticizer or super-plasticizer are used in high strength concretes to improve the workability at low water-cement ratio. Super-plasticizer is a must in the high performance concrete. Selection of minimum water-cement ratio to achieve good strength and then super-plasticizer to obtain the desired workability is the best strategy in good concrete practice.

Workability of concrete is quantified and measurable by a number of tests. Some tests are very simple but may or may not integrate the totality of the workability of the concrete. That is mixing, placing, compacting, finishing and obtaining good homogeneity. The following are the tests that measure the workability of concrete:

(a) Slump test,
(b) Compaction factor,
(c) Vee-bee (or *V-B or Vebe test*) consistometer,
(d) Flow test,
(e) Remolding test,
(f) Ball penetration or Kelley ball test, and
(g) Two-point test.

Slump test is the simplest and most extensively used test to measure the workability of the concrete. A frustum of a conical tube of 300 mm high, 200 mm diameter at bottom, 100 mm at top with two handles attached to the surface at about the middle height is used as a mould. Concrete is placed in three layers, each layer being consolidated by 25 tamping of a rod and then the top surface is leveled. The mould is pulled up and the wet molded concrete is allowed to slump or slide down. The magnitude of the slump of the concrete from the 300 mm height is called *slump of the concrete*. Fig. 2.2 illustrates the slump cone and the slump of the concrete. More slump obviously means more workable. It is considered to be reasonably good for low slump or high slump but not very good for the middle range of the slump. The fresh concrete is classified into four broad groups based on slump, and they are listed in table 2.8.

Table 2.8 Classification of Fresh Concrete Based on Slump.

Class >	Low	Medium	High	Very high
Slump (mm)	0-35	40-75	80-150	160 & above

The concrete gets stiffened with time even in a small duration of 10 minutes transit. The slump of the concrete at time of placing is likely to be less than that measured at mixing time. The phenomenon of reduction of slump with time is called *slum loss*. Ready mixed concrete is transported for longer distance and duration when compared to the site mixed concrete.

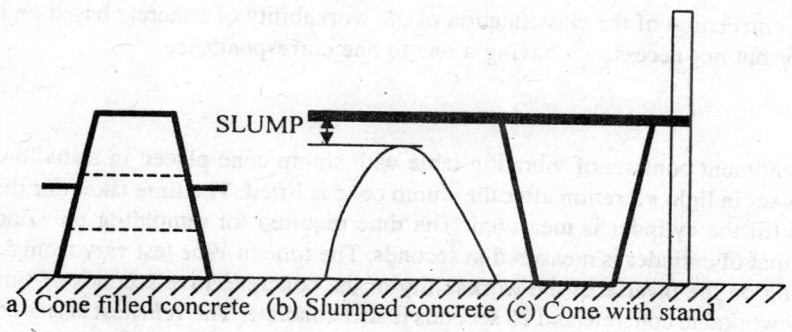

a) Cone filled concrete (b) Slumped concrete (c) Cone with stand

Fig. 2.2 Slump Cone & Slump Measure

The required slump is always expected to be at the time of placing the concrete. The normally recommended slump for different types of concreting environment in dry weather condition is listed in table 2.9.

Table 2.9 Recommended Concrete Slump for Construction.

Type of work	Slump (mm)
PCC, Shallow slabs & beams with nominal reinforcement	Low (20 to 40)
Beams and columns with reasonable reinforcement	Medium (40 to 75)
Heavily reinforced concrete members	High (75 to 125)
Deep members with heavy reinforcement (pumped concrete)	Very high (125 to 200)

Compacting Factor Test

The compacting factor test is based on the degree of compactness achieved in placing (or dropping) the concrete from a specific height. The ratio of the density of the concrete actually achieved to the density that is possible in fully compacted condition is called *compacting factor*. Two conical hoppers with doors at bottom are placed one over the other at a specified distance. The top hopper is bigger than the lower one. A cylindrical mold of 150 mm by 300 mm is placed below the hoppers. The top hopper is filled with fresh concrete without any compaction. On opening of the bottom door of the top hopper, the concrete falls into the lower hopper. The concrete fills up the lower hopper by gravity and then overflows. Similarly the bottom door of the lower hopper is released and the concrete is allowed to fall and fill the bottom cylinder. The density of the concrete in the cylinder is calculated. The ratio of this density to that of the fully compacted density of the concrete in the cylinder is called the compacting factor. The workability of the concrete based on the compacting factor can be divided into four classes and they are listed in table 2.10. Fig. 2.3 illustrates the general arrangement of the compacting factor test equipment.

Table 2.10 Classification of Workability of Fresh Concrete Based on Compacting Factor.

Class	Low	Medium	High	Very high
Compacting factor	0.75	0.85	0.92	0.95

There is some correlation of the classification of the workability of concrete based on the slump and compaction factor but not necessarily having a one to one correspondence.

Vebe Test

The Vebe test equipment consists of vibrating table with slump cone placed in a shallower cylindrical tube. The table is set in light vibration after the slump cone is lifted. The time taken for the freestanding concrete cone to fill the cylinder is measured. The time required for remolding the concrete from the conical shape to that of cylinder is measured in seconds. The time in *Vebe* test vary from 5 seconds to 25 seconds depending on the inverse of the workability of the concrete. Five seconds of time of *Vebe* test represents highly workable concrete and 25 seconds is stiff concrete. The *vebe* test is less commonly used when compared to the two tests mentioned earlier. Fig 2.4 illustrates the principles of *Vebe* test.

2.13 SEGREGATION AND BLEEDING OF CONCRETE

Concrete mix consists of particles of different density and sizes. The heavier particles tend to settle down or move to bottom layer. There is a tendency of the segregation of the particles while placing and during vibration of concrete. Well-mixed concrete has lesser tendency to segregate. If concrete is messed instead of mixed or simply dumped instead of placed properly, the segregation of the particle takes place. The

Fig. 2.3 Compacting Factor Test

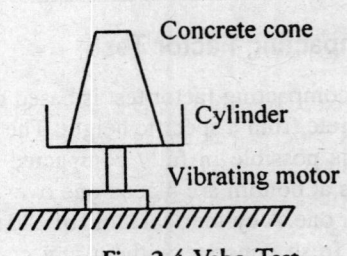

Concrete cone

Cylinder

Vibrating motor

Fig. 2.4 Vebe Test

segregation of concrete is the property of green concrete to get separated into particles and lose the homogeneity of the concrete. Well-graded aggregate and uniformly mixed concrete has the least chance of segregation while placing and in vibration. The natural vibration frequency of the particles depends on the size, shape and the mass. The right amount of vibration, not too much of it is to be applied depending on the workability of the concrete. There is no prescribed test to measure segregation, but visual observation indicates the segregation of concrete. Optimal mixing, right delivery and compaction, and careful vibration avoids segregation of concrete. The segregated concrete has voids and honeycombing. It is not durability.

Water being the lightest of the components of the concrete, it tends to float to the surface. The phenomenon of rising of water to the top surface of the concrete during compaction or vibration is called *bleeding*. Over vibration or inappropriate compaction of concrete results in bleeding. The bleeding of concrete results in less homogeneous concrete and soft-top layer. Cement or similar fine particles are carried by water to the top surface at times. Such fine particles with water forms a soft crust at the top of the concrete. This phenomenon of rising of the water along with fine particles of cement is called *laitance*. The laitance leaves a soft-top crust on the concrete thus reducing the strength and durability of the concrete.

2.14 CURING OF CONCRETE

Curing of concrete is a misleading word for a common man. Curing means to cure from a problem or decease. Curing of concrete has altogether a different meaning. Cement when comes in contact with moisture, reacts chemically on the surface of the cement particles. Bonding with the surrounding particles of cement is developed during the chemical reaction of cement. The process of formation of compounds of cement with water is called hydration. The hydration of cement generates heat and the heat is called heat of hydration. The hydration process is initiated at about eighty percent of humidity environment. Uncontrolled heat of hydration causes micro cracking in concrete. Further, moisture is needed continuously for some period in and around cement particles to continue with the formation of hydrates of cement. Therefore moist environment is a necessity for the hardening concrete. Supply of moisture or water to the hardening concrete to dissipate the heat of hydration and to aid further chemical reaction of cement is called *curing of concrete*. The chemical reaction gets accelerated with increase of the temperature of the water or moisture.

The concrete gains strength with increase in duration of curing. An approximate strength of concrete at 90 days with different periods of curing is given table 2.11. Concrete that is allowed to gain strength by exposing to air is called *air curing*. It is not really any curing but given a notation for no curing with water and for lack of better terminology.

Table 2.11 Strength of Concrete with Curing.

Days of water curing >	0	3	7	28
Ratio of strength on 90 days to that at 28 days curing >	0.5	0.75	1.0	1.2

Moist curing: The concrete is considered green until about the final setting time of cement. The final setting time of cement is not necessarily the final set of the concrete. The final setting time of most cements is about 10 hours but the concrete after placing in position under normal temperature of the order 20 degrees Celsius is likely to set in about six hours. Five to ten hours of pre-curing time is adequate in hot weather conditions. The hardening of concrete in warm weather of about 30+ degrees Celsius is faster so the curing with moisture can start after eight hours of casting. After about 6 to 10 hours of casting, the exposed surfaces of hardening concrete must be kept continuous under damp or wet condition by sprinkling of water continuously. Or the surface can be covered with jute sacks that are kept under wet condition. The concrete can also be kept under submerged condition wherever feasible for curing. This is probably done in pre-cast concrete construction. The concrete test samples are always cured in water submerged condition. But such a situation is hardly available for most structures. Commonly accepted practice is to cover the concrete with wet jute bags and sprinkling of water as frequently as needed to keep the bags wet all the time. Most asked question is how frequently the water must be sprinkled? The frequency of spraying water depends on the weather conditions. In rainy season when the humidity is high, the sprinkling of water can be once in four hours. But in the dry summer, sprinkling has to be more frequent. The aim is to keep the gunny bags wet all the time. In any case, the surface of the concrete should be wet or under 80 percent humidity condition. Another question that is asked is how long the curing of concrete be continued without interruption? 28 days curing is the best and desirable to result into a strong and durable concrete. Under no circumstances the period of curing should be less than 7 days. A canvas or polyethylene sheet placed over the wet surface of the concrete helps in reducing loss of moisture from the concrete. Seven days of water curing will atleast give strength of that of 28 days cured value on 90th day. However one must remember the durability of the concrete will be affected even if the strength is achieved if the curing time is cut down.

Curing of many concrete structures is given the lowest importance in the unorganized sector in India. The sprinkling of water over the concrete is usually handed over to a small boy or a woman worker or a watchman. Further, the construction is in progress while some portion requiring curing. Many parts of the structure are not easily accessible to the helpless boy who is supposed to maintain the wetness of the surface of the concrete. The engineer and the supervisor must pay more attention to the curing of the concrete. Curing of slabs is about the easiest, and curing of the concrete in columns is about the most difficult one. The slabs being relatively thin and are flat, ponding of water on the top face of the slab is adequate. However the bottom face of the slab is sprayed with water now and then. The concrete slabs have built-in higher factor of safety because of continuity in two directions when compared to that available to columns. The slabs collapse only after developing a yield line mechanism. Columns are the most critical elements and least compacted and least cured in common practice. Covering the columns with wet jute bags that are kept under wet condition is a must. The side faces of the beams are the other items that are not properly cared/cured.

Curing of mass concrete: Thick concrete members such as raft slabs, pile caps and other thick concrete sections need special attention in curing. Concrete raft slabs of the order 1000 mm to 1500 mm or even more thicker are common in large constructions. Since the bottom face is resting on lean concrete, loss of moisture on that surface is almost none. Concrete sections thicker than 1500 mm need special curing arrangements. The temperature of the concrete at the time of laying need to be about 15 degrees Celsius or even less in mass concrete. Ice cold water can be used in mixing the concrete so that the interior of the concrete is at a lower temperature to start with. Surface curing doesn't reach the interiors of thick members. If the thickness of the concrete is more than 2000 mm, cooling pipelines have to be laid in the concrete so that cool water can be circulated through such lines for curing. Rigid perforated tubes can be embedded in mass concrete at the time of laying the concrete. Cold water is pumped through the pipes to provide heat absorption and to maintain 80 percent humidity inside. Too much of un-dissipated heat of hydration can cause micro cracking and damage to the mass concrete.

Steam curing: Concrete can be cured by controlled steam. The concrete must be allowed to harden for about 3 hours, then it is covered under tarpaulin or put in steam chambers. Pre-cast concrete elements can be placed in steam chambers. Steam at about 80° to 100° C is allowed over the concrete, gradually raising temperature of the concrete chamber upto 70° C in about 2 hours time. The steam curing cycle consists of four parts. Pre-heating period, heating period, steaming period and post-steam period. The pre-steam period is the period the concrete is allowed to set partly and it may be two to three hours. Too early exposure of concrete to the steam can cook the concrete resulting in bubbling or boiling surface. The concrete shouldn't be subjected to sudden rise in temperature. The heating period is the period in which the steam is allowed into the chamber at atmospheric pressure to heat the chamber. This period depends on the mass of the concrete and the size of the member and the arrangements of steaming such as chamber or tarpaulin. Steam curing for about 8 to 10 hours can be considered required to obtain an equivalent 7 days strength of water cured concrete. Steam curing of 18 to 24 hours gives strength equal to that of 28 days water cured one. The entire cycle of steam curing can be about 18 to 24 hours. The steam is cut off slowly in about an hour's time and the concrete is allowed to cool in the chamber for another hour. Improperly done steam curing can cause cooking and spalling of the surface concrete. Steam curing is normally done for prestress concrete or pre-cast concrete members. It is a matter of economics of storage and supply of the product, reuse of prestressing equipment and the bed for quick turn over. The steam curing appears to be economical in producing prestressed concrete pole, railway sleepers, pipes etc.

2.15 SHRINKAGE AND CREEP OF CONCRETE

Shrinkage of concrete is the decrease in volume due to evaporation of moisture from concrete and hardening of the concrete. The shrinkage is divided into three types, and they are *plastic shrinkage, drying shrinkage and thermal shrinkage*. There is some settlement or subsidence of concrete even before the concrete gets hardened, and this reduction in volume of fresh concrete is called *plastic shrinkage*. Stiffening of the top layer of the concrete may cause settlement of the inside concrete. The plastic shrinkage may be caused by bleeding, absorption of water by subgrade soil, settlement of formwork, rapid evaporation of moisture from the surface etc. Plastic shrinkage produces horizontal cracking on the top surface. All precautions should be taken to minimize the loss of moisture from the surface.

Fresh concrete when exposed to ambient humidity undergoes volume change due to change in the moisture content in the concrete. The humidity in the fresh concrete is normally more than the ambient humidity. Therefore a reduction in the volume of the concrete takes place. The decrease in volume of the concrete due to change in the moisture caused by ambient condition is called *drying shrinkage*. Thermal shrinkage is the one that is caused by the cooling of the concrete. Decrease in volume of the concrete takes place as it is exposed to the ambient temperature that is lower than that of the concrete. The decrease in the volume of concrete due to plastic or dry shrinkage or thermal shrinkage results in cracking, sometime visible and or invisible. The drying shrinkage can be reversed to some extent by induction of moisture into the concrete. Curing of the concrete at right time and for the right period minimizes the shrinkage.

The shrinkage of concrete is proportional to the amount of water added at the time of mixing and the amount of cement content. Some aggregate characteristics will also influence the shrinkage. Finer cement reacts faster and develops heat of hydration at faster rate thus resulting in shrinkage of the concrete. Temperature at the time of placement of the concrete influences the thermal shrinkage. Higher the temperature at placement, higher is the shrinkage. Pre-cooling of the concrete either with cold water or cooled aggregate reduces the shrinkage. The shrinkage of concrete takes place for a long period. In some constructions, the elements may be cast at different periods. There may be a reasonable time difference between the casting of concretes in different parts of the structure. The concrete cast at different periods shrinks at different rates. The difference in the shrinkage of the different parts is called the differential shrinkage. Part of the shrinkage is recoverable by wetting of the concrete. Structural engineer is primarily interested in the total and differential shrinkage. The shrinkage is a function of the following factors:

❖ Shrinkage decreases with increase in the aggregate-cement ratio,
❖ Shrinkage increases with increase in the cement content,
❖ Shrinkage increases with increase in the water-cement ratio,
❖ Shrinkage decreases with increase in the curing period.

The plastic shrinkage is irrecoverable, whereas the drying shrinkage is partly recoverable by adding moisture to the dried concrete. Half of the shrinkage takes place in the first one-month, and seventy five per cent in the first six months after commencement of drying of concrete. The plastic shrinkage can be reduced considerably by protecting the surface of the concrete immediately after casting of the concrete. The best principle to reduce the shrinkage is to reduce cement and water-cement ratio, and then add an admixture or plasticizers to improve the workability. The shrinkage of concrete can be eliminated by use of expansive cement, the cement that expands while setting.

Shrinkage is basically a strain without stress, and yet produces cracking in the concrete. Restraint on the shrinkage will cause stresses in the concrete. The shrinkage starts at the surfaces of the concrete and extends into the interior. Approximate shrinkage strain for a water-cement ratio of 0.7 can be as much as 0.0007 while it is about 0.0003 for a water-cement ratio of 0.45.

The total free shrinkage for most reinforced concrete construction that is cured for 28 days is taken as 0.0003.

$$\epsilon_{sh} = 0.0003 \tag{2.7}$$

In practice, the columns are cast on one day and the beam and slab are cast few days later. Or the slab including the beams is cast in different stages. The members cast on different days will have different rates of shrinkage. In bridge building, the beams may be pre-cast members and the concrete slab is placed at site over the pre-cast beams. The difference in of shrinkage of the two or more members is called *differential shrinkage*. Differential shrinkage produces interface forces in the members and even cracking.

Creep of Concrete

Concrete when subjected to loads undergoes elastic strain to start with. The stress in the member produces strain and the strain increases if the stress is maintained for a longer duration. The time depended strain under constant stress at 100 percent humid concrete is called *basic creep*. Basic creep of concrete takes place in mass concrete structures. The time dependent strain under constant stress at normal conditions is called *creep strain*. The creep strain per unit stress is called *specific creep*. The ratio of creep strain to the elastic strain is called *creep coefficient*. On unloading, most of the elastic strain is recovered and only part of the creep strain is recovered with time. Creep is a function of relative proportions of aggregate, water, cement, type of aggregate, porosity of concrete, curing period, level of stress and age of loading.

Creep of concrete increases with increase in cement content and water-cement ratio. It also increases with decrease in the age of concrete at loading. Creep also increases with increase in the exposed temperature of the concrete. Seventy percent of the creep strain takes place in the first one month and the balance may take place in the next three years period time.

Creep strain relation is:

$$\epsilon_{cc} = C_c \, \epsilon_{se} \tag{2.8}$$

Where ϵ_{cc} = creep strain, C_c = creep coefficient, and ϵ_{se} = elastic strain

The total strain at any given period is equal to the sum of creep and elastic strains. The creep coefficients are given in table 2.12.

Table 2.12 Creep Coefficients.

Age of concrete at loading	Creep coefficient
7 days	2.2
28 days	1.6
1 year	1.1

As the creep strain increase, deflections in the beams and slabs also increase. The deflection caused by the dead load is to be multiplied by the creep coefficient to obtain the creep deflection. However because of the reinforcement doesn't creep at the same rate as that of concrete, the creep deflection of the reinforced concrete members is only a portion of the creep coefficient. Some redistribution of the stresses takes place indirectly due to creep. In case of columns, the concrete undergoes creep and transfers part of the load to the reinforcement. Therefore the reinforcement in columns under direct compression are subjected to higher levels of stress than that estimated by simple compatibility condition. The creep can cause failure of a structure because of excessive strain in the concrete. This can happen when the dead load is much higher than the live load. In such situations, the permanent load dominates the total strain.

2.16 STRENGTH OF CONCRETE

Strength of concrete is its ability to resist load before collapse. The coarse aggregate in concrete is like a bone skeleton in human structure. The bone primarily resists compression, so much so the aggregate in concrete is the prime source to resist the compression. Aggregate can also resist tension to some extent but in the concrete matrix, there are other weak links in tension and hence the aggregate doesn't control tension problem. The cement-sand mortar in concrete is like muscle in human body. The mortar fills the voids in the aggregate and binds it together to form in to a homogeneous mass. Water in the concrete is like the life giving blood of human being. Water by itself may not resist load, but when combined with cement, it results into compounds of strength and durability. The strength of concrete is measured by its ability to resist compressive stress. 150 mm concrete cube is taken as basic specimen to measure the compressive strength. 200 mm cube is considered as a standard testing specimen in some parts of Europe. 150 mm by 300 mm cylinder is the compression test specimen in North America. Therefore the strength of concrete is referred with respect to a test specimen and it bears a relation with the concrete prism. The compressive strength of prism in direct compression is different from the under bending compression. The strength of concrete is associated with the hardened concrete and depends on several factors as listed below.

1. *Strength decreases with increase in water-cement ratio*. Water-cement ratio is one of the most important factors in determining the strength of concrete. This can be compared to the ratio of the blood content to the weight of a person. There is always an optimal value of this ratio. Too less blood causes anemic and too much of it can cause high blood pressure or other effects. The decrease in the water-cement ratio decreases the workability of the concrete. The advantage of the low water-cement ratio is obtained only if good consolidation of the concrete is achieved. The water-cement ratio shouldn't be indiscriminately decreased leading to poor workability of concrete Workability agents such as super-plasticizers have to be added to achieve required workability and consequently good consolidation and strength. Some of the water that is added into the concrete mix is directly consumed for chemical reactions, and the water available beyound remains as free water in the concrete.

This free water in the hardened concrete evaporates leaving micro-voids in the concrete. The voids in concrete decrease the strength of concrete. More water added at mixing of concrete results in more shrinkage and porosity. One has to strike a balance between strength and workability.

2. *Strength increases with increase in cement content.* Cement acts as a binder in the presence of moisture. The binding material helps in binding aggregate in to a homogeneous solid mass. More cement may mean, more binding material. Cement content beyond a point is of no use and in fact it may cause bad side effects. Protein is good for human body but if one eats too much of proteins, the body will go sick, therefore an optimal cement content must be added for a given situation. More cement also means more heat of hydration and more shrinkage. Further, cement is the expensive material among the basic components of the concrete. It is therefore necessary to make an optimal use of cement in making concrete.

3. *Strength of concrete increases with increase in the size of aggregate.* Strength of the concrete depends on the strength, shape and size of the aggregate. As already mentioned, the aggregate is the filler and the strength giver. Basic strength of the aggregate is important and reflects in the final strength of the concrete. Every thing else being same, the strength of the concrete increases marginally with increase in the size of the aggregate. The type of construction, size of the members and intensity of reinforcement and cover requirement control the size of the aggregate that can be used in the concrete. In plain and mass concrete, one would like to select larger aggregate. 20 mm aggregate is the most commonly used aggregate in reinforced and prestressed concretes. Aggregate crushing value less than 45 is recommended in ordinary plain and reinforced concrete. A value of 30 is preferred in prestressed concrete and reinforced concrete exposed to very severe environment. 30 or 25 mm size aggregate is recommended in road pavement with aggregate crushing value not less than 30.

4. *Strength of concrete depends on the grading, shape and texture of aggregate.* Well-graded aggregate results into compact homogeneous mass therefore better strength. About equal angles of angular shaped aggregate gives better interlocking therefore gives higher strength. Round aggregate also gives good strength. The texture of the aggregate needs to be rough and not smooth to provide good interlocking. Flaky or oblong aggregate should be avoided in making concrete.

5. *Strength increases with curing period of concrete.* As mentioned earlier, cement generates heat of hydration during its reaction with water. This heat of hydration if not absorbed by external agency, produces cracking in the concrete thus leading to a poor product. The heat of hydration can be absorbed by water that is sprinkled on the surface of the concrete. The process of dissipating the heat of hydration and also assisting in saturated chemical reaction of hardening concrete is called *curing of concrete*. In the process curing, the heat of hydration is dissipated and calcium silicates are formed and the concrete gains strength with time. Curing of concrete normally starts after about 10 hours of casting of concrete. The methods of curing may differ but the essentiality of curing is to protect the concrete surface from the loss of moisture and maintenance of wet surface around the concrete. The strength of concrete increases with curing. The suggested standard curing period is 28 days. All standard test specimens must be cured for twenty-eight days. Similarly the concrete structures need to be cured for 28 days to gain full potential strength. Concrete with ordinary Portland cement gains about two-thirds of the 28 days strength in seven days of curing. The rate of gain in strength of concrete with curing increases with curing period but its gain is rapid to start with and slows down with time. The gain of strength beyond 28 days of curing is considered to be marginal and may be about 30 percent in a year period. If the curing is stopped after 7 days, then the concrete will not give its full potential. The gain in strength in such a case will be about eighty to eight-five percent of its full potential.

6. *Concrete gains strength faster with increase in the temperature at curing.* Temperature helps in accelerating in the formation of chemical compounds. It doesn't mean hot water should be poured on concrete. The temperature of the curing water needs to be increased slowly. Hot water is not recommended in curing but faster curing is achieved with warmer water when compared with cold water. Cold water is used in curing the mass concrete. Curing of concrete with steam is also practiced for very quick results. One day of steam curing is considered to be equal to 28 days of water curing. However there is a method of applying the steam curing.

Approximate strength in relation with the age of concrete with respect to that of 28 days curing is given in Table 2.13. However the revised code of practice on plain and reinforced concrete doesn't permit correction to the strength of the concrete with age.

Table 2.13 Strength of Concrete with Age.

Age of concrete	Relative strength with respective to 28 days
7 days	0.65 to 0.7
28 days	1.0
3 months	1.10 to 1.15
6 months	1.15 to 1.20
12 months	1.20 to 1.25

Strength and Stress-strain Behaviour

Strength and stress-strain behaviour of concrete is measured by uni-axial compressive test performed on concrete cube or cylinder specimen. 150 mm concrete cube is the standard specimen accepted in India and some countries. The mould of the specimen must be either cast iron or steel, machined to high dimensional accuracy. Concrete is placed in the mould and vibrated by a specified method and cured in water for 28 days. After 28 days, the cube surface is dried and tested in a compression-testing machine at a prescribed rate of loading. Typical stress-strain behavior of concrete cube is indicated in Fig. 2.5.

The testing is under controlled rate of loading rather than the strain controlled. The rate of loading is 2.5 N/mm^2/sec. The stress-strain curve of concrete in compression is linear upto 30 percent of the strength,

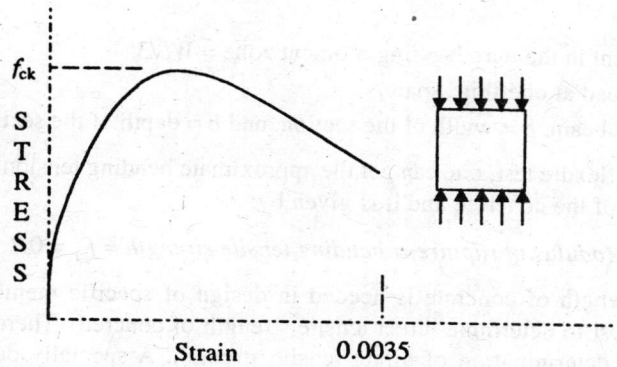

Fig. 2.5 Stress-strain of Concrete.

then an increase in the rate of strain from 30 to 50 percent of the strength, and from there to about 85 percent of the strength, the rate of increase in strain is faster. It is difficult to control the rate of loading after the load reaches 75 percent of the strength, and the crushing of the concrete takes place suddenly. The maximum bending compressive strain in concrete is idealized to 0.0035. The limiting strain can be higher in case of strain controlled tests. The crushing strain in direct compression is limited to 0.002. The instantaneous loading on concrete indicates an increase in the strength. This loading may be termed as impact loading and the strength reading of the concrete with impact will be higher. Repeated cyclic loading on concrete cube can reduce the strength to 70 percent. Similarly sustained load can decrease the strength to about 80 percent of the normal test load. The cylinder strength of concrete is about 85 percent of the cube strength. It is therefore seen that the procedure in testing of the concrete has to be standardized to aim at consistency in understanding of the strength of the concrete. There are other strengths such as prism, flexure, tension, split, shear, bond, etc., which are generally inter-related with the cube strength.

The tensile strength in bending, which is called strength in flexure of concrete is obtained by a flexure test of plain concrete beam specimen. The codes of practice specify a standard specimen subjected to two concentrated loads at one-third span points as shown in Fig. 2.6.

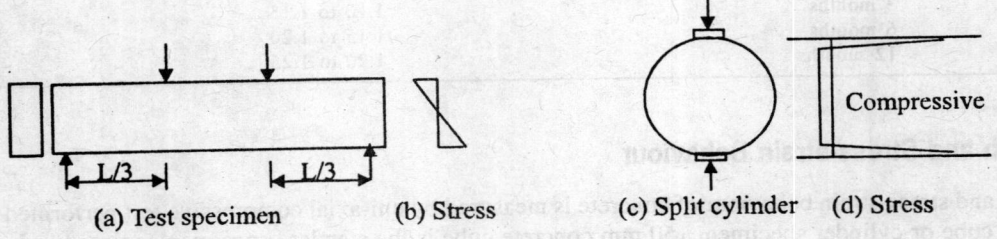

(a) Test specimen   (b) Stress   (c) Split cylinder   (d) Stress

Fig. 2.6 Flexure and Split Cylinder Tests.

The bending tensile strength of concrete from the flexure test is also called as modulus of flexure and it is:

$$f_{cr} = \frac{M}{Z} = \frac{2WL}{bh^2} \tag{2.10}$$

Where
  M = bending moment in the pure bending moment zone = $WL/3$
  W = concentrated load at one-third span
  L = Span of the test beam, b = width of the section, and h = depth of the section

In the absence of a flexure test, one can get the approximate bending tension of the concrete from the compression strength of the concrete and it is given by:

$$Modulus\ of\ rupture = bending\ tensile\ strength = f_{cr} = 0.7\ \sqrt{f_{ck}} \tag{2.11}$$

The direct tensile strength of concrete is needed in design of specific members. Unfortunately it is difficult to design a test to determine direct tensile strength of concrete. There used to be a test called *briquette test* for the determination of direct tensile strength. A specially designed specimen called briquette with gripping ends was suggested at one time. But it is found that it is almost impossible to apply axial tension load without causing bending on the specimen. Therefore the test was considered to be not good enough to determine the real tensile strength of the concrete. An empirical expression is often used to get an approximate direct tensile strength of concrete and it is:

$$Direct\ tensile\ strength = f_{ct} = 0.35\ \sqrt{f_{ck}} \qquad (2.12)$$

Where f_{ct} = direct tensile strength of the concrete.

Another test called *split cylinder* test was developed to estimate tensile strength of the concrete. The test is also called *a Brazilian test*. A 150 mm by 300 mm concrete solid cylinder is laid horizontal between the two platens of a compression-testing machine and subjected to compression load on diagonally opposite faces of the cylinder. Fig. 2.6(c) illustrates the test and Fig. 2.6(d) illustrates the stress distribution across the depth of the cylinder. The split tensile strength can be computed as:

$$f_{ct} = \frac{P}{\pi DL} \qquad (2.13)$$

where P = compressive force, L = length of the cylinder, D = diameter of the cylinder.

The direct tensile strength of the concrete is approximately equal to about sixty percent of the flexure strength of the concrete. The cracking strength of concrete is normally referred as that equal to the split cylinder strength. The tensile strength of the concrete in bending is equal to the modulus of rupture in case of beams subjected to bending. The allowable tensile stress is obtained by dividing the tensile strength by factor of safety.

Table 2.14 Modulus of Rupture and Bond Strength (stress in N/mm²).

Concrete strength = f_{ck} =	10	15	20	25	30	35	40	45
Modulus rupture	2.2	2.7	3.1	3.5	3.8	4.1	4.4	4.7
Bond stress for 0.25 mm slip								
Plain bars	—	1.9	2.4	2.9	3.3	3.4	3.5	3.5
Deformed bars	—	3.5	4.4	5.2	5.8	6.3	6.5	6.6
Bond stress for no slip:								
Plain bars	—	1.5	1.7	2.0	2.2	2.5	2.7	2.8
Deformed bars	—	1.9	2.1	2.5	2.8	3.1	3.4	3.5

The reinforcement bars are embedded in concrete and the bars should not slip from concrete when subjected to pull or push. The bonding capacity of the concrete with the bar is called *bond strength* of the concrete. The bond strength of the concrete is determined by *pullout* test. A bar is embedded in a concrete cylinder and pulled from one end. Pulling force is applied at one end of the bar and slip at the other end of the bar is measured. The bond strength is determined in two stages. One that corresponds to no slip condition and the other is slip condition of the bar. The load at which the slip is initiated is called the no slip bond strength. The ultimate bond strength is the bond stress corresponding to load at which 0.25mm slip occurs. The bond stress is equal to the pulling force divided by the surface area of the embedded bar. Table 2.14 gives different strengths of the concrete recommended for design.

Cube test: Concrete cube under compression has its top and bottom faces in contact with the compression platens. As the cube is subjected to uni-axial compression, the cube will get shortened and the lateral dimensions tend to expand by the Poisson's effect. However the contact top and bottom faces of the cube are constrained against the lateral expansion. The fracture of the cube is influenced by the frictional force on the top and bottom faces of the cube in addition to the axial compression. The failure is not by uni-axial compression but by combined action of the forces on the cube. The failure or the crushing pattern of the test cube is like a double symmetrical pyramid. Figure 2.7(a) illustrates the failure pattern of a concrete cube. The failure is brittle and is due to the combined effect of compression and friction on the faces. On the other hand when a concrete prism is subjected to uni-axial compression, the surface friction of the platens at the top and bottom faces damp out in short length and the middle portion of the prism experiences direct compression or uniaxial compression. The compressive strength of the concrete prism has a relation with that of the cube.

Table 2.15. Compressive Strength of Prism with Respect to 150 mm Cube.

Length/width	0.5	1.0	2.0	3.0	4.0	5+
Relative strength	1.5	1.0	0.8	0.72	0.68	0.67

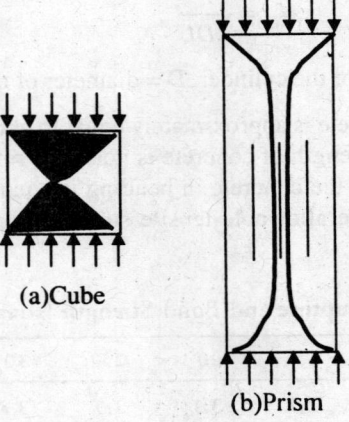

(a)Cube

(b)Prism

Fig. 2.7 Failure Pattern of Concrete Cube and Prism under Uni-axial Compression.

The strength of a prism specimen decreases with increase in height-to-side ratio and converges to a value when this ratio is 4 or 5. The compressive strength of prismatic element where the length to thickness ratio is five or more, is 0.67 times that of the 150 mm cube. Similarly, the 150 by 300 mm cylinder strength is about 0.80 times that of the 150 mm cube.

Table 2.16 Strength of Concrete Cubes with Respect to that of 150 mm Cube.

Cube size (in mm) =	100	150	200	300
Relative strength wrt. 150 cube	105	1.0	0.95	0.87

Strength of the concrete determined by cube sample varies with the size of the cube. The strength of the specimen increases with decrease in size of the cube. Table 2.16 gives approximate distribution of strength of concrete based on the size of the cube.

The strength that is indicated by a cube specimen is estimated to be more than that by the cylinder specimen. The size of the cube and the diameter of the cylinder are to be same. Its relation is:

$$f_{cu} = (0.75 \text{ to } 0.80) f_{cy} \qquad (2.14)$$

where

f_{cy} = cylinder strength having height to diameter ratio equal to two, and

f_{cu} = strength of concrete cube having the same size as that of the diameter of the cylinder.

The lower value applies to smaller sizes.

Moisture content in concrete provides lubrication effect and reduces the strength when compared with a dry sample.

$$\text{Strength of a dry sample} = (1.1 \text{ to } 1.20) \text{ times the strength of the saturated sample} \qquad (2.15)$$

Modulus of Elasticity of Concrete

Modulus of elasticity of a material is defined as the ratio of the uni-axial stress to the corresponding strain with in the elastic limit. The stress-strain relation in concrete is almost linear upto 30 percent of the crushing strength and then it is non-linear. Therefore the measurement of elastic modulus in concrete is not as clear as that of steel. Three different moduli are mentioned in such non-linear materials.

Tangent modulus: The slope of the tangent drawn at a point to the stress-strain curve is called tangent modulus. This value depends on the point chosen on the curve. It decreases with increase in the stress level chosen. It is useful in the study of non-linear behavior of concrete structure.

Scent modulus: The slope of a line drawn from the origin of the stress-strain curve to a point at 40 percent of the ultimate stress is called the scent modulus. This probably is realistic in the determination of elastic deflection of concrete structures.

Chord modulus: As the name indicates, it is the slope of the chord drawn between two points on the stress-strain curve. This is of very little practical usefulness.

Dynamic modulus: The tangent modulus measured at the starting stress-strain curve is considered to be the dynamic modulus of elasticity. This can be obtained from the compression test of a cube or a cylinder. This value is likely to be atleast 20 to 40 percent higher than the scent modulus.

Flexural modulus of elasticity: A simply supported plain concrete beam is subjected to concentrated load at mid-point and the deflection of the beam is measured for incremental loads. Flexural modulus of elasticity is computed from the load deflection relation of a simply supported beam. The test beam has to be slender so that the deflection due to shear is negligible. The flexural modulus of elasticity is given by:

$$E_c = \frac{WL^3}{48Iv} \qquad (2.16)$$

The modulus of elasticity of concrete is essential to compute deflections of structures. The modulus of elasticity of concrete is taken close to the flexural modulus of elasticity. The code of practice suggests the modulus of elasticity in terms of the strength of the concrete. The Indian code of practice on plain and reinforced concrete suggests the following. The modulus of elasticity of the concrete E_c (in N/mm^2) can be taken as

$$E_c = 5000 \sqrt{f_{ck}} \qquad (2.17)$$

A linear stress distribution across the depth of a beam is assumed for a beam subjected to bending moment. As the bending moment increase, the plain concrete beam fails when the tensile stress concrete at the extreme fibre reaches the flexural strength. All the tensile force is resisted by the reinforcement in reinforced concrete beam under bending. Redistribution of stresses in concrete compression zone takes place as the load tends to failure level. Consequently the strength of concrete in bending is higher than that in direct compression. There is no such redistribution of stress in columns under compression.

2.17 DURABILITY OF CONCRETE STRUCTURES

The property of concrete structure to withstand the environmental effects along with loads is the *durability* property. The durability is measured on qualitative scale rather than a specific quantitative scale. The internal and external factors that influence the durability and life of a structure are listed and explained briefly in this section. The degree of the effect is different for different factors. A list of parameters that influence the durability is given and explained later.

Physical and mechanical factors:
Chemical factors:
Biological and environmental factors
Non-structural factors

1. *a)*   *Physical internal factors*:
   Compaction of concrete, Porosity and permeability of concrete, Surface finish of concrete, Duration of curing of concrete; Cover to the reinforcement and Surface cracking to start with.
   b)   *External factors*:
   Abrasion and erosion (wear & tear),
   Exposure to wetting and drying (moisture) and Freezing and thawing
2. *Chemical factors*:
   Chemical composition of Aggregate, Quality of cement and water used, Exposure to acidic environment, Sulphate attack, Aggregate alkaline reaction, Other chemical aggressive actions.
3. *Biological and environmental factors*:
   Biological growth, such a moss, algae etc on the surface of concrete
4. *Non-structural factors*:
   Shape and size of structural elements, Drainage from the structure,
   Joints, Inserts, Bearings, Railings, Anchorage and Fixtures.

There is interdependence between the factors that are listed above.

Physical Factors

The main physical factors that influence the durability of concrete are explained in brief:

❖ **Compaction of concrete and Moisture transport:** Production of good concrete should be the aim of any builder. Good workable concrete leading to dense and well-compacted concrete minimizes the transport of moisture within. The porosity or permeability of the concrete is a direct result of poor compaction of concrete. Transportation of moisture induces physical and chemical reactions that are responsible for the deterioration of the concrete. Diffusion of air, carbon dioxide and chloride ions etc. into concrete combined with moisture add to the problem. The carbonation of concrete leads to increase in volume and cracking. Similarly the presence of moisture and air leads to corrosion of reinforcement.

❖ **Surface Finish:** Water in contact with the concrete surface is transported through capillary action even in well-compacted concrete. Uneven and rough surface retains moisture and transports. Micro-cracked surfaces of concrete permit the movement of water freely into the concrete. An impermeable concrete surface is the most ideal finish even though not possible, but even such surfaces develop micro cracking in due course because of changes in temperature and chemical actions. Adequate precautions must be taken to maintain an even and un-cracked surface of concrete.

❖ **Wetting and drying:** Water carries dissolved salts such as sulphates, chlorides, carbonates etc. into the pores of the concrete under wet condition. The moisture in the concrete evaporates during drying leaving dissolved salts in crystallized state. The accumulation of such crystals increases the volume leading to cracking of the concrete. Further the concentrated crystallized chemical act on the aggregates and reinforcement causing carbonation and corrosion as the case may be. Efflorescence of crystallized salts at the surface of concrete is also a result of wetting and drying.

❖ **Freezing and Thawing:** Water or moisture transported into the pores of the concrete when exposed to freezing and thawing causes volume changes in addition to the acceleration of chemical degradation. This leads to cracking of the concrete and deterioration.

❖ **Chemical Factors**

Chemical reactions take place due to internal structure or external exposure environment. Some of the important chemical attacks on concrete are briefly mentioned. Moisture transportation aids the chemical attacks.

❖ **Carbonation, Acid formation and attack:** Moisture along with environmental gases, liquids and solids lead to formation of acids that react with hydrates of cement. The chemical attack breaks the chemical bonds in the calcium silicates and hydrates thus destroying the strength of the concrete. The extent of deterioration depends on the density and porosity of the concrete, cement and aggregate properties and the exposure conditions.

❖ **Sulphate attack:** The sulphate ions from soil or water or even from cement react with calcium hydroxide in the concrete forming sulphates. The increase in volume due to formation of sulphates within the pores of the concrete result in cracking and further degradation of the concrete. The presence of aluminates in cement has a tendency to expand in volume on reaction with sulphates.

❖ **Alkali and chemical attacks:** Concrete is initially saturated with lime therefore has an alkali environment. The sodium and potassium ions in the alkali solutions attack the silica of the aggregate. The rate of attack depends on the active silica in the aggregate and dust particles, porosity of the concrete and moisture transport. Concrete surface when exposed to acidic environment reacts with aggregates resulting in increase of volume and consequent cracking.

❖ **Biological and Environmental Factors**

Continuous presence of moisture on the surface of concrete promotes biological growth such as moss, algae and small plants. Penetration of plant roots and biological products into concrete results in cracking and deterioration.

❖ **Corrosion of Reinforcement**

Oxides of iron and steel are formed when the reinforcement is exposed to moisture and oxygen. Similarly the chloride ions accelerate the formation of oxides of iron. The alkali environment in the concrete with pH value in the range of 12 provides good protection against formation of iron oxides. The diffusion of carbon dioxide into concrete reacts with calcium hydroxide forming calcium carbonate. Therefore the pH value in concrete decreases to 9 and the protection to the reinforcement decreases. The dissolution of positively charged iron ions and the combination of the released electrons with moisture and oxygen results into hydroxyl ions. The ion activates the formation of ferric oxide. The formation of ferric oxide on the surface of the reinforcement results into increase in volume of rusted surface. The increase can be as much as 50 to 200 percent of the iron content.

The high alkalinity and relatively high electrical resistivity of Portland cement under moist condition protects the embedded reinforcement in concrete. Factors that are likely to cause corrosion to the reinforcement are:

(a) Inadequate cover to reinforcement,
(b) Cracking of the concrete,
(c) Honey combing of concrete,

(d) Carbonation of hydrates,

(e) Electrolysis.

Good concrete with well-finished surface and having adequate cover to the reinforcement provide good protection to reinforcement against corrosion. Cracks or voids in the concrete provide an access to moisture, air and other environmental chemical agents. Rusted steel will expand in volume thus causing cracking along the reinforcement. The hydrated cement when exposed to carbon dioxide gets carbonated and results in increased shrinkage and cracking. The alkali protection to the reinforcement is lessened. Similarly passage of electricity or development of static potential difference around the reinforcement causes oxidation and corrosion. The following are recommended to protect the reinforcement against corrosion. The main aim is to minimize the permeability.

❖ The most important is to place good concrete with excellent surface finish and cure it well.
❖ Select lowest feasible water-cement ratio.
❖ Minimize the use of admixtures containing calcium chloride, or soluble chlorides. Restrict the entry of salts through the water used either in the concrete mix or in curing.
❖ At least 14 days of uninterrupted curing be done so that good hydration takes place without micro-cracking.
❖ Adequate cover to the reinforcement be provided.
❖ **Stress corrosion in pretension steel**
    The oxidation of stressed steel wires results in sudden splitting of the wires. This phenomenon is present in stressed steels of prestressed concrete constructions.
❖ **Non-structural component factors**
    The discontinuities in construction, construction and expansion joints are liable for problems. Corrosive and even non-corrosive fixtures and inserts exposed to moisture are liable for causing cracking in the concrete. Such non-structural elements must be planned, positioned, placed, protected and maintained carefully.
    As per the IS: 456:2000, the code of practice of plain and reinforced concrete, the factors influencing the durability of concrete structures include:

❖ The environment,
❖ The cover to the embedded steel,
❖ Type and quality of construction material,
❖ Cement content and water-cement ratio of concrete,
❖ Workmanship to obtain full compaction and efficient curing,
❖ The shape size of the member,

2.18 DURABILITY DESIGN CONSIDERATIONS

The exposure conditions of a structure decide the selection of structural and architectural materials, design and detailing of the elements. The basic five exposure conditions for the durability design considerations are:

1. Mild exposure,
2. Moderate exposure,
3. Severe exposure,
4. Very severe exposure, and
5. Extreme exposure.

Mild exposure: The mild exposure condition may be called as protected environment. Interiors of dwellings, offices, commercial complexes and workshops of non-aggressive environment and other structures protected against weathering, wetting and drying are classified in this group.

Moderate exposure: Foundations buried under non-aggressive soils and ground water, structures exposed to rain and not under frequent wetting and drying, outside non-aggressive environment, high humidity halls, running water (canals, dams, weirs etc.) bath rooms and water tanks come under this class.

Severe exposure: Structures exposed to frequent wetting and drying of ordinary water, partially submerged under water, occasional freezing situations, structures in coastal regions, structures subjected to constant vibrations, foundations in aggressive soils are considered to be under this classification.

Very severe exposure: Structures exposed to sea water spray, extreme freezing, chemical fumes, colour dying halls, partially submerged sea water (all harbor structures in contact with sea water), septic tanks come under this classification.

Extreme exposure: Concrete roads; structures under constant abrasion in wet and dry conditions, direct contact with aggressive chemicals, floors of chemical plants are classified in this category.

The important parameters are listed here. Table 2.17 lists the desirable specifications for the different materials etc for the five exposure conditions.

a) Materials

Quality and Quantity of aggregate,
Type and Quantity of cement,
Water Quality, Water-Cement ratio,
Admixtures if any to be used,
Concrete mix design and its workability,
Reinforcement and
Inserts and Fixtures.

b) Detailing and workmanship

Surface finish, shape and texture,
Cover to the reinforcement and precautions,
Finishing of joints,
Drainage of water,
Curing of concrete,
Accessibility for maintenance and
Quality assurance scheme.

c) Maintenance

Inspection regulations,
Systematic maintenance scheme and implementation,
Immediate attention to damages,
Replacement and Repairs,
Renovation and
Prevention of accumulation of water.

Table 2.17 Recommended Limits for Durability Considerations (unless otherwise stated it is by weight).

Material / Property	Exposure Classification			
	Mild (1)	Moderate (2)	Severe (3)	Very Severe & extreme (4)
Aggregate	Normal	Normal	Good	Good
Grading	40	40	30	30
Crushing value (Max.)				
Aggregate density	2000	2200	2200	400
(kg/cum) (minimum)				
Amount of Silts/clay/fine				
(less than 0.15mm)				
(Maximum %)				
In Coarse	2	2	1	1
In Sand	4	3	2	2
Sulphates Admissible (max %)	1	1	0.5–1.0	0.5–1.0
2. **Cement (OPC)**				
Minimum(kg/cum)				
PCC	220	2400	250	260-280
RCC	300	300	320	340-360
PSC	350	350	400	400-400
Maximum content (kg/cum)	← 450 →			
	(unless special considerations are given)			
3. Maximum Water-Cement ratio				
PCC	0.60	0.60	0.50	0.45-0.40
RCC	0.55	0.50	0.45	0.45-0.40
PSC	0.50	0.50	0.45	0.45-0.40
	(Better avoid PSC in highly aggressive environment)			
4. Minimum grade of concrete				
PCC	M10	M155	M20	M20, M25
RCC	M20	M25	M30	M35, M40
PSC	M35	M40	M45	M45
5. Water quality for mixing				
minimum. pH value	4.5	6.0	6.0	6.5
6. Maximum chloride content Maximum				
acid soluble Chloride content as % of				
chloride ion by mass of concrete				
PCC			1%	
RCC			0.15%	
PSC			0.10%	
7. Minimum cover (in mm) to reinforcement bars in RCC				
(The cover specification applies to main and secondary reinforcement)				
Slab	20	20	25	30 to 35
Walls	20	20	25	30 to 35
Beams	25	30	45	50 to 60
Columns	40	40	45	50 to 75

(Cont.)

(Cont.)

*	(1) A wearing or maintenance protective coat to be provided to the concrete in extreme exposure condition (Reinforcement be coated with protective coating)			
	(2) Cover should not be less than the maximum size of bar			
8.	Inserts projecting beyond the concrete surface should be made of:	Steel	GI	GI	Stainless steel
9.	Minimum clear spacing of reinforcement	Diameter of the bar, or Maximum size of aggregate + 5 to 10 mm,			
10.	Minimum cover (in mm) to prestressed concrete wires or cables Slabs Beams	20 25	25 30	25 45	25–30 45–75
11.	Recommended maximum crack widths (in mm)	0.3	0.2	0.1	0.1
12.	Minimum clear spacing of cables ducts in *PSC*	←———— 1.5 to 2 times the diameter of the ducts (preferably 2 times the dia of duct.) ————→			

Structural Concrete: The constituents of concrete, namely aggregate, cement, water and admixtures must satisfy the standard requirements. Further there are limits on qualities and quantities of the basic materials to suit the exposure requirements. Similarly the mix proportions, methods of mixing, laying, consolidation, finishing, formwork and curing are important. Testing methods and quality assurance must be given adequate importance.

Aggregate: Normal aggregate well graded and containing least amount of fine particles (less than 0.15 mm) should be selected. The crushing value of the aggregate reflects the grading and angularity. Some recommended limits on aggregate are given in Table 2.17. Some variations in the limits are permitted based on the type of construction such as Plain concrete, Reinforced concrete etc.

Cement and Water: Portland cements of grades 33, 43 and 53 are suitable for structural concrete. Portland composite (Pozzolana) cements can also be used depending on the type of construction conditions. 53-grade cement is faster in hardening when compared with the 33 grade. A minimum quantity of cement content, especially in reinforced concrete is required to provide adequate alkaline environment to protect the steel from corrosion. Similarly maximum limit is suggested to minimize shrinkage and heat of hydration effects. Shrinkage of concrete increase with water contents hence a reduction in water-cement ratio enables higher cement content. Water used in concrete making should confirm to the standards and should not contain oil, organic matter, humic acid etc. The water-cement ratio should be as low as possible not only for strength consideration but also to minimize the porosity of the concrete. The maximum water-cement ratio admissible for different exposure conditions is given in table 2.17.

Admixtures: Admixture added to cement or concrete to improve the properties such as workability, setting time or to decrease the permeability should not normally exceed 5 percent and preferably in the range of 2 percent or as recommended by the manufacturer. Admixtures or additives should not contain any chlorides in any form in reinforced or prestressed concrete constructions.

Strength of concrete: Strength of concrete need not necessarily indicate the durability, but yet it is considered as a desirable index for durability. Concrete must have a minimum strength for durability considerations. It indirectly reflects the quality of production and porosity. Minimum acceptable grades of concrete under different exposure conditions in the opinion of the author are listed in Table 2.17. The ones suggested by the code is listed separately.

Reinforcement: Some suggestions on cement and water contents were already explained and the limits are indicated in the Table 2.17 to protect the reinforcement from corrosion. The other suggestions and design parameters are:

a) Uninterrupted curing to avoid micro-cracking and complete hydration,
b) Blended cements with slag or fly ash or silica fumes can be used,
c) Limit the total sodium oxide to a maximum of 0.6%
d) Use Low water – cement ratio and
e) Provide adequate cover to the reinforcement and not less than that mentioned in Table 2.17.

Proper placing of reinforcement to ensure the cover to the reinforcement is ensured. Unfortunately cover to the reinforcement is the most neglected factor as the workers trample during construction. Table 2.17 suggests the minimum cover to be provided to the reinforcement. Similarly the clear spacing of the bars and cables should be adequate enough to achieve sound concrete around the bars. A special protective treatment to the concrete surface or to the reinforcement should be provided in aggressive environment as of extreme exposure.

Inserts: Steel or galvanized iron or stainless steel inserts can be embedded in concrete depending on the exposure conditions. Wrongly placed or wrong inserts exposed to moisture or retention of moisture cause surface damages, which leads to deterioration of the structure. Table 2.17 recommends desirable type of inserts.

The minimum cement contents and the water-cement ratio as suggested by the code are listed in table 2.18.

Table 2.18 Minimum Quantity of Cement or Cementetious Material Required, and Maximum W/C Ratio Permitted as per IS: 456-2000.

Exposure	Plain Concrete		Reinforced Concrete		Minimum Grade Concrete	
	Min. Cements (kg/m³)	Max. free w/c	Min. Cements (kg/m³)	Max. free w/c	Plain Concrete	Reinforced Concrete
Mild	220	0.6	300	0.55	–	M20
Moderate	250	0.60	300	0.50	M15	M25
Severe	260	0.50	350	0.45	M20	M30
Very Severe	280	0.45	375	0.45	M20	M35
Extreme	300	0.40	375	0.40	M25	M40

Note: Cement content specified is irrespective of grade of cement,
It is inclusive of additions made to concrete such as flyash, blast furnace slag etc., with respect to water cement ratio,
Minimum grade of concrete for plain concrete is not specified for mild exposure conditions
Aggregate nominal size is 20 mm

Table 2.19. Adjustment to Minimum Cement Contents for Aggregate other than 20 mm Nominal Size.

Nominal size of aggregate (mm)	Adjustment to minimum cement content of Table 2.18 (kg/m³)
10	+40
20	0
40	−30

Maximum cement content: The code on practice of plain and reinforced concrete specifies maximum cement content. Cement paste shrinks on drying, more cement means more shrinkage. Further, heat of hydration is also increases with increase in cement content. Heat of hydration leads to micro cracking. Micro cracking further leads to ingress of water and ultimately deterioration of concrete, corrosion of reinforcement. It is therefore desirable to limit the maximum quantity of cement content. The maximum cement content recommended by the Indian code of practice is 450 kg/m^3 unless otherwise approved by the engineer in-charge.

Chlorides in concrete: The chloride ion may be present in the components of concrete, such as water, aggregate, sand, admixtures, and cement. Or chloride may be diffusing from the environment into the exposed surface of the concrete. The chloride ion causes corrosion of reinforcement at an accelerated rate. The code recommends limits of chlorides content in concrete as given in table 2.20.

Table 2.20. Maximum Admissible Limit of Chloride Content in Concrete.

Type or use of concrete	Maximum total acid soluble chloride content in kg/m^3
Concrete containing metals and steam cured, prestressed concrete	0.4
Plain concrete with metals or reinforced concrete	0.6
Plain concrete	3.0

Maximum limit in sulphate: Sulphates are present in many aggregates, and even in cement. The soil on which the structure stands may also contain sulphates. Even the environment contains some forms of sulpher and sulphates. Sulphates combining with water results in compounds of higher volume. It is necessary to limit the total sulpher trioxide (SO$_3$) content in the concrete to 4 percent by mass of the cement. This limit doesn't apply to supersulphated cement complying to IS: 6909.

Alkali-aggregate reaction: Almost all aggregates contain silica as the main element. Alkalis such as sodium (Na$_2$O) and potassium (K$_2$O) react with silica in the aggregate producing expansive compound. This reaction is produced if the concrete is exposed to high moisture. The cement or the aggregate may contain alkali reactive constituents. Increase in the volume of the reacted aggregate will cause cracking and deterioration of the concrete. The following precautions should be taken.

- Use non-reactive aggregate,
- Use fly ash conforming to IS: 3812 or blast furnace slag conforming to IS: 12089. Or use Portland Pozzolana cement etc.,
- Protect the concrete from constant exposure to moisture,
- Limit the alkalis to 1.1% in cement.

2.19 FIRE PROTECTION SPECIFICATIONS

Domestic facilities such as cooking gas and electric power lines are provided in most homes. A liability of fire hazards to homes exists. Modern buildings have large amount of combustible material such as clothes, wooden furniture, books and paper, plastic items and wall and curtain hangings etc. The permanent buildings are built with brick and concrete. These building materials resist the fire well. The strength of the material deteriorates fast under fire. The increase in vertical heights of the buildings and compactness of the multistory buildings make the fire fighting a complex problem. The buildings must be built to resist fire to desirable safe levels. This section is only an introduction to the fire protection specifications of simple buildings. Fire protection measures to tall buildings are not dealt in this book. The time period for which a building can withstand fire without serious damages is called *fire rating*. The basic fire ratings and the protection specifications to the structural materials based on the code suggestions are mentioned in this section. Minimum dimensions and minimum cover required to the reinforcement for different periods of fire exposure are listed in tables 2.21 and 2.22.

Table 2.21 Minimum Required Dimensions for Members (in mm) for Fire Rating in Hours.

	Beam & floor dimensions			Column dimension		Wall thickness	
Rating hours	Beam width	Rib width	Floor thickness	Exposed fully	One face exposed	0.4%<p<1.0%	p>1%
0.5	200	125	75	150	100	100	100
1.5	200	125	110	250	140	140	100
2.0	200	125	125	300	160	160	100
4.0	280	175	170	450	240	240	180

p = percentage of reinforcement

It can be seen that the code has come with precise minimum dimensions and minimum covers to the reinforcement. In the present, practice the minimum thickness of ribs of waffle slabs are made in the range of 100 mm as against 125 mm now recommended. The thickness of such waffle floor slabs may even start from 60 mm now as against 75 mm thickness suggested for half an hour fire rating.

Table 2.22. Nominal Cover Requirement to Reinforcement (in mm).

Rating	Beams		Slabs		Ribs		Columns
In hours	Ss	Cont.	Ss	Cont.	Ss	Cont.	
1.0	20	20	20	20	20	20	40
1.5	20	20	25	20	35	20	40
2.0	40	30	35	25	45	35	40(45)*
3.	60	40	45	35	55	45	40(55)
4.	70	50	55	45	65	55	40(65)

Ss = simply supported, Cont. = continuous, *the values given in the brackets recommended by the author on the assumption that if the ribs need 45mm cover the columns that are exposed to the same order of heat should have atleast that much cover. Columns at about the middle height of the floors are seen to have had maximum damaged in fire.

The waffle slabs are usually provided in large halls for public gatherings and also in tall buildings. The minimum fire rating of such public halls is 1.5 hours and above. If that is the case, the waffle slab thickness can't be less than 110 mm. The minimum cover to the reinforcement in ribs for 1.5 hours of fire rating is 35 mm in simply supported waffle slabs. The minimum width of the rib will be atleast 160 mm. Further the cover requirement as per the present code applies to main as well as secondary reinforcement.

2.20 QUALITY ASSURANCE IN CONCRETE STRUCTURES

Concrete is a powerful material leading to innovative structures and methods of construction. The innovative structures require innovative and quality controls to give high reliability. The concrete construction uses a wide spectrum of skilled and semi-skilled workers. Experienced engineers and construction companies, build outstanding structures with excellent quality controls; and on the other end of the spectrum, skilled and semi-skilled workers build concrete houses for common man with little or no supervision. The quality assurance of concrete structures varies from reliability level to uncertainty level. There are many non-engineered buildings being built in India. There is a need to educate the semi-skilled and engineers on the quality assurance of concrete structures.

A number of questions do arise on the quality assurance programme. The questions are how and what exactly is to be assured? In the final analysis, it is the quality of the structure to be assured and not only the quality of the materials as dominantly observed today. Components that control the quality of a structure are:

1. Quality of the basic materials such as aggregate, cement, water, Admixtures, etc.
2. Mix proportions of concrete and its production,
3. Construction and fabrication of the structure, formwork,
4. Transportation, placing, compaction and finishing of concrete,
5. Tolerances in quantities and qualities, and detailing,
6. Curing of concrete,
7. Strength and serviceability design of the structure,
8. Durability specifications and maintenance,
9. Life expectance.

At the moment, the quality assurance is leaning heavily towards items one and two of the above list because of the historical and technological background. The idea that the quality control in the basic materials and the production of the concrete ensures the quality of the final structure is a necessary condition but not sufficient. Input of excellent material doesn't guarantee an excellent product. There are a number of inter links between the quality of basic materials and the final structure. Between the cup and the lip, there are a number of slips.

Code Provisions on Quality Control

The clauses on the quality control of concrete structures given in the code are primarily on materials and production of concrete and not on the final structure. Characteristic strength of the concrete is the centre of focus in the design and even construction of concrete structures. The characteristic strength is defined as the strength of the material below which not more than five percent of the test results fall. Strength of concrete, as a matter of fact, strength of any material will follow a normal distribution. Figure 2.8 illustrates a normal distribution curve in which the five-percent cumulative value is indicated.

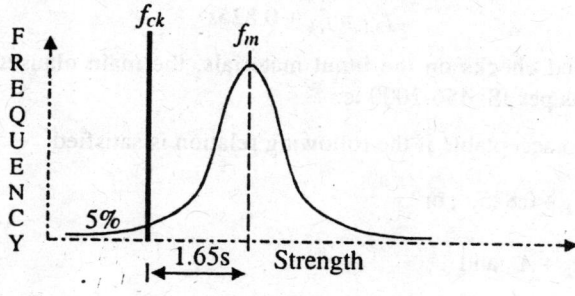

Fig. 2.8 Normal Distribution Curve.

The characteristic strength in terms of probability of failure definition can be expressed as:

$$P(f_c < f_{ck}) = p_f \tag{2.18}$$

For a normally distributed strength of concrete where p_f is the accepted probability of failure. The above probability acceptance can be expressed as:

$$\phi(f_{ck} - f_m)/s = p_f \tag{2.19}$$

where

f_m and s are the population mean value and standard deviation of the normal distribution, f_{ck} characteristic strength, ϕ is the cumulative normal distribution function.

The expression can be inverted to establish the relation between the mean and characteristic strengths, and it is:

$$\phi^{-1}(p_f) = (f_{ck} - f_m)/s = -k \tag{2.20}$$

Where k is an index that signifies the acceptable probability of failure. The minus sign is assigned as the five percent is on the negative end of the curve. The value of k is 1.65 for five-percent acceptability of failure.

The inverse of the equation can be rearranged as:

$$f_m = f_{ck} + 1.65s \tag{2.21}$$

The minimum size of the sample needs to be about fifty for a normal distribution, however the code accepts the size of thirty. Even that is not a practicable size for day to day quality control of concrete. Three concrete cube specimens form a sample. In most constructions, thirty-sample size is not practicable in day-to-day control, therefore four consecutive non-overlapping samples are considered to be a practicable size to test the acceptability of the concrete. As the size is smaller than the minimum population size, a modification to the acceptability expression is suggested. An expression, which satisfies the five percent acceptability criterion with a sample smaller than thirty, is given by:

$$f_{ml}' = f_{ck} + 1.65\,s\,(1-1/\sqrt{n}) \tag{2.22}$$

in which n is the size of the sample and f_{ml} is the mean value of the smaller size sample. The equation reduces to the earlier one for n tending to infinity and the mean value comes out to be same as characteristic strength for a single sample.

The above equation for four non-overlapping samples reduces to:

$$f_{ml} = f_{ck} + 0.825s \tag{2.23}$$

Besides the pre-assigned checks on the input materials, the main clauses on the quality control acceptability of concrete as per IS: 456-2000 is:

a)   The concrete is said to acceptable if the following relation is satisfied

$$f_{ml} \geq f_{ck} + 0.825s \; ; \text{ or}$$

$$f_{ml} \geq f_{ck} + A \, , \text{ and} \tag{2.24}$$

$$f_i \geq f_{ck} - B \tag{2.25}$$

Where f_{ck} = characteristic strength,

$\quad\quad f_{m1}$ = mean strength of any four consecutive non-overlapping samples,

$\quad\quad f_i$ = strength of a sample,

$\quad\quad A$ = 3 MPa for $M15$ concrete and 4 MPa for $M20$ grade concrete and above

$\quad\quad B$ = 3 MPa for $M15$ concrete and 4 MPa for $M20$ grade concrete and above

The concrete is liable for rejection if it is porous or honeycombed, improper construction joints and tolerances on size of the members. It can also be rejected for improper placement of reinforcement and inadequate cover, and not following the specifications.

The mean strength requirement set up by the code is on the liberal side especially for concretes of $M35$ and above. The mean strength of four consecutive samples needs to be four Newtons per square mm or 0.825 times the standard deviation is easy to satisfy. The value 0.825 derived from the normal distribution of the statistics is acceptable if the sample size is large. Consider a case of eight sets of four consecutive non-overlapping samples. Let each of the samples satisfies the criterion of the mean value greater than the characteristic strength by 0.825 times standard deviation. But when all the thirty two samples are considered as a large sample, the global quality control expression demands that the mean value should be greater than the characteristic strength by 1.65 time the standard deviation of the total sample. The standard deviation of the total sample will be more than the standard deviation of any one of the eight samples. Further the multiplier of the standard deviation is 1.65. The chances of satisfying the acceptance criterion of the large population of thirty-two samples are not high.

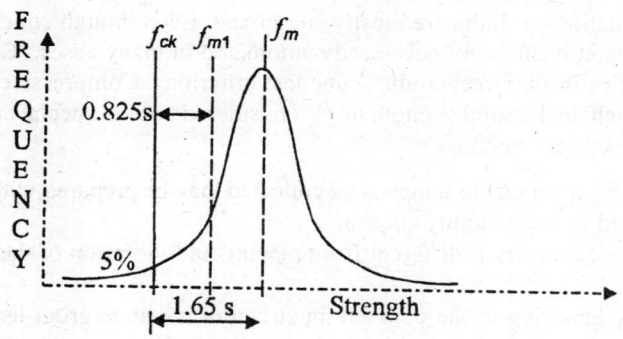

Fig. 2.9 Mean Value for 4 Sample Acceptance.

The acceptability criteria are supposed to be valid for all concretes upto 60 MPa. That may mean even if the concrete is of $M60$ grade, the mean value of the order 64 MPa of four consecutive samples is acceptable. In that sense the section is liberal. To say that strength of every sample should have a value greater than characteristic strength minus 4 MPa is more stringent. For example that for $M60$ concrete, no sample strength should be less than 56 MPa. This could cause some problem because the margin for individual sample strength is rather small. For example, the Indian Road Congress (IRC) permits strength of a sample not less than 80 percent of the characteristic value, and the earlier version also permitted this eighty-percent.

Concrete Cube Test

As per Indian building code of practice, a concrete sample consists of three 150 mm-cube concrete specimens, cured for 28 days in water and tested under moist surface dry condition. The sample is subject to the following conditions:

1. Sample should to be selected randomly,
2. The mould of the cube is well-finished steel or cast iron,
3. The concrete is placed and vibrated under careful control conditions,
4. Cured under good moist condition,
5. Tested on 7th or 28th day of curing the sample.

The concrete CUBE TEST has become a demi-god in quality assurance of concrete structures. It is a hope that the concrete in the structure has the same qualities of the sample! The code further states the following on the Inspection and Testing of the structure:

Care should be taken to see that:
a) Design and details are being capable of being executed,
b) Clear instructions on the inspection standards,
c) Clear instructions on the permissible deviations,
d) " Elements critical to workmanship, structural performance, durability and appearance are identified"
e) To verify the quality is satisfactory in the individual parts, especially the critical ones.

Other Deficiencies of Acceptable Criteria in Cube Test

A concrete sample consists of 3 specimens to be tested on 7th day, another 3 specimens to be tested on 28th day after casting and curing. Standard also specifies, how the samples to be collected, cast, cured and tested. The small constructions in India are mostly site mixed. Even though concrete is made by weigh batching, the batching plant itself is not necessarily automated in many cases. Consequently, there are some inherent deficiencies in the acceptability cube test criterion. Compressive and flexural strength tests are specified of which the flexural strength test is considered only in special conditions. Some of the weak points in the acceptance criteria are:

(1) The concrete batch from which the samples are collected may be prepared with care even though the samples are supposed to be randomly chosen,
(2) Filling, compaction of samples is different from placing and vibration of the concrete in the structure,
(3) The formwork and shuttering to the concrete in structure is not as grout leak proof like the cube mould,
(4) The time difference in transportation and laying of the concrete in position could be different from that of the preparation of cube specimens,
(5) The congestion of reinforcement that might cause honey combing is not reflected in making of the cubes,
(6) The curing of cubes is done systematically when compared to the curing of the concrete in structure,
(7) Testing of 3 specimens as a sample and taking an average strength of the sample is not dependable. Even if more number of samples is considered but the total number in a batch of construction is not high for a statistically acceptable level,
(8) There is a likelihood of less quality control in real concrete structure, detailing of reinforcement, formwork and curing as the actual concrete is not subjected to any testing.

In case where the tensile strength of the concrete plays an important role, flexure test is recommended by the code.

2.21 NON-DESTRUCTIVE TESTING OF CONCRETE STRUCTURES

Methods of non-destructive testing are developed during the last four decades. Reliable equipment for such non-destructive testing is manufactured only in recent times. Further a number of non-destructive tests is available. Ultra sonic non-destructive testing has been accepted as a reliable method of testing of welding in all-important steel structures. At the moment, rebound hammer test, ultrasonic test, and core-cutting tests are considered to be dependable tests in concrete construction. Of these, ultrasonic test has its own advantages and considered being efficient in locating honeycomb and porous concrete. The core cutting sampling test is adapted to a limited extent because of practical difficulties associated with core drilling, specimen preparation and drilling in thin elements etc. Rebound hammer test that indicates the hardness of concrete surface reflects the strength of the concrete indirectly. This test is used quite extensively in assessing the quality of actual concrete in the structure. The test can be carried out quite extensively and a reliable statistical approach can be applied for acceptance of the test. Some of the strong points in quality control through rebound hammer test are:

(1) A very large number of test readings can be taken in short duration,
(2) Large number of locations on the structure can also be selected,
(3) A statistical approach with high degree of confidence can be applied,
(4) Actual structure is tested, therefore the test result reflects the totality of the final product indicating the concrete mix, its consolidation, surface finish, formwork and curing,
(5) The builder/contractor or even the supervisor will be careful during the real concreting because the actual structure is under testing.

An indiscriminate application of the non-destructive testing without proper correlation factors can lead to misleading results. Number of precautions and correction factors to the results of concrete rebound hammer test are needed. These special aspects are:

(1) Surface texture preparation,
(2) The size or the thickness of the element under testing,
(3) Moisture content in the concrete,
(4) Level of maturity of the concrete with respect to 28 days strength,
(5) Accuracy of the test hammer,
(6) Stability of the supporting base used during testing.

The results may have the marginal variation if the method of testing and speed of impact is not uniform, so while calibrating the equipment, the person actually testing can be involved to reflect the method of testing.

The following are some of the observations regarding the concrete test hammer results:

(1) Moist concrete surface will show a lower strength when compared with a dry one,
(2) Thin elements such as slabs, reflect higher strength when compared with thick elements and mass concrete,
(3) Rough surface will give a lower reading when compared with smooth one,
(4) Grout coated or plastered surface will also show lower strength when compared with the original concrete,
(5) A direct reading on aggregate surface will indicate a higher value.

Since very large sampling can be done even on a single element without much effort, data will reflect the final quality of the concrete. A rationalized statistical approach to obtain the strength of the concrete is dependable when compared to the results of standard cube tests. Many Engineers feel that the results of the concrete rebound hammer gives the strength of the concrete on the surface of the structure but does not reflect the quality of concrete inside. Testing of the surface extensively should give much more confidence when compared to cube strength randomly selected and tested. As per present practice, a cube test consisting of 3 to 6 cubes is expected to represent the quality of the concrete of the structure irrespective of where and from which portion of the structure the concrete cube is supposed to have been taken as a sample. If one can accept such a concrete cube as representative sample, there is no need to doubt the reliability of statistically arrived result based upon concrete rebound hammer test.

Acceptable Criteria of non-destructive Rebound Hammer test

Sampling and Acceptable criteria is:
(1) Minimum number of locations should be 10 for an element or for a given pour of concrete,
(2) At least 10-hammer reading should be taken at each location. A location means a spot of about 75 mm square.
(3) All readings beyond 20% of the average value should be rejected. If the number of readings to be rejected exceeds 20% of the readings, then the reliability of the location should be examined. There could be a special problem such as honey combing or porous concrete at such locations.
(4) The total number of readings should be atleast 100 for given element or of a pour of concrete.
(5) The co-efficient of variation of the readings should be within 15%. The characteristic strength of the concrete should be calculated after converting each of the readings to the equivalent strengths with appropriate correction factors.
(6) The characteristic strength of the concrete can be calculated by the formula given below:

$$f_{ck} = f_m - 1.65s \qquad (2.26)$$

In which f_m = mean value of the strength; s = standard deviation

This implies that the probability of the strength falling below the specified value is not more than Five percent.
(7) In case 20% of the locations have fallen under the rejected, then the quality of concrete is not acceptable.

Example

A simple illustration of results of non-destructive test undertaken on roof slab of a building is given in this example.

The example gives statistics of the data. The most probable characteristic strength predicted by the computer program design exclusively for NDT test is also listed. A histogram created by the software on the NDT hammer test. However, this in not presented here. There is a considerable scope in predicting a reliable strength of the concrete based upon of the statistical analysis of the test results.

Statistics of non-destructive test (Data in N/mm²)

Total number of samples	= 80;	Max. Value of the data	= 36.30
Min. value of the data	= 0.70,	Number of interval	= 7
Class interval width	= 3.71,		
MEAN	= 21.64,	MEDIAN	= 21.63,
MODE	= 21.76,	Standard Deviation	= 5.80

SKEWNESS Coefficient	= 0.596,	KURTOSIS Coefficient	= 2.203
CHI-SQUARE	= 38.3	CONFIDENCE Level	= 0.000
Correlated STD DEV	= 4.64	Coefficient of Variation	= 0.214

Specified Strength of the Concrete   = 20.0
Most probable Characteristic Strength = 14.8

OBJECTIVE QUESTIONS

A number of objective questions are listed here. These questions help to understand the subject better. The candidate is expected to tick the one most appropriate among the four listed under each question. There appears to be more than one solution, that may appear correct, but the most appropriate and applicable in general must be chosen. The star marked questions are more difficult ones that require more experience in the field.

(1) *Concrete Aggregates*

2-1.1   As per Indian standard specifications a coarse aggregate used in concrete contains particles of about:
a) 10% passing through 4.75 mm IS sieve   b) 20% passing through 10 mm IS sieve
c) 10% passing through 10 mm IS sieve   d) 10% passing through 2.36 mm IS sieve

2-1.2   As per Indian standard specifications a fine aggregate used in concrete is the one which has 85% to 100% passing through IS sieves Number:
a) 2.36 mm   b) 4.75 mm   c) 1.18 mm   d) 600 micron

2-1.3*   As per Indian standard specifications a 40 mm single-size aggregate used in concrete contains about:
a) 85% to 100% retained on 40 mm sieve   b) 20% retained on 40 mm sieve
c) 85 to 100% passing through 40 mm IS sieve   d) 0 % retained on 40 mm sieve

2-1.4*   As per Indian standard specifications a 40 mm size graded aggregate used in concrete contains about:
a) 95 to 100% retained on 40 mm sieve   b) 95 to 100% passing through 40 mm sieve
c) 80 to 100% passing through 40 mm sieve   d) 80 to 100% retained on 40 mm sieve

2-1.5*   As per Indian standard specifications a 20 mm graded aggregate used in concrete contains about:
a) 25 to 50% retained on 10 mm sieve   b) 10 to 20% passing through 10 mm sieve
c) 25 to 50% passing through 10 mm sieve   d) 5 to 20% retained on 20 mm sieve

2-1.6   The following fine aggregate is not recommended for use in concrete if it contains:
a) 80% passing through 600 micron sieve   b) 10% passing through 300 micron sieve
c) 80% retained on 300 micron sieve   d) 80% retained on 600 micron sieve

2-1.7 **   20 mm nominal size all-in-one aggregate contains about:
a) 30% retained on 20 mm sieve   b) 30% passes through 20 mm sieve
c) 80% passes through 20 mm sieve   d) 95% passes through 20 mm sieve

2-1.8**   The percentage of 20 mm nominal all-in-one size aggregate passing through 4.75 mm sieve is:
a) 95   b) 70   c) 45   d) 20

2-1.9*   Maximum percentage of clay lumps permitted in the aggregate used in concrete-making is about:
a) 10   b) 7   c) 4   d) 1

2-1.10*   Maximum percentage of coal permitted in the aggregate used in concrete-making is about:
a) 8   b) 5   c) 3   d) 1

2-1.11*   Maximum percentage of soft fragments permitted in the aggregate used in concrete is about:
a) 5   b) 3   c) 1   d) 0

2-1.12*   Maximum percentage of total deleterious material permitted in aggregate is about:
a) 10   b) 7   c) 5   d) 2

2-1.13   Specify gravity of the fine aggregate used in concrete is about:
a) 3.1   b) 2.8   c) 2.1   d) 2.5

2-1.14   Bulk density of loose coarse aggregate used in concrete is about (in kN per cum):
a) 26   b) 20   c) 16   d) 12

2-1.15   Bulk density of compacted coarse aggregate used in concrete is about (in kN per cum):
a) 21   b) 18   c) 16   d) 12

2-1.16* Void ratio of coarse sand used in concrete is about:
a) 0.55   b) 0.50   c) 0.45   d) 0.40

2-1.17 Moisture content and percentage absorption of water associated with aggregate are:
a) Same   b) Directly connected   c) Unrelated   d) Indirectly connected

2-1.18 Bulking of fine aggregate is due to:
a) Absorbed moisture   b) Moisture content,   c) Less compaction   d) Voids

2-1.19* Maximum bulking factor of sand is about:
a) 1.1   b) 1.2   c) 1.4   d) 1.5

2-1.20 Maximum bulking of sand is likely to occur at percentage moisture content of:
a) 4   b) 8   c) 12   d) 16

2-1.21 Aggregate crushing value of the aggregate used for road-making concrete should be:
a) Less than 30   b) Less than 45   c) More than 30   d) More than 45

2-1.22 Aggregate crushing value of the aggregate that is used in concrete should be:
a) More than 45   b) More than 30   c) Less than 45   d) Less than 30

2-1.23 Aggregate crushing value of the aggregate for runway concrete should be:
a) Less than 45   b) Less than 30   c) More than 30   d) More than 45

2-1.24 Aggregate impact value of an aggregate to be used in building concrete should be:
a) More than 30   b) More than 45   c) Less than 30   d) Less than 45

2-1.25 Abrasion value of the aggregate that can be used for concrete should less than:
a) 40   b) 30   c) 20   d) 0.0

2-1.26 Flakiness of an aggregate has the following effect on the strength of the concrete in which it is used:
a) Increases the strength   b) Does not effect the strength
c) Decreases the strength   d) Decreases the soundness

2-1.27 Bulk density of an aggregate can be defined as:
a) The ratio of weight of the aggregate to the volume of its solids
b) The ratio of the weight of the aggregate to its volume
c) The ratio of the solid weight of the aggregate to the solid volume
d) Specific gravity of the aggregate multiplied by the unit volume

2-1.28 Void ratio of an aggregate can be defined as:
a) Ratio of the volume voids to that of the solids
b) Ratio of solid weights to the specific weight
c) Ratio of volume of voids to the total volume of the aggregate
d) None of the above

2-1.29 Best shape of an aggregate for making good concrete is:
a) Rounded   b) Irregular   c) Angular   d) Flaky

2-1.30 Best texture of an aggregate for making concrete is:
a) Smooth   b) Crystalline   c) Granular   d) Glossy

2-1.31 Angular aggregate should have the following characteristic:
a) Sharp edges   b) Flat edges   c) Irregular surface   d) Flaky surface

2-1.32 Flakiness of an aggregate can be defined as the ratio of:
a) The thickness to length of the aggregate
b) The thickness to mean size of the aggregate
c) The thickness to the width of the aggregate
d) The thickness to the average length and width of the aggregate

2-1.33* Maximum percentage of flaky aggregate permitted in an acceptable concrete aggregate is:
a) 5   b) 15   c) 25   d) 35

2-1.34 Flakiness index of an aggregate is defined as:
a) Ratio of the thickness to the length of aggregate particles
b) Ratio of the thickness to the mean size of aggregate particles
c) Percentage of aggregate passing through standard angular gauge sieves by weight
d) Percentage of aggregate retained on standard angular gauge sieves by weight

2-1.35* Flakiness index of an aggregate to be used in making of concrete should be:
a) Less than 25   b) Greater than 25   c) Less than 50   d) Greater than 50

2-1.36* Aggregate can be defined as elongated if the ratio of its length to its mean size is greater than:
a) 2.5   b) 1.8   c) 1.2   d) 1.0

2-1.37* Elongation index can be defined as:
a) Percentage of material retained on standard length gauge sieves
b) Ratio of the material retained on the standard angular gauge sieves to the original one

c) Ratio of the length to its mean size of the aggregate

d) Percentage of material passing through standard length gauges

2-1.38 **Percentage of elongated aggregate permissible in concrete should be:

a) Less than five   b) Less than fifteen   c) Less than twenty   d) Less than fifty

2-1.39   Use of flaky aggregate has the following effect on the concrete:

a) Increases the density

b) Decreases the strength

c) Increases the workability

d) Decreases the water-cement ratio demand.

2-1.40   The use of flaky aggregate has the following effect on the concrete:

a) Decreases the workability

b) Decreases the compaction

c) Increases the strength

d) Decreases the flexural strength

2-1.41   Use of elongated aggregate has the following effect on the concrete:

a) Increases the workability

b) Increases the strength

c) Decreases the compaction

d) Increases the compaction

2-1.42   Fineness modulus of an aggregate represents:

a) Gradation of particle size

b) Young's modulus of the aggregate

c) Fine particles in the aggregate

d) None of the above

2-1.43   Total number of standard sieves needed to determine the fineness modulus of any aggregate is:

a) 1   b) 5   c) 8   d) 11

2-1.44   Standard sieves in mm required to determine the fineness modulus of any aggregate are:

a) 40, 20, 10, 4.75, 2.36, 1.18, 0.60 (600 micron), 0.30, 0.15

b) 80, 40, 20, 10, 4.75, 2.36, 1.18, 0.60

c) 150, 80, 40, 20, 10, 4.75

d) 150, 80, 40, 20, 10, 4.75, 2.36, 1.18, 0.60 (600 micron), 0.30, 0.15

2-1.45*   Standard sieves in mm needed to determine the fineness modulus of sand are:

a) 4.75, 2.36, 1.18, 0.60, 0.30, 0.15

b) 2.36, 1.18, 0.60, 0.30, 0.15

c) 1.18, 0.60, 0.30

d) 4.75, 2.36, 1.18, 0.60

2-1.46**   Normal range of fineness modulus of 20 mm single size aggregate is:

a) 6 to 8   b) 5 to 7   c) 4 to 10   d) 3 to 5

2-1.47   Fineness modulus of an aggregate can be defined as:

a) Percentage of cumulative material retained on eleven standard sieves

b) Percentage of cumulative material passing through eleven standard sieves

c) Percentage of cumulative material retained on eight standard sieves

d) Percentage of finer particles passing through 4.75 mm sieves

2-1.48   Fineness modulus of sand usable in concrete is in the range of:

a) 4 to 6   b) 3 to 5   c) 2 to 4   d) 2 to 3

(2) Cement

2-2.1   The two main basic materials in making of the Portland cement are:

a) Chalk and lime

b) Limestone and silica

c) Chalk and alumina

d) Lime and alumina

2-2.2   The three main raw materials used in making of the Portland cement are:

a) Limestone, sandstone and clay

b) Lime, silica and clay

c) Lime, clay and gypsum

d) Silica, alumina and gypsum

2-2.3   The approximate temperature in kiln in which the granules of Portland cement material are formed is:

a) 800 to 1000 degree Celsius

b) 1000 to 1200 degree Celsius

c) 1500 to 1700 degree Celsius

d) 1300 to 1500 degree Celsius

2-2.4   Gypsum is added in manufacturing of the Portland cement:

a) At the beginning of the grinding the raw materials, b) At the time of grinding into powder

c) In the kiln at the time of formation of the granules, d) After grinding the granules

2-2.5   Gypsum is added in cement manufacturing to obtain the property of:

a) Cementing

b) Quick setting of the cement

c) Retarding the setting of the cement

d) To absorb heat of hydration of the cement

2-2.6   The cementing property in Portland cement is primarily due to the base material:

a) Lime   b) Silica   c) Alumina   d) Gypsum

2-2.7   The main chemical compound in Portland cement is:

a) Tricalcium aluminate

b) Dicalcium silicate

c) Tricalcium silicate

d) Tetracalcium silicate

2-2.8 The two main chemical compounds of Portland cement are:
a) Tricalcium silicate and dicalcium silicate   b) Dicalcium silicate and tricalcium aluminate
c) Tricalcium aluminate and silicate   d) Tricalcium silicate and tricalcium aluminate

2-2.9 Appropriate percentage of tricalcium silicate present in Portland cement is:
a) 70 to 80   b) 50 to 60   c) 30 to 40   d) 15 to 40

2-2.10* Approximate percentage of tricalcium aluminaferrite present in Portland cement is:
a) Less than 10   b) 5 t0 10   c) 10 to 20   d) 20 to 30

2-2.11 The following compound of cement is responsible for the development of the early strength of Portland cement paste:
a) Tricalcium aluminaferrite   b) Dicalcium silicate
c) Gypsum   d) Tricalcium silicate

2-2.12 The following cement compound is responsible for the development of later stage strength of Portland cement paste:
a) Gypsum   b) Tricalcium aluminate
c) Dicalcium silicate   d) Tricalcium silicate

2-2.13** Total percentage of sulphur present in Portland cement should not be more than:
a) 3%   b) 6%   c) 10%   d) 15%

2-2.14** Total percentage of magnesia present in the Portland cement should not be more than:
a) 2%   b) 6%   c) 10%   d) 15%

2-2.15 The development of strength of cement paste and its fineness are related as follows:
a) Inversely proportional   b) Directly proportional
c) Not related at all   d) Remotely related

2-2.16 The setting of Portland cement may be defined as:
a) Setting of cement paste
b) Change of cement paste from fluid to hardened state
c) Gain of strength of cement paste
d) None of the above

2-2.17 The flash set of Portland cement paste is:
a) Surface hardening only   b) Premature hardening
c) Hardening without developing heat of hydration   d) Flashing of heat of hydration

2-2.18** Flash setting of Portland cement paste can be prevented by the addition of:
a) Calcium chloride   b) Calcium aluminate
c) Gypsum   d) Pozzolana

2-2.19** The percentage of the residue left after sieving good Portland cement in 90 micron sieve should not be more than:
a) 30   b) 20   c) 10   d) 0

2-2.20 Specific surface of cement is:
a) The ratio of the surface area to its volume   b) The surface area of one gram of cement particles
c) The ratio of the specific gravity to its surface   d) none of the above

2-2.2 * Minimum specific surface of a good Portland cement is about (in sqm/kg):
a) 1000   b) 500   c) 200   d) 50

2-2.22 The fineness of cement is tested by:
a) Specific surface - air permeability method   b) Le Chatelier method
c) Normal consistency apparatus   d) Vicat needle

2-2.23 The soundness of Portland cement can be tested by:
a) Vicat needle   b) Le Chateliers apparatus
c) Sieve analysis   d) Specific surface analysis

2-2.24 The expansion of Portland cement is indicated through the following test:
a) Sieve analysis   b) Normal consistency   c) Setting time   d) Soundness

2-2.25* Minimum compressive strength of mortar of grade 33 ordinary Portland cement-standard sand mortar after 72 hours of curing should not be less than (in Newton per sq mm):
a) 16   b) 20   c) 33   d) 20

2-2.26* Minimum compressive strength of mortar of the ordinary Portland cement-standard sand mortar after 7 days of curing should not be less than (in Newton per sq mm):
a) 16   b) 20   c) 33   d) 20

2-2.27 * Minimum compressive strength of mortar of the ordinary Portland cement-standard sand mortar after 28 days of curing should not be less than (in Newton per sq mm):

a) 16   b) 20   c) 33   d) 20

2-2.28 * Specific gravity of Portland cement is about:
   a) 3.50   b) 2.65   c) 3.15   d) 2.20

2-2.29 Bulk density of ordinary Portland cement is about (in kN per cubic metre):
   a) 28   b) 23   c) 18   d) 14

2-2.30 Percentage void ratio of Portland cement is about:
   a) 60   b) 50   c) 40   d) 30

2-2.31 Weight of a normally supplied Portland cement bag is about (in kg):
   a) 100   b) 75   c) 50   d) 25

2-2.32 Following compound causes the expansion of Portland cement paste:
   a) Free silica   b) Iron oxide   c) Voids   d) Free lime

2-2.33 Soundness of Portland cement can be improved by:
   a) The addition of lime
   c) The addition of gypsum
   b) The addition of calcium chloride
   d) Fine grinding of clinker

2-2.34 Rapid-hardening cement can be obtained by:
   a) The addition gypsum
   c) Addition of calcium sulphate
   b) Fine grinding of clinker
   d) Higher content of lime

2-2.35 White cement is produced in:
   a) Electrical furnace kiln
   c) Coal powder fired kiln
   b) Oil fired kiln
   d) Fire brick refractory kiln

2-2.36 Fineness of cement is determined by:
   a) Fineness modulus test
   c) Grain size apparatus
   b) Chemical analysis
   d) Air permeability apparatus

2-2.37** Particle size of ordinary Portland cement is less than:
   a) 90 micron   b) 150 micron   c) 50 micron   d) 10 micron

2-2.38 The 'unsoundness' of cement is indicated by:
   a) Late setting of cement paste
   c) Quick setting of cement paste
   b) Cracking of cement paste
   d) Large increase in volume of cement paste

2-2.39 The expansion of cement paste can be due to:
   a) Poor grinding of cement clinker
   c) Presence of uncombined calcium hydroxide
   b) Presence of Calcium Oxide
   d) Presence of hydrated calcium sulphate

2-2.40 The presence of the following compounds cause expansion of hardened cement paste:
   a) Calcium Carbonate, Calcium Hydroxide
   b) Calcium Oxide, Calcium Hydroxide, Calcium Carbonate
   c) Calcium Oxide, Magnesium Oxide, Calcium Sulphate
   d) Potassium Oxide, Calcium Silicate, Iron Oxide

2-2.41 The 'unsoundness' of ordinary Portland cement may be caused by:
   a) Excess quantity of clay in the raw material mixing
   b) Excess quantity of lime in the raw material mixture
   c) Presence of free oxides of calcium
   d) Presence of moisture in cement

2-2.42 Le Chatelier test on cement is carried out to determine:
   a) Soundness of cement
   c) Flexural strength of cement
   b) Normal consistency of cement
   d) Final setting time of cement

2-2.43** Cement paste in Le Chatelier apparatus is boiled in water for the following reason:
   a) To cause expansion of the cement paste
   b) To cause quick shrinkage of the cement paste
   c) To slow down the chemical reactions and hydration
   d) d) To accelerate the hydration of the cement paste

2-2.44* Autoclave test is performed to determine the following property of cement:
   a) Normal consistency
   c) Chemical composition
   b) Heat of hydration
   d) Soundness

2-2.45 Autoclave test determines the presence of the following compounds:
   a) Ca O, Ca SO$_4$
   c) Ca SO$_4$, Fe$_2$O$_3$
   b) Ca O, Mg O
   d) K$_2$O, Fe$_2$O$_3$

2-2.46* Complete soundness of Portland cement is determined by the following tests:
   a) Autoclave or Le Chatelier test
   c) Le Chatelier test only
   b) Brazilian split cylinder
   d) Autoclave only

2-2.47   The grades of Portland cements manufactured are:
a) 33, 43 and 53
b) 250, 350 and 400
c) 200, 300 and 400
d) There is no such grade in Portland cement

2-2.48   The number 330 in grade 330 cement signifies:
a) Compressive strength of cement paste on 28 days of curing
b) Compressive strength of standard cement mortar cube in kg/cm^2 at 28 days of curing
c) Compressive strength of standard concrete cube in kg/cm^2 at 28 days of curing
d) Specific surface of cement

(3) Admixtures

2-3.1   Pozzolana is a material used:
a) In stabilizing foundations
b) As an admixture to cements
c) As a distempers
d) In the water-proof treatment

2-3.2   Main component of Pozzolana is:
a) Calcium silicate
b) Siliceous material
c) Calcium chloride
d) Gypsum

2-3.3   Finely-divided Pozzolana reacts with lime producing:
a) Tricalcium silicate
b) Dicalcium silicate
c) Calcium silicate
d) Calcium chloride

2-3.4   Addition of Pozzolana to cement results in:
a) Increased curing time
b) Decreased curing time
c) Increase in early-setting lime
d) Increase in strength

2-3.5   Addition of Pozzolana to cement causes:
a) Less heat of hydration
b) More heat of hydration
c) Decrease in workability
d) Increase in strength

2-3.6   Addition of Pozzolana to cement causes:
a) Reduced permeability
b) Increased permeability
c) Increase in heat of hydration
d) Reduction in curing time

2-3.7   Addition of Pozzolana to Portland cement causes:
a) Increase in bleeding
b) Decrease in curing time
c) Reduction in bleeding
d) Increase in heat of hydration

2-3.8*   Addition of Pozzolana to Portland cement causes:
a) Decrease in shrinkage
b) Increase in shrinkage
c) Increase in bleeding
d) Increase in heat of hydration

2-3.9   The addition of Pozzolana to Portland cement may cause:
a) Increase in early strength
b) Decrease in curing time
c) Increase in permeability
d) Decrease in early strength

2-3.10   The following material contains Pozzolana properties:
a) Dolomite   b) Limestone   c) Flyash   d) Clay

2-3.11*   Following material contains Pozzolana properties:
a) Black cotton clay   b) Diatomaceous clay   c) Dolomite   d) Limestone

2-3.12*   Early setting of concrete can be attained by the addition of the following admixtures:
a) Hydrogen peroxide   b) Calcium chloride   c) Calcium sulphate   d) Gypsum

2-3.13*   Setting time of cement can be increased by the addition of:
a) Calcium chloride   b) Hydrogen peroxide   c) Gypsum   d) Sodium

(4) Workability of Concrete

2-4.1   The workability of concrete can be improved by the addition of the following admixture:
a) Calcium carbonate
b) Calcium sulphate
c) Hydrated lime
d) Hydrogen peroxide

2-4.2   The workability of concrete can be improved by the addition of the following admixtures:
a) Calcium sulphate
b) Bentonite
c) Calcium sulphate
d) Dolomite

2-4.3   The workability of concrete can be improved by the addition of the following admixture:
a) Calcium sulphate
b) Hydrogen peroxide
c) Flyash
d) Lime

2-4.5 An air-entraining agent when added to cement concrete improves:
a) Strength of concrete
b) Workability of concrete
c) Density of concrete
d) Durability of concrete

2-4.6 A foaming agent when added to concrete improves the following quality:
a) Strength
b) Durability
c) Density
d) Workability

2-4.7 Oily agents when added to concrete improve the following quality:
a) Strength
b) Durability
c) Density
d) Workability

2-4.8 Compaction factor test helps in the determination of the following quality of concrete:
a) Strength
b) Porosity
c) Workability
d) Compaction

2-4.9 Slump test helps in the determination of the following quality of concrete:
a) Strength
b) Settlement
c) Shrinkage
d) Workability

2-4.10 Workability of concrete can be improved by:
a) Increasing the size of the aggregate
b) Decreasing the aggregate content
c) Increasing the fine aggregate content
d) Increasing the flaky aggregate content

2-4.11 The property of the workability of concrete reflects:
a) Ease in mixing and placing of the concrete
b) Ease in mixing, placing and compacting of the concrete
c) Ease in mixing of the ingredients of the concrete
d) Ease in mixing of coarse and fine aggregates

2-4.12 Compaction factor of concrete really reflects:
a) Density of the concrete
b) Quality of the cement paste
c) The energy required for good compaction
d) Compaction of concrete

2-4.13 Workability of concrete can best be associated with:
a) Ease of placement
b) Resistance to segregation
c) Level of compaction
d) Internal work needed to produce compactness

2-4.14 Rational method of testing workability of wet concrete is:
a) Slump test
b) Hardness test
c) Compacting factor test
d) Flow test

2-4.15 Some commonly field tests for workability of wet concrete are:
a) Vebe test, Slump test
b) Slump test, Remoulding test
c) Flow test, Compactness test
d) Kelly ball test, Slump test

2-4.16 High void ratio in concrete is the result of:
a) Poor workability of concrete
b) Large void ratio in coarse aggregate
c) Poor quality of cement
d) Low water - cement ratio

2-4.17 Slump of wet concrete is:
a) More for rounded aggregate when compared with the angular one
b) Less for rounded aggregate when compared with the angular one
c) Decreases with the increase in size of aggregate
d) More for flaky aggregate when compared with angular one

2-4.18 Segregation of concrete while placing it in position reflects:
a) Good workability of concrete
b) Poor workability of concrete
c) Large size aggregate
d) Low cement content

2-4.19 Segregation of concrete can be minimized by:
a) Low water-cement ratio
b) Vibration of concrete
c) Proper positioning of concrete
d) Using well graded aggregate

2-4.20 Bleaching of concrete is associated with:
a) Seepage of water from concrete
b) Flow of cement paste to one location
c) Surfacing of cement to the top face of fresh concrete
d) Rising of water to the top face of the fresh concrete

2-4.21 Longer mixing period of concrete mix in mixers has the following effect:
a) Improves the consistence of concrete
b) Segregates the particles
c) Tends to decrease the strength of concrete
d) Increases the strength of concrete

2-4.22* Volume of coarse aggregate (with 40% void ratio) required to make one cubic metre of Reinforced Concrete of mix 1:2:4 by volume is (in cum) (Use w/c ratio of 0.65 by weight and select the nearest value):
a) 1.0
b) 0.78
c) 0.65
d) 0.56

2-4.23   The number of bags of cement required to make one cubic metre of concrete 1:2:4 (with 40% void ratio in aggregates) is (Use w/c ratio of 0.65 and select the nearest value):
a) 4.5   b) 6.2   c) 7.2   d) 8.5

2-4.24   The weight of cement required in kg to make one cubic metre of Reinforced Concrete of mix 1:2:4 by weight (Assume w/c ratio of 0.65) is:
a) 320   b) 360   c) 400   d) 450

2-4.25   The amount of water required for a workable Reinforced Concrete of mix 1:2:4 by weight is about (in litres) (Assume w/c ratio of 0.65):
a) 150   b) 200   c) 250   d) 300

(5) *Strength of concrete*

2-5.1   The main compound of cement responsible for Strength of concrete:
a) Calcium aluminate   b) Dicalcium silicate   c) Tricalcium silicate   d) Gypsum

2-5.2   Strength of concrete increases with the increase in:
a) Water-cement ratio   b) Cement-water ratio
c) Sand-cement ratio   d) Water-aggregate ratio

2-5.3   Strength of concrete increases with increase in:
a) Water-cement ratio   b) Aggregate-cement ratio
c) Size of the aggregate up to a point   d) Sand content

2-5.4   Strength of cement concrete increases with:
a) Decrease in the water-cement ratio
b) Increase in the water-cement ratio
c) Decrease in the size of the aggregate.
d) Decrease in the temperature of water used in curing

2-5.5   Rate of increase of hardening of the cement concrete increases with:
a) Increase in the temperature of water of curing   b) Decrease in the temperature of water of curing
c) Increase in the water-cement ratio   d) Increase in the moisture content

2-5.6*   Approximate ratio of Strength of cement concrete at 3 months to that at 28 days of curing is:
a) 0.85   b) 1.0   c) 1.15   d) 1.35

2-5.7   Approximate ratio of Strength of cement concrete at 7 days to that at 28 days curing is:
a) 0.40   b) 0.67   c) 0.90   d) 1.15

2-5.8   Approximate ratio of Strength of cement concrete at one year to that at 28 days of curing is about:
a) 0.85   b) 1.00   c) 1.10   d) 1.30

2-5.9   Approximate ratio of direct tensile strength to its cube the compressive strength is:
a) More than 0.25   b) 0.1 to 0.15   c) 0.05 to 0.10   d) 0.01 to 0.05

2-5.10*   Approximate ratio of bond stress at 0.25 mm slip of plain bars to its cube compressive strength is:
a) 0.40   b) 0.20   c) 0.10   d) 0.05

2-5.11*   Approximate ratio of bond stress at 0.25mm slip of deformed bars to its cube compressive strength is:
a) 0.35   b) 0.25   c) 0.14   d) 0.10

2-5.12   Approximate ratio of strength of 150 mm cube of concrete to that of 150 mm by 300 mm cylinder is:
a) 1.50   b) 1.30   c) 0.85   d) 0.67

2-5.13   Approximate ratio of strength of 150 mm by 150 mm by 300 mm square prism to that of a 150 mm cube is:
a) 0.80   b) 1.00   c) 1.15   d) 1.30

2-5.14   Approximate ratio of strength of 150 mm by150 mm by 600 mm prism to that of a 150 mm cube of concrete is:
a) 1.15   b) 1.00   c) 0.85   d) 0.68

2-5.15   Approximate ratio of strength of concrete of a 100 mm cube to that of a 150 mm cube is:
a) 1.25   b) 1.08   c) 0.76   d) 0.65

2-5.16   Approximate ratio of strength of a 300 mm concrete cube to that of a 150 mm cube is:
a) 1.25   b) 1.15   c) 1.00   d) 0.75

2-5.17*   Approximate ratio of strength of a dry sample to that of a saturated sample of cement concrete is:
a) 1.15   b) 1.05   c) 0.85   d) 0.75

2-5.18*   Approximate ratio of axial compressive of concrete to its bending compressive strength is:
a) 0.65   b) 0.80   c) 1.00   d) 1.15

2-5.19   Split strength of concrete can be determined by:
a) Brazilian (cylinder test) b) Cube test   c) Briquettes test   d) Vicat's apparatus

2-5.20   Strength of concrete shows an increase with:
a) Increase in the rate of loading of the test specimen
b) Decrease in the rate of loading of the test specimen
c) Unaffected by the rate of loading of the test specimen
d) None of the above

2-5.21   Approximate standard rate of loading of a 150 mm concrete cube for the determination of its compressive strength (in Newton per sq mm per min) is:
a) 10 to 20   b) 20 to 40   c) 40 to 60   d) 60 to 80

2-5.22   Durability of Portland cement concrete is proportional to:
a) Water-cement ratio   b) Aggregate-cement ratio
c) Cement-aggregate ratio   d) Sand content

2-5.23** Admissible maximum water-cement ratio for durable structural concrete exposed to mild environment is:
a) 1.0   b) 0.85   c) 0.65   d) 0.4

2-5.24*   Admissible maximum water-cement ratio for a durable structural concrete exposed to rain and sun is:
a) 1.0   b) 0.8   c) 0.6   d) 0.5

2-5.25*   Admissible maximum water-cement ratio for a durable structural concrete exposed to sea water is:
a) 1.0   b) 0.7   c) 0.5   d) 0.4

2-5.26   Durability of cement concrete is usually improved by the addition of:
a) More cement   b) More granite aggregate
c) High vibration   d) More coarse sand

2-5.27*   Minimum quantity of cement content required in kg per cum of durable plain concrete is about:
a) 160   b) 220   c) 330   d) 400

2-5.28*   Minimum cement content required for durable reinforced cement concrete is about (in kg per cum):
a) 150   b) 200   c) 250   d) 300

2-5.29*   Minimum cement content required for prestressed concrete for durability is (in kg per cum):
a) 200   b) 250   c) 300   d) 350

2-5.30** Approximate ratio of bond strength of plain bars to that of the deformed bars at no slip in concrete is:
a) 0.6   b) 0.8   c) 0.75   d) 1

2-5.31*   Young's modulus of concrete (E) is given by:
a) $E = 1000 f_{ck}$ b) $E = 5000$ times square root of f_{ck}
c) $E = 5700 f_{ck}$ d) $E = 10000$ times square root of f_{ck}
where f_{ck} = cube strength of concrete in Newton per sq mm.

2-5.32   Density of concrete:
a) Increases with a decrease in the size of the aggregate
b) Is independent of the size of the aggregate
c) Increases with an increase in the size of the aggregate
d) None of the above

2-5.33*   Weight of concrete with 10 mm size aggregate is about (in kN per cum):
a) 18   b) 21   c) 24   d) 26

2-5.34*   Unit weight of concrete with 40 mm size aggregate is about (in kN per cum):
a) 20   b) 23   c) 26   d) 28

2-5.35   Concrete gets disintegrated due to the action of:
a) Oxalic acid   b) Phosphoric acid   c) Humic acid   d) Sulphuric acid

2-5.36   Concrete gets disintegrated slowly due to the action of:
a) Acetic acid   b) Oxalic acid   c) Nitric acid   d) Hydrochloric acid

2-5.37*   Following acid does not cause the disintegration of concrete:
a) Carbolic   b) Oxalic   c) Acetic   d) Sulphurous

2-5.38*   Following salt solutions have no effect on concrete:
a) Sodium carbonate   b) Calcium Chlorides   c) Fluorides   d) Ammonium nitrite

2-5.39*   Following products have no affect on the durability of the concrete:
a) Most Petroleum oils   b) Phenol   c) Fish oil   d) Vinegar

2-5.40** Following affect the durability of concrete:
a) Molasses   b) Alcohol   c) Cider   d) Petroleum oils

2-5.41*   Following may cause only surface effects on concrete:
a) Hydrochloric   b) Groundnut oil   c) Milk products   d) Baking soda

(6) *Shrinkage and Creep in Concrete*

2-6.1 Shrinkage in concrete is primarily due to:
a) Constant load on the member
b) Depletion of moisture from the concrete
c) Restraint in concrete
d) Increase in the moisture content

2-6.2 Shrinkage in concrete produces cracking if the concrete is:
a) Allowed to move freely
b) Cured improperly
c) Restrained by an external or boundary condition
d) Restrained by the embedded reinforcement

2-6.3 The shrinkage of concrete is directly proportional to:
a) Water content at the time of mixing
b) Sand content in the concrete
c) Coarse aggregate content
d) Aggregate to cement ratio

2-6.4 The shrinkage of concrete is proportional to:
a) Cement content
b) Sand content
c) Aggregate content
d) Temperature of water

2-6.5 The shrinkage of concrete increases with:
a) Increase in the age of concrete
b) Decrease in the age of concrete
c) Increase in the moisture content
d) None of the above

2-6.6* 50% of the shrinkage of concrete occurs in:
a) 7 days of curing
b) 28 days of curing
c) 1 month of drying
d) 6 months of drying

2-6.7 Most of the shrinkage in concrete takes place in:
a) 28 days of curing
b) 1 month of natural drying
c) 6 months of natural drying
d) 12 months of drying

2-6.8 Shrinkage in cement concrete decreases with:
a) Increase of moisture content in the concrete
b) Removal of moisture content in the concrete
c) Addition of load on the concrete
d) Removal of load on the concrete

2-6.9* Approximate shrinkage of cement concrete is:
a) 0.001
b) 0.005
c) 0.0003
d) 0.0001

2-6.10 Shrinkage of ordinary concrete is about (in mm/m):
a) 0.1 to 0.3
b) 0.3 to 0.6
c) 0.6 to 10
d) None of the above

2-6.11* Shrinkage of light weight concrete is about (in mm/m):
a) 0.1 to 0.3
b) 0.3 to 0.5
c) 0.5 to 0.8
d) 0.8 to 10

2-6.12* Shrinkage of aerated concrete is about (in mm/m):
a) 0.1 to 0.3
b) 0.3 to 0.5
c) 0.5 to 0.8
d) 0.8 to 10

2-6.13 Maximum differential shrinkage in concrete elements cast in a span of one month can be:
a) 0.0002
b) 0.0006
c) 0.001
d) 0.003

2-6.14 Units of shrinkage of concrete are:
a) Strain
b) mm/mm
c) Rate of stress
d) Rate of strain

2-6.15 Shrinkage in concrete is a:
a) Load and time dependent process
b) Time dependent drying process
c) Temperature dependent process
d) Curing dependent process

2-6.16 Actual shrinkage of reinforced concrete decreases with:
a) Decrease in reinforcement
b) Independent of reinforcement
c) Increase in reinforcement
d) Decrease in high strength steel

2-6.17 Shrinkage in concrete shows up:
a) During curing
b) Immediately after cracking
c) Immediately after loading the member
d) With time after curing

2-6.18 Shrinkage in concrete is likely to decrease with:
a) Increase in the designed strength of the concrete
b) Decrease in the designed strength of the concrete
c) Independent of the designed strength of the concrete
d) Decrease in curing time

2-6.19 Force caused by shrinkage of a member:
a) Increases with the increase in the size of the members
b) Decrease with the increase in the size of the members
c) Independent of the size of the member
d) Increases with increase in the percentage of reinforcement

2-6.20 Shrinkage in plain concrete produces:
a) Limited number of cracks at critical locations
b) Wide cracks at middle span
c) Thin but many cracks at about equal spacing
d) Thin and limited cracks at random

2-6.21 Shrinkage cracks in reinforced concrete members can be controlled by:
a) The percentage of reinforcement
b) Independent of the amount of reinforcement
c) Member size
d) Restrained boundary condition

2-6.22 Shrinkage cracking in reinforced concrete decrease with:
a) Increase in the diameter of the reinforcement bars
b) Decrease in the diameter of the reinforcement bars
c) Increase in the strength of the reinforcement
d) Decrease with smaller bars at closer spacing

2-6.23 Shrinkage cracking in reinforced concrete decrease with:
a) Well distributed reinforcement across the section
b) Higher reinforcement at the tension face of the member
c) Reasonable reinforcement at the compression
d) Increase in the shear reinforcement

2-6.24 Shrinkage cracking in columns is:
a) Never occurs
b) Will always be there but cannot be seen
c) Will occur in lightly loaded framed structures
d) Dominant in axially loaded columns

2-6.25 For a given length of a RC beam, the shrinkage cracking:
a) Is more in simply supported case
b) Is independent of the boundary conditions
c) Is more in restrained boundary conditions
d) Is less in restrained boundary conditions

2-6.26 The width of the shrinkage crack in RC beams:
a) More at tension face of the member
b) More at the compression face of the member
c) Same on all faces of the member
d) Less at the tension face and more at the middle depth

2-6.27 The requirement of shrinkage reinforcement in RC beams:
a) Can be taken care if the percentage of tension reinforcement is more than that minimum for shrinkage
b) Should be provided on four faces and it can be a part of compression or tension reinforcement
c) Should always be over and above the tension and compression reinforcement
d) Should be met over and above the main reinforcement

2-6.28* Approximate width of shrinkage crack visible to the naked eye is about:
a) 0.2 mm   b) 0.1 mm   c) 0.04 mm   d) 0.02 mm

2-6.29* If the shrinkage coefficient of a concrete is 0.0003 mm then the desired size of the plain concrete floor panel to avoid cracking of the panels should be about:
a) 2000 mm
b) 1200 mm
c) 600 mm
d) There is no relation at all

2-6.30 Flooring cast-in-situ plain concrete is to be laid continuously:
a) To avoid shrinkage cracking in the floor
b) To minimize the shrinkage cracking in the floor
c) But thin shrinkage cracking is caused in the floor
d) To allow continuous curing

2-6.31 Creep strain in concrete is:
a) Time dependent under constant load
b) Temperature dependent under constant load
c) Moisture dependent under constant load
d) None of the above

2-6.32 Creep in concrete is associated with:
a) Removal of moisture from the concrete
b) Removal of load from the concrete
c) Addition of load on the concrete
d) Time

2-6.33* Creep in concrete increases with:
a) Decrease in curing time,
b) Decrease in the age of concrete at the time of loading,
c) Decreases with cement content,
d) None of the above.

2-6.34 The total creep strain in concrete is directly proportional to:
a) Elastic strain
b) Moisture content in the concrete
c) Time of loading
d) Duration of loading

Key Solutions to the Objective Questions

The solutions are listed in the following tables under each sub-heading.

(1) Concrete Aggregates

	1	2	3	4	5	6	7	8	9	10
0	a	b	c	b	c	a	d	c	d	d
10	a	c	b	c	a	c	d	b	c	b
20	a	c	b	d	c	c	d	c	c	b
30	a	b	b	d	a	b	a	c	b	a
40	b	a	d	d	a	b	a	c		

(2) Cements

	1	2	3	4	5	6	7	8	9	10
0	b	c	d	b	c	c	a	a	b	b
10	d	c	a	b	d	b	a	c	c	b
20	c	a	b	d	a	a	c	c	b	c
30	b	d	d	b	b	d	c	d	a	c
40	b	a	d	d	b	d	a	b		

(3) Concrete Admixtures

	1	2	3	4	5	6	7	8	9	10
0	b	d	c	a	b	d	c	d	d	
10	b	d	c							

(4) Workability of Concrete

	1	2	3	4	5	6	7	8	9	10
0	c	d	c	b	d	d	c	c	d	a
10	b	c	b		b	a	a	b	c	a
20	b	b	b		b					

(5) Strength and Durability of Concrete

	1	2	3	4	5	6	7	8	9	10
0	c	b	c	a	a	c	b	d	a	c
10	a	b	a	d	b	d	b	b	d	b
20	a	c	c	d	c	a	c	d	c	a
30	b	c	a	d	b	a	a			a
40	c									a

(6) Shrinkage and Creep in Concrete

	1	2	3	4	5	6	7	8	9	10
0	b	c	a	d	a	c	d	a	c	b
10	c	d	d	b	d	d	d	b	a	c
20	a	d	c	b	c	d	b	a	b	c
30	a	b	b	d						

Strength Design of Beams of Rectangular Section

3.1 BASIC CONCEPTS

Structure or any part of the structure should have reasonable margin of safety and indicate warning before failure. Strength of a member is its capacity to withstand loads. Strengths are associated with bending, shear, axial, combined force resistance. In statically determinate members such as cantilever or simply supported beam, failure of a section leads to failure of the member. But in statically indeterminate structure, failure of a section leads to redistribution of forces to the other sections and further loading leads to formation of mechanism and collapse. The redistribution of forces in reinforced concrete structures is limited by its ductility, and therefore the total redistribution forces is not accounted in most cases. Only a limited redistribution of forces is allowed in the present day practice of limit state design of reinforced concrete structures. The limit state design ensures satisfactory serviceability and adequate safety against failure. Two types of limit states are considered by the Indian code of practice.

- *Limit state of strength and*
- *Limit state of serviceability*

Suitable implementing specifications and detailing reinforcement provides the durability of the structure. Designer specifies quality of materials and the builder must be able to supply such materials; and the supervisor must enforce quality control in construction.

3.2 STATE OF STRESSES IN CONCRETE AND STEEL

The characteristic stress-strain curve for concrete is obtained normally from testing of concrete cube under direct compression. Concrete cylinder specimen is also used in some countries. The character of the specimen is extended to prism with modification on the limit of strength. Figure 3.1 illustrates the stress-strain curve of concrete under compression. Line *abc* is the stress-strain curve for concrete cube specimen with characteristic strength as f_{ck}. The prism strength of concrete is taken as 0.67 times that of the cube and it is indicated by curve ade. The design strength of concrete is obtained by dividing the prism strength by a partial safety factor. For the purpose of structural design, the stress-strain behavior of the concrete is idealized as indicated in curve *afg*. The curve is idealized as parabolic from starting point of zero strain up to 0.002 strains, and then by a straight line up to the crushing strain. The crushing strain of concrete is normally taken as 0.0035.

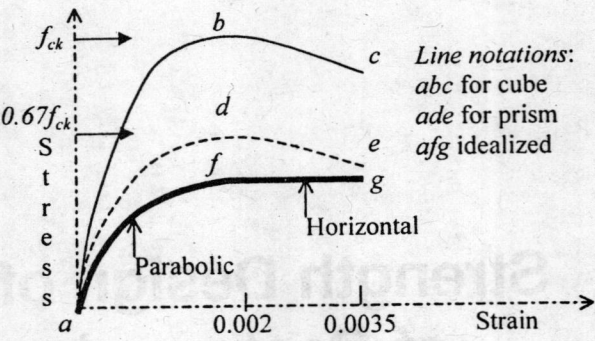

Fig. 3.1 Stress-strain for Concrete and Idealized Curve.

The notations in the stress-strain curve of the concrete are:

f_{ck} = characteristic strength of concrete (cube specimen)

$0.67f_{ck}$ = prism strength of concrete

$0.67f_{ck}/\gamma_c$ = design strength of concrete = $0.445f_{ck}$ the factor 0.445 is normally rounded to 0.45

γ_c = partial safety factor applied to concrete = 1.5; this is applied to the crushing value of the concrete

0.0035 = crushing strain of concrete; (the maximum admissible strain)

0.002 = level of strain of concrete where the parabolic curve ends and straight line starts

The idealized stress-strain curve of concrete is parabolic from 0.00 to 0.002 strains and then linear up to the ultimate strain of 0.0035. Figure 3.2 illustrates the stress-strain curve for mild steel and high yield steel deformed bars. The notations of the stress-strain curve are listed here.

f_y = characteristic strength of steel. It is equal to yield stress in mild steel; proof stress for *HYSD at a* corresponding 0.2% residual strain

abd = curve for mild steel, yield stress at point *b*, breaking at point *d*; for steel with yield at 250 MPa.

abce = stress-strain curve for high yield steel deformed bars with proof stress is in the range of 415, 500 and 550 MPa. Commonly used grade of steel at the moment is Fe415,

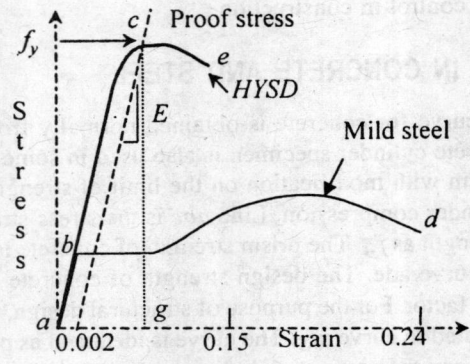

Fig. 3.2 Stress-strain Curve for Steel.

0.002 = residual strain corresponding to proof stress

   $E =$ modulus elasticity of steel and it is 200 MPa. (200,000 N/mm^2)

The dotted inclined line is parallel to the elastic curve with residual strain of 0.2 percent.

   ε_y = strain corresponding to the proof stress = $0.002 + f_y/E$;
f_y/γ_s = design proof stress
   γ_s = partial safety factor applied to steel in bending tension = 1.15, applied to the proof stress.

   The design strength of a material is obtained by dividing the characteristic strength by a partial safety factor associated with the material. The characteristic strength is obtained from control specimen used in testing. Suitable correction factors have to be applied to the to obtain the strength of the member in relation to the control specimen. In the present case the concrete cube is the control specimen so to obtain the prism strength one has to multiply by 0.67. On the other hand, the control specimen for steel is the reinforcement bar itself therefore no correction is needed. The partial safety factors applied to materials differ for different types of members. The factor that is applied to bending members is not the same as that applied to compression members. The bending members are subjected to maximum stress at the extreme fiber to start with, and there is a redistribution of stress within the section.

3.3 DESIGN BENDING MOMENT CAPACITY OF RECTANGULAR SECTION

The limit state design of reinforced concrete section in bending is based on the following assumptions:
   The material is assumed to be homogeneous and isotropic. At a microscopic level this is not a valid assumption but acceptable at macroscopic level. This implies that the section as a whole behaves like a homogeneous material for all practical purposes even though concrete is matrix of aggregate and mortar with embedded reinforcement. Concrete resists compression only while the reinforcement resists all the tensile force, plus part of compression if steel is placed in the compression zone.
   Concrete is brittle and has an ultimate compressive strain of 0.0035. Ultimate tensile strain may be in the order of 0.00015, which is less than one twentieth of that of the compressive strain. The tensile stress is often taken equal to zero. Even though concrete is capable of resisting a small tension, but at the macroscopic level, steels resists tension in total. Further the member may have micro-cracks and not capable of resisting any tension. Plane sections in a reinforced concrete beam before bending will remain plane even after bending. This assumption holds good till collapse and acceptable for all slender members. The deformation of slender members is dominated by the bending. The compatibility of deformations and strains is controlled by this assumption. The embedded reinforcement is bonded with concrete and even if the section is treated as cracked one. Adequate bond length must be available at all the critical sections.

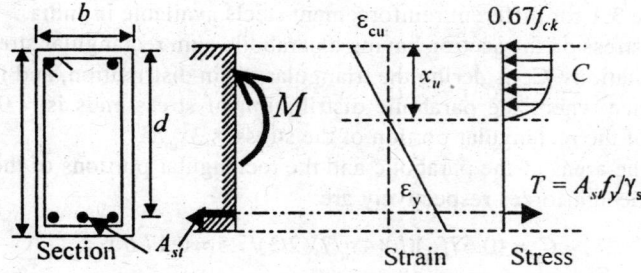

Fig. 3.3 Section under Bending, Notations.

The bending resistance of a section for simultaneous occurrence of the limiting strains in concrete and steel is called *balanced moment capacity*. Such a section is called *balanced section*. The area of reinforcement corresponding to it is called *balanced reinforcement*.

Consider a rectangular section of width b and depth h. Figure 3.3 illustrates the notation. The notations are:

A_{st} = area of tension reinforcement,

A_{stb} = area of tension reinforcement for a balanced section,

ε_{cu} = crushing strain in concrete = 0.0035,

ε_y = yield strain in steel = $f_{st}/\gamma_s = f_{st}/1.15$,

d = depth of centroid of reinforcement from extreme compression fibre. It is called *effective depth of reinforcement*,

T = design tensile capacity of reinforcement,

C_d = design compressive capacity of concrete section,

x_u = depth of the compression zone or it is also called as depth of compression block,

x_c = distance of the centroid of the total compression force on concrete from top fibbers,

E_s = modulus of elasticity of steel and it is taken = 200,000 N/mm^2,

f_y = yield or proof strength (stress) of steel reinforcement,

The *neutral axis* distance in a balanced cross section can be obtained from the similar triangles of the strain as shown in Fig. 3.3.

$$\frac{\varepsilon_{cu}}{\varepsilon_y} = \frac{x_u}{d - x_u} \tag{3.1}$$

$$x_u = \frac{\varepsilon_{cu}\,d}{\varepsilon_{cu} + \varepsilon_y} = \frac{0.0035d}{0.0035 + 0.002 + f_y / 1.15E_s}$$

$$x_u = \frac{805d}{1265 + f_y} = k_u\,d \tag{3.2}$$

Where

$$k_u = \frac{805}{1265 + f_y}$$

The Eq. (3.2) gives the neutral axis distance in a balanced section. The value of the neutral axis coefficient is given in table 3.1 for different reinforcement steels available in India.

The compressive stress on the section having a parabolic cum rectangular stress distribution is shown in Figure 3.4 with notations. Considering the triangular strain distribution, and from similar triangles of the strain, the distance where the parabolic distribution of stress ends is = $0.002x_u/0.0035 = 4x_u/7$. Similarly the depth of the rectangular portion of the stress is $3x_u/7$.

The considering the areas of the parabolic and the rectangular portions of the stress distribution, the corresponding total design forces respectively are:

$$C_p = (0.67f_{ck})(b)(4x_u/7)(2/3)/1.5 = 0.17\,b\,x_u f_{ck}$$

$$C_r = (0.67f_{ck})(b)(3x_u/7)/1.5 = 0.19\,b\,x_u f_{ck}$$

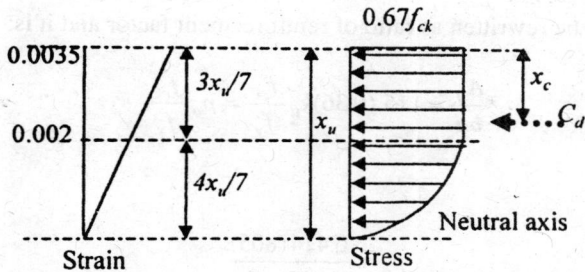

Fig. 3.4 Strain & Stress Distribution in Compression Zone.

The total design compression force, which is equal to the sum of the above two is:

$$C_d = C_p + C_r = 0.36\, b\, x_u\, f_{ck}$$

The distance (x_c) from the top fibre is obtained by taking the moment of the forces about the top fibre.

$$C_d(x_c) = C_p((\,5/8)(4\, x_u\, /7) + 3\, x_u/7) + C_r(3\, x_u/14)$$

This will give the distance of the centroid of the compression force as

$$x_c = 0.42\, x_u$$

The distance between the design compressive and tensile forces is called *lever arm* and it is:

$$= d\text{-}x_c = d - 0.42 x_u = (\,1 - 0.42 k_u)\, d = jd$$

$$(3.3)$$

Where

$j =$ is the ratio of the lever arm distance to the effective depth of steel and it is:

$$j = 1 - 0.42 k_u$$

$$(3.4)$$

The values of lever arm distance coefficient, j are listed in table 3.1.

Table 3.1 Design Moment & Neutral Axis Coefficients.

f_y (N/mm²)	K	j	k_u	p_o	% steel for M20-M50
250	0.149	0.78	0.531	0.220	1.79 – 4.48
360	0.141	0.79	0.495	0.205	1.14 — 2.85
415	0.138	0.80	0.479	0.198	0.95 – 2.38
500	0.133	0.80	0.456	0.189	0.76 – 1.89

The corresponding design tensile force from the steel reinforcement is:

$$T = A_{st} f_y / \gamma_s = A_{st} f_y / 1.15 = 0.87\, A_{st} f_y$$

The equilibrium of forces on the section in the horizontal direction gives: $T = C_d$

$$A_{st} f_y / 1.15 = 0.36\, b x_u\, f_{ck} = 0.36\, k_u\, bd\, f_{ck}$$

The above equation can be rewritten as ratio of reinforcement factor and it is:

$$\frac{A_{st}}{bd} = 1.15\,(0.36)k_u\,\frac{f_{ck}}{f_y} = p_o\,\frac{f_{ck}}{f_y} \tag{3.5}$$

Where

$$p_o = \frac{0.414\,(805)}{1265 + f_y} \tag{3.6}$$

The value of p_o is an indication of percentage of reinforcement required in a balanced section. These values are listed in table 3.1. The value is in the range of 0.189 to 0.220, a narrow band. The table 3.1 also the range of percentage of reinforcement for concrete of grades $M20$ and $M50$ in which percentage of reinforcement with respective the effective area of cross section is:

$$p = \frac{100A_{st}}{bd} = 100p_o\,\frac{f_{ck}}{f_y} \tag{3.7}$$

It may be noted that the approximate percentage of reinforcement of Fe 415 grade steel is 0.95 for $M20$ grade concrete and 2.38 for $M50$ concrete.

The balanced moment capacity of the rectangular section is given by:

$$M_{rb} = C_d jd = Kbd^2 f_{ck} \tag{3.8}$$

Where
$$K = 0.36jk_u$$

3.4 DESIGN LOADS AND COLLAPSE BENDING MOMENT

Loads on a structure are specified by the code or by the user. Characteristic load is considered as the base line. For all purposes, the characteristic load is same as that specified by the codes. However, in case specified loads are not available, the concept of characteristic load is to be used. *Characteristic load* is defined as that load which has a 95 percent probability of not being exceeded during the life of the structure. *Design load* is obtained by multiplying the specified or characteristic load by partial safety factor. The design load consists of combinations of different loads that act on the structure. For example, a slab is subjected to its' own weight, other fixed loads, live loads and wind or earthquake load etc. A design load has built in safety margins through partial safety factors and combination of different types of loads. The design loads are placed on the structure and it is analyzed for internal forces of the members. The design load in general terms is given by:

$$W_r = \Sigma\,\gamma_i\,W_i$$

Where
  W_r = design load,
  $W_i = i^{th}$ possible load that is likely to occur on the structure,
  γ_i = partial safety factor applied to the load.

The partial safety factor applied to the load is 1.5 for strength limit and about 1.0 for serviceability state. The dead loads are always present, whereas full live loads, wind loads act at times. Therefore the design load combinations are termed as: live load combination, wind load combination, earthquake load combination etc. Some of the commonly used strength limit state combinations are listed here.

Live load combination, say combination 1:

Design load = 1.5 DL + 1.5 LL

Wind (or earthquake) load combination, there are two possible combinations; combination 2 and 3
Design load = 1.5 DL +1.5 (WL or EL) or ,
Design load = 1.2 DL +1.2 LL +1.2 (WL or EL); in which the dead load and wind load act cumulatively,
Design load = 0.9 DL + 1.5 (WL or EL), in case the dead load has compensating effect on the wind load.

Where
  DL = dead load,
  LL = live load,
  WL = wind load,
  EL = earthquake load.

The wind load and the earthquake loads are not expected to act at the same time. Either wind or earthquake acts and not together. The wind or earthquake loads are the most critical specified to the zone of the structure. In some cases, the wind load has the over turning effect whereas the dead load has stabilizing effect. In such a case, a partial safety factor of 0.9 is applied to the dead load to be on the safe side. The design load mentioned here is with reference to the limit state of strength. This load is also called as collapse load or some times called as ultimate load. The term collapse load is misleading; it need not necessarily imply that the structure will collapse at that load. The collapse load is applicable to more for plastic analysis of structures, however it is used in the limit state design for convenience of referring to the strength limit.

The structure should be analyzed for the combinations listed above plus any other similar load combination such as snow, blast, impact, shrinkage, etc. as the case may be. One can multiply the loads and analyze the structure, or analyze the structure for each basic load and then suitably multiply with the appropriate partial safety factor and then combine. The combination could be at basic load level or at the resultant forces for the basic load. In computer-based analysis, the structural analysis is carried for each of the basic loads then the combination is suitably chosen. This is probably better when there are many load combinations.

The partial safety factors applied to the loads in the service limit state are different from those in the strength state. These load combinations for *serviceability limit state* is:

Design load = DL + (WL or EL) or,
Design load = DL +0.8 LL +0.8 (WL or EL); in which the dead load and wind (or Earthquake) load act cumulatively,
Design load = 0.9 DL + (WL or EL), in case the dead load has compensating effect on the wind load.

3.5 DESIGN FOR BENDING MOMENT

The real member design involves design for bending, shear, bond and torsion if any etc. The method of design is developed stage by stage for the sake of simplicity and clarity. The following steps are recommended for moment design:

1. Calculate or estimate the basic loads such as dead, live, wind etc. on the structure. The size of the member may not be known to start with so one has to estimate the self weight of the member or members,

2. Select the concrete grade based on the exposure condition and economics, and reinforcement grade,
3. Analyze the structure for internal forces and deformations by elastic linear analysis for each load combination after applying the partial safety factors.
4. Select the critical moment (M_c) and apply the condition that the moment capacity is more than the design moment. That is $M_{rb} \geq M_c$, Where M_c is the collapse moment, M_{rb} is the balanced moment capacity of the section. Normally one has to assume the width of the section and compute the effective depth of the section.
5. Round off the effective depth to a convenient figure and then compute the area of reinforcement. Select suitable number and diameter of the bars to match the required are of reinforcement.

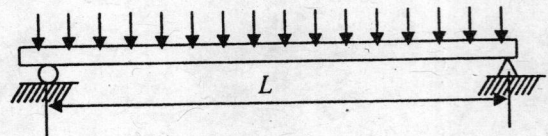

Fig. 3.5 Simply Supported Beam.

Example 3.1 Design for balanced moment capacity

A five metre effective span simply supported beam in mild exposure condition is subjected to a live load of 24 kN/m. It is to be designed with $M20$ concrete and $HYSD$-$Fe415$ bars.

Solution

The given data is: $L = 5$ m, $f_{ck} = 20$ N/mm^2, $f_y = 415$ N/mm^2, $w_l = 24$ kN/m, the design coefficients for the present data can be obtained from table 3.1 and they are: $K = 0.138$, $j = 0.80$, $k_u = 0.479$

Assume the width of the beam as $b = 250$ mm and depth (h) of the beam in the range of $L/12$ for simply supported beams, say $h = 450$ mm for the purpose of dead weight. Fig. 3.5 illustrates the simply supported beam.

Self weight of the beam = $w_g = 0.25(0.45)25 = 2.81$ kN/m, use 3 kN/m
Total design load at collapse state (strength limit state) = $w_t = 1.5(3 + 24) = 40.5$ kN/m
Maximum bending moment at collapse =

$$M_c = \frac{w_t L^2}{8} = \frac{40.5(25)}{8} = 126.56 \text{ kNm}$$

The balanced moment capacity must be more than the collapse moment, that is:

$$M_{rb} \geq M_c, \text{ This leads to:}$$

$$Kbd^2 f_{ck} = 0.138(250)(20)d^2 \geq 126.56(10^6)$$

$$\text{or, } d = \sqrt{126.56(10^6)/(0.138)(250)(20)} = 428 \text{ mm}$$

Let the effective depth be 430 mm. The minimum cover to the reinforcement is 25 mm or the diameter of the bar. Let the diameter of the bar be assumed 25 mm. Therefore the total depth of the beam is = $h = 430 + 25/2 + 25 = 467.5$ mm. Now revise the assumed depth to 470 mm. The assumed self-weight is close to the final value.

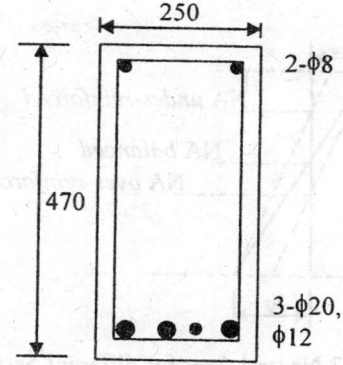

Fig. 3.6 Example 3.1.

The area of the tension reinforcement needed is:

$$A_{st} = \frac{1.15M_c}{jdf_y} = \frac{1.15(126.56)10^6}{0.8(430)415} = 1020 \text{ mm}^2$$

Use 3 numbers of 20 mm bars and one number of 12 mm bar.

The area of reinforcement provided is
A_{st} (provided) = 3*314 + 113 = 1055 mm² > 1020 mm² needed
Provide 2 numbers of 8 mm hanger bars at the top.
These bars are used to tie up the shear reinforcement.

3.6 UNDER-REINFORCED AND OVER-REINFORCED SECTIONS

Once the width of the section is selected, the depth is uniquely determined for a balanced section. The depth of the section is rounded off to a rational value, even after deciding the balanced section. If the depth of the beam is different from the balanced one, then the crushing strain in concrete and the yield tension strain in steel don't occur simultaneously.

If the depth of the beam selected is more than that of a balanced section, then the steel reaches yield strain earlier than the crushing strain in concrete. The strain in steel increases rapidly and the strain in concrete reaches the crushing value at failure. The failure initiated by the yielding of steel gives a warning before the failure. The failure is called *ductile failure*.

Figure 3.7 illustrates the location of neutral axes for different possible combinations of strains in steel and concrete. A reinforced concrete section in which the yielding of steel is initiated first before the concrete in compression zone is subjected to crushing is called *under-reinforced section*. The cost of reinforcement steel is about seventy times that of concrete by volume. The designer would like to use minimum quantity of steel as far as possible. The cost of a beam depends on cost of materials, cost of formwork and labour, and other space limitations. In under-reinforced concrete section, the strength of steel is fully utilized therefore it is likely to be economical. If the depth of a beam is made smaller than that of a balanced section, then the extreme concrete fiber will reach its crushing strain earlier when compared to the yield strain in reinforcement. The neutral axis distance in such a section is larger than that of a balanced section. A reinforced concrete section in which the concrete in extreme fiber in

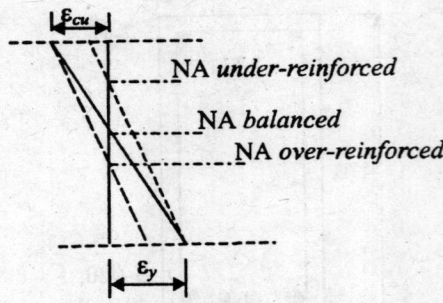

Fig. 3.7 Neutral Axes for different Section.

compression is subjected to crushing before yielding of steel is called *over-reinforced section*. This simply means that the steel in the beam is not fully utilized. Fig. 3.7 illustrates the relative location of the neutral axis of over-reinforced beam. The two advantages in an under-reinforced section are:

● The expensive steel is not fully utilized, therefore not economical, and
● The failure of the section occurs with warning.

The moment capacity of an under-reinforced beam is controlled by steel, while that of an over-reinforced is by concrete.

The neutral axis location is obtained by equating the compressive force on concrete to the tensile force in the reinforcement. The compressive force on the concrete is limited to the maximum. The equilibrium condition gives

$$A_{st}(0.87f_y) = 0.36b(k_u d)f_{ck} \qquad (3.9)$$

The neutral axis distance is given by

$$k_u = \frac{0.87 A_{st} f_y}{0.36 b d f_{ck}} \qquad (3.10)$$

Moment capacity of under-reinforced rectangular section is controlled by steel and it is equal to the area of tension reinforcement multiplied by the lever and strength of steel with partial safety factor of 0.87

$$M_r = 0.87 A_{st} f_y (1 - 0.42 k_u) d \qquad (3.11)$$

The lever arm is likely to be less than the balanced section that is given above but close to it. For all practical purposes, the lever of the balanced section is considered in computing the moment capacity. The error is marginal and on the safer side.

For an over-reinforced rectangular section, the moment capacity is same as that of a balanced section:

$$M_r = C_d jd = Kbd^2 f_{ck} \qquad (3.12)$$

Most of such sections are provided with reinforcement in the compression zone. Such section is called doubly reinforced section.

Example 3.2. Design of under-reinforced section

A 5 m effective span simply supported beam is subjected to a live load of 24 kN/m and is designed with *HYSD-Fe*415 reinforcement. The beam is exposed to mild environment and must have an over all depth of 550 mm for architectural consideration. Design the beam for bending moment.

Solution

The given data is: $L = 5$ m, $h = 550$ mm, mild exposure so $f_{ck} = 20$ N/mm², $f_y = 415$ N/mm²; $w_l = 24$ kN/m, the design coefficients obtained from table 3.1 are: $K = 0.138$, $j = 0.80$, $k_u = 0.479$
   Assume the width of the beam as $b = 250$ mm,
   Self weight of the beam $= w_g = 0.25(0.55)25 = 3.4$ kN/m, use 3.5 kN/m
   Total design load $= w_t = 1.5(3.5 + 24) = 41.25$ kN/m
   Maximum bending moment at collapse =

$$M_c = \frac{w_t L^2}{8} = \frac{41.25(25)}{8} = 128.9 \text{ kNm}$$

The balanced moment is compared with the collapse moment to determine the depth of a balanced section:

$$M_r \geq M_c, \text{ which leads to:}$$

$$Kbd^2 f_{ck} = 0.138(250)(20)d^2 \geq 128.9(10^6)$$

$$\text{or, } d = \sqrt{128.9(10^6)/(0.138)(250)(20)} = 432 \text{ mm}$$

The depth of a balanced section is 432 mm while that provided is equal to $550 - 35 = 515$ mm, for 20 mm reinforcement bars. The section is under-reinforced as the depth provided is more than that of a balanced section. The area of tension reinforced required be obtained from the moment capacity of the under-reinforced section.
   The area of the tension reinforcement needed is:

$$A_{st} = \frac{1.15 M_c}{jd f_y} = \frac{1.15(128.9)10^6}{0.8(515)415} = 869 \text{ mm}^2$$

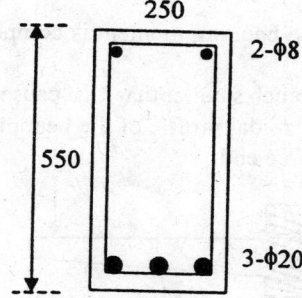

Fig. 3.8 Example 3.2.

Use 3 numbers of 20 mm bars.

The area of reinforcement provided is

A_{st} (provided) $= 3*314 = 942$ mm^2 > 869 mm^2 needed.

Provide 2 numbers of 8 mm hanger bars at the top.

These bars are used to hang up the shear reinforcement.

The lever arm in this example is taken equal to that of the balanced section. The area of reinforcement computed is marginally more than that is really needed. The actual neutral axis distance is:

$$k_u = \frac{0.87 A_{st} f_y}{0.36 b d f_{ck}} = \frac{0.87(942)415}{0.36(250)(515)(20)} = 0.366$$

The corresponding lever arm ratio is

$$j = (1 - 0.42 k_u) = 0.846$$

The approximation is 0.8 as against 0.846 actual.

Example 3.3 Design of cantilever beam

A 3 m cantilever span beam of is subjected to a concentrated load of 16 kN at 500 mm from free edge. The beam is in open non-aggressive environment.

Solution

Effective span in cantilever beam built in masonry is equal to the clear span plus the effective depth of the beam at support. This is as per the Indian code of practice IS:456-2000. The bending moment is zero at the free end and increases to maximum at support. Therefore the beam can be made tapering with the maximum design depth at the support and nominal depth at free end. Select concrete grade M25 as the beam is exposed to moderate condition.

Fig. 3.9 illustrates the beam profile. Select HYSD-Fe415 reinforcement steel. The corresponding design coefficients are: $K = 0.138, j = 0.80, k_u = 0.479$. The effective depth of the beam at support is assumed as = clear span/7 = 3/7 = 0.4 m. Let the width of the beam = b = 0.25 m. For the purpose of self-weight assume the average depth of the beam as 350 mm.

The effective cantilever span is = L = clear span + effective depth of the beam = 3.4 m.

Let self-weight of the beam is = $w_g = 0.35(.25)25 = 2.2$ kN/m.

Even though the beam is tapering the bending moment is computed assuming a uniform depth. The error is marginally on the safer side.

The self-weight bending moment is much smaller than that caused by the concentrated load. For more accurate moment calculations, the trapezoidal profile of the beam is considered. The concentrated load is acting at a distance of 500 mm from free end.

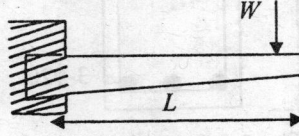

Fig. 3.9 Cantilever Beam.

The maximum bending moment that occurs at a distance d beyond the support is:

$$M_c = \frac{1.5w_g(3.4)^2}{2} + 1.5W_l(3.4-0.5) = 0.75(2.2)11.56 + 24(2.9) = 19.1 + 69.6 = 88.7 \text{ kNm}$$

The design criterion is:

$$M_r = Kbd^2f_{ck} > 88.7 = M_c, \text{ this leads to:}$$

$$d \geq \sqrt{88.7(10)^6 / 0.138(250)(25)} = 321 \text{ mm}$$

The cover required for moderate exposure condition is 30 mm, so the overall depth required is $= 321 + 10 + 30 = 361$ mm, so provide $h = 370$ mm and $d = 330$ mm, then the area of reinforcement required is:

$$A_{st} = \frac{1.15M_c}{jdf_y} = \frac{1.15(88.7)(10)^6}{0.8(330)(415)} = 932 \text{ mm}^2$$

The lever arm used is that of the balanced section. Provide 3 numbers of 20 mm bars at top, the area of steel provided is 942 mm² as against 932 mm². Adequate development length should be provided and the development length requirement is discussed later. Provide 2 numbers of 8 mm bars at bottom as hanger bars. The depth of the beam can be made nominal at the free end and equal to 200 mm.

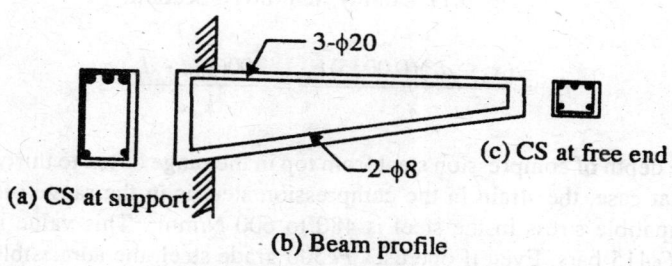

(a) CS at support   (b) Beam profile   (c) CS at free end

Fig. 3.10 Reinforcement Details Example 3.3.

3.7 DOUBLY REINFORCED SECTION

There are situations in which the depth of a beam is chosen less than that of a balanced one. In such a case, the area of concrete is not adequate to carry the collapse moment, therefore additional reinforcement is provided in the compression zone. This reinforcement is called compression reinforcement. A design with compression reinforcement along with tension steel is called *doubly reinforced section*.

The bending moment to be resisted by the compression reinforcement is the difference of the collapse moment and the balanced moment capacity of the section, and it is:

$$M_{rd} = M_c - M_{rb} \tag{3.13}$$

The following notations are used with reference to the doubly reinforced concrete section.

M_{rd} = moment capacity corresponding to compression reinforcement,
A_{sc} = area of compression reinforcement,
d_c = distance of centroid of compression reinforcement from extreme compression fibre (say top fibre),

f_{sc} = compressive stress in compression steel compatible with strain in concrete at that level subject to maximum of yield strength of steel,

A_{st2} = area of tensile reinforcement corresponding to that matching the compression reinforcement, usually it is equal to the compression steel provided d_c is nominal and steels are of same quality,

The actual strain in compression steel can be calculated from the compatibility of strains in the compression zone. Fig 3.11 illustrates the strains at different depths and other notations used in doubly reinforced concrete. The stress in the compression steel is:

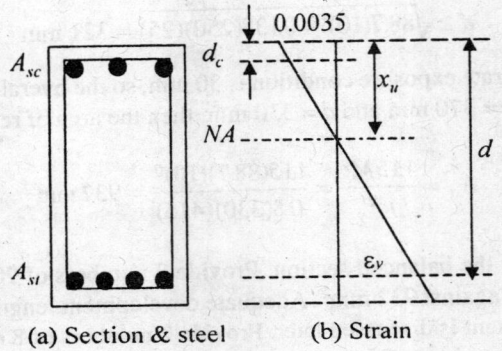

(a) Section & steel   (b) Strain

Fig. 3.11 Doubly Reinforced Section.

$$f_{sc} = \frac{(x_u - d_c)(0.0035)E_s}{x_u} = \frac{700(k_u d - d_c)}{k_u d} \le f_y \qquad (3.14)$$

One can assume the depth of compression steel from top in the range of ten to thirty percent of the neutral axis distance. In that case, the strain in the compression steel is in the range of 0.0024 to 0.003. The corresponding compatible stress in the steel is 480 to 600 N/mm^2. This value is more than the yield strength of *HYSD-Fe*415 bars. Even if one uses *Fe*500 grade steel, the admissible stress is equal to the yield stress. In any case, a check has to be made based on the cover distance. The area of compression steel can be obtained by equating the excess moment above the balanced one to that resisted by the steel, and it is:

$$A_{sc} = \frac{\gamma_s M_{rd}}{(d - d_c)f_{sc}} = \frac{1.15 M_{rd}}{(d - d_c)f_{sc}} \qquad (3.15)$$

The partial factor of safety is applied to stress or strength of steel in the above expression.

Similarly the corresponding tensile reinforcement, this is in addition to the balanced reinforcement is:

$$A_{st2} = \frac{1.15 M_{rd}}{(d - d_c)f_y} \qquad (3.16)$$

The total area of tensile reinforcement is:

$$A_{st} = A_{sb} + A_{st2} \qquad (3.17)$$

Example 3.4 Design of doubly reinforced beam

A 5 m effective span simply supported beam is subjected to a live load of 24 kN/m. The beam is exposed to mild environment and must have an over all depth of 400 mm for architectural consideration.

Solution

The given data is: $L = 5$ m, $h = 400$ mm, $f_{ck} = 20$ N/mm^2, $f_y = 415$ N/mm^2, $w_l = 24$ kN/m, the design coefficients for the present data can be obtained from table 3.1 and they are: $K = 0.138$, $j = 0.80$, $k_u = 0.479$

Assume the width of the beam as $= b = 250$ mm,
Self weight of the beam $= w_g = 0.25(0.4)25 = 2.5$ kN/m,
Total design load $= w_t = 1.5(2.5 + 24) = 39.75$ kN/m, use 40 kN/m

Maximum bending moment at collapse $=$

$$M_c = \frac{w_t L^2}{8} = \frac{40(25)}{8} = 125 \text{ kNm}$$

The effective depth of the section is:

$$d = 400 - 35 = 365 \text{ mm}$$

The balanced moment capacity of the section is:

$$M_{rb} = Kbd^2 f_{ck} = 0.138(250)(365^2)(20)/10^6 = 91.5 < 125 \text{ kNm}$$

The balanced moment capacity of the section is less than the collapse moment; therefore the beam has to be designed as doubly reinforced concrete section.

The area of the tension reinforcement needed for balanced section portion is:

$$A_{sb} = \frac{1.15 M_{rb}}{jdf_y} = \frac{1.15(91.5)10^6}{0.8(365)415} = 868 \text{ mm}^2$$

The area of compression and tension reinforcement for the remaining portion of the bending moment is;

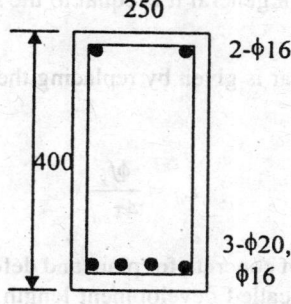

Fig. 3.12 Example 3.4.

$$A_{sc} = A_{st2} = \frac{1.15(M_c - M_{rb})}{(d - d_c)f_y} = \frac{1.15(125 - 91.5)10^6}{(365 - 35)415} = 281\,\text{mm}^2$$

The total area of tension reinforcement provided is:

$$A_{st} = A_{sb} + A_{st2} = 868 + 281 = 1149\,\text{mm}^2$$

Provide 3 numbers of 20 mm bars and 1 of 20 mm bar.

A_{st} (provided) = 3*314 + 201 = 1143 mm²; Actually required is 1149 mm².

Provide 2 numbers of 16 mm to bars at 35 mm below the top fibre. These bars can also be used as hanger bars for shear reinforcement.

3.8 DEVELOPMENT LENGTH, ANCHOR LENGTH, CURTAILMENT OF BARS, SPLICES AND LAP LENGTHS

The reinforcement bars in concrete are expected to have perfect bond with the concrete. There must be adequate concrete all round the surface of the bar. Cover of concrete around the bar must not less than the diameter of the bar or the size of the aggregate for sound bonding. The bond strength between the bar and the concrete depends on the type of surface of the bar and the strength of the concrete. The deformed bars have higher bonding strength. Length of a reinforcing bar required to develop its full capacity without slipping from the concrete is called *development length*. The perimeter of the bar of bond length multiplied by bond strength should be equal or more than the strength of the bar. This criterion can be expressed as:

$$\pi \phi L_d \tau_{bd} = \frac{\pi \phi^2 \sigma_s}{4}, \text{ or}$$

$$L_d = \frac{\phi \sigma_s}{4 \tau_{bd}} \tag{3.18}$$

Where

ϕ = diameter of the bar

τ_{bd} = bond strength of concrete with respect to bar

σ_s = stress in the reinforcement, in general it is equal to the strength of steel

L_d = development length

The full development length of a bar is given by replacing the stress by the strength of steel. The full development is given by:

$$L_d = \frac{\phi f_y}{4 \tau_{bd}} \tag{3.19}$$

Table 3.2 gives the bond strengths of concrete for plain and deformed reinforcement. The length of the bar that will hold against any slip is called development length. Table 3.3 gives the full development length.

Table 3.2 Bond Stress in Limit State Design for Bars.

Grade of concrete	M20	M25	M30	M35	M40 & above
Plain bars (tension)	1.2	1.4	1.5	1.7	1.9
Deformed bars (tension)	1.9	2.2	2.4	2.7	3.1
Deformed bars (compression)	2.4	2.7	3.0	3.4	3.9

Bending of bars at ends increases the anchor capacity. Every 45° bent of a bar provides a bond capacity four times the diameter of the bar. A bar bent at 90° gives anchor length of 8 times the diameter; similarly, a bar bent in 180⁰ gives an anchor length of 16 times the diameter of the bar. The total additional bond length that can be accounted from the bends of bars is limited to a maximum of 16 times the diameter of the bar.

Anchor length in shear stirrups or in ties: Stirrups are provided to resist shear stress in beams. The stirrups are bent around the hanger bars. Adequate anchor length is deemed to have been provided if the end of the stirrup is bent at 90 degrees and further continued to length of 8 times the diameter of the stirrup. In case of 135 degrees bent, the required extension length beyond is 6 times the diameter of the bar. If the ends of the stirrups are bent at 180 degrees, then the bar must extend by 4 times the diameter. In most case the stirrups are bent at 135 or 90 degrees.

Anchor in bundled bars: The development length of 2 bundled bars is 10 percent more than that of a single bar. The development length for three and four bundled bars is 20 and 33 percent more than that of a single bar.

Splice and lap lengths: Joint or splice in reinforcement bar is provided because the required length is not available. The length of over lap between two successive bars is called *lap length*. The lap length must allow full transfer of force from one bar to another. The can be at locations where the bending moment is less critical. It is desirable to limit the number of bars spliced at one location to about one third of the bars, under any circumstance, not more than half of the bars can be spliced. Most desirable location of splice is at one-third span in simply supported beam or slab, and about quarter span in continuous beam. Similarly the splices in column bars can be at about one third the height of the column. The tendency of splice all bars in column at one location must be avoided. The actual stress in the bar at lap location can be considered in computing the development length. The laps are considered as staggered if the centre-to-centre distance of the splices is 1.3 times the development length. The recommended lap length are listed in table 3.3.

Table 3.3 Full Development Lengths of Bars.

Type of Deformed bars	M20	M25	M30	M35	M40 & above
Bars in tension	54φ	47φ	43φ	38φ	33φ
Bars in compression	43φ	38φ	34φ	30φ	27φ

φ = diameter of the bar

- Lap length = development length (L_d) subject to a minimum of 30φ for flexural tension,
- Lap length = 2 L_d subject to a minimum of 30φ for direct tension members,
- The straight length of the bar beyond hook shall not be less than 15φ subject to a minimum of 200 mm,
- Lap length = 1.4 L_d if the lap occurs in tension bar at top with a cover less than twice the diameter of the bar,

- Lap length = 1.4 L_d if the lap is at corner of a section and the minimum cover is less than twice the diameter of the bar, or the clear distance between the laps is less than 75 mm or 6 times the diameter of the bar,
- Lap length = 2 L_d for top corner bar where the cover is less than 2 times the diameter of the bar, or the distance between the laps is less than 75 mm or 6 times the diameter of the bar.
- In case of hanger and shrinkage bars, the lap length is atleast 30 times the diameter of the bars,
- Lap length = L_d in bar under compression subject to minimum of 24 times the diameter of the bar,
- Where two different diameters are lapped, smaller diameter is considered for lap length,
- Laps in the bundled bars is to be done for one bar at a time and not all the bars at one location,
- Lap splices should be avoided for bars larger than 36 mm diameter, If the laps of bars larger than 36 mm diameter, additional spiral should be provided around the lap.

The laps provided at the bottom face of beam have full consolidation of concrete. Therefore, the lap length at bottom face of the beams is same as the development length. However, the concrete at top face of a beam or at corners of the section may not have adequate concrete consolidation. The concrete has a tendency of settling down leaving a possible separation of the bar and the concrete under the bar. Therefore the lap length provided at top face of the beam or at corners of the section is given extra lap length.

Welded splices: Welded splices are recommended for bars larger than 36 mm diameter. The strength of the weld material should be equal or greater than the strength of the material of the bar. The weld must be designed such that its strength is equal or more than the strength of the bar in compression splice, about 25 percent more in case of tension bar. Butt or lap welding can be used in the splices.

Curtailment of bars: The bending moment decreases away from the critical section. Some bars of the critical section can be curtailed (cut off) where they are not required. The following points be considered while curtailing the bars:

- A curtailed bar should extend beyond the point where it is not required to resist the force for a distance equal to *effective depth of the beam or 12 times the diameter of the bar*, which ever is larger. However the shear at the cut of section should not exceed two-third of the shear strength of the section.
- At least *one-third of the positive* reinforcement in simply supported beam and *one-fourth of the positive* reinforcement of continuous beam must be continued over the support. Such bars should extend over the support by length equal to one-third of the development length.
- The diameter of the main bar at support or at the point of inflection should be limited to such that the full development length of the bar shall not exceed $M_{ri}/V + L_0$. Where M_{ri} is the bending moment capacity of the section, V is the design shear force at the section; L_0 is equal to the sum of the anchorages beyond the centre of the simple support, and the effective depth or 12 times the diameter of the bar in case of point of contraflexure.
- At least one-third of the total reinforcement of the negative bending moment at the support must extend beyond point of inflection for a distance not less than effective depth or 12 times the diameter of the bar, or one-sixth of the clear span which ever is larger.
- Bars in bundled bars must not terminate at one point. The bars must terminate at different points separated by distance not less than 40 times the diameter of the bar. However the bundle of bars can terminate on the support.

Example 3.5 Curtailment of bars

A simply supported beam of 5 m effective span is subjected to a live load of 24 kN/m. The beam is exposed to mild environment and must have an over all depth of 400 mm for architectural consideration. The beam was designed as doubly reinforced with two of 16 mm bars at top and 3 of 20 mm plus one 16 mm bar at bottom at mid span. Determine at what location one 20 mm bar and 16 mm bar can be terminated (see example 3.4).

Solution

The given data is: $L = 5$ m, $h = 400$ mm, $b = 250$ mm, $f_{ck} = 20$ N/mm2, $f_y = 415$ N/mm2, $w_l = 24$ kN/m, the design coefficients for the present data are: $K = 0.138$, $j = 0.80$, $k_u = 0.479$;

Self weight of the beam = $w_g = 0.25(0.4)25 = 2.5$ kN/m,
Total design load = $w_t = 1.5(2.5 + 24) = 39.75$ kN/m, use 40 kN/m
Maximum bending moment at collapse =

$$M_c = \frac{w_t L^2}{8} = \frac{40(25)}{8} = 125 \, kNm$$

The reinforcement at bottom is 3 of 20 mm bars and one of 16 mm bar. Of this two bars are continued till the support and the remaining are curtailed at distance x from support. Design total load is = 40 kN/m. A schematic representation of the external design moment diagram and the resisting moment diagram for curtailed reinforcement is shown in Fig. 3.13.

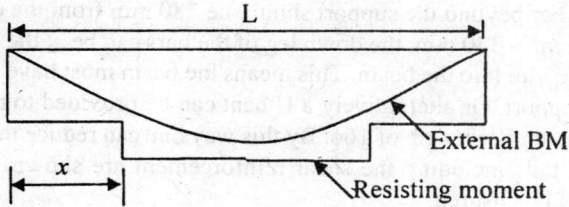

Fig. 3.13 Bending & Resisting Moments.

Area of total reinforcement at mid section = $A_{st} = 3(314) + 201 = 1143$ mm^2
Area of reinforcement left after curtailing 1 of 16 mm and 1 of 20 mm = $A_{st1} = 2(314) = 628$ mm^2
The bending moment at a distance x from the support is = $M_{cx} = w_t \, x(l - x)/2 = 20 \, x(l - x)$
Proportioning the area of tension reinforcement with respect to the bending moments is:

$$\frac{M_{cx}}{M_c} = \frac{A_{st1}}{A_{st}} = \frac{628}{1143} = 0.549 \text{ , or}$$

$$M_{cx} = 0.549(125) = 68.625 \, kNm = 20x(5 - x)$$

The above equation results in to: $x^2 - 5x + 3.8625 = 0$, the solution of the quadratic is:

$$x = \frac{5 \pm \sqrt{25 - 4(3.8625)}}{2} = \frac{5 - 3.1}{2} = 0.95 \, m$$

The bars can be curtailed at a distance of effective depth or 12 times the diameter of the **bar beyond the** point where they are not required. The effective depth is 365 mm and 12 times diameter of the larger bar is 240 mm, so select the larger of the two. The two bars can be curtailed at 950 – 365 = 585 mm from support.

Example 3.6 Anchor length in cantilever

A cantilever beam of 3 m clear span is subjected to a load of 16 kN at 500 mm from free end. The beam is exposed to open non-aggressive environment. See example 3.3 in which the beam is designed. The design data is: $f_{ck} = 25$, $f_y = 415$ N/mm^2, $b = 250$ mm, $h = 370$ mm, $d = 330$ mm and the main top reinforcement is 3 of 20 mm bars.

Solution

The cantilever beam is extended in to the wall for stability. The length of the beam embedded in to the wall must accommodate the adequate anchor length. The top main reinforcement bar can be bent at 90 degrees to have a adequate anchor length. The maximum bending moment occurs at the support where the bars are anchored. It is therefore necessary a careful anchor length be designed. The full development length for $M25$ concrete taken from table 3.3 is $L_d = 47\phi$. The 90 degrees bent of the bar provides an anchor length of 8ϕ, therefore the net anchor length required is:

Anchor length = $47\phi - 8\phi = 39\phi = 780$ mm.

The L bent length of the bar beyond the support should be 780 mm from the effective support. Since the effective depth of the beam is 330 mm, the down leg of the bars can be at the most 300 mm. The bars must extend 780 – 300 = 480 mm into the beam. This means the beam must have an anchor length of 480 mm beyond the effective support. Or alternatively a U bent can be provided to reduce the beam anchor length. The bent provides an anchor length of 16ϕ. By this way one can reduce the anchor of the beam to 160 mm. Reinforcement details including the shear reinforcement are shown in Fig. 3.14. The static stability of the beam should be ensured.

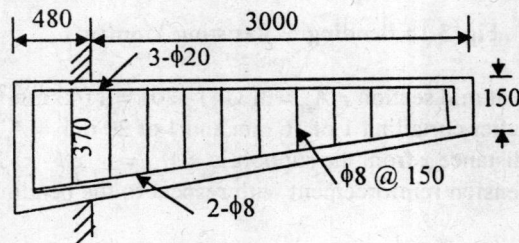

Fig. 3.14 Reinforcement Details of Cantilever Beam.

3.9. DESIGN FOR TRANSVERSE SHEAR IN RECTANGULAR BEAMS

The bending moment is the dominant force in slender members. A reinforced concrete beam is usually designed for bending moment first and then designed for shear force. The size of the beam is selected primarily based on the bending moment. The shear reinforcement is then designed for the size in most cases. However in some cases of high shear force, the size of the section chosen for bending capacity may not be adequate.

A simply supported beam subjected to uniformly distributed load is shown in Fig.3.15 (a). The corresponding bending moment and shear force diagrams are illustrated in the figure. Bending moment is maximum at mid span and shear force is maximum at support.

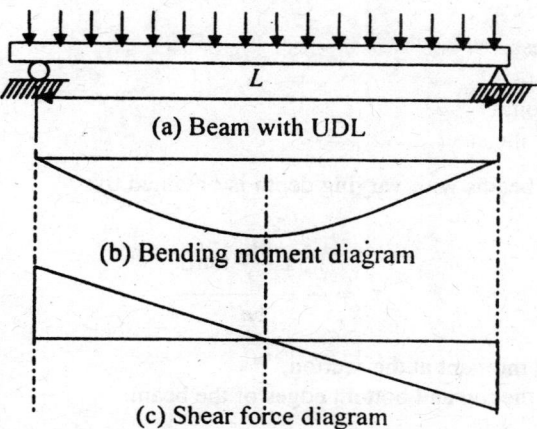

(a) Beam with UDL

(b) Bending moment diagram

(c) Shear force diagram

Fig. 3.15 BMD and SFD for Simply Supported Beam.

Figure 3.16 (a) illustrates a reinforced rectangular cross section with the effective depth and neutral axis distances indicated. The shear stress distribution of an uncracked cross section is parabolic along the depth of the section as shown in Fig. 3.16 (b). The average shear stress accepted by the code as an empirical value is also shown by a rectangular distribution. The maximum shear stress on the section occurs at the centroid of the section is =

$$\tau_m = \frac{1.5V_u}{bh}$$

In case of cracked section, the variation of the shear stress up to the neutral axis is parabolic with zero stress at the top and then it is uniform till the level of the tension reinforcement. This variation is shown in Fig. 3-16 (c). The figure also indicates the average shear stress used for design purpose. For the sake of design convenience, an average shear stress also called as *nominal shear stress* is given by:

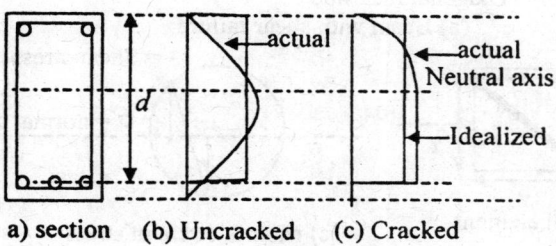

a) section   (b) Uncracked   (c) Cracked

Fig. 3.16 Shear Stress Distribution.

$$\tau_v = \frac{V_u}{bd} \tag{3.20}$$

Where
τ_v = nominal shear stress,
V_u = shear force,
b = width of the section,
d = effective depth of the steel

The nominal shear stress in beams with varying depth is obtained by:

$$\tau_v = \frac{V_u \pm \dfrac{M_u}{d} \tan \beta}{bd} \tag{3.21}$$

Where M_u = design bending moment at the section,
β = angle between the top and bottom edges of the beam

Shear stress produces diagonal tension, and when the diagonal tensile stress reaches the limiting tension in concrete, the section is considered to have failed in shear. Consider an infinitesimal element that is subjected to shear and compression in the shear zone.

Consider a simply supported beam shown in Fig. 3.17 (a) in which the shear force is a maximum near the support. A crack stating from the support and inclined at about 40° with horizontal generates going up to the top of the beam. Consider an infinitesimal element in this zone, and is subjected to shear stress and axial stress. The axial stress can be either compressive or tensile. Fig. 17 (b) illustrates the state of stress on the small element. The maximum principal stress is tensile stress and at a plane at 45° with the horizontal for pure shear stress condition. For an compressive and shear stresses, the principal plane may be inclined at an angle less than 45°. Fig. 3.17 (c) shows the Mohr's circle of stresses. If the principal stress reaches the limiting tensile stress of concrete, then the section will crack. Further the tensile stress capacity of concrete is moderated by the amount of tensile reinforcement. Higher the tensile reinforcement, higher is the tensile capacity of the concrete. Allowable shear capacities of the concrete are a

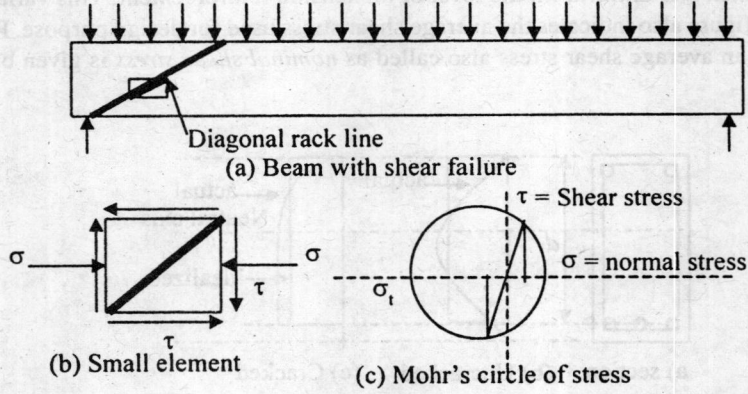

(a) Beam with shear failure

(b) Small element   (c) Mohr's circle of stress

Fig. 3.17 Tensile Stress a Failure Concept in Shear Design

function of the tensile reinforcement and these values are given in table 3.4. The shear reinforcement is provided to resist the shear force that is over and above that resisted by the concrete. The shear reinforcement is called transverse reinforcement, and provided in the form of ties connecting the top and bottom longitudinal bars. The shear reinforcement is also referred as stirrups. Further, the shear capacity of concrete even with shear reinforcement is limited to a maximum value. Under any circumstances, the concrete should not be subjected to a shear stress more than that specified by the code. The absolute maximum admissible shear strength of concrete is given in table 3.4. If the nominal shear stress on a section does not exceed the shear strength of the concrete, only nominal shear reinforcement need to be provided. Shear reinforcement must be provided in case the shear stress exceeds the shear strength of the concrete, subject to the condition that it is within the maximum admissible with the shear reinforcement. When the actual shear stress on the section is more than the maximum admissible with the shear reinforcement, the section must be redesigned such that the nominal shear stress is within the maximum limit. The design criterion for shear can, stated in three cases:

Table 3.4 Design Shear Strength and the Absolute Maximum Admissible on Concrete (N/mm²).

% tension steel	M15	M20	M25	M30	M35	M40 + above
		Design shear strength of concrete without shear reinforcement				
0.15	0.28	0.28	0.29	0.29	0.29	0.30
0.25	0.35	0.36	0.36	0.37	0.37	0.38
0.50	0.46	0.48	0.49	0.50	0.50	0.51
0.75	0.54	0.56	0.57	0.59	0.59	0.60
1.00	0.60	0.62	0.64	0.66	0.67	0.68
1.25	0.64	0.67	0.70	0.71	0.73	0.74
1.50	0.68	0.72	0.74	0.76	0.78	0.79
2.00	0.71	0.79	0.82	0.84	0.86	0.88
2.50	0.71	0.82	0.88	0.91	0.93	0.95
3.00 & above	0.71	0.82	0.92	0.96	0.99	1.01
		Absolute Maximum shear strength admissible with shear reinforcement				
$\tau_{cmax.}$	2.5	2.8	3.1	3.5	3.7	4.0

Note: Correction factor to the shear strength of slabs is applied. There is no change in the shear strength for slabs of 300 mm thick and above. As the thickness of the slab decreases from 300 mm to 150 mm, strength is enhanced upwards. The enhanced factor is given below

Overall depth(mm)	300	275	250	200	175	150
Enhanced multiplying factor	1.0	1.05	1.10	1.20	1.25	1.30

The shear strength of axially compressed members is enhanced by a factor is:
Enhanced Multiplying factor = $1 + 3P_u/A_g f_{ck}$

Case 1. Nominal shear stress is less than the shear strength of concrete.

$$If \ \tau_v \leq \tau_c$$

Provide only nominal shear reinforcement. And this nominal shear reinforcement is:

$$s_{max} = s_v = \frac{A_{sv} f_{yu}}{0.4b}$$

$$(3.22)$$

Where

s_v = spacing of vertical stirrups,
A_{sv} = area of the stirrup steel,
f_{yv} = proof strength of stirrup steel,

This expression implies that the reinforcement is designed to resist a shear stress of 0.4 N/mm^2.
The spacing of the stirrups is subjected to a maximum spacing of 0.75d or 300 mm.

$$s_{max} \leq 0.75d \text{ , or}$$

$$s_{max} \leq 300\,\text{mm} \tag{3.23}$$

Case 2: In case, the nominal shear stress is greater than the shear capacity of the concrete,

$$If\ \tau_v \leq \tau_c \leq \tau_{cmax} \tag{3.24}$$

Where

$\tau_{c\,max}$ = maximum shear strength of concrete with shear reinforcement.

Shear reinforcement must be provided to resist the shear force over and above that resisted by the concrete portion. The shear capacity of the concrete is given by:

$$V_{rc} = bd\tau_c \tag{3.25}$$

The shear force to be resisted by the reinforcement is: $V_{su} = V_u - V_{rc}$
The stirrup capacity of vertical stirrups is given by:

$$V_{rs} = \frac{A_{sv} f_{yv}\, d}{1.15 s_v} \tag{3.26}$$

In case the stirrups are set at an inclination, the shear capacity of the inclined stirrups is given by:

$$V_{rs} = \frac{A_{sv} f_{yv}\, d}{1.15 s_v}(\sin\alpha + \cos\alpha) \tag{3.27}$$

Where

α = angle between the axis of the beam and the stirrup. This angle should not be less than 45°.

Sometimes the main reinforcement is bent up (called cranked bars) when not required. However bent up bar is taken to the top face of the beam. In continuous beams, the maximum bending moment normally occurs at the intermediate supports. This bending moment is *called negative moment*. The moment that causes compression on the top face is called *positive bending moment* and the corresponding reinforcement that is placed at the bottom is called *positive reinforcement*. Therefore the reinforcement provided to resist the positive bending moment is bent up at the appropriate location at about 45° with the axis of the beam. The bent part of the bar contributes to the shear strength of the beam. The shear strength of the beam is given by:

$$V_{rs} = 0.87 A_{sv} f_{yv} \sin\alpha \tag{3.28}$$

The spacing of the vertical stirrup reinforcement to resist the shear force over and above that resisted by the concrete can be obtained by equating the capacity steel to the balance of the shear force to be resisted by steel, that is $V_{rs} = V_{us}$. The rearrangement of the equation leads to:

$$S_s = \frac{0.87 A_{sv} f_{yv}\, d}{V_{su}} = \frac{0.87 A_{sv} f_{yv}\, d}{V_u - V_{rc}} = \frac{0.87 A_{sv} f_{yv}}{b(\tau_v - \tau_c)} \tag{3.29}$$

The spacing of the inclined stirrup reinforcement is given by:

$$s_s = \frac{0.87 A_{sv} f_{yv}\, d}{V_{su}(\sin\alpha + \cos\alpha)} = \frac{0.87 A_{sv} f_{yv}\, d}{(V_u - V_{rc})(\sin\alpha + \cos\alpha)} = \frac{0.87 A_{sv} f_{yv}}{b(\tau_v - \tau_c)(\sin\alpha + \cos\alpha)} \tag{3.30}$$

The spacing is subject to the maximum spacing listed in case 1. The spacing of the stirrups should not be more than 300 mm.

Case 3: The shear force exceeds the capacity of the concrete with stirrup reinforcement.

$$If\ \tau_v \geq \tau_{c\max} \tag{3.31}$$

Table 3.4 gives the absolute maximum shear strength of concrete with shear reinforcement. If the nominal shear stress on the cross section exceeds that of the absolute maximum with shear reinforcement, the size of the section has to be upgraded. The size of the section, either width or depth of the section, preferably the depth, must be increased to a value such that the nominal shear stress becomes less than the absolute maximum.

Example 3.7: Design for shear force

A 5 m effective span simply supported beam is subjected to a live load of 24 kN/m and is designed with *M*20 concrete and *HYSD-Fe*415 bars. The beam is exposure to mild environment. Design for shear.

Solution

The given data is: $L = 5$ m, $f_{ck} = 20$ N/mm^2, $f_y = 415$ N/mm^2, $w_l = 24$ kN/m. Fig. 3.5 illustrates the simply supported beam. The beam is designed in example 3.1 to resist the bending moment. The size of the beam is 250 mm by 470 mm. The reinforcement provided at mid span is three numbers of 20 mm bars and one of 12 mm bar. It is assumed that two bars are curtailed and the reinforcement at support is only two of 20 mm bars.

Total design load at collapse state is $= w_t = 1.5(3 + 24) = 40.5$ kN/m

Even though the maximum shear force is next to the support, because of the diagonal tension, the critical shear failure is at a distance *d* from the support in simply supported beam. The effective depth of the section is $= d = 435$ mm. The design shear force at the section is:

$$V_u = \frac{w_t L}{2} = w_t d = 40.5(2.5 - 0.435) = 83.63 \text{ kN}$$

The nominal shear stress is =

$$\tau_v = \frac{V_u}{bd} = \frac{83630}{250(435)} = 0.78 \text{ N/mm}^2$$

Assuming only two 20 mm tension bars are available at the support,
The percentage of steel is = 2*314*100/(250*435) = 0.58%
The shear strength of concrete for 0.58 percent of reinforcement, and the absolute shear strength of concrete with stirrup steel taken from table 3.4 are:

$$\tau_c = 0.50 \text{ N/mm}^2 \text{ and } \tau_{cmax} = 2.8 \text{ N/mm}^2$$

The shear strength of concrete is 0.50 N/mm^2, which is less than the nominal shear stress. Select two legged 8 mm for stirrups, the area of two legs of the stirrup is: 100 mm^2. The spacing of the stirrup reinforcement is:

$$s_v = \frac{0.87 A_{sv} f_{yv}}{b(\tau_v - \tau_c)} = \frac{0.87(100)(415)}{250(0.78 - 0.50)} = 516 \text{ mm}$$

The maximum spacing of the stirrups is controlled by:

$$s_{max} \leq \frac{A_{sv} f_{yv}}{0.4b} = \frac{100(415)}{0.4(250)} = 415 \text{ mm} \text{ and}$$

$s_{max} = d = 435$ mm or 300 mm, which ever less. Provide two legged 8 mm stirrups at 300 mm spacing.

The total length of the beam is 5 000 + 450 = 5450 mm. Fig. 3.18 illustrates the reinforcement details including curtailed bars. The notation 2L-ϕ8@300 is two legged 8 mm bars spaced at 300 mm.

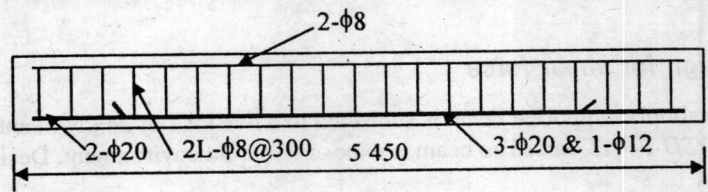

Fig. 3.18 Reinforcement Details of Simple Beam Example 3.7.

Example 3.8: Design for shear of an under-reinforced section

A simply supported 5 m effective span beam is subjected to a live load of 24 kN/m. The beam is exposed to mild environment and has an over all depth of 550 mm. The beam was designed for bending moment in example 3.2 and the details are:

The data obtained from example 3.2 is: $L = 5$ m, $f_{ck} = 20$ N/mm^2, $f_y = 415$ N/mm^2, $w_l = 24$ kN/m, $b = 250$ mm, $h = 550$ mm, $d = 515$ mm. Top reinforcement is 2 of 8 mm and bottom bars are 3 of 20 mm and only two bars are continued to end of the beam. The design total load is 41.25 kN/m, and the gross length of the beam is 5,520 mm.

Solution

The design for bending moment was covered in example 3.2 and the reader is advised to refer to the example. The beam is simply supported, so the critical shear occurs at a distance of effective depth from the face of the support. As the support details are not given here, one can assume the section for shear design at a distance from the effective support. The shear force at the section is:

$$V_u = w_t(0.5L - d) = 41.25(2.5 - 0.515) = 81.9 \text{ kN}$$

The nominal shear stress is $\tau_v = \dfrac{V_u}{bd} = \dfrac{81900}{250(515)} = 0.64 \, \text{N/mm}^2$

Only two 20 mm tension bars are available at the support, so

The percentage of steel is = 2*314*100/(250*515) = 0.49%

The shear strength of concrete for 0.49 percent of reinforcement and the absolute shear strength of concrete with stirrup steel taken from table 3.4 are:

$$\tau_c = 0.47 \, \text{N/mm}^2 \text{ and } \tau_{c \max} = 2.8 \, \text{N/mm}^2$$

The shear strength of concrete is 0.47 N/mm², which is less than the nominal shear stress. Select two legged 8 mm bar as shear stirrup, the area of two legs of the stirrup is: 100 mm². The spacing of the stirrup reinforcement is:

$$s_v = \frac{0.87 A_{sv} f_{yv}}{b(\tau_v - \tau_c)} = \frac{0.87(100)(415)}{250(0.64 - 0.47)} = 895 \, \text{mm}$$

The maximum spacing of the stirrups is controlled by:

$$s_{\max} \le \frac{A_{sv} f_{yv}}{0.4b} = \frac{100(415)}{0.4(250)} = 415 \, \text{mm} \quad \text{and}$$

$s_{\max} = d = 515$ mm or 300 mm, which ever least. Provide two legged 8 mm stirrups at 300 mm spacing. The maximum spacing controls the shear reinforcement in this example and it is same as that in example 3.7

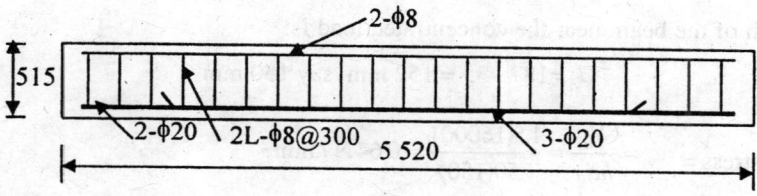

Fig. 3.19 Reinforcement of Simple Beam Example 3.8.

Example 3.9: Design of shear reinforcement in cantilever beam

A 3 m clear span cantilever beam is subjected to a concentrated load of 16 kN at 500 mm from free end. The beam is exposed to non-aggressive open environment. The beam was designed for bending moment in example 3.3. Reinforcement of 3 numbers of 20 mm bars at top, and 2 numbers of 8 mm bars at bottom as hanger bars are provided. The depth of the beam is nominal at the free end and it is 150 mm.

Solution

The average self-weight of the beam is 2.2 kN/m. The design loads are obtained by multiplying the loads by 1.5 partial safety factors. The maximum shear force in a cantilever beam occurs at the face of the support.

The design shear force at support is:

$$V_u = 1.5\, w_g\,(L) + 1.5\, W_l = 1.5(2.2) + 1.5(16) = 27.3 \text{ kN}$$

The nominal shear stress is $= \tau_v = \dfrac{V_u}{bd} = \dfrac{27300}{250(330)} = 0.33$ N/mm^2

Three 20 mm tension bars are available at the support,

The percentage of steel is $= 3*314*100/(250*330) = 1.1\%$

The shear strength of concrete for 1.1 percent of reinforcement and the absolute shear strength of concrete with stirrup steel taken from table 3.4 are:

$$\tau_c = 0.67 \text{ N/mm}^2 \text{ and } \tau_{cmax} = 3.1 \text{ N/mm}^2$$

The shear strength of concrete is more than the nominal shear stress. Therefore minimum shear reinforcement is provided.

The maximum spacing of the stirrups is controlled by:

$$s_{max} \le \frac{A_{sv} f_{yv}}{0.4b} = \frac{100(415)}{0.4(250)} = 415 \text{mm} \text{ or it is also limited to:}$$

$s_{max} = d = 330$ mm or 300 mm, which ever least. The maximum value of spacing of the stirrup steel is 300 mm at the support but the depth of the beam at free end is only 150 mm, one has to check for the shear capacity just next to the concentrated load. The size of the section next to the concentrated load is:

$$h_o = 150 + (370 - 150)(500)/3000 = 187 \text{ mm}$$

The effective depth of the beam near the concentrated load is:

$$d_o = 187 - 35 = 152 \text{ mm, say } 150 \text{ mm}$$

Nominal shear stress $= \tau_v = \dfrac{V_u}{bd_0} = \dfrac{1.5(16000)}{250(150)} = 0.64 \text{ N/mm}^2$

Even this nominal shear stress is less than the shear strength of the concrete. Therefore only nominal shear reinforcement is provided. As the beam is short one, provide a uniform spacing of 150 mm throughout the length of the beam. Provide two legged 8 mm stirrups at 150 mm spacing. Vide Fig. 3.14 for reinforcement details of the beam.

Example 3.10: Complete design for balanced section

A simply supported beam of 6.5 m effective span in severe environment is subjected to a live load of 20 kN/m and superimposed dead load of 6 kN/m. Design the beam for moment and shear.

Solution

The data of the beam is:

The effective span = L = 6.5 m,

Superimposed dead load = w_s = 6 kN/m

Superimposed live load = w_l = 20 kN/m

Exposure condition = severe,

so the minimum grade of concrete is $M25$

f_{ck} = 25 N/mm^2, select steel grade: f_y = 415 N/mm^2,

Minimum cover for severe exposure is = 45 mm

The design coefficients can be obtained from table 3.1, and they are: K = 0.138, j = 0.80, k_u = 0.479

Design for bending moment

Assume the width of the beam b = 300 mm and depth (h) of the beam in the range of $L/12$ for simply supported beams, say h = 550 mm for the purpose of dead weight. Fig. 3.20 illustrates the simply supported beam.

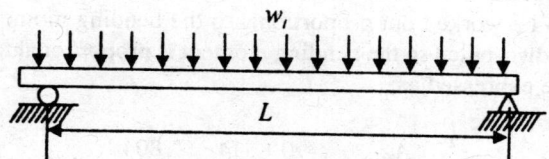

w_l

L

Fig. 3.20 Simply Supported Beam.

Self weight of the beam = w_g = $bh\gamma c$ = 0.3(0.55)25 = 4.2 kN/m, use 5.0 kN/m

Total design load at collapse state (strength limit state) =

$$w_t = \gamma_1 (w_g + w_s + w_l) = 1.5(5 + +6 +20) = 46.5 \text{ kN/m}$$

Maximum bending moment occurs at mid-span =

$$M_c = \frac{w_t L^2}{8} = \frac{46.5(6.5)^2}{8} = 245.6 \text{ kNm}$$

The balanced moment capacity must be more than the collapse moment, that is:

$$M_{rb} \geq M_c, \text{ This leads to:}$$

$$Kbd^2 f_{ck} = 0.138(300)(25)d^2 \geq 245.6(10^6)$$

$$\text{or, } d = \sqrt{245.6(10^6)/(0.138)(300)(25)} = 488 \text{ mm}$$

Let the effective depth be 490 mm. The minimum cover to the reinforcement is 45 mm or the diameter of the bar. Let the diameter of the bar be assumed 25 mm. Therefore the total depth of the beam is = h = 490 + 25/2 +45 = 547.5 mm. Now revise the assumed depth of 550 mm for dead weight purpose. The self-weight is just right.

The area of the tension reinforcement needed is:

$$A_{st} = \frac{1.15 M_c}{jd f_y} = \frac{1.15(245.6)10^6}{0.8(490)415} = 1737 \, \text{mm}^2$$

Use 3 numbers of 25 mm bars and one of 20 mm bar at mid span.
The area of reinforcement provided is

$$A_{st} \text{ (provided)} = 3*490 + 314 = 1784 \, \text{mm}^2 > 1737 \, \text{mm}^2 \text{ needed}$$

Provide 2 numbers of 10 mm hanger bars at the top to support the stirrups.
Minimum width required to accommodate four bars at one level is

b (min) = four diameter of bars + 3 spacing + 2 end covers =
b (min) = $(3*25 + 20) + 3*25 + 2*45 = 260 \, \text{mm} < 300 \, \text{mm}$ provided

Curtailment of bars

Out of the four bars provided at the bottom face, one 25 mm bar and one 20 mm bar can be curtailed. The length of curtailment can be worked out proportional to the bending moment.

The load is uniformly distributed so the bending moment is proportional to the square of the span. The curtailment length can be expressed as:

$$\frac{x^2}{L^2} = \frac{A_{st1}}{A_{st}} = \frac{490 + 314}{3*490 + 314} = \frac{804}{1784} = 0.45$$

where
x = effective length of the curtailed bars and
A_{st1} = area of the curtailed bars. The above expression gives:

$$x = L\sqrt{0.45} = 4.36 \, \text{m} = 4360 \, \text{mm}$$

Minimum length bar that should extend beyond required length is equal to effective depth of the beam or 12 times the diameter of the bar whichever is larger.
The minimum bar extension on each side is = 490 mm or 12*25 =300 mm. Use 490 mm
The length of the curtailed bars = x + 2*490 = 4360 +980 = 5340 mm

Design for shear

The critical shear at a distance equal to the effective depth from the face of the support is:

$$V_u = w_t(0.5L - d) = 46.5(6.5/2 - 0.490) = 128.34 \, \text{kN}$$

The nominal shear stress is $= \tau_v = \dfrac{V_u}{bd} = \dfrac{128340}{300(490)} = 0.87 = \text{N/mm}^2$

Only two 25 mm tension bars are available at the support,
The percentage of steel is = 2*490*100/(300*490) = 0.67%

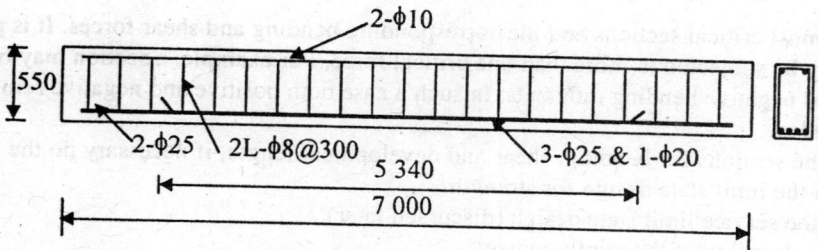

Fig. 3.21 Reinforcement Details of Example 3.10.

The shear strength of concrete for 0.67 percent of reinforcement and the absolute shear strength of concrete with stirrup steel taken from table 3.4 are:

$$\tau_c = 0.54 \text{ N/mm}^2 \text{ and } \tau_{c\,max} = 3.1 \text{ N/mm}^2$$

The shear strength of concrete is 0.54 N/mm², which is less than the nominal shear stress. Select two legged 8 mm bar as shear stirrup, the area of two legs of the stirrup is: 100 mm². The spacing of the stirrup reinforcement is:

$$s_v = \frac{0.87 A_{sv} f_{yv}}{b(\tau_v - \tau_c)} = \frac{0.87(100)(415)}{300(0.87 - 0.54)} = 364 \text{ mm}$$

The maximum spacing of the stirrups is controlled by:

$$s_{max} \leq \frac{A_{sv} f_{yv}}{0.4b} = \frac{100(415)}{0.4(300)} = 345 \text{ mm},$$

further the maximum spacing is limited to:
$s_{max} = d = 490$ mm or 300 mm, which ever less;
Provide two legged 8 mm stirrups at 300 mm spacing.

Example 3.11 Complete Design of overhang Beam

A overhang beam in moderate environment with 6 m middle span and 2 m cantilever on one side is subjected to a superimposed dead load of 8 kN/m and a live load of 22 kN/m. Design the beam with HYSD-Fe500 steel.

Solution

So far only simply supported or cantilever beams have been designed. The maximum bending moment in simply supported beam is $wL^2/8$. In the experience of the author during the last four and half decades, some students have a tendency to use the bending moment of $wL^2/8$ for all types of beams in hurry in examinations. The student must first analyze the beam for design forces and then design the section. The normal steps recommended in the design of a member are:

- Select materials and specifications for the exposure condition,
- Perform load computation, usually this part is small unless wind or earthquake loads have to be obtained,
- Perform structural analysis for different load combinations, this part depends on the complexity of the structure, a number of computer programs are available to obtain critical forces and displacements,

- Select the most critical sections and the corresponding bending and shear forces. It is possible that a section may be subjected to more than one critical force. For example, a section may be subjected to positive and negative bending moments. In such a case both positive and negative reinforcement has to be provided,
- Designed the sections for bending, shear and development length, if necessary do the curtailment of the bars, in the limit state design for strength,
- Check for the service limit state design (discussed later),
- Perform the detailing of the reinforcement,
- Draw or sketch the reinforcement details

Design date

Simple span $= L = 6$ m, and overhang cantilever $= a = 2$ m,
Superimposed dead load $= w_s = 8$ kN/m, live load $= w_l = 22$ kN/m,
Concrete grade for moderate exposure $= f_{ck} = 25$ N/mm^2,
Reinforcement grade *HYSD* bars of $= f_y = 500$ N/mm^2,
Cover to the reinforcement in moderate exposure in beams is $= 30$ mm,
The design coefficients from table 3.1 are: $K = 0.133$, $j = 0.80$,

Load Analysis

The uniformly distributed live load must be placed on the beam so as to cause most adverse effects. There are three possible critical load locations, and they are: (1) load over the entire beam as shown in Fig. 3.22

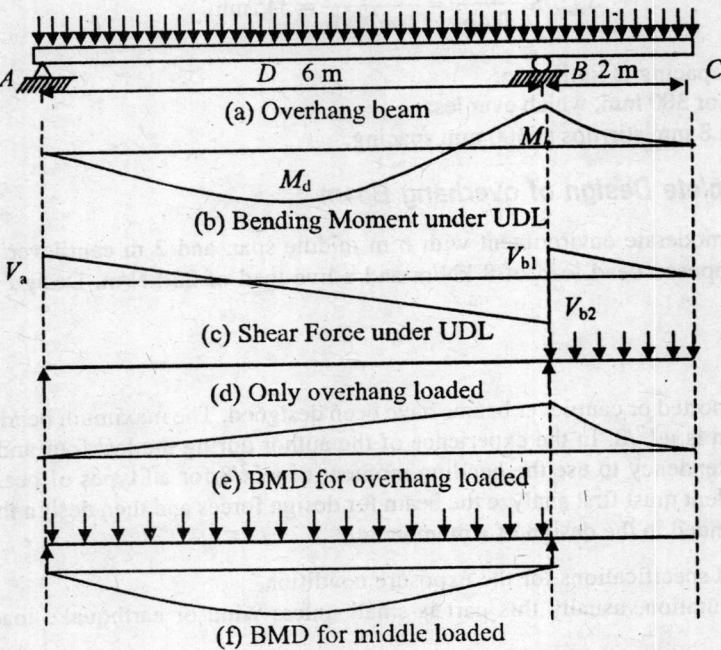

Fig. 3.22 Overhang Beam with Load Conditions, **Example 3.11.**

(a), the corresponding bending and shear force diagrams are shown in Fig. 3.22(b) and (c). The second critical load location is the uniformly distributed load (UDL) on the overhang portion as shown in Fig 3.22(d), and the corresponding BMD is shown in Fig. 3.22(e). The third load location is that the UDL on middle span as shown in Fig. 3.22(f), and the corresponding BMD is in Fig. 3.22(f). The dead load must be placed over the entire length of the beam. It can be seen that the mid span of the simply supported beam is subjected to positive BM in load cases 1 and 3, and negative BM in live load case 2. The maximum positive bending moment occurs at mid span in the third load location. The critical location is obtained from the extremum principle.

Let the depth of the beam be assumed as $L/12$ for the purpose of computing the self-weight. That is $h = 0.5$ m. Let the width of the section be $= b = 0.3$ m.

Self weight $= w_g = 0.3(0.5)25 = 3.75$ kN/m, use $w_g = 4$ kN/m,

Dead load $= w_d = 4 + 8 = 12$ kN/m

Live load $= w_l = 22$ kN/m,

Structural Analysis

The live load is almost twice of the dead load, so it is possible that the maximum bending moment is controlled by the live load location.

Load case 1 when the entire span is under UDL (dead plus live load together)

The supports are at A and B, and the free end of the beam is C, the middle point is at D as shown in Fig. 3.22(a). The reaction at point A, (R_a), under UDL is obtained by taking the moment about the support point B.

$$R_a = w\,(6 + 2)*(6-8/2)/6 = 8w/3 = 8*(12 + 22)/3 = 90.7 \text{ kN}$$

The reaction of support at B is:

$$R_b = 8w - R_a = 8(34) - 90.7 = 181.3 \text{ kN}$$

The bending moment at a distance x from the end support is:

$$M_x = R_a x - w\frac{x^2}{2} = 90.7x - 17x^2$$

The maximum bending moment is obtained by equating derivative of the moment with respect to x to zero.

$$\frac{\partial M_x}{\partial x} = 90.7 - 34x = 0$$

This leads to $x = 90.7/34 = 2.67$ m and the maximum bending moment is:

$$M_x = 90.7x - 17x^2 = 90.7(2.67) - 17(2.67) = 121 \text{ kNm}$$

The bending moment at the mid point, that is 3m from the support is:

$$M_d = 90.7x - 17x^2 = 90.7(3) - 17(3)(3) = 119.7 \text{ kNm}$$

The difference between the two bending moments is less than 2 percent. As the bending moment due to live load is critical at the mid-span, and the difference is only marginal, one can consider the mid-span as the critical section. The factored positive bending moment at about the mid span in the span AB is:

$$M_{ld} = 1.5 \ (121) = 181.5 \text{ kNm} \tag{3.32}$$

The factored absolute maximum negative bending moment that occurs at the support B is:

$$M_{lb} = 1.5 \ w \ a^2/2 = 1.5(34)(4)/2 = 102 \text{ kNm} \tag{3.33}$$

The factored shear force at support A is same as the reaction at that point multiplied by partial safety factor

$$V_{la} = 1.5 \ (90.7) = 136 \text{ kN} \tag{3.34}$$

The factored shear force to the right of the support B is:

$$V_{1rb} = 1.5(34)2 = 102 \text{ kN} \tag{3.35}$$

The factored shear force to the left of the support B is:

$$V_{1lb} = 1.5R_b - V_{1rb} = 1.5(181.3) - 102 = 170 \text{ kN} \tag{3.36}$$

Load case 2, the live on the overhang only

Figure 3.22(f) indicates the load location and the bending moment diagram. The factored negative bending moment at support B due to live load is:

$$M_{2bl} = 1.5(22) \ 2*2/2 = 66 \text{ kNm}$$

The corresponding negative bending moment at mid span is:

$$M_{2dl} = M_{2bl} /2 = 33 \text{ kNm}$$

The bending moment due to live load at the mid span is negative as compared to that due to the dead load. The factored bending moment can be calculated from the positive BM from the dead load and negative bending moment due to live load. The total load intensity is 34 kN/m whereas the dead load intensity is 12 kN/m. Therefore the factored bending moment due to dead load at mid span can be obtained by proportioning the 181.5 kNm by 12/34. The net positive factored bending moment at mid-span can be obtained by the superposition of the two.

$$M_{2d} = (181.5)(12/34) - 33 = 64 - 33 = 31 \text{ kNm} \tag{3.37}$$

The net bending moment at mid span is still positive. As far as the negative bending moment at support B is concerned, both the cases are exactly the same. The shear is not critical.

Load case 3. The live load is on the mid-span

The factored bending moment at about the mid-span due to dead and live loads is obtained by adding the that caused by the dead and live load over only on the simple span. And it is:

$$M_{3d} = 181.5(12)/34 + 1.5(22) \ L^2/8 = 64 + 148.5 = 212.5 \text{ kNm} \tag{3.38}$$

The reaction due to live load acting on the simple beam alone is:

$$R_{3la} = wL/2 = 22(6)/2 = 66 \text{ kN}$$

The reaction due to dead load is proportioned from case 1 by 12/34. The factored shear force at support A is:

$$V_{3a} = 1.5 (90.7)12/34 + 1.5(66) = 147 \text{ kN} \tag{3.39}$$

The reaction at B is smaller in this case when compared to case 1 and hence the shear on either side is less.

Design moments and shear forces
Design for bending moment

The two critical sections of the beam are: 1. At about the mid-span (D) of AB and at support B. The critical factored bending moments are:

$$M_{1d} = 1.5 (121) = 181.5 \text{ kNm (+ve)}$$
$$M_{2d} = (181.5)(12/34) - 33 = 64 - 33 = 31 \text{ kNm (+ve)}$$
$$M_{1b} = 1.5 \, w \, a^2/2 = 1.5(34)(4)/2 = 102 \text{ kNm (- ve)}$$

The beam is to be prismatic up to in segment AB and then tapering to the free end. Therefore design the beam section based on the absolute maximum bending moment. The design criterion for bending is:

$$M_r = Kbd^2 \, f_{ck} \geq M_u = M_{1d} = 181.5 \text{ kNm}$$

$$d \geq \sqrt{\frac{181.5(10)^6}{0.133(300)25}} = 426.6 \text{ mm}$$

The overall depth of the beam required is = 426.6 + cover + diameter of bar/2 = 426.6 + 30 + 12.5 = 469.1 mm

Use the overall depth of the beam as 475 mm and the effective depth as 430 mm

The area of positive reinforcement required at mid span is:

$$A_{std} = \frac{1.15M_{1d}}{jdf_y} = \frac{1.15(181.5(10)^6}{0.8(430)(500)} = 1214 \text{ mm}^2$$

Provide 4 numbers of 20 mm diameter bars at the bottom, then the area provided is = 4*314 = 1242 mm²

Two of these bars continued from support to support and the remaining two can be curtailed.

Design for negative bending moment that occurs at support B

The depth of the beam is same as that at the mid-span. The negative reinforcement required is:

$$A_{std} = \frac{1.15M_u}{jdf_y} = \frac{1.15M_{1b}}{jdf_g} = \frac{1.15(102)(10)^6}{0.8(430)(500)} = 682 \text{ mm}^2$$

Provide 2 numbers of 12 mm bars from end to end and 1 number of 25 mm bar extra over the support.

The 25 mm bar is continued for 2 m beyond the support. The actual curtailment of the bar is worked out but not indicate in this example for the sake of simplicity.

Design for Shear force

The maximum shear force 170 kN occurs to the left of the support B and that at the support A is 136 kN. The critical shear for design is at a distance d from the support. The rate at which the intensity is decreasing is 34 kN/m The design shear force at distance 0.43 m from the support is:

$$V_u = 170 - 1.5(34)*0.43 = = 148 \text{ kN}$$

The nominal shear stress at left of support B is:

$$\tau_v = \frac{V_u}{bd} = \frac{148,000}{300(430)} = 1.13 N / mm^2$$

The percentage of reinforcement at the support B is:

$$p = \frac{A_{st}(100)}{bd} = \frac{691(100)}{300(430)} = 0.5\%$$

Shear strength of concrete with 0.5% reinforcement and without stirrup steel taken from table 3.4 are:

$$\tau_c = 0.47 \text{ and } \tau_{c \, max} = 3.1 \text{ N/mm}^2$$

Select two legged 8 mm stirrups, then the area of stirrup steel is 100 mm², the spacing of the stirrups is:

$$s_v = \frac{0.87 \, A_{sv} \, f_{yv}}{b(\tau_v - \tau_c)} = \frac{0.87(100)(500)}{300(1.13 - 0.47)} = 220 \text{ mm}$$

The maximum spacing of the stirrups is controlled by:

$$s_{max} \leq \frac{A_{sv} \, f_{yv}}{0.4b} = \frac{(100)(500)}{0.4(300)} = 416 \text{ mm}$$

$s_{max} = d = 430$ mm or 300 mm, which ever less; The maximum value of spacing of the stirrup steel is 220 mm.

Provide two legged 8 mm stirrups at 220 mm spacing at the critical section. The minimum stirrup requirement is 8 mm bars at 300 mm spacing. The maximum spacing will control for most of the span except for one meter on to the left of the support B. The depth of the beam at free end of the section is taken as 250 mm. The stirrups on the overhang portion are provided at 220 mm. Fig. 3.23 illustrates the reinforcement details.

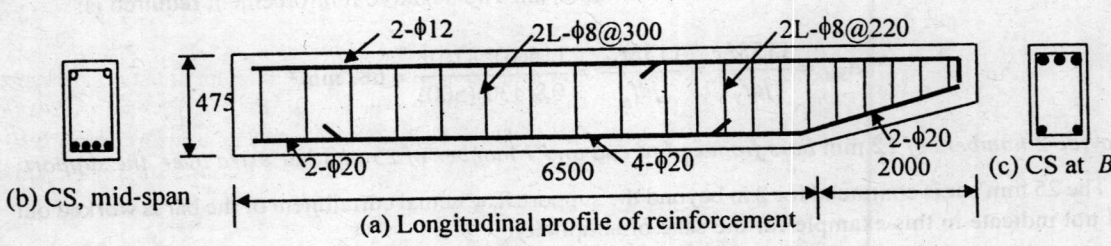

Fig. 3.23 Reinforcement Details of Example 3.11.

3.10 DESIGN FOR TORSION

Most beams are subjected to transverse loads passing through the centroidal axis. However if the load acts away from centroidal plane, a twisting moment about the axis of the member is generated. The twisting moment is called *Torsion*. Loads transferred through brackets attached to the beams away from centroid produce torsion. The supports must resist the torsion. If adequate support restraints are not provided, the member will fail in stability.

Torsion at a section is resisted by in-plane shear stresses. The magnitude of the shear stress caused by the torsion is proportional to the distance of the fibre from the centroid of the section in circular sections. In case of rectangular or any non-circular sections, the variation of the shear stress even though increases with the distance from the centroid, but the corners of the section will warp and are stress free. The warping of the section is dominant in oblong cross sections. Figure 3.24 illustrates the variation shear

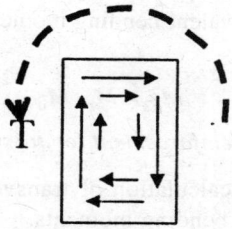

Fig. 3.24 Shear Stress Variation in Resting Torsion.

stress against torsion. Most of the cross sections in reinforced concrete are either rectangular or flanged or tee shapes. Reinforcement must be provided to resist the torsion. The torsion is converted into equivalent shear force and bending moment and the reinforcement is provided to resist the equivalent forces.

The equivalent shear force of the torsion is given by:

$$V_e = V_u + 1.6 \frac{T_u}{b} \qquad (3.40)$$

Where

V_e = equivalent shear force,
V_u = factored shear force,
T_u = factored torsion.

The equivalent nominal shear stress is given by the same expression give earlier and it is:

$$\tau_{ve} = \frac{V_e}{bd} \qquad (3.41)$$

The minimum shear reinforcement conditions continue to holds good. The stirrups must be around the corner bars. In case of pure shear force, the stirrups can be hooked with the hanger-bars. Additional longitudinal reinforcement must be provided to resist the torsion. The longitudinal reinforcement is to be designed for an equivalent bending moment. The equivalent bending moment for torsion along with the moment is:

$$M_{el} = M_u + M_t \qquad (3.42)$$

Where

M_{el} = equivalent bending moment for which the longitudinal reinforcement to be provided,

M_u = Factored bending moment,

M_t = equivalent moment torsion and it is:

$$M_t = T_u \left[\frac{1 + D/b}{1.7} \right]$$ (3.43)

$D = h$ = overall depth of the section

The tensile reinforcement must be provided to resist the equivalent bending moment. However, if the equivalent torsion M_t is equal or more than the factored bending moment, an additional longitudinal reinforcement must be provided on the compression face of the beam. The amount of reinforcement on the compression face must meet an equivalent bending moment that is equal to M_{e2}

Where

$$M_{e2} = M_t - M_u$$ (3.44)

Design of transverse and longitudinal reinforcement for torsion

The expressions derived earlier for calculation of transverse and longitudinal reinforcement will holds good for the equivalent shear and bending moments.

The area of shear reinforcement for equivalent shear is given by:

$$A_{sv} = \frac{1.15 T_u s_v}{b_1 d_1 f_y} + \frac{1.15 V_u s_v}{2.5 d_1 f_y}$$ (3.45)

Subject to a minimum of: $$A_{sv} = \frac{1.15(\tau_{ve} - \tau_c) b s_v}{f_y}$$ (3.46)

Where b_1 = centre to centre distance of the corner bars in the width direction

d_1 = centre to centre distance of the corner bars in the depth direction

Torsional failure causes a skew bending plane, and the compression face depends on the relative magnitudes of bending and torsion. If the bending moment is dominant, the compression face is at top fibre. Typical end simply supported and torsion constrained condition is shown in Fig. 3.25

Example 3.12: Design for Torsion in simple beam

A 5 m effective span simply supported beam is subjected to a live load of 24 kN/m and torsion of 30 kNm from end to end. The beam is exposure to mild environment. Design the beam for torsion. (The beam was designed for bending in examples 3.1 and 3.7)

Solution

The given data is: $L = 5$ m, $f_{ck} = 20$ N/mm^2, $f_y = 415$ N/mm^2, $w_l = 24$ kN/m, $T = 30$ kNm, $b = 300$ mm

Let the total depth of the beam is = $h = 520$ mm, then self weight is = $0.3*0.52*25 = 3.9$, use 4 kN/m

Total design load at collapse state is = $w_t = 1.5(4 + 24) = 42$ kN/m

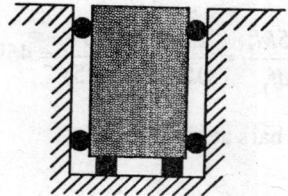

Fig. 3.25 Torsion Restrained Support.

Design for bending and torsion

Maximum factored bending moment that occurs at the mid support is =

$$M_u = \frac{w_t L^2}{8} = \frac{42(25)}{8} = 131.25 \text{ kNm}$$

Factored torsion on the beam from end to end is = $T_u = 1.5(30) = 45$ kNm
The moment equivalent torsion is:

$$M_t = T_u \left[\frac{1 + h/b}{1.7} \right] = 45 \left[\frac{1 + 520/300}{1.7} \right] = 64.3 \text{ kNm}$$

The equivalent bending moment is:

$$M_{el} = M_u + M_t = 131.25 + 64.3 = 195.55 \text{ kNm}$$

The balanced moment capacity must be equal to more than the collapse moment, that is:

$$M_{rb} \geq M_{el}, \text{ This leads to:}$$

$$Kbd^2 \, f_{ck} = 0.138(300)(20)d^2 \geq 195.55(10^6)$$

$$or, d = \sqrt{195.55(10^6)/(0.138)(300)(20)} = 486 \text{ mm}$$

The total depth of the section is = $h = 486 + 25 + 10 = 521$ mm; Use $d = 485$ mm
    The assumed depth is close to the final design depth of a balanced section. Therefore, the assumption is OK.
    The area of the tension reinforcement needed is:

$$A_{st} = \frac{1.15 M_{el}}{jdf_y} = \frac{1.15(195.55)10^6}{0.8(485)415} = 1397 \text{ mm}^2$$

Use 4 numbers of 20 mm bars and one number of 16 mm bar.
    The area of reinforcement provided is = A_{st} (provided) = 4*314 + 201 = 1457 mm² > 1397 mm² that is needed
    Since the equivalent torsion moment is less than the factored bending moment, the top face reinforcement can be nominal. However the author recommends a top reinforcement close to that required for the equivalent bending moment. Top reinforcement needed is:

$$A_{sc} = \frac{1.15M_t}{jdf_y} = \frac{1.15(64.3)10^6}{0.8(486)(415)} = 459 \text{ mm}^2$$

Provide 2 numbers of 16 mm hanger bars at the top.

Design for Equivalent Shear force

The critical section for shear failure is at a distance d from the support. The critical shear failure occurs at a distance d from the support in simply supported beams. The effective depth of the section is $= d = 485$ mm and the percentage of tension reinforcement at the end section is taken as two numbers of 20 mm bars. The design shear force at the section is:

$$V_u = \frac{w_t L}{2} - w_t d = 42(2.5 - 0.485) = 84.6 \text{ kN}$$

The equivalent shear force of the torsion is given by:

$$V_e = V_u + 1.6\frac{T_u}{b} = 84.6 + 1.6\frac{45}{0.3} = 324.6 \text{ kN}$$

The nominal shear stress is =

$$\tau_{ev} = \frac{V_e}{bd} = \frac{324600}{300(485)} = 2.23 \text{ N/mm}^2$$

The shear strength of concrete for 0.91 percent of reinforcement, and the absolute shear strength of concrete with stirrup steel taken from table 3.4 are:

$$\tau_c = 0.61 \text{ N/mm}^2 \text{ and } \tau_{c\,max} = 2.8 \text{ N/mm}^2$$

Assume a spacing of the stirrups at 200 mm, then compute the area of stirrup.

$$\text{Let } s_v = 200 \text{ mm}$$

Area of stirrup steel is subject to a minimum of:

$$A_{sv} = \frac{1.15(\tau_{ve} - \tau_c)bs_v}{f_y} = \frac{1.15(2.23 - 0.61)300(200)}{415} = 270 \text{ mm}$$

The area of shear reinforcement for equivalent shear is given by:

$$A_{sv} = \frac{1.15T_u s_v}{b_1 d_1 f_y} + \frac{1.15V_u s_v}{2.5d_1 f_y} = \frac{1.15(45)(10^6)200}{(300-70)(520-70)415} + \frac{1.15(84600)200}{2.5(520-70)415}$$

$$= 241 + 42 = 284 \text{ mm}^2$$

Provide 2 legged 16 mm stirrup, then the area of stirrup steel is 402 mm². The spacing of the stirrup can be prorated with the area of steel provided and that required:

The stirrup spacing is $= s_v = 402(200)/284 = 283$ mm.

Provide 2 legged 16 mm bars at 280 mm spacing.

Example 3.13: Design of L bent cantilever beam

A 3 m cantilever beam of is subjected to a concentrated load of 16 kN at 500 mm from the free end. The load is on L bent portion of the cantilever as shown in Fig. 3.27. The beam is exposed to open non-aggressive environment. The L bent span is 500 mm. The beam is fixed into masonry support wall.

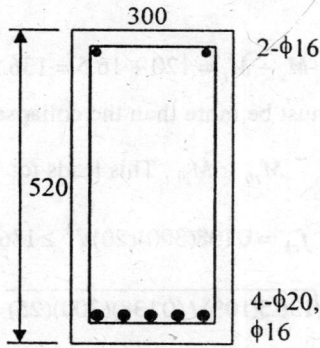

300

2-ϕ16

520

4-ϕ20,
ϕ16

Fig. 3.26 Example 3.12.

Solution

Effective span in cantilever beam is equal to the clear span plus the effective depth of the beam at support. The bending moment is zero at the load end and increases to maximum at support. The beam is also subjected to torsion as the load is on the L bent. Select concrete grade $M25$.

Select $HYSD$-Fe415 reinforcement steel. The design coefficients are: $K = 0.138$, $j = 0.80$, $k_u = 0.479$. f_{ck} = 25 N/mm², $f_y = 415$ N/mm². Let the width of the beam = b = 0.30 m. For the purpose of self-weight assume the average depth of the beam as 400 mm.

The effective cantilever span is = L = clear span + effective depth of the beam =3.4 m.

Self-weight of the beam is = w_g = 0.4(0.3)25 = 3.0 kN/m.

The maximum factored bending moment that occurs at a distance d beyond into the support is:

$$M_u = 1.5 w_g (3)(3.4)/2 + 1.5 w_g (1)(3.4) + 1.5 W_l (3.4) =$$

$$1.5((3.0)5.1 + 3(3.4)) + 24(3.4) = 1.5(15.3 + 10.2) + 81.6 = 120 \text{ kNm}$$

The factored torsion on the beam is constant on the beam, and it is:

$$T_u = 1.5(16)(0.5) = 12 \text{ kNm}$$

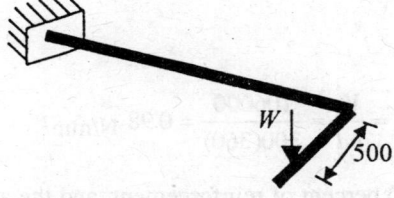

W

500

Fig. 3.27 Cantilever L Beam Example 3.13.

The equivalent moment for torsion is:

$$M_t = T_u \left[\frac{1 + h/b}{1.7} \right] = 12 \left[\frac{1 + 400/300}{1.7} \right] = 16.5 \text{ kNm}$$

The design bending moment is:

$$M_{el} = M_u + M_t = 120 + 16.5 = 136.5 \text{ kNm}$$

The balanced moment capacity must be more than the collapse moment, that is:

$$M_{rb} \geq M_{el}, \text{ This leads to:}$$

$$Kbd^2 f_{ck} = 0.138(300)(20)d^2 \geq 136.5(10)^6$$

$$or, \ d = \sqrt{136.5(10^6)/(0.138)(300)(25)} = 363 \text{ mm}$$

The total depth of the section will be $= h = 363 + 30 + 10 = 403$ mm; Use $d = 400$ mm
The assumed depth is same as the final design depth of section.
Provide $h = 400$ mm and $d = 360$ mm, then the area of reinforcement required is:

$$A_{st} = \frac{1.15 M_{el}}{jdf_y} = \frac{1.15(136.5)(10^6)}{0.8(360)(415)} = 1313 \text{ mm}^2$$

Provide 5 numbers of 20 mm bars at top, the area of steel provided is 1560 mm². Adequate development length should be provided and the development length requirement is discussed later. Provide 2 numbers of 10 mm bars at bottom as hanger bars.

Design for equivalent shear force

The critical section for shear is at support. The design shear force at the section is:

$$V_u = 1.5 w_g (L+1) + 1.5 W_l = 1.5((3)(3+1) + 16) = 42 \text{ kN}$$

The equivalent shear force of the torsion is given by:

$$V_e = V_u + 1.6 \frac{T_u}{b} = 42 + 1.6 \frac{12}{0.3} = 106 \text{ kN}$$

The nominal shear stress is =

$$\tau_{ev} = \frac{V_e}{bd} = \frac{106000}{300(360)} = 0.98 \text{ N/mm}^2$$

The shear strength of concrete for 1.0 percent of reinforcement, and the absolute shear strength of concrete with stirrups taken from table 3.4 are:

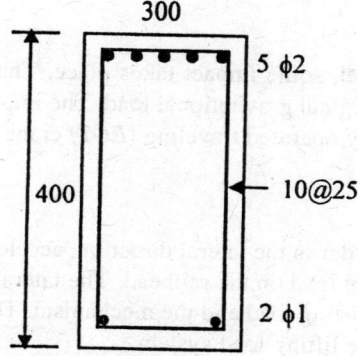

Fig. 3.28 Example 3.13.

$$\tau_c = 0.64 \ N/mm^2 \ \text{and} \ \tau_{cmax} = 3.1 \ N/mm^2$$

The area of shear reinforcement for equivalent shear is computed assuming spacing of stirrups $s_v = 250$ mm:

$$A_{sv} = \frac{1.15 T_u s_v}{b_1 d_1 f_y} + \frac{1.15 V_u s_v}{2.5 d_1 f_y} = \frac{1.15(12)(10^6)250}{(300-70)(400-70)415} + \frac{1.15(42000)250}{2.5(400-70)415} = 110 + 35 = 145 \ mm^2$$

Area of stirrups is subject to a minimum of:

$$A_{svm} = \frac{1.15(\tau_{ve} - \tau_c)bs_v}{f_y} = \frac{1.15(0.98-0.64)300(250)}{415} = 70 \ mm^2$$

Provide 2-legged 10 mm stirrup, then the area of stirrup steel is 157 mm². The spacing of the stirrup can be prorated with the area of steel provided and that required:

Provide 2-legged 10 mm bars at 250 mm spacing.

3.11 DESIGN OF GANTRY GIRDER

Gantry girder is a beam that carries overhead traveling cranes on rails. Large beams in bridges and industrial structures are usually referred as girders and they are designed as members subjected to bending, shear and torsion. The crane lifts or lowers load, moves laterally with the help of a trolley and transports along the longitudinal and lateral directions. Direction parallel to the axis of the girder is called longitudinal direction and perpendicular to the axis is called lateral direction. This section is only for the design of the reinforced concrete girders; so not much of explanation or data on the traveling cranes is given. The total crane system consists of:

- Overhead traveling crane girder,
- Overhead traveling trolley, a small trolley to move in the lateral direction,
- Lifting hook and its mechanism, and
- Lifting weight.

Impact Load

When the crane lifts or lowers a load, some impact takes place. The *impact factor* is the ratio of the additional load experienced to the original gravitational load. The impact factor in hand-operated cranes is about 10 percent and in electrically operated traveling *(EOT)* crane is 20 to 25 percent.

Lateral Load

As the trolley moves on the crane girder in the lateral direction, acceleration or braking takes place. The braking of the trolley results in lateral load on the railhead. The lateral load is measured as a percentage of the trolley weight along with the lifting load and the mechanism. The percentage of the lateral load is about 5 percent of the trolley and the lifting load system.

Wheel Load

Crane girder may have four or eight wheels. When the trolley is close to the wheels of the crane, the load on those wheels is maximum and the load on the other pair of wheels is minimum. *Wheel load* refers the maximum possible load that is likely to come on the wheel. The gantry girder has to be designed for the maximum wheel loads.

Longitudinal Load (Drag force)

The crane may brake or accelerate as it moves on the rails. Such a change in movement results into a load in the longitudinal direction. This load is often referred as *drag force*. The drag force is proportional to the wheel load and it is about 10 percent in case of *EOT* cranes.

Fatigue

Fatigue on a structure reduces the life of the structure or may even cause fracture. The failure due to fatigue is usually sudden and brittle. A fatigue factor is normally assigned to a structure and the allowable stresses are reduced depending on the frequency of the loads. At the moment, no clear guidelines are given by the code for fatigue of reinforced concrete girders. The author recommends a reduction factor of about 5 to 10 percent to the strength of the gantry girder due to fatigue.

Example 3.14: Design of Gantry girder

A gantry girder of 8 m simply supported span has the following data (notations are indicated);

Span of the girder = L = 8 m
Spacing of the wheels = a = 3.6 m,   height of rail = 0.05 m,
Wheel maximum load including impact factor = W_w = 240 kN,
Lifting load = W_l = 250 kN,   Hook weight = W_k = 30 kN,
Weight of trolley = W_r = 50 kN,
Impact factor = 20%, Longitudinal load = 10%, Lateral load = 5%.

Figure 3.29 illustrates the girder with wheel loads. The maximum bending moment occurs under one of the two loads, when that load and the centroid of the loads are at equal distances from the middle point of the beam. In the present case, the loads are equal so the centroid of the loads is at half of the distance of the wheel. Therefore the load is at a distance one fourth of the wheel spacing ($a/4$). Let A be the left

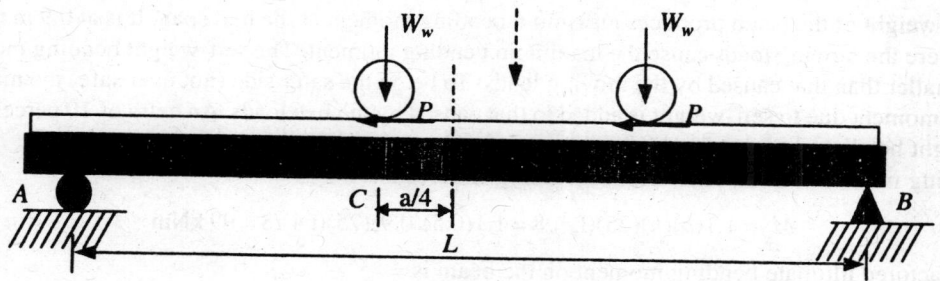

Fig. 3.29 Gantry Grider with Wheel Load at Critical Location for BM.

support and B is the right support and C is the location of the load under which the maximum bending moment is developed for the moving loads.

Maximum bending moment computation

Wheel load with impact factor is: $= W_w = 240$ kN

Distance of load centroid from $A = AC = L/2 - a/4 = 4 - 0.9 = 3.1$ m

The reaction at support $A = R_a = 2(240)(3.1)/8 = 186$ kN

Bending moment under the load is: $M_1 = R_a(3.1) = 576.6$ kNm

The longitudinal force due to drag is $10\% = P = 200/10 = 20$ kN

Where 200 kN is the wheel load without impact

Let the width of the beam be about $L/16$, use $b = 0.5$ m

Let the depth of the girder about $L/10$, use $h = 0.9$ m

Fig. 3.30 Gantry Grider with Rail & Wheel.

The height of the drag force that is acting at the top of the railhead is:

$$= 0.9/2 + 0.15 = 0.6 \text{ m}$$

The horizontal drag force causes equal and opposite reactions at the two supports and it is:

$$R = 2P(0.6)/L = 2(40)(0.6)/8 = 60 \text{ kN}$$

The bending moment on the beam at the load location by drag force is:

$$M_2 = R(3.1) = 186 \text{ kNm}$$

The self-weight of the beam produces maximum bending moment at the mid span. It is at 0.9 m from the point where the moving loads cause the maximum bending moment. The self-weight bending moment is much smaller than that caused by the moving loads. To be on the safer side (not over safe) the maximum bending moment due to self-weight is added to that caused by the live loads. An extra of 10 percent of the self-weight bending moment is added to account for the rail weight and other fixtures.

Bending moment due to self-weight including fixtures is:

$$M_g = 1.1(b)(h)(25)(L^2)/8 = 1.1(0.5)(0.9)(25)(64)/8 = 99 \text{ kNm}$$

The factored ultimate bending moment on the beam is =

$$M_u = \gamma_f(M_g + M_1 + M_2) = 1.5(99 + 576.6 + 186) = 1307.4 \text{ kNm}$$

The lateral load causes torsion on the beam. The lateral load is 5 percent and it is:

$$\text{Lateral load} = 0.5 \text{ (Lifting weight + kook weight)} = 0.05(500 + 30) = 26.5 \text{ kN}$$

The height at which it is acting above the centroid of the beam is = 0.45 + 0.15 = 0.6 m
The factored torsion on the beam is:

$$T_u = 1.5(26.5)(0.6) = 23.9 \text{ kNm}$$

The equivalent torsion on the beam is:

$$M_t = T_u\left[\frac{1+h/b}{1.7}\right] = 23.9\left[\frac{1+900/500}{1.7}\right] = 39.4 \text{ kNm}$$

The equivalent design bending moment is:

$$M_{el} = M_u + M_t = 1307.4 + 39.4 = 1346.8 \text{ kNm}$$

Giving a five percent fatigue reduction factor, and equating the reduced balanced moment capacity to the collapse moment gives:

$$0.95M_{rb} \geq M_{el} \text{ This leads to:}$$

$$0.95Kbd^2 f_{ck} = 0.138(500)(25)d^2 \geq 1346.8(10^6)$$

$$or, d = \sqrt{1346.8(10^6)/0.95(0.138)(500)(25)} = 905 \text{ mm}$$

Assume two rows of 25 mm bars as the main steel. The cover to the reinforcement is 30 mm and the clear space between the two rows as 40 mm, then the total depth of the section is = h = 905 + 20 +25 + 30 = 980 mm;

$$\text{Use } h = 980 \text{ mm and } d = 980 - 30 - 25 - 20 = 905 \text{ mm}$$

The assumed depth is 900 mm and the final design depth of section is 980 mm. The increase in depth is 8 percent. There is a marginal increase in the torsion that is in the ratio of (490 +150)/(450+150) = 1.06. The bending moment due to drag and that due to torsion will increase by about 8 percent. The bending moment due to self-weight also increases by about 9 percent. There is no change in the main bending moment due to wheel loads. The final bending moment after correction is:

$$M_{el} = 1380 \text{ kNm}$$

The area of reinforcement required is:

$$A_{st} = \frac{1.15M_{el}}{(0.95)jdf_y} = \frac{1.15(1380)(10)^6}{(0.95)0.8(905)(415)} = 5560 \text{ mm}^2$$

Provide 12 numbers of 25 mm bars at bottom in two rows. The area of steel provided is 5880 mm² as against 5560 mm² that is needed. Adequate development length should be provided. Provide a nominal reinforcement at mid height of the section. Provide one of 12 mm bar at mid height on each side face. Provide 8 numbers of 25 mm bars at the bottom lower row and 4 numbers of 25 mm bars in the second row. The percentage of tension steel is one. The side cover is 30 mm so the clear spacing between the lower row bars is:

$$\text{Clear spacing of lower bars} = (600 - 30 - 30 - 8*25)/7 = 48 \text{ mm}.$$

The minimum required spacing of the bars is the bar diameter 25 mm or the size of the aggregate plus 5 mm. The beam is subject to torsion even though not high.

Design for Shear Force

The maximum shear force occurs at a distance d from the support when one of the concentrated loads is at that point as shown in Fig. 3.31.

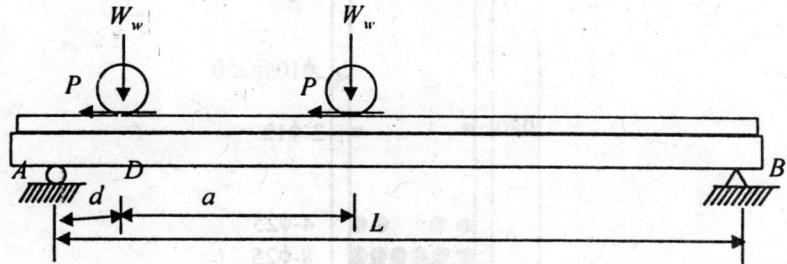

Fig. 3.31 Location of Loads for Maximum Shear Force Example 3.14.

The reaction at A is obtained by taking moment about the other support at B.

$$R_a = W_k(L - d + L - d - a)/L + w_gL/2 + 2P(h/2 + 0.15)/L$$

$$R_a = 240(16 - 2*0.905 - 3.6)/8 + 13.6*4 + 2*40(0.48 + 0.15)/8$$

$$= 342 + 54.4 + 6.3 = 402.7 \text{ kN}$$

The factored shear force at distance d from support

$$V_u = R_a - w_g d = 402.7 - 13.6(0.905) = 390.4 \text{ kN}$$

The equivalent shear force of the torsion is given by:

$$V_e = V_u + 1.6\frac{T_u}{b} = 390.4 + 1.6\frac{23.9}{0.6} = 455 \text{ kN}$$

The nominal shear stress is = $\tau_{ev} = \dfrac{V_e}{bd} = \dfrac{455000}{600(905)} = 0.84$ N/mm^2

The shear strength of concrete for 1.0 percent of reinforcement, and the absolute shear strength of concrete with stirrup steel taken from table 3.4 are:

$$\tau_c = 0.64 \text{ N/mm}^2 \text{ and } \tau_{cmax} = 3.1 \text{ N/mm}^2$$

Let $s_v = 200$ mm
Area of stirrup steel is subject to a minimum of:

$$A_{sv} = \frac{1.15(\tau_{ve} - \tau_c)bs_v}{f_y} = \frac{1.15(0.84 - 0.64)600(200)}{415} = 67 \text{ mm}^2$$

The area of shear reinforcement for equivalent shear is given by:

$$A_{sv} = \frac{1.15 T_u s_v}{b_1 d_1 f_y} + \frac{1.15 V_u s_v}{2.5 d_1 f_y} = \frac{1.15(23.9)(10^6)200}{(600-70)(980-70)415} + \frac{1.15(390400)200}{2.5(980-70)415}$$

$$= 28 + 96 = 124 \text{ mm}^2$$

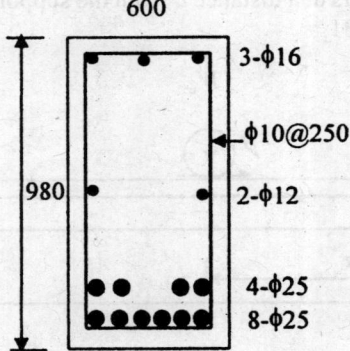

Fig. 3.32 Example 3.14.

Provide 2-legged 10 mm stirrup, then the area of stirrup steel is 157 mm^2. The spacing of the stirrup can be prorated with the area of steel provided and that required:
The stirrup spacing is $s_v = 157(200)/124 = 253$ mm.

Provide 2-legged 10 mm bars at 250 mm spacing.

Provide 3 numbers of 16 mm bars at the top face so that the spacing between the bars is not more than 300 mm.

PROBLEMS

3.1   Determine the neutral axis of a cross section for *M*30 grade concrete and *HYSD-Fe*550 grade reinforcement. The limit of compression strain in concrete is 0.004 and that in reinforcement is 0.002 + $f_y/1.15E_s$.

3.2. A rectangular section is made of *M*30 grade concrete and *HYSD-Fe*550 reinforcement. The limit compressive strain in concrete is 0.004, and yield strain in steel is $0.002 + f_y/1.15E_s$. Determine the balanced moment capacity of the section.

3.3. A 5 m span simply supported beam is subjected to a uniformly distributed live load of 20 kN/m. Design a balanced reinforced concrete cross section beam with width equal to 60% of the effective depth. Sketch the reinforcement details. Use *M*25 grade concrete and *HYSD-Fe*415 reinforcement.

3.4. A 5 m span simply supported beam, 400 mm square is subjected to a uniformly distributed live load of 20 kN/m. Design the reinforcement details using *M*20 concrete and *HYSD-Fe*500 reinforcement bars.

3.5. A 5 m span simply supported beam, 250 by 400 mm is subjected to a uniformly distributed live load of 20 kN/m. Design the reinforcement details using *M*20 concrete and *HYSD-Fe*500 reinforcement bars.

3.6. A 4 m cantilever beam with 400 mm width is subjected to a uniformly distributed live load of 20 kN/m. Design a balanced reinforced concrete rectangular section using *M*20 concrete and *HYSD-Fe*415 steel bars.

3.7. A 4 m cantilever beam 200 by 300 mm is subjected to a uniformly distributed live load of 20 kN/m. Design reinforcement detail of the section for *M*25 grade concrete *HYSD-Fe*415 reinforcement bars.

3.8 A 4 m cantilever beam is subjected to a concentrated load of 35 kN at the free end. Design a balanced reinforced concrete cross section using *M*25 concrete and *HYSD-Fe*415 reinforcement. Use *b* = 300 mm.

3.9. A staircase is formed with independent steps 1.2 m cantilevering from an upright reinforced concrete wall. it Live load is 1.5 kN placed on 1.1 m from the support. The width of the step is 300 mm. Design a balanced reinforced concrete cross section using *M*25 concrete and *HYSD-Fe*500 bars.

3.10. A staircase slab spans between two walls separated by a distance of 4.8 m. The rise and tread of the staircase are 150 and 300 mm respectively. The staircase slab is 1.25 m width and has two landings of 1.25 m at each end. Design the slab for a live load of 4 kN/m using *M*25 concrete *HYSD-Fe*500 bars.

3.11. A 8 m simply supported beam is subjected to uniformly distributed load of 20 kN/m. The section is 400 mm by 800 mm and having 4 numbers of 20 mm bars as bottom reinforcement. Determine the moment capacity of the section. The concrete is *M*25 grade and reinforcement is *HYSD-Fe*415.

3.12. A reinforced concrete cross section of 250 mm by 450 mm is provided with 4 numbers of 20 mm bars in the tension and 3 numbers of 20 mm bars in the compression at 50 mm from the extreme compression fibre. The concrete is *M*25 grade and reinforcement *HYSD-Fe*415. Determine the moment capacity of the section.

3.13 Double over-hang beam of total length 8 m with over-hang span of 1.6 m on each end is subjected to a UDL of 20 kN/m. Design the beam with a depth of 450 mm, *M*25 grade concrete, and *HYSD-Fe*500 reinforcement. Sketch the reinforcement details.

3.14. An electrical light pole of 8 m total length is embedded into ground to a depth of 1.5 m. The wind load acting on the pole is 3 kN at 0.5 m from the top of the pole. Design a reinforced concrete square cross section pole using *M*25 concrete and *HYSD-Fe*415 reinforcement.

3.15 A 2 m high compound wall is subjected to wind force of 1.5 kN/m. Design a reinforced concrete wall with foundation depth 1 m below ground level. Use *M*25 grade concrete and *HYSD-Fe*415 reinforcement bars. Check for stability of the wall.

3.16. A boundary wall of 3.5 m high above ground level is subjected to 1.5 kN/m wind load. The wall is designed with square pilasters spaced at 3.0 m. Design the wall as well as the pilasters in reinforced concrete using *M*25 concrete and *HYSD-Fe*415 reinforcement bars. The depth of the foundation is 1200 mm. Assume that foundation has adequate stability.

3.17 A precast concrete pole of 7 m length and having square cross section of 300 mm size is provided with 4 numbers of 20 mm *HYSD-Fe*415 reinforcement bars placed one at each corner of the section. Clear cover to the bars is 35 mm. The member is lifted from the ground holding it at its middle. Determine whether the moment capacity is adequate for handling.

3.18 Determine the moment capacity of a rectangular sectioned reinforced concrete beam for the following data; *b* = 200 mm, *d* = 600 mm, A_{st} = 603 mm². Concrete is *M*25 grade and the reinforcement is *HYSD-Fe*500. State whether the section is over or under-reinforced.

3.19 Tension reinforcement in a singly reinforced rectangular section is 120% of that balanced steel. Determine the enhanced moment capacity of the cross section as a function of balanced moment capacity, assuming all other properties remaining the same. (*M*20, *HYSD-Fe*415).

3.20 Tension reinforcement in a balanced rectangular section is 80 percent of that of the balanced reinforcement. Determine the moment capacity of the section as a function of balanced moment capacity assuming all other data remaining the same. (*M*20, *HYSD-Fe*415).

4

Strength Design of Non-Rectangular Section and T Beams

4.1 INTRODUCTION

Rectangular and flanged beams are commonly used in buildings and bridges. This chapter presents design of non-rectangular and *T* section beams. The sections are assumed to be symmetrical about the vertical (load) axis. The action the external loads is assumed to pass through shear centre thus avoiding torsion and warping of the section. The shear centre of the symmetrical sections coincides with the centroid of the section. A section is said to be balanced section if the limiting strains in reinforcement and concrete occur simultaneously at the time of collapse. The idealization of the stress-strain curve, limits of strains and strengths of materials are independent of the shape of the cross section. Non-rectangular sections discussed in this chapter include:

- Triangular section,
- Trapezoidal section and
- *T* section.

The three shapes listed have architectural or constructional value. The triangular section can be upright or inverted. The upright triangle section is not suitable in simply supported beam since small area of cross section is available at top to resist compression. Moreover, the large area available in the tension zone has no structural use and only adds up to dead weight. Fig. 4.1(a) illustrates this aspect. Similarly, the diamond shape is also not an efficient shape for beams. The inverted triangle has larger area of concrete at top face therefore more efficient in resisting the compressive force. However very little width is available at the bottom layer to house the reinforcement. The reinforcement may have to be arranged in two or more rows.

Trapezoidal section discussed in this chapter refers to a section having the top and bottom faces parallel to each other. An upright trapezoid is the one in which the width of the top face is smaller than that of the bottom face. Tapered foundation slabs are in trapezoidal shape with smaller width at top. Such tapered slabs are becoming less common because of practical reasons.

4.2 UPRIGHT TRIANGULAR SECTION

An upright triangular has its base horizontal and apex at top. Let b the width of the base of the triangle and h is the total height. d is the effective depth of the reinforcement from the apex of the triangle. Fig 4.1(a) illustrates the triangular cross section. The upper triangular portion above the neutral axis (*NA*) is

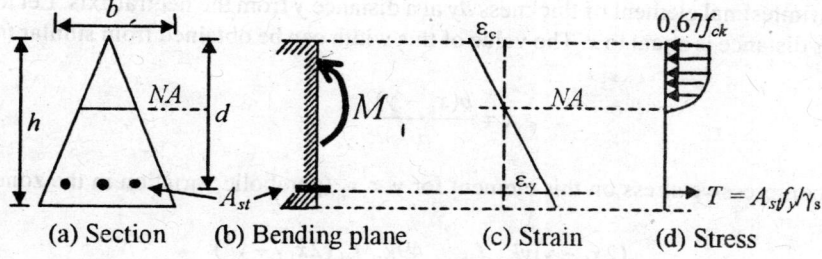

(a) Section   (b) Bending plane   (c) Strain   (d) Stress

Fig. 4.1 Triangular Section under Bending, Notations.

under compression. The line of action of the load passes through the centroid of the section. Fig. 4.1(b) indicates the plane on which the bending moment is acting. Plane section remains plane even after bending. Fig.4.1(c) illustrates the strains and Fig.4.1 (d) illustrates the stress distribution along the depth of the section at collapse. The notations (vide Fig. 4.2) are:

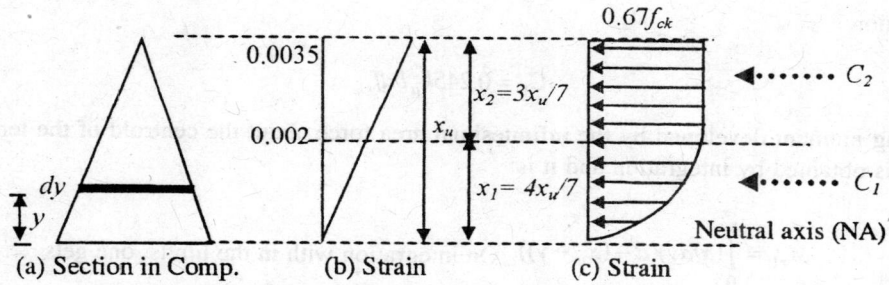

(a) Section in Comp.   (b) Strain   (c) Strain

Fig. 4.2 Strain & Stress Distribution in Compression Zone.

A_{st} = area of tension reinforcement,
b = width of the base of the triangle,
C_1 = compressive force contributed from the portion of the parabolic distribution of the stress on concrete,
C_2 = compressive force contributed from the portion of the rectangular distribution of the stress on concrete,
d = depth of the reinforcement from the apex,
h = overall depth of the section,
k_u = balanced section neutral axis distance coefficient,
M_r = moment capacity of the section,
x_u = neutral axis distance from top fiber = $k_u d$,
x_1 = depth of parabolic portion of the stress on concrete = $4x_u/7$,
x_2 = depth of rectangular portion of the stress on concrete = $3x_u/7$.

Design for Bending

The width of the section in the compressive zone varies linearly so the forces produced by the parabolic and rectangular portions of the stress block are derived in two stages by integration. Further, the stress distribution varies along the depth of the section. Many steps in the derivation and integration are omitted smooth reading.

Consider an infinitesimal element of thickness dy at a distance y from the neutral axis. Let the width of the element at this distance is equal to x. The value of this width can be obtained from similar triangles as:

$$x = \frac{b(x_u - y)}{d} \tag{4.1}$$

The intensity of compressive stress on this element for $y < x_1$ (parabolic variation in the zone) is:

$$f = \frac{(2x_1 - y)yk_p f_{ck}}{x_1^2} = \frac{49k_p f_{ck}(2x_1 y - y^2)}{16x_u^2} \tag{4.2}$$

The total force coming from the parabolic stress distribution on the segment is:

$$C_1 = \int_0^{x_1} xf dy = \frac{49(0.67)bf_{ck}}{16x_u^2 d} \int_0^{x_1} (x_u - y)(2x_1 y - y^2)dy,$$

On integration

$$C_1 = 0.245k_u^2 bdf_{ck} \tag{4.3}$$

The bending moment developed by the infinitesimal area force about the centroid of the tension reinforcement is obtained by integration and it is:

$$M_{r1} = \int_0^{x_1} (xf dy)(d - (x_u - y)), \text{ On integration with in the limits, one gets}$$

$$M_{r1} = 0.164(1 - 0.67k_u)k_u^2 bd^2 f_{ck} \tag{4.4}$$

The compressive force acting on the upper portion where the rectangular stress distribution is acting is:

$$C_2 = \frac{1}{2}\left(\frac{x_u b}{d}\right)x_u(0.67)f_{ck} = 0.0615k_u^2 bdf_{ck} \tag{4.5}$$

The bending moment developed by the rectangular portion of the stress block about the centroid of steel is:

$$M_{r2} = C_2(d - \frac{2x_2}{3}),$$

It works out to be

$$M_{r2} = 0.0615(1 - 0.286k_u)k_u^2 bd^2 f_{ck} \tag{4.6}$$

The total design compressive force on the cross section can be obtained by the sum of the two components, that is $(C_1 + C_2)$ and then dividing the sum by the partial safety factor of 1.5.

$$C = (C_1 + C_2)/\gamma_{mc} = 0.204k_u^2 bdf_{ck} \tag{4.7}$$

The design moment capacity of the section can be obtained by the sum of the two components, M_{r1} and M_{r2} and then divided it by the partial factor of safety:

$$M_{rb} = (M_{r1} + M_{r2})/\gamma_{mc} = K_2 bd^2 f_{ck} \tag{4.8}$$

where

$$K_2 = 0.15(1 - 0.565k_u)k_u^2 \tag{4.9}$$

The area of tension reinforcement required can be obtained by equating the compression force to the tensile force with partial safety factor of 1.15 applied to the force in steel:

$$T = A_{st} f_y / 1.15 = C = 0.204k_u^2 bdf_{ck}$$

The rearrangement of the above equation gives the tension reinforcement for balanced section as:

$$A_{st} = 0.23k_u^2 bd \left[\frac{f_{ck}}{f_y} \right] \tag{4.10}$$

The reinforcement ratio (not real ratio because bd is not the gross area of the section) is:

$$\frac{A_{st}}{bd} = P_2 \left[\frac{f_{ck}}{f_y} \right] \tag{4.11}$$

Where

$$P_2 = 0.234k_u^2 \tag{4.12}$$

The design coefficients for triangular section are listed in table 4.1 for ready reference.

Table 4.1 Design Coefficients for Triangular Section.

f_y	k_u	j	K_2	P_2
250	0.531	0.514	0.030	0.066
415	0.479	0.536	0.025	0.053
500	0.456	0.548	0.023	0.049

The liver arm distance in the case of upright triangle is:

$$jd = \frac{M_{rb}}{C} = \frac{K_2 bd^2 f_{ck}}{0.204k_u^2 bdf_{ck}} = \frac{K_2 d}{0.204k_u^2} = \frac{0.15(1 - 0.565k_u)d}{0.204} = 0.735(1 - 0.565k_u)d \tag{4.13}$$

Or

$$j = 0.735(1 - 0.565k_u) \tag{4.14}$$

Design for Shear

The shear stress distribution due to transverse load on a triangular section is complex when compared with that of the rectangular section. The shear stress is maximum at the centroid of an uncracked section. In reinforced cracked section the distribution is uniform in the cracked zone till the tension reinforcement. In the absence of well-defined theory of shear distribution on the cracked triangular section, an empirical approach is applied in the design. The nominal shear stress developed in the cross section is obtained by dividing the design shear force by the effective area of cross section and that is equal to half of the base width multiplied by the effective depth. The nominal shear stress computed is on the lower side, rather on the unsafe side. One has to use the width of the section at the reinforcement level, rather than that at the base.

The approximate nominal shear stress is:

$$\tau_v = \frac{V_u}{(b/2)d}$$

The stirrup is placed along the side inclined faces even though the stirrup is in the vertical plane. So the effective resistance is proportional to the sine of the inclination of the stirrup leg.

The spacing of the stirrups can be obtained from the previous chapter with the correction and it is given by:

$$s_v = \frac{0.87 A_{sv} f_{yv} \sin\alpha}{(b/2)(\tau_v - \tau_c)}$$

where α = angle of inclination of the stirrup legs with the horizontal axis (assumed symmetrical legs)

The average width of the section is assumed to be equal to the half of the base width. This is on the liberal side leading to a safer side.

Example 4.1: Simply supported beam of triangular cross section

A simply supported beam of effective span 5 m in mild exposure is subjected to a live load of 24 kN/m. It is to be designed with $M30$ concrete and $HYSD$-Fe415 bars. The cross section of the beam is triangular.

Solution

The given data is: span = L = 5 m, strength of concrete = f_{ck} = 30 N/mm^2, strength of reinforcement = f_y = 415 N/mm^2, live load = w_l = 24 kN/m, the design coefficients for the present data can be obtained from table 4.1 and they are: K_2 = 0.025, k_u = 0.479, p_2 = 0.053

Assume the width of the beam section as b = 350 mm and depth (h) of the beam in the range of $L/8$ for simply supported beams, let say h = 700 mm for the purpose of dead weight. Fig. 4.3 illustrates the simply supported beam.

Self weight of the beam = w_g = 0.35(0.7)25/2 = 3.1 kN/m, use 3.5 kN/m

Total design load at collapse state (strength limit state) = w_t = 1.5(3.5 + 24) = 41.25 kN/m

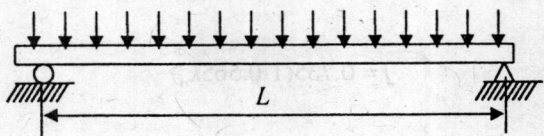

L

Fig. 4.3 Simply Supported Beam, Example 4.1.

Maximum bending moment at collapse =

$$M_c = \frac{w_t L^2}{8} = \frac{41.25(25)}{8} = 128.9 \text{ kNm}$$

The balanced moment capacity of the section must be more than the collapse moment, that is:

$$M_{rb} \geq M_c \text{ , This leads to:}$$

$$K_2 bd^2 f_{ck} = 0.025(350)(30)d^2 \geq 128.9(10^6)$$

$$or, \ d = \sqrt{128.9(10^6)/(0.025)(350)(30)} = 700 \text{ mm}$$

Let the effective depth be 700 mm. The minimum cover to the reinforcement must be 25 mm or the diameter of the bar. Let the diameter of the bar be assumed 25 mm. Therefore the total depth of the beam is = h = 700 + 25/2 +25 = 737.5 mm. Now revise the depth to 740 mm. The weight assumed is slightly higher than the actual one. The revised self-weight is = w_g = 3.3 kN/m, the design load was 3.5 kN/m. The design is acceptable.

The area of the tension reinforcement needed is:

$$A_{st} = p_2 \frac{f_{ck}(bd)}{f_y} = 0.053 \frac{30(350)(700)}{415} = 939 \text{ mm}^2$$

Use 3 numbers of 20 mm bars.
The area of reinforcement provided is: A_{st} (provided) = 3*314 = 942 mm² > 939 mm² needed
Provide one of 10 mm hanger bar at the top to tie up the shear reinforcement.

Design for Shear

The design procedure for shear for non-rectangular shape of reinforced concrete sections is not spelled out by the code. In the absence of the guidelines, an empirical procedure is suggested. A nominal shear stress is computed using the half width of the base of the triangle.

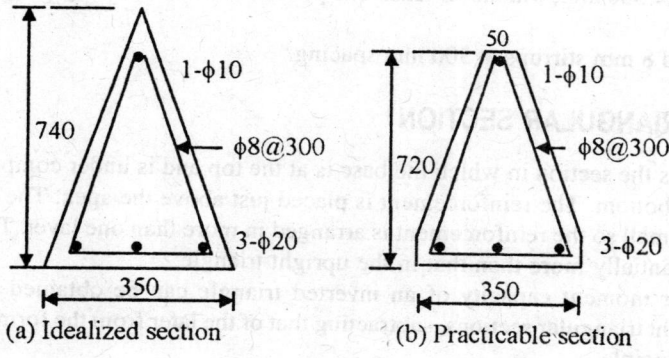

Fig. 4.4 Cross Section, Example 4.1.

The shear force at a distance d from the support where the shear force is critical is:

$$V_u = w_t(L - d) = 41.25(5/2 - 0.7) = 74.25 \text{ kN}$$

The nominal shear stress on the section is:

$$\tau_v = \frac{V_u}{bd/2} = \frac{74250}{(350)(350)} = 0.61 \text{ N/mm}^2$$

Percentage of reinforcement is:

$$p = \frac{342(100)}{0.5(350)700} = 0.3\%$$

The admissible shear stresses without and with shear reinforcement are:

$$\tau_c = 0.39 \text{ N/mm}^2 \text{ and } \tau_{cmax} = 3.5 \text{ N/mm}^2$$

The shear strength of concrete is 0.39 N/mm², which is less than the nominal shear stress. Select two legged 8 mm bar as shear stirrup, the area of two legs of the stirrup is: 100 mm².
The slope of the stirrup leg is given by:

$$\tan\alpha = (740 - 45 - 25)/175 = 670/175 = 3.83 \text{ ,or}$$

$$\alpha = 75 \text{ degrees}$$

The spacing of the stirrup reinforcement is:

$$s_v = \frac{0.87A_{sv}f_{yv}\sin\alpha}{(b/2)(\tau_v - \tau_c)} = \frac{0.87(100)(415)(0.96)}{175(0.61 - 0.39)} = 900 \text{ mm}$$

The maximum spacing of the stirrups is controlled by:

$$s_{max} \leq \frac{A_{sv}f_{yv}}{0.4b/2} = \frac{2(100)(415)}{0.4(350)} = 590 \text{ mm and,}$$

$s_{max} = d = 700$ mm or 300 mm, whichever least. The permissible maximum spacing of the stirrups is 300 mm.

Provide two legged 8 mm stirrups at 300 mm spacing.

4.3 INVERTED TRIANGULAR SECTION

An inverted triangle is the section in which the base is at the top and is under compression. The apex of the triangle, is at the bottom. The reinforcement is placed just above the apex. The sectional area available at apex point is small so the reinforcement is arranged in more than one layer. The area available for compression is substantially more than that in the upright triangle.

The derivation for moment capacity of an inverted triangle can be obtained by superposition of rectangular and upright triangular sections, subtracting that of the later from the former. Fig 4.5 illustrates the superposition principle

The notations used are same as those used earlier:

A_{str} = area of reinforcement in rectangular section,
A_{stt} = area of reinforcement in triangular section,
A_{st} = area of reinforcement in inverted triangular section,

Subtract the triangular section from the rectangular section, then results the inverted triangle as shown in Fig. 4.5(c). This is mathematically expressed as:

$$M_r \text{ (Inverted triangle)} = M_r \text{ (Rectangular)} - M_r \text{ (Triangular)} \tag{4.15}$$

$$M_{rb} = Kbd^2 f_{ck} - K_2 bd^2 f_{ck} = (K - K_2)bd^2 f_{ck} \tag{4.17}$$

Wherever a triangle word is mentioned, it is the upright triangle. The inverted triangle is indicated by Inv. triangle.

The final moment capacity of an inverted triangle is:

$$M_r = K_3 bd^2 f_{ck} \tag{4.18}$$

Where
$$K_3 = K - K_2 = 0.36\, k_u - 0.151 k_u^2 + 0.085 k_u^3$$

The values of K and K_2 can be obtained from the earlier expressions or tables The compressive force due to bending moment on the inverted triangle can be obtained by subtracting the compressive force of the triangle from that of the rectangular section.

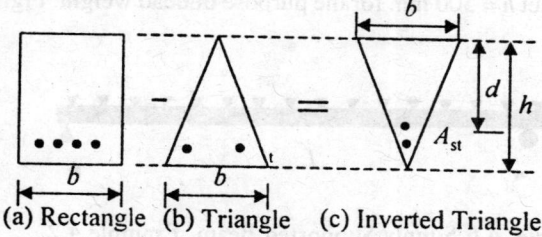

Fig. 4.5 Superposition to Get Inverted Triangle.

C (inverted triangle) = C (rectangular) – C (triangle)

$$C = 0.36\, k_u bd f_{ck} - 0.204\, k_u^2 bd f_{ck} = (0.36 - 0.204 k_u)k_u bd f_{ck} \tag{4.19}$$

The effective lever arm of the resisting moment is obtained by dividing the moment capacity expression by the net compressive force.

$$j = \frac{M_{rb}}{Cd} = \frac{K_3}{(0.36 - 0.204 k_u)k_u} \tag{4.20}$$

Similarly the area of tension reinforcement is obtained.

A_{st} (Inverted triangle) = A_{st} (Rectangle) - A_{st} (Triangle)

$$\frac{A_{st}}{bd} = (p - p_2)\frac{f_{ck}}{f_y} = p_3 \frac{f_{ck}}{f_y} \tag{4.21}$$

Table 4.2 gives the values of K_3, j and the percentage of reinforcement of inverted triangle.

Table 4.2 Design Coefficients for Inverted Triangular Section.

f_y	k_u	K_3	j	p_3
250	0.531	0.119	0.85	0.154
415	0.479	0.113	0.90	0.145
500	0.450	0.110	0.91	0.140

Example 4.2: Simply supported beam of inverted triangular cross section

A simply supported beam of 5 m effective span in mild exposure is subjected to a live load of 24 kN/m and it is to be designed with $M30$ concrete and $HYSD$-$Fe415$ bars. The cross section of the beam is an inverted triangular.

Solution

The given data is: $L = 5$ m, $f_{ck} = 30$ N/mm^2, $f_y = 415$ N/mm^2, $w_l = 24$ kN/m, the design coefficients for the present data can be obtained from table 4.2 and they are: $K_3 = 0.113$, $k_u = 0.479$, $j = 0.90$, $p_3 = 0.145$

Assume the width of the beam as $= b = 350$ mm and overall depth (h) of the beam in the range of $L/10$ for simply supported beams, let $h = 500$ mm for the purpose of dead weight. Fig. 4.6 illustrates the simply supported beam.

Fig. 4.6 Simply Supported Beam, Example 4.2.

Self weight of the beam $= w_g = 0.35(0.5)25/2 = 2.2$ kN/m, use 3.0 kN/m
Total design load at collapse state (strength limit state) $= w_t = 1.5(3.0 + 24) = 40.5$ kN/m
Maximum bending moment at collapse $=$

$$M_c = \frac{w_t L^2}{8} = \frac{40.5(25)}{8} = 126.6 \text{ kNm}$$

The balanced moment capacity must be more than the collapse moment, that is:

$$M_{rb} \geq M_c \text{ This leads to:}$$

$$K_3 bd^2 f_{ck} = 0.113(350)(30)d^2 \geq 126.6(10^6)$$

$$or, d = \sqrt{126.6(10^6)/(0.113)(350)(30)} = 327 \text{ mm}$$

Let the effective depth is about 330 mm. The minimum cover to the reinforcement is 25 mm or the diameter of the bar. Let the diameter of the bar be assumed 20 mm and the bars are placed in two layers.

The spacing between the bars be assumed as 30 mm. Therefore the total depth of the beam is $= h = 330 + 20/2 + 30 + 25 = 395$ mm. The reinforcement is placed at the corner of the triangle. A small flatness at the bottom of the section is provided to avoid the sharp bottom corner. Let the cover provided is 35 mm and then smoothened the corner. Now revise the assumed depth to 475 mm and then the effective depth is $(475 - 65 - 25) = 390$ mm. The assumed depth of the beam for the self-weight purpose is 500 mm. The design is acceptable.

The area of the tension reinforcement needed is:

$$A_{st} = \frac{1.15M_c}{jdf_y} = \frac{(1.15)126.6(10^6)}{0.9(390)415} = 999 \text{ mm}^2$$

Use 1 of 25 mm and 2 numbers of 20 mm bars.

The area of reinforcement provided is: A_{st} (provided) $= 490 + 2*314 = 1118$ mm^2
Provide two numbers of 10 mm hanger bars at the top to tie the shear reinforcement

Design for Shear

The design for shear of non-rectangular reinforced concrete section is not spelled out by the code. The effective width of the section chosen is half of the width of the section. A nominal shear stress is computed using the half width of the base of the triangle.

The shear force at a distance d from the support where the shear force is critical is:

$$V_u = w_u(L - d) = 40.5(5/2 - 0.36) = 86.67 \text{ kN}$$

The nominal shear stress on the section is:

$$\tau_v = \frac{V_u}{bd/2} = \frac{86670}{0.5(350)(390)} = 1.28 \text{ N/mm}^2$$

Percentage of reinforcement is:

$$p = \frac{1118(100)}{0.5(350)390} = 1.6\%$$

The allowable shear stresses without and with shear reinforcement are obtained from table 3.4:

$$\tau_c = 0.78 \text{ N/mm}^2 \text{ and } \tau_{cmax} = 3.5 \text{ N/mm}^2$$

The shear strength of concrete is 0.78 N/mm^2, and is less than the nominal shear stress. Select two legged 8 mm bar as stirrup, so the area of two legs of the stirrup is: 100 mm^2.

Slope of the stirrup leg, $\tan \alpha = (450 - 50 - 30)/175 = 370/175$
$$\alpha = 65 \text{ degrees}$$

The spacing of the stirrup reinforcement is:

$$s_v = \frac{0.87A_{sv}f_{yv}\sin\alpha}{(b/2)(\tau_v - \tau_c)} = \frac{0.87(100)(415)(0.9)}{175(1.28 - 0.78)} = 370 \text{ mm}$$

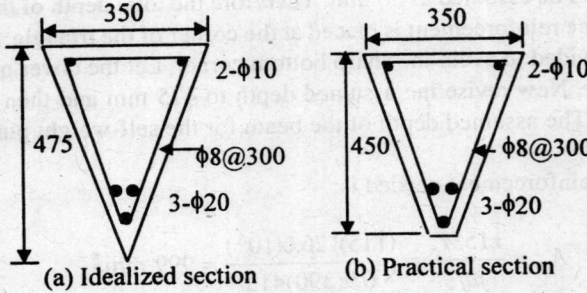

(a) Idealized section   **(b) Practical section**

Fig. 4.7 Cross Section, Example 4.2.

The maximum spacing of the stirrups is controlled by:

$$s_{max} \leq \frac{A_{sv} f_{yv}}{0.4b/2} = \frac{2(100)(415)}{0.4(350)} = 590 \text{ mm and,}$$

s_{max} is to be limited to d or 300 mm, whichever less; The permissible maximum spacing of the stirrup is 300 mm. Provide two legged 8 mm stirrups at 300 mm spacing.

Example 4.3: *Simply supported beam of inverted triangular section, under-reinforced section*

A simply supported beam of 5 m effective span in mild exposure is subjected to a live load of 24 kN/m and is to be designed with M30 concrete and HYSD-Fe415 reinforcement bars. The cross section of the beam is an inverted triangle with total depth of the beam as 600 mm and width 350 mm.

Solution

The given data is: $L = 5$ m, $f_{ck} = 30$ N/mm², $f_y = 415$ N/mm², $w_l = 24$ kN/m, the design coefficients for the present data can be obtained from table 4.2 and they are: $K_3 = 0.113$, $k_u = 0.479$, $j = 0.90$, $p_3 = 0.145$

Width of the beam is $= b = 350$ mm and depth of the beam is $= h = 600$ mm
Self weight of the beam $= w_g = 0.35(0.6)25/2 = 2.7$ kN/m, use 3.0 kN/m
Total design load at collapse state $= w_t = 1.5(3.0 + 24) = 40.5$ kN/m

Maximum bending moment at collapse $=$

$$M_c = \frac{w_t L^2}{8} = \frac{40.5(25)}{8} = 126 \text{ kNm}$$

The balanced moment capacity must be more than the collapse moment, that is:

$M_{rb} \geq M_c$, This leads to:

$$K_3 bd^2 f_{ck} = 0.113(350)(30)d^2 \geq 126.6(10^6)$$

$$or, \ d = \sqrt{126.6(10^6)/(0.113)(350)(30)} = 327 \text{ mm}$$

The effective depth for a balanced section is 327 mm, and the total depth provided is 600 mm. This means that an effective depth of about $600 - 65 = 535$ mm is available. The minimum cover to the reinforcement is 25 mm or the diameter of the bar. Let the diameter of the bar be assumed 20 mm and the bars are placed in two layers. The spacing between the bars can be assumed as 30 mm. Therefore the effective depth of the beam is = $h - distance\ from\ apex\ to\ centre\ of\ steel = 600 - 20/2 - 30 - 25 = 535$ mm. The reinforcement is placed at the bottom corner of the triangle and to avoid the sharp corner at bottom, let the cover be taken in the order of 35 mm plus the corner is smoothened with a small flatness at the bottom. Now revise the effective depth to 535 mm.

The area of the tension reinforcement required is obtained by dividing the collapse moment by the lever arm and the strength of the steel, and multiplied by the partial safety factor.

$$A_{st} = \frac{1.15 M_c}{jd f_y} = \frac{1.15(126.6)(10^6)}{0.9(535)415} = 729 \text{ mm}^2$$

Use 2 numbers of 20 mm and one of 12 mm bars.

The area of reinforcement provided is: A_{st} (provided) = 741 mm^2
Provide two numbers of 10mm hanger bars at the top to tie the shear reinforcement

Design for Shear

A nominal shear stress is computed using the half width of the base of the triangle. The shear force at a distance *d* from the support where the shear force is critical is:

$$V_u = w_f(L - d) = 40.5(5/2 - 0.535) = 79.6 \text{ kN}$$

The nominal shear stress on the section is:

$$\tau_v = \frac{V_u}{bd/2} = \frac{79600}{(350)(535)/2} = 0.85 \text{ N/mm}^2$$

Percentage of reinforcement is:

$$p = \frac{1118(100)}{0.5(350)535} = 1.2\%$$

The admissible shear stresses without and with shear reinforcement are obtained from table 3.4:

$$\tau_c = 0.7 \text{ N/mm}^2 \text{ and } \tau_{cmax} = 3.5 \text{ N/mm}^2$$

The shear strength of concrete is 0.7 N/mm^2, which is less than the nominal shear stress. Select two legged 8 mm bar as shear stirrup, the area of two legs of the stirrup is: 100 mm^2.

The slope of the stirrup leg , tan α = 530/175
$$\alpha = 72 \text{ degrees}$$

The spacing of the stirrup reinforcement is:

$$s_v = \frac{0.87 A_{sv} f_{yv} \sin \alpha}{(b/2)(\tau_v - \tau_c)} = \frac{0.87(100)(415)(0.72)}{175(0.85 - 0.7)} = 1275 \text{ mm}$$

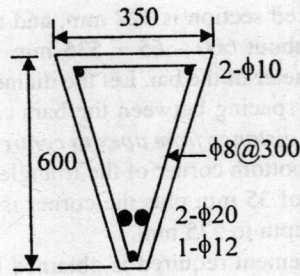

Fig. 4.8 Cross Section, Example 4.3.

The maximum spacing of the stirrups is controlled by:

$$s_{max} \leq \frac{A_{sv} f_{yv}}{0.4b/2} = \frac{2(100)(415)}{0.4(350)/2} = 1185 \text{ mm and,}$$

s_{max} is to be limited to the effective depth or 300mm, whichever least. The permissible maximum spacing of the stirrup is 300 mm.

Provide two legged 8 mm stirrups at 300 mm spacing.

4.4 TRAPEZOIDAL SECTIONED BEAM

A trapezoidal shape is developed by combining a rectangular section with a triangular one. The top and bottom faces of the section are parallel to each other, and the load action must pass though the centroid of the section. The beam is assumed to be prismatic and straight. In case the load acts eccentric with the centroidal axis, torsion is generated on the section. Fig. 4.9 illustrates the notations. The wider top face of a trapezium is advantageous in simply supported beams. The bending moment capacity and the other design parameters of the trapezoidal section can be obtained by superposition of the properties of the rectangular and triangular sections.

The notations are:

b_1 = width of the trapezium at top,

b_2 = width of the trapezium at bottom, assumed to be larger than the top width for derivation, it can be smaller also.

A_{st} = total area of tension reinforcement,

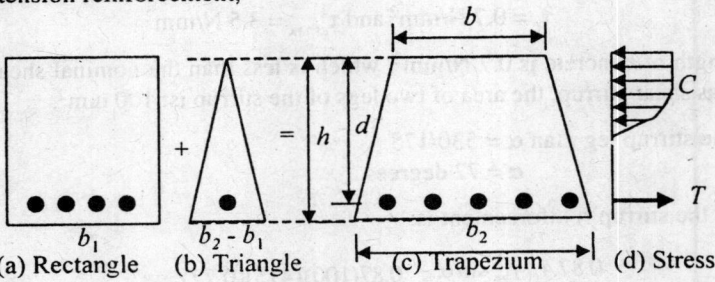

Fig. 4.9 Trapezoidal Section from Rectangular and Triangular Sections.

h = total depth of the section,

d = effective depth of the reinforcement from top.

The section is divided into a rectangle with width b_1 and a triangle with base with as $b_2 - b_1$. The moment capacity of the trapezium is the sum of the moment capacities of rectangle and the triangle, and it is given by:

$$M_{rb} = Kb_1 d^2 f_{ck} + K_2(b_2 - b_1)d^2 f_{ck}$$

It can be rearranged as:

$$M_{rb} = [(K - K_2)b_1 + K_2 b_2]d^2 f_{ck} \qquad (4.22)$$

Similarly the compressive force on the section is:

$$C\,(trapezium) = C\,(rectangle) + C\,(triangle)$$

$$C = (0.36 k_u b_1 + 0.204 k_u^2 (b_2 - b_1))df_{ck} \qquad (4.23)$$

The lever arm coefficient can be obtained from the above two expressions.

$$j = \frac{(K - K_2)b_1 + K_2 b_2}{0.36 k_u b_1 + 2.204(b_2 - b_1)k_u^2} \qquad (4.24)$$

The area of tension reinforcement required can be obtained from the lever arm ratio.

$$A_{st} = \frac{1.15 M_c}{jdf_y} \qquad (4.25)$$

The derivation is obtained with wider side at the bottom face. The trapezoid with wider top face may result in $b_2 - b_1$ as negative. The design for shear force is similar to that of the design of triangular section.

Example 4.4: Design of trapezoidal-sectioned beam

A simply supported beam of 5 m effective span in moderate exposure is subjected to a live load of 24 kN/m and it is to be designed with *HYSD-Fe*415 reinforcement bars. The cross section of the beam is an inverted trapezium.

Solution

The given data is: $L = 5$ m, $f_y = 415$ N/mm², $w_l = 24$ kN/m, select $f_{ck} = 25$ N/mm². The design coefficients

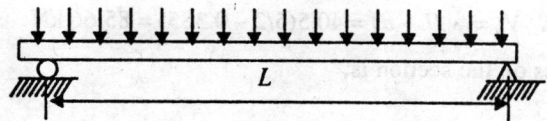

Fig. 4.10 Simply Supported Beam, Example 4.4.

for the present data can be obtained from table 4.1 and they are: $k_u = 0.479$, $K = 0.138$, $K_2 = 0.025$, $p_2 = 0.053$

Assume the widths of the beam as $b_1 = 250$ mm, and $b_2 = 350$ mm, and depth (h) of the beam in the range of $L/10$ for simply supported beams, say $h = 500$ mm for the purpose of dead weight. Fig. 4.10 illustrates the simply supported beam.

Self weight of the beam = $w_g = (0.35 + 0.25)(0.5)25/2 = 1.8$ kN/m, use 3.0 kN/m

Total design load at collapse state (strength limit state) = $w_t = 1.5(3.0 + 24) = 40.5$ kN/m

Maximum bending moment at so called collapse load =

$$M_c = \frac{w_t L^2}{8} = \frac{40.5(25)}{8} = 126.6 \text{ kNm}$$

The balanced moment capacity (Eq. 4.22) must be more than the collapse moment, that is:

$$M_{rb} \geq M_c$$

$$M_{rb} = [(K - K_2)b_1 + K_2 b_2]d^2 f_{ck} = ((0.138 - 0.025(250) + 0.025(350))(25)d^2 = 925 \ d^2$$

$$or, \ d = \sqrt{126.6(10^6) / (925)} = 370 \text{ mm}$$

Let the effective depth is 370 mm. The minimum cover to the reinforcement is 30 mm or the diameter of the bar. Let the diameter of the bar be assumed 20 mm and the bars are placed in one layer Therefore the total depth of the beam is = $h = 370 + 20/2 + 30 = 410$ mm. Now revise the assumed depth to 425 mm and the effective depth as $(425 - 40) = 385$ mm. The assumed depth of the beam for self weight purpose is 500 mm and that is close to the final value. The design is acceptable. The lever arm ratio from Eq. 4.24 is:

$$j = \frac{(K - K_2)b_1 + K_2 b_2}{0.36 k_u b_1 + 0.204(b_2 - b_1)k_u^2} = \frac{(0.138 - 0.025((250) + 0.025(350))}{0.36(0.479)(250) + 0.204(100)(0.479)^2} = 0.77$$

The area of the tension reinforcement needed is:

$$A_{st} = \frac{1.15 M_c}{jdf_y} = \frac{1.15(126.6)(10)^6}{0.77(385)(415)} = 1184 \text{ mm}^2$$

Use 4 numbers of 20 mm bars at bottom as the main reinforcement.

The area of reinforcement provided is A_{st} (provided) = 4*314 = 1256 mm^2

Provide two numbers of 8mm hanger bars at the top to tie up the shear reinforcement

Design for Shear

The shear force at a distance d from the support where the shear force is critical is:

$$V_u = w_t(L - d) = 40.5(5/2 - 0.385) = 85.66 \text{ kN}$$

The nominal shear stress on the section is:

$$\tau_v = \frac{V_u}{b_1 d} = \frac{85660}{250(385)} = 0.89 \text{ N/mm}^2$$

Percentage of reinforcement is:

$$p = \frac{1256(100)}{0.5(350 + 250)385} = 1.1\%$$

The admissible shear stresses without and with shear reinforcement are obtained from table 3.4:

$$\tau_c = 0.64 \text{ N/mm}^2 \text{ and } \tau_{cmax} = 3.1 \text{ N/mm}^2$$

The shear strength of concrete is 0.64 N/mm², which is less than the nominal shear stress. Select two legged 8 mm bar as shear stirrup, the area of two legs of the stirrup is: 100 mm².

The slope of the legs of the shear stirrups is close to 90 degrees and the nominal shear stress is computed using the smaller of the widths.

Slope of the stirrup leg, tan α = (425 –50)/50 and α = 82 degrees

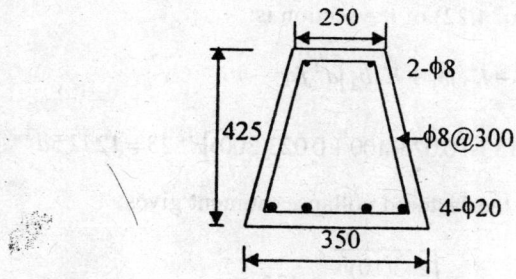

Fig. 4.11 Example 4.4.

The spacing of the stirrup reinforcement is:

$$s_v = \frac{0.87 A_{sv} f_{yv} \sin \alpha}{((b_1 + b_2)/2)(\tau_v - \tau_c)} = \frac{0.87(100)(415)(0.99)}{300(0.89 - 0.64)} = 476 \text{ mm}$$

The maximum spacing of the stirrups is controlled by:

$$s_{max} \leq \frac{A_{sv} f_{yv}}{0.4 b_1} = \frac{(100)(415)}{0.4(250)} = 415 \text{ mm, and}$$

s_{max} is to be limited to = d = 385 mm or 300 mm, whichever less. The maximum spacing of the stirrup is 300 mm.

Provide two-leg 8mm stirrups at 300mm spacing.

Example 4.5: Design of inverted trapezoidal sectioned beam

A simply supported beam of 6 metres in span is subjected to an imposed load of 30 kN/m and exposed to moderate environment. The architect has specified top and bottom widths of the trapezium as 400 mm and 300 mm.

Solution

The beam is in moderate exposure environment so select $M25$ grade concrete and $HYSD$-$Fe500$ steel reinforcement. Minimum cover to reinforcement is $= 30$ mm. And $b_1 = 0.4$ m, $b_2 = 0.3$ m. This is an inverted trapezium. The coefficients for rectangular section are $j = 0.8$, $K = 0.133$, and for triangular section; $K_2 = 0.023$, $j_2 = 0.548$, $k_u = 0.456$.

Let the depth of the section $= h =$ about $L/10 = 600$ mm. This is used for the self-weight of the section.
The self-weight $= w_g = (0.4 + 0.3)(0.6)25/2 = 5.25$ kN/m, use 6 kN/m
The design load for strength is $= w = 1.5(w_g + w_l) = 54$ kN/m

Design for Moment

Factored design bending moment is $= M_c = \dfrac{wL^2}{8} = \dfrac{54(36)}{8} = 243$ kNm

The resisting moment capacity (Eq. 4.22) of the section is:

$$M_r = \left[(K - K_2)b_1 + K_2 b_2\right]d^2 f_{ck}$$

$$= \left[(0.133 - 0.023)400 + 0.023(300)\right]d^2 25 = 1272.5d^2$$

Equating the moment capacity to the factored collapse moment gives:

$$d = \sqrt{\frac{243(10)^6}{1272.5}} = 436 \text{ mm}$$

Select the total depth of the section as $= h = 500$ mm, this gives
Effective depth $= d = 500 - 40 = 460$ mm
The lever arm ratio of the beam from Eq. 4.24 is:

$$j = \frac{(K - K_2)b_1 + K_2 b_2}{0.36k_u b_1 + 0.204(b_2 - b_1)k_u^2} = \frac{(0.133 - 0.023)400 + 0.023(300)}{0.36(0.456)(400) + 0.204(-100)0.456^2} = 0.83$$

The area of tension reinforcement required is:

$$A_{st} = \frac{1.15M_c}{jdf_y} = \frac{1.15(243)10^6}{0.83(360)500} = 1464 \text{ mm}^2$$

Provide 5 numbers of 20 mm bars at the bottom, and 2 of 10 mm bars at top as hanger bars.
The area of tension provided is $= A_{st} = 1570$ mm^2

Design for Shear

The maximum shear force at a distance d from the support is:

$$V_u = w(L/2 - d) = 54(3 - 0.46) = 137.16 \text{ kN}$$

The nominal shear stress is obtained by taking the average width of the section.

$$\tau_v = \frac{V_u}{bd} = \frac{137160}{350(460)} = 0.86 \text{ N/mm}^2$$

Let two of the five bars are curtailed, then the percentage of reinforcement three continued bars is 0.6%
The admissible shear strengths of concrete without and with shear reinforcement from table 3.4 are:

$$\tau_c = 0.53 \text{ and } \tau_{cmax} = 3.1 \text{ N/mm}^2$$

The slope of the legs of the stirrup is 83 degrees. Select two leg 8 mm stirrups, then the spacing of the stirrups is:

$$s_v = \frac{0.87 A_{sv} f_{yv} \sin \alpha}{(b)(\tau_v - \tau_c)} = \frac{0.87(100)(500)(0.99)}{350(0.86 - 0.53)} = 372 \text{ mm}$$

The maximum spacing of the stirrup is only 300 mm, *so provide the 8mm stirrups at 300 mm spacing.*

Range of Lever Arm for Trapeziodal Sections

Equation 4.24 that gives the lever arm ratio can be modified be dividing the numerator and the denominator by b_1 and let the ratio of the bottom and the top sides be denoted as: $b_2/b_1 = r$. The lever arm ratio of Eq. 4.24 reduces to:

$$j = \frac{(K - K_2)b_1 + K_2 b_2}{0.36 k_u b_1 + 0.204(b_2 - b_1)k_u^2} = \frac{K + K_2(r-1)}{0.36 k_u + 0.204(r-1)k_u^2} \qquad (4.26)$$

For an upright trapezium with topside as half of the size of the bottom face, the value of *j* is 0.74, and that for an inverted trapezium with bottom side as half of the top face, the value is 0.84. The value reduces to that of rectangular one for the ratio of the sides equals to 1. Normally the inverted triangles and inverted trapeziums are more commonly used. Therefore the lever arm ratio is more than that of a rectangular one.

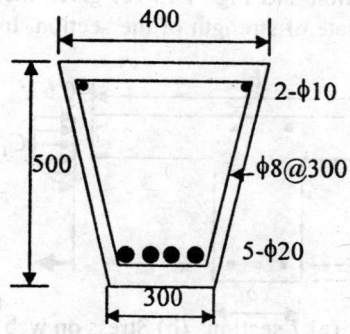

Fig. 4.12 Example 4.5.

4.5   T SECTIONED BEAMS

The slab is cast integral with the beam in most cases. The shear reinforcement of the beam projects into the slab and the shear hanger bars are invariably at the top face of the slab. Therefore the beam behaves as a T (tee) beam. The top slab portion is referred as top flange. In simply supported beams, the top face is under compression, so the flange portion is either partially or fully under compression. The neutral axis distance from top is in the range of 0.45 times the effective depth of the section. The slab portion is likely to be under compression. In cantilever beams, the top face is under tension, and if the slab is built on the top face of the cantilever beam, then the slab is under tension. Continuous beam and framed beam slab construction, the portion at the intermediate supports is under negative bending moment. This means that the top face of the beam is under tension thus the beam at intermediate supports is not strictly a T beam. A beam section having flanges both at top and bottom is called I beam. Such flanged sections are uncommon in reinforced concrete construction but very common in prestressed concrete construction.

Effective Flange Width

The slab at the top face of the beam bends along with the beam. However, only a portion of the slab bends to the same curvature as that of the beam. The shear force on the beam influences upto a limited distance of the slab width. This effect is called shear lag effect. The beam may support a 4 m wide slab but all the 4 metre width may not bend to the to the same curvature of the beam. The width of the slab that acts together with the beam is called *effective width*. The effective width of the flange is subject to the following:

b_f = average of the widths of the slab on either side of the beam, or
  $= b_w + 6t + L_o/6$

where

b_f = effective width of the flange,
b_w = width of the beam section, it is also called *web width*,
L_o = distance between two successive points of zero bending moments of the beam, = span in simply supported beam, or = twice the span of a cantilever,
t = thickness of the slab.
h = overall depth of the section

Figure 4.13(a) illustrates the T beam cross section along with the notations. Fig. 4.13(b) indicates the stress distribution on the web portion and Fig. 4.13 (c) gives the stress distribution on the overhang portion of the flange at the limit-state of strength of the section. In most cases the entire flange area is

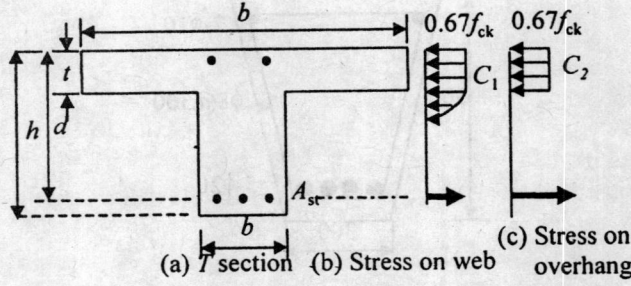

(a) T section   (b) Stress on web   (c) Stress on overhang

Fig. 4.13 T **Beam Section and Notations.**

subjected to compression and the stress distribution is idealized as rectangular shape. The web portion is much thinner than the flange and it is subjected to parabolic or rectangular cum parabolic distribution of the stress. Idealization of the stress distribution of rectangular pattern on the total flange width is normally assumed. This is subject to the condition that the neutral axis falls beyond the flange thickness. If the neutral axis falls in the flange, the section is considered as a rectangular one with the width of the section equal to the width of the flange. The moment capacity expression of the rectangular section will then be extended to the *T* beam.

The maximum strains in concrete and steel at the time of failure and the neutral axis distance are:

$$\varepsilon_{cu} = 0.0035$$

$$\varepsilon_y = 0.002 + f_y/1.15$$

$$\frac{x_u}{d - x_u} = \frac{\varepsilon_{cu}}{\varepsilon_y}$$

where:

x_u = neutral axis distance

d = distance of steel reinforcement from the extreme compression fibre,

f_y = proof stress in steel and 1.15 is the partial safety factor applied to steel.

The neutral axis distance from the above equation for simultaneous occurrence of the limiting strains, reduces to:

$$x_u = \frac{\varepsilon_{cu} d}{\varepsilon_{cu} + \varepsilon_y} = \frac{0.0035d}{0.0055 + f_y/1.15} = k_u d \qquad (4.27)$$

The compressive force on the flanged section can be divided into two portions. The web width that can be considered as rectangular and the second portion is the overhang portions of the flange beyond the width of the web. The compressive force on the web rectangular portion of the section is same as that of the rectangular sectioned beam with width equal to that of the web. And the compressive force is equal to:

$$C_1 = 0.36 b_w k_u b_w d f_{ck} \qquad (4.28)$$

The compressive force on the overhang portions of the flange is obtained by considering a rectangular stress distribution over the flange thickness. The compressive force on the overhang flange is:

$$C_2 = (b_f - b_w) t (0.67 f_{ck}) / \gamma_m = 0.45 (b_f - b_w) t f_{ck} \qquad (4.29)$$

Taking the moment of the above two forces about the centre of the reinforcement gives:

$$M_{rb} = K b d^2 f_{ck} + 0.45 (b_f - b_w) t f_{ck} (d - 0.5t) \qquad (4.30)$$

Where

K = the value obtained for rectangular section in chapter 3

Equating the total compressive force to the tensile strength (equilibrium of forces) gives the area of tension steel.

$$A_{st} = \frac{C_1 + C_2}{f_y / \gamma_s} = \frac{1.15 p_0 b_w d f_{ck}}{f_y} + 0.518(b_f - b_w)t \frac{f_{ck}}{f_y} \qquad (4.31)$$

Close Approximation and Reasonable Idealization

A uniform rectangular stress distribution is taken over the flange neglecting the projection of the web below the web flange. The parabolic portion of the stress block is acting on small width and the lever arm distance of this to the steel is small when compared to the compression on the flange potion. This idealization may be considered as on the safer side and estimates the capacity of the section slightly lesser than the real value. With this idealization, the compressive force on the section (same as on the flange) and the moment capacity of the section are:

$$C_d = 0.45 \, b_f t f_{ck} \qquad (4.32)$$

$$M_r = 0.45 b_f t (d - 0.5t) f_{ck} \qquad (4.33)$$

The effective depth of the section is obtained by equating the limit moment to the moment capacity of the section:

$$d = 0.5t + \frac{M_c}{0.45 b_f t f_{ck}} \qquad (4.34)$$

And the corresponding area of tension reinforcement is:

$$A_{st} = \frac{1.15 M_c}{(d - 0.5t) f_y} \qquad (4.35)$$

The above two expressions holds good if the neutral axis falls in the web. In case the thickness of the flange is large, then the neutral axis lies in the flange and the sectional moment capacity is equal to that of rectangular section with width of the section equal to the width of the flange.

$$M_r = K b_f d 2 f_{ck}, \; for \; x_u < t \qquad (4.36)$$

An approximate relation of the neutral axis with respect to the thickness of the flange can be established by considering one of the expressions for moment capacity.

Let $\qquad\qquad x_u = k_u d > 1.15 \, t$

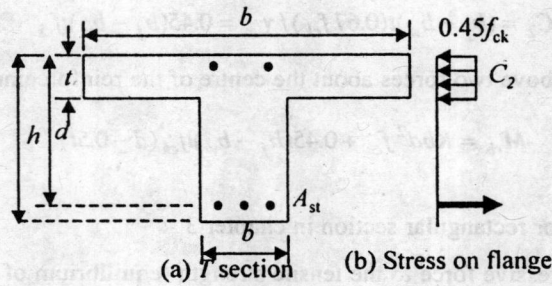

(a) *T* section   (b) Stress on flange

Fig. 4.14 T Beam; Approximate Stress Profile.

For reinforcement *HYSD-Fe*415, the value of the neutral axis coefficient is 0.479; then the relation between the thickness and the effective depth is:

$$d > 2.4\,t \tag{4.37}$$

This relation gives only a guideline about the possibility of neutral axis falling in the web in flanged sections.

Example 4.6: *T* Section Simply Supported Beam

A 8.4 m effective span simply supported beam is supporting a slab of 120 mm thickness. The beams are spaced at 4 m apart. The slab carries a uniformly distributed load of 4 kN/m². The beam is in protected environment.

Solution

The given data and the assume information for preliminary calculation are:
Effective span = $L = 8.4$ m, $t = 120$ mm, $w_l = 4$ kN/m², $f_{ck} = 20$ N/mm²
The superimposed load from the weight of the slab is: $w_s = 0.125*4*25 = 12.5$ kN/m
The superimposed live load from the slab is: $w_1 = 4*4 = 16$ kN/m

The width of the flange is the minimum of the following:

b_f = Centre to centre distance of the beams = 4 m = 4000 mm, or
$b_f = L_o/6 + b_w + 6t = 2.32$ m

Assume the size of the web, excluding the thickness of the slab as 300 by 500 mm for self-weight purpose.

The assumed self-weight = $w_g = 0.3(0.5)*25 = 3.75$ kN/m, use 4 kN/m as self-weight.
Total design load on the beam is = $w_t = 1.5(w_g + w_s + w_l) = 47.75$ kN/m, use 48 kN/m
Maximum bending moment that occurs at mid span is

$$M_c = \frac{w_t L^2}{8} = \frac{48(8.4)^2}{8} = 436.36$$

The effective depth of the section that is required is given by:

$$d = 0.5t + \frac{M_c}{0.45 b_f t f_{ck}} = 0.06 + \frac{436.36(1000)}{0.45(2.3)(0.12)(20)(10)^2} = 0.24 \text{ m}$$

Even though 240 mm effective depth is adequate from the compression force point of view. It is better use the effective depth in the range of span/16 to minimize the reinforcement requirement, the deflections and crack width. Let the overall depth including the thickness of the slab be 600 mm and the effective depth as 560 mm.

$$\text{Use } h = 600 \text{ mm and } d = 560 \text{ mm}$$

The area of tension reinforcement needed is:

$$A_{st} = \frac{1.15 M_c}{(d - 0.5t) f_y} = \frac{1.15(436.36)10^6}{(560 - 60)415} = 2418 \text{ mm}^2$$

Provide 5 numbers of 25 mm bars at the bottom as the main reinforcement. And 2 numbers of 10 mm at top.

The area of reinforcement provided is $A_{st} = 5*490 = 2450$ mm^2

Two out of the five bars are curtailed. Assume 25 mm cover on at the side faces and all the five bars are arranged in a row, then the spacing between the bars is = [(300 –50) –5*25]/4 = 31.5 mm. The minimum clear spacing between the bars should be equal to the diameter of the bar or size of the aggregate plus 6 mm. The size of the aggregate is 20 mm, therefore, the clear spacing is adequate. It is advisable to provide about 40 mm spacing for better concreting. The percentage of reinforcement with respect to the web is:

$$p = \frac{2450*100}{300*560} = 1.45\%$$

Design for Shear Force

The design for shear is similar to that of rectangular beam. The critical shear force occurs at a distance of effective depth of the reinforcement from the support. The design shear force at a distance d from the support is:

$$\text{Nominal shear stress} = \tau_v = \frac{V_u}{b_w d} = \frac{1.5(w_t)(L/2-d)}{b_w d} = \frac{1.5(48000)(4.2-0.56)}{300(560)} = 1.56 \text{ N/mm}^2$$

The percentage of reinforcement of three bars that are continued is 0.8%
The admissible shear strengths of concrete without and with shear reinforcement from table 3.4 are:

$$\tau_c = 0.58 \text{ and } \tau_{cmax} = 2.8 \text{ N/mm}^2$$

Select two leg 10 mm stirrups, then the spacing of the stirrups is:

$$s_v = \frac{0.87 A_{sv} f_{yv}}{(b_w)(\tau_v - \tau_c)} = \frac{0.87(157)(415)}{300(1.56-0.58)} = 141 \text{ mm}$$

Provide the 10 mm stirrups at 140 mm spacing.

The design was based on the approximate moment capacity of concrete section. However the design for shear is for the total collapse of the section. The actual moment capacity of the concrete section can be obtained from the Eq. (4.30) and it is:

$$M_{rb} = Kbd^2 f_{ck} + 0.45(b_f - b_w)tf_{ck}(d - 0.5t)$$

$$M_{rb} = 0.138(0.3)(0.56)^2(20) + 0.45(2.3 - 0.3)0.12(20)(0.56-.06) = 0.26 + 1.08 = 1.34 \text{ MNm}$$

The effective depth required is only 0.24 m while the that selected is 0.56 m, therefore the concrete capacity of the section works out to be far higher than that really needed. The failure is controlled by the tension in the reinforcement. Providing larger concrete capacity than that of the reinforcement doesn't enhance the capacity of the section. In case one uses the effective depth as 0.24 m, the moment capacity based on the more accurate expression is = 0.047 + 0.39 = 0.437 MNm as against 0.4364 MNm from the approximate expression. Fig. 4.15 illustrates the section.

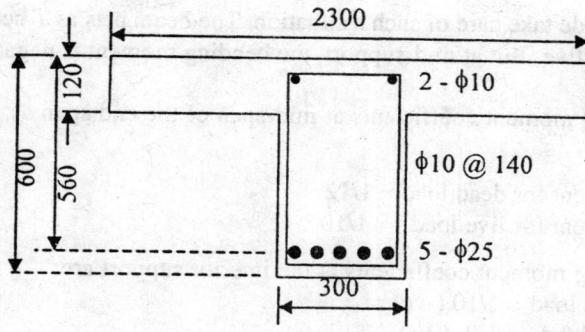

Fig. 4.14 T Section, Example 4.6.
(All dimension in mm)

Example 4.7: Continuous beam design

A two-span continuous beam of 12 m span each spaced at 3.5 m is supporting a slab of 150 mm thickness. The slab is designed to carry a live load of 4 kN/m². The beam is exposed to severe environment. (Fig. 4.16)

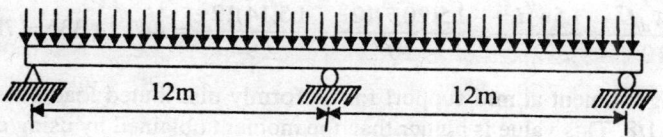

Fig. 4.16 Two-span Continuous Beam, Example 4.7.

Solution

The data and the other information on the problem is:

Two-span continuous beam, each span = L = 12 m, slab thickness = t = 150 mm, spacing of beams = 3.5 m

Live load = w_l = 4 kN/m², exposure condition = severe, therefore use $M30$ concrete, f_{ck} = 30 N/mm², Reinforcement used = $HYSD$-$Fe415$; f_y = 415 N/mm²

Select the the width of the web as 400 mm, and

depth of web as = $L/16$ = 0.75 m (for self weigth purpose only)

Self weight of the web = 0.4*0.75*25 = 7.5 kN/m

Weight of the slab on each beam = 0.15*3.5*25 = 13.2 kN/m

Live load on the slab to the beam = w_l = 3.5*4 = 14 kN/m

Total dead load = w_d = 7.5 + 13.2 = 20.7 kN/m

The two-span continuous beam has maximum negative bending moment (magnitude) at mid-support. The maximum positive *BM* is not exactly at mid-span but close enough to idealize at the mid-span. One can perform structural analysis of two-span continuous beam and come out with the actual moments and shear forces on the beam. However, the code recommends design bending moment and shear force coefficients. During the construction stage, the continuity of the beam may not exist so the moments recommended by the code take care of such a situation. The beam acts as T beam at mid-span where the bending moment is positive. But at mid-support, the bending moment is negative so beam behaves as a rectangular section.

The *positive bending* moment coefficients at mid-span of the end span as given by the *IS:456-2000* are:

Moment coefficient for dead load = 1/12
Moment coefficient for live load   = 1/10

The *negative bending* moment coefficients at the interior support are:
That due to dead load = 1/10 (-ve)
That due to live load = 1/9 (-ve)

Design of the Section at Middle Support

The design bending moment at mid-span after multiplying with the partial load factor of 1.5 is:

$$M_b = \frac{1.5w_dL^2}{12} + \frac{1.5w_lL^2}{10} = \frac{1.5(20.7)12^2}{12} + \frac{1.5(14)12^2}{10} = 372.6 + 302.4 = 675 \text{ kNm}$$

Maximum negative (numerical value only) bending moment at the middle support is:

$$M_c = \frac{1.5w_dL^2}{10} + \frac{1.5w_lL^2}{9} = \frac{1.5(20.7)12^2}{10} + \frac{1.5(14)12^2}{9} = 447.1 + 336 = 783.1 \text{ kNm}$$

The actual bending moment at mid support for uniformly distributed load over the span is negative and the coefficient is 1/8. This value is higher than the moment obtained by using coefficients suggested by the code. This implies that certain redistribution of the moments is incorporated; otherwise it appears on the unsafe side at the support.

The section at the middle support behaves like a rectangular section and so equating the factored moment to the balanced moment capacity of a rectangular section gives:

$$M_{rb} = Kbd^2f_{ck} \geq 783.1 = M_c , \text{This leads to}$$

$$d \geq \sqrt{\frac{783.1(10)^6}{0.138(400)(30)}} = 688 \text{ mm}$$

The effective depth of balanced section is 688 mm. It is desirable to make the beam as an under-reinforced section. Further, the shear stress at the intermediate support is also to be reasonable. Therefore select the effective depth of the beam as 720 mm and the overall depth as 800 mm. It is anticipated that the reinforcement will be placed in two layers at the top face. Use $d = 720$ mm, $h = 800$ mm, then the web depth is 800-150 = 650 mm. This value is close to the one that is assumed for self-weight purpose.

The reinforcement required at the support to be placed on the top fibre is:

$$A_{st} = \frac{1.15M_c}{jdf_y} = \frac{1.15(783.1)(10)^6}{0.8(720)(415)} = 3768 \text{ mm}^2$$

Provide 7 numbers of 25 mm bars and 2 numbers of 16 mm bars. The total area provided is 3832 mm². Two numbers of 16 mm bars are continued from end to end and the remaining bars can be curtailed at the appropriate location. The bars should extend to a length of 0.33 times the span beyond the face of the support. The bar can be 4 m on either side of the support.

The shear force coefficients at the interior support next to the outer support are: 0.6 for both live and dead loads.

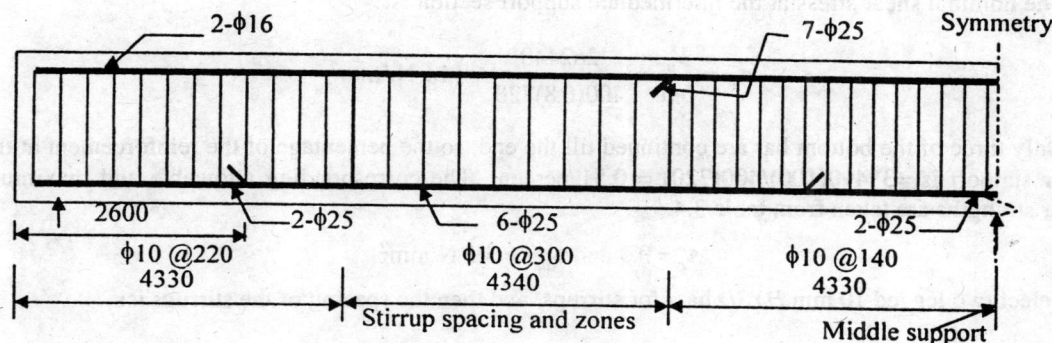

Fig. 4.17 Longitudinal Section of Reinforcement Details, Example 4.7.

The design shear force at the middle support is:

$$V_u = 0.6(1.5)(w_d + w_l)L = 0.6(1.5)(20.7 + 14) = 374.76 \text{ kN}$$

The nominal shear stress at the intermediate support section is:

$$\tau_v = \frac{V_u}{bjd} = \frac{376760}{400(0.8)720} = 1.64 \text{ N / mm}^2$$

The percentage of reinforcement is 1.33, therefore the allowable and maximum admissible shear strengths taken from table 3.4 are:

$$\tau_c = 0.72 \text{ and } \tau_{max} = 3.5 \text{ N/mm}^2$$

Select two legged 10 mm *HYSD-Fe*415 bars for stirrups, then the spacing of the stirrups is:

$$s_v = \frac{0.87 A_{sv} f_{yv}}{(b)(\tau_v - \tau_c)} = \frac{0.87(157)(415)}{400(1.64 - 0.72)} = 144 \text{ mm}$$

The maximum spacing of the stirrup is only 300 mm; *so provide the 10 mm stirrups at 140 mm spacing in the one third of span on either side of the middle support.*

Design of the Section at the Mid-span of the Beam

The area of tension reinforcement at mid-span computed for the *T*-beam is:

$$A_{st} = \frac{1.15 M_b}{(d - 0.5t) f_y} = \frac{1.15(675)10^6}{(720 - 75)415} = 2900 \text{ mm}^2$$

Provide 6 numbers of 25 mm bars at the bottom as the main reinforcement. The area of reinforcement provided is 2940 mm². Continue 3 of the six bars from end to end and curtail the remaining 3 bars at 0.2L from each end.

The design shear force at the outer support (including the partial safety factor) from the code is:

$$V_u = 1.5(0.4\,w_d + 0.45\,w_l)\,L = 1.5(0.4\,(20.7) + 0.45(14))12 = 262.44\ \text{kN}$$

The nominal shear stress at the intermediate support section is:

$$\tau_v = \frac{V_u}{bjd} = \frac{262440}{400(0.8)720} = 1.14\ \text{N/mm}^2$$

Only three of the bottom bar are continued till the end, so the percentage of the reinforcement at the outer support is $=3*490(100)/400(720) = 0.51$ percent. The corresponding allowable and maximum shear strengths are taken from table 3.4.

$$\tau_c = 0.5\ \text{and}\ \tau_{max} = 3.5\ \text{N/mm}^2$$

Select two legged 10 mm *HYSD* bars for stirrups, and then the spacing of the stirrups is:

$$s_v = \frac{0.87 A_{sv} f_{yv}}{(b)(\tau_v - \tau_c)} = \frac{0.87(157)(415)}{400(1.14 - 0.5)} = 221\ \text{mm}$$

Provide the 10 mm stirrups at 220 spacing up to one third of span from the end support. Provide the same stirrups at 300 mm spacing in the middle one third.

The longitudinal reinforcement details of one span are shown in Fig. 4.17. The effective span of the beam is 12 m so the total gross length of the span of the beam is 13 m. The reinforcement details are given for 13 m span, where the effective span is only 12 m

The reinforcement is set symmetric with respect to the mid-support. The cross sectional details at mid span and at the-mid support are shown in Fig. 4.18.

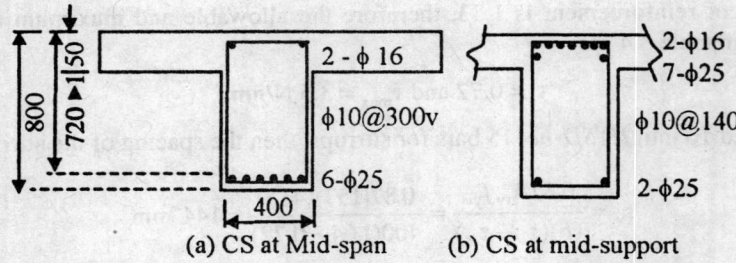

(a) CS at Mid-span   (b) CS at mid-support

Fig. 4.18 Cross Section of Beam Example 4.7.

PROBLEMS

4.1   Determine the limit moment capacity of a diamond shaped cross section with its side *a*, and included angle 90°. The depth of reinforcement from the apex is 80% of the overall depth. The beam is exposed to moderate environment, so use *M25* concrete and *HYSD-Fe415* reinforcement. The strain limits are same as given by the code.

4.2 A square cross section has a notch cut out in the compression flange as shown in Fig.1. The size of the notch is 1/3 rd of the size of the section. Derive the limit moment capacity of the section for *M25* grade concrete and *HYSD-Fe415* reinforcement. The strain limits are same as given by the code.

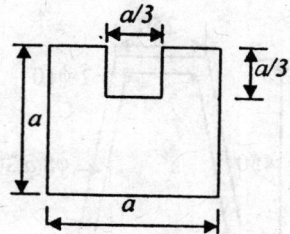

Fig. 1 Problem 4.1.

4.3   Derive the moment capacity of a circular cross section with reinforcement at a distance of 0.8 times the diameter from the top compression fibre. The strain limits are same as given by the code. Use *M*25 grade concrete and *HYSD-Fe*415 reinforcement. Calculate the percentage of reinforcement needed for a balanced cross section.

4.4   A simply supported 4 m beam span is subjected to a live load of 16 kN/m. Design a reinforced concrete inverted triangular cross section beam with top width as 250 mm and the vertical depth as 350 mm. Use *M*30 grade concrete and *HYSD-Fe*415 reinforcement. The exposure condition is severe. Sketch the reinforcement details.

4.5   A 6 m effective span simply supported beam is subjected to a live load of 16 kN/m. Design a reinforced concrete inverted triangular cross section beam with top width as 200 mm and the vertical depth as 500 mm. Use *M*25 grade concrete and *HYSD-Fe*415 reinforcement. The exposure condition is moderate Sketch the reinforcement details.

4.6   A four metre span cantilever is subjected to a live load of 20 kN/m. Design an upright triangular balanced reinforced concrete using *M*20 concrete and HYSDFe-415 reinforcement bars both for bending and shear. Use base width as *b* = 400 mm.

4.7   A 1.6 m wide staircase is made of beam and slab reinforced concrete section. The slab thickness is 100 mm and the web size is 200 mm by 300 mm. The going is four metres and the landings at both ends are 1.6 m each. The live load on the staircase is 4 kN/m². Design the staircase using *M*20 grade concrete and *HYSD-Fe*415 reinforcement bars. The exposure condition is mild.

4.8   A double over-hang beam of total length 10 m and overhangs on each side is 2 m. The beam is supporting a 120 mm thick slab of 4 m width, and the load on the slab is 4 kN/m². Design the reinforcement concrete beam for a web size of 300 mm by 450 mm. Use *M*25 concrete and *HYSD-Fe*500 reinforcement bars. The beam is exposed to mild environment.

4.9   Calculate the limit moment of a *T*-beam with 100 mm slab thick and width 4500 mm. The web size is 250 mm by 400 mm. It is made of *M*25 grade concrete and six numbers of 25 mm bars of *HYSD-Fe*500 grade reinforcement arranged in two layers. The clear spacing between the layers of the reinforcement is 35 mm.

4.10 A reinforced concrete beam section shown in Fig. 2 is made of *M*30 grade concrete and *HYSD-Fe*500 reinforcement bars. Determine the limit moment capacity of the section.

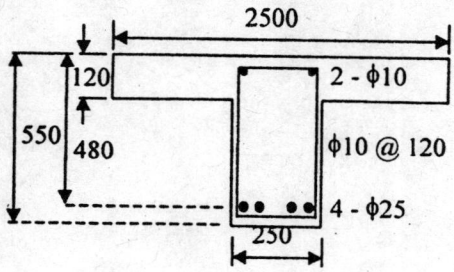

Fig. 2 Problem 4.10.
(All dimension in mm)

4.11 Trapezoidal section shown in Fig. 3 is made of *M*25 concrete and *HYSD-Fe*415 reinforcement. The cover to the reinforcement is 35 mm. Determine the limit moment capacity of the section.

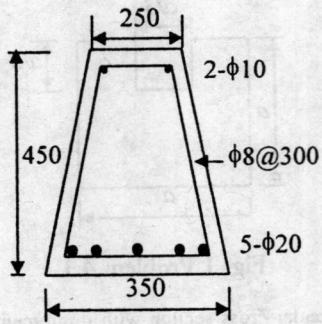

Fig. 3 Problem 4.11.

4.12 An inverted trapezoidal section shown in Fig. 4 is made of *M*25 concrete and *HYSD-Fe*415 reinforcement. The cover to the reinforcement is 35 mm. Determine the limit moment capacity of the section.

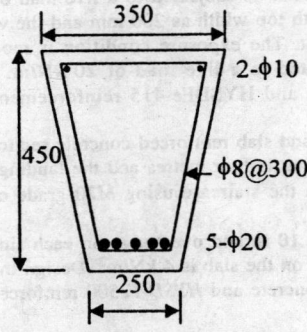

Fig. 4 Problem 4.12.

Design of Prestressed Concrete Beams

5.1 INTRODUCTION

Prestressed concrete construction developed during nineteen-twenties came into practice in forties of last millennium in India. Prestresses are introduced in a member such that the stresses due to loads on the structure are compensated either partly or fully. Concrete is weak in tension but resists the compression. A beam under bending develops tensile stress on one face and compressive on the other. The tensile force in the member is resisted by the embedded reinforcement in reinforced concrete structures. Pre-compression is introduced to compensate the tension in member. The pre-compression can be introduced by different means. Pre-stretched cables at suitable locations in the member are placed and anchored such that the force of the cable when transferred to the member introduces stresses in the member. Pre-strain of the order 0.004 is introduced in stressing cables. The derivation of the moment capacity of the section is similar to that of the reinforced concrete. The crushing strain of concrete is taken as 0.0035 for all grades of concretes except for high performance concrete. The strength of the concrete in prestressed concrete is in the range of 35 N/mm^2 to 60 N/mm^2. The crushing strain limit in the high performance concrete of $M60$-$M120$ is in the order of 0.0040 to 0.0045.

Post-tensioned and *Pre-tensioned* constructions are the two main methods of prestressed concrete. The *post-tensioned* concrete is the one in which the concrete is cast and cured to gain a reasonable strength and then prestressed. Cable ducts are provided in the member to insert (*thread*) high tensile steel cable or wires. The cables are stretched against the hardened concrete and then anchored to the member. Suitable anchors are provided to transfer the cable force to the concrete. The minimum strength of concrete in such construction is 35 N/mm^2. However, strength of 45 N/mm^2 and above is preferred. The *pre-tensioned* concrete is the one in which the high tensile steel wires are pre-stretched against rigid abutments first to a desired value at the appropriate locations. The concrete is placed on the stretched wires in the molds, cured and then the force from the wires are transferred to the hardened concrete. The wires force is transferred to the concrete through bond between the cable and the concrete. The minimum acceptable strength of the concrete in pre-tensioned concrete construction is 45 N/mm^2. Each of the constructions has its merits of application depending on the type of member, available facilities, time etc. The proof strength of high tensile steel wire is in the range of 1500 N/mm^2 to 2000 N/mm^2. The proof strength corresponds to 0.2 percent residual strain. The proof strength of the high tensile bars is in the range of 1100 kN/mm^2. The wires are usually drawn from bars and proof strength of the wire increases with decrease in the diameter of the wire. The wires come in diameters of 8 mm, 7 mm, 5 mm, 4 mm, 3 mm,

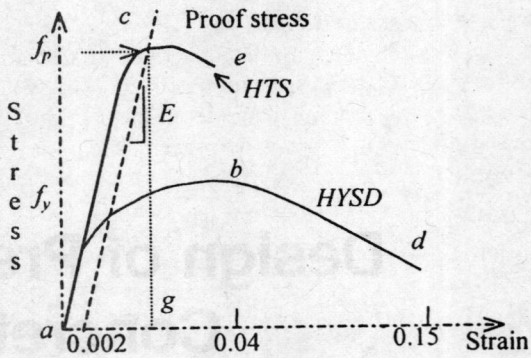

Fig. 5.1 Stress-Strain Curve for Steel.

1.6 mm etc. The wires are bundled into cables and strands of different load capacities. Strands having a capacity 3000 kN are produced for large construction. The transfer of prestress is in the range of 70 percent of the proof strength, and the jacking force is up to 80 percent of the proof strength. Loss of prestress takes place from jacking to transfer. Further loss of prestress takes place from transfer of prestress to the effective prestress at the working load. The loss of prestress is due to the following reasons.

- Loss due to anchorage take up,
- Loss due to elastic shortening,
- Loss or gain due to bending,
- Loss due to creep in concrete,
- Loss due to shrinkage of concrete,
- Loss due to relaxation of steel,
- Loss due to friction and wobble of the cable,

The loss of prestress affects the stresses and deflections of the members and has very little influence on the ultimate strength. The fracture of the high tensile wires is sudden. The failure of prestressed concrete structures is sudden whether it is initiated by steel or by concrete. Nominal non-tensioned reinforcement is provided in most prestressed concrete structures to minimize the secondary effects and reduce bursting failures. The non-tensioned reinforcement is high yield steel. Use of Fe500 or Fe550 as nominal reinforcement or stirrups is not efficient.

Long span girders, Folded plates, shell structures, water tanks, nuclear reactors etc. are commonly built in prestressed concrete. The cables in the post-tensioned construction are not inherently bonded with the concrete. However, grouting of the ducts is done invariably resulting in some bonding of the cables. The primary aim of the grouting of the ducts is to protect the high tensile cables from corrosion. Figure 5.1 illustrates the stress-strain curve of high yield steel deformed (HYSD) and high tensile steel (HTS).

5.2 DESIGN MOMENT CAPACITY OF RECTANGULAR SECTION

The concrete is assumed homogeneous and isotropic. At a microscopic level this is not a valid assumption but, acceptable at macroscopic level. This implies that the section as a whole behaves like a homogeneous material for all practical purposes even though concrete is matrix of aggregate and mortar. Concrete resists compression while the reinforcement resists all the tensile force. Concrete is brittle and has a maximum compressive strain in bending as 0.0035.

Plane sections before bending remain plane even after bending. This assumption holds well up to collapse of the section in slender members. The deformation of slender members is dominated by the bending. Adequate bond or anchorage is assumed available in all the sections.

The bending capacity of a section for simultaneous occurrence of the limiting strains in concrete and steel is called *balanced moment capacity*. Such a section is called *balanced section*. The area of rein-forcement corresponding to it is called *balanced steel*.

Consider a rectangular section of width b and depth h. Fig. 5.2(a) illustrates the notation. The nota-tions are:

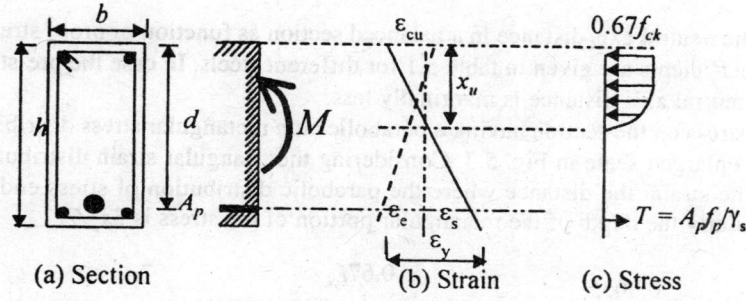

| (a) Section | (b) Strain | (c) Stress |

Fig. 5.2 Section under Bending, Notations.

A_p = area of tensioned steel,

A_{pb} = area of tensioned steel of a balanced section,

ε_{cu} = crushing strain in concrete = 0.0035,

ε_e = strain in steel caused prestressing, it is normally taken as = 0.004

ε_s = strain in steel caused by bending moment at failure,

ε_y = stain at proof stress in steel, (yield strain in steel) =$0.002 + f_p/E_s \gamma = 0.002 + f_p/E(1.15)$,

d = depth of centroid of steel from extreme compression fibre. It is simply called *effective depth of steel*,

T = design tensile capacity of tensioned steel,

C_d = design compressive capacity of concrete section,

x_u = depth of the compression zone or it is also called as depth of compression block,

x_c = distance of the centroid of the total compression force on concrete from top fibers,

E_s = modulus of elasticity of steel and it is normally taken = 200,000 N/mm^2

f_p = yield or proof strength (stress) of steel

$f_{p\gamma_s}$ = partial safety facted applied to steel.

The *neutral axis* distance in a balanced cross section can be obtained from the similar triangles of the strain as shown in Fig. 5.2(b).

$$\frac{\varepsilon_{cu}}{\varepsilon_s} = \frac{\varepsilon_{cu}}{\varepsilon_y - \varepsilon_e} = \frac{x_u}{d - x_u} \qquad (5.1)$$

The rearrangement of the above equation gives:

$$x_u = \frac{\varepsilon_{cu} d}{\varepsilon_{cu} + \varepsilon_y - \varepsilon_e} = \frac{0.0035 d}{0.0035 + 0.002 + f_p / 1.15 E_s - 0.004}$$

The expression after substitution of the modulus of elasticity of steel reduces to:

$$x_u = \frac{805d}{345 + f_p} = k_u d \qquad (5.2)$$

Where

$$k_u = \frac{805}{345 + f_p}$$

The Eq. (5.2) gives the neutral axis distance in a balanced section as function of proof stress. The values of the neutral axis coefficients are given in table 5.1 for different steels. In case the pre-strain in steel is less than 0.004, the neutral axis distance is marginally less.

The compressive stress on the section having a parabolic cum rectangular stress distribution is shown in Fig. 5.2(c) and to enlarged scale in Fig. 5.3. Considering the triangular strain distribution, and from similar triangles of the strain, the distance where the parabolic distribution of stress ends is $0.002\, x_u / 0.0035 = 4x_u/7$. Similarly the depth of the rectangular portion of the stress is $3x_u/7$.

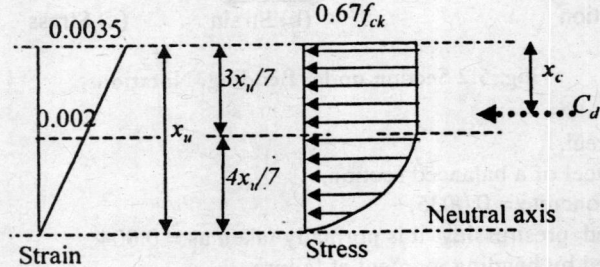

Fig. 5.3 Strain & Stress Distribution in Compression Zone.

The considering the areas of the parabolic and the rectangular portions of the stress distribution, the corresponding total design forces are:

$$C_p = (0.67f_{ck})(b)(4x_u/7)(2/3)/1.5 = 0.17\, b\, x_u f_{ck}$$

$$C_r = (0.67 f_{ck})(b)(3x_u/7)/1.5 = 0.19\, b\, x_u f_{ck}$$

The total design compression force, which is equal to the sum of the above two is:

$$C_d = C_p + C_r = 0.36\, b\, x_u f_{ck}$$

Let (x_c) = distance of the compressive force from the top fibre. It is obtained by taking the moment of the force about the top fibre.

$$C_d(x_c) = C_p((5/8)(4\, x_u/7) + 3\, x_u/7) + C_r(3\, x_u/14)$$

This gives the distance of the compression force as:

$$x_c = 0.42\, x_u$$

The distance between the design compressive and tensile forces is called *lever arm* and it is:

$$jd = d - x_c = d - 0.42x_u = (1 - 0.42k_u)\,d \tag{5.3}$$

Where $j =$ is the ratio of the lever arm distance to the effective depth of steel and it is:

$$j = 1 - 0.42k_u \tag{5.4}$$

The values of lever arm distance coefficient, j are listed in table 5.1.
The corresponding design tensile force from the steel reinforcement is:

$$T = A_p f_p / \gamma_s = A_p f_p / 1.15 = 0.87\, A_p f_p$$

The equilibrium of forces on the section in the horizontal direction gives: $T = C_d$

$$A_p f_p / 1.15 = 0.36\, bx_u f_{ck} = 0.36\, k_u\, bd\, f_{ck}$$

Table 5.1 Design Moment & Neutral Axis Coefficients.

f_p (N/mm²)	K	j	k_u	P_o
1400	0.133	0.80	0.46	0.19
1500	0.128	0.81	0.44	0.18
1600	0.123	0.83	0.41	0.17
1650	0.119	0.83	0.40	0.16
1700	0.116	0.84	0.39	0.16

Note: All fractions rounded off to the second decimal. Moment capacity rounded down and lever arm rounded up to be on the safer side of design.

The above equation can be rewritten as ratio of reinforcement factor for balanced section and it is:

$$\frac{A_{pb}}{bd} = 1.15(0.36)k_u\,\frac{f_{ck}}{f_p} = p_o\,\frac{f_{ck}}{f_p} \tag{5.5}$$

Where

$$p_o = \frac{333.27}{345 + f_p} \tag{5.6}$$

The value of p_o is an indication of percentage of reinforcement required in a balanced section. These values are listed in table 5.1. The value is in the range of 0.165 to 0.182, a narrow band. The percentage of steel with respective the effective area of cross section is:

$$p = \frac{100\,A_{pb}}{bd} = 100\,p_o\,\frac{f_{ck}}{f_y} \tag{5.7}$$

The Indian code of practice for prestressed concrete on structures recommends maximum amount of high tensile steel so that the beams are not over-reinforced. The admissible maximum reinforcement ratio is given by:

$$\frac{A_p}{bd} \le 0.24\,\frac{f_{ck}}{f_p} \tag{5.8}$$

The balanced moment capacity of the rectangular section is:

$$M_{rb} = C_d jd = Kbd^2 f_{ck} \qquad (5.9)$$

Where

$$K = 0.36 k_u j \qquad (5.10)$$

Practical or architectural considerations may require the depth of the beam to be different from the balanced depth. If the depth provided is more than the balanced depth of a section, then the steel will reach the yield strain before the concrete reaches the crushing strain. Such a section is called under-reinforced section. The *under-reinforced section* is the one in which the yielding of the steel takes place before the strain in the concrete reaches its crushing value. The failure of the prestressed concrete structures is mostly due to the crushing of the concrete. The crushing of concrete due to excess curvature caused by large strain in steel is called secondary failure. A section in which the crushing of concrete occurs before the yielding of steel, is called *over-reinforced section*. Failure of a structure must be avoided at all costs, and yet there exist a probability of failure so it is desirable to have warning before a failure. Therefore most structures are designed as under-reinforced.

Influence of the pre-strain of steel on the sectional capacity. Effective pre-strain in steel in prestressed concrete is of the order of 0.004. In most cases, the concrete fibres may be subjected to negligible pre-compression in comparison with that of the crushing strain. Except in axially prestressed members, the pre-compression in the concrete has little influence on the limiting strength of the section. It is desirable to account for the pre-compressive strain in concrete in axially pressed concrete sections. There may be some situations where the pre-strain in concrete at the ultimate compression zone can be tension. There is an extra margin of compression strain limit in computing the limiting strength of the section. The design of prestressed concrete beam for bending is illustrated first. Design of a beam for shear, anchorage, serviceability design is discussed later.

Example 5.1: Simply Supported Beam-balanced Section Design for Moment

A 8 m effective span simply supported beam in moderate environment is subjected to a superimposed uniformly distributed load of 30 kN/m over the span. The beam is pretensioned type. Design a rectangular section beam with width of the beam as 300 mm. 7 mm *HTS* wires have proof strength of 1500 MPa.

Solution

The minimum grade of concrete for pretensioned type concrete beam is *M*45 and is adequate for moderate exposure condition. The design data is: $L = 8$ m, $f_{ck} = 45$ N/mm^2, and $w_l = 30$ kN/m. The design coefficients from table 5.1 are: $K = 0.128$, $j = 0.81$

Let the total depth of the beam is 600 mm for the purpose of estimating the self-weight.
The self-weight is $= w_g = 0.3*0.6*25 = 4.5$ kN/m, use 5 kN/m to start with.
The factored limit load of collapse on the beam is $= w_t = 1.5(w_g + w_l) = 1.5(5 + 30) = 52.5$ kN/m
The maximum bending moment occurs at the mid-span, and it is

$$M_c = \frac{w_t L^2}{8} = \frac{52.5(64)}{8} = 420 \text{ kNm}$$

The moment capacity of the section must be more than the collapse moment. This leads to:

$$M_r = Kbd^2 f_{ck} \geq M_c = 420 \text{ kNm}$$

Let 7 mm *HTS* wires are used, the proof strength of the wires is 1500 N/mm^2.

$$d = \sqrt{\frac{M_c}{Kbf_{ck}}} = \sqrt{\frac{420(10)^6}{0.128(300)(45)}} = 493 \text{ mm}$$

Let the effective depth of the beam be 500 mm. So the self-weight assumed is close and on the safer side. The actual depth of the beam is finalized after the choice of the wires and detailing. The area of tensioned steel is:

$$A_p = \frac{1.15 M_c}{jdf_c} = \frac{1.15(420)^6}{0.81(500)(1500)} = 795 \text{ mm}^2$$

Provide 21 numbers of 7 mm wires in three rows. Area of steel provided is 808 mm^2. The minimum spacing of the wires is 25 mm for practical consideration of anchoring. Provide 8 wires in bottom two rows, spaced at 30 mm. The side cover available to the centre of the 7 mm wire is (300-7*30)/2 = 45 mm. The minimum clear cover required in the exposed environment is 30 mm. Arrange the balance of the (21-16) = 5 wires in the third row from the bottom. Each row is spaced at 30 mm. The centroid of the three rows of the wires can be obtained by taking the moment of the area of the wires about the bottom row. And its distance from the centre of the bottom row is = (5*60 + 8*30)/(21) = 26 mm. The minimum cover required to the wires in the lowest row of wires is 30 mm. Provide a cover of 40 mm to the last row of wires. The minimum overall depth required for the beam is equal to the effective depth + distance to the outer most row of wires + half the diameter of the wire + the clear cover.

The overall depth = 500 + 26 + 7/2 + 40 = 569.5 mm. Use $h = 570$ mm.

Fig. 5.4 illustrates the cross section of the beam that is required to resist the bending moment acting on the section. The design of the beam for shear is discussed later.

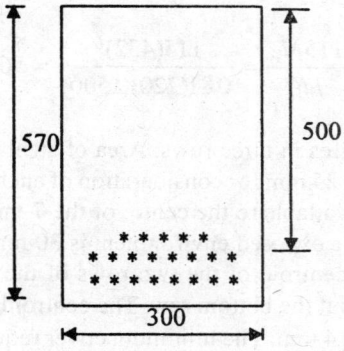

Fig. 5.4 Cross Section Example 5.1.

Example 5.2: Design Simply Supported Under-reinforced Beam Section

Eight metre simply supported beam is subjected to a superimposed uniformly distributed load of 30 kN/m over the entire span. The beam is exposed to moderate environment and pretensioned. The size of the beam is prefixed as 300 mm by 800 mm. Design the section with 7 mm *HTS* wires having a proof strength of 1500 MPa.

Solution

The minimum grade of concrete for pretensioned concrete construction is *M*45 and adequate for moderate exposure. The design data is: $L = 8$ m, $f_{ck} = 45$ N/mm^2, and $w_l = 30$ kN/m. The design coefficients from table 5.1 are: $K = 0.128$, $j = 0.81$

$$\text{The self-weight is} = w_g = 0.3*0.8*25 = 6 \text{ kN/m},$$

The limit load of collapse on the beam is = $w_t = 1.5(w_g + w_l) = 1.5(6 + 30) = 54$ kN/m
The maximum bending moment occurs at the mid-span at collapse, and it is

$$M_c = \frac{w_t L^2}{8} = \frac{54(64)}{8} = 432 \text{ kNm}$$

The moment capacity of the section must be more than the collapse moment. This leads to:

$$M_r = Kbd^2 f_{ck} \geq M_c = 432 \text{ kNm}$$

Let 7 mm *HTS* wires are used in the design and the proof strength of the wires is 1500 N/mm^2.

$$d = \sqrt{\frac{M_c}{Kbf_{ck}}} = \sqrt{\frac{432(10)^6}{0.128(300)(45)}} = 500 \text{ mm}$$

The effective depth for a balanced section is 500 mm whereas the overall depth of the beam is already fixed at 800 mm. The effective depth can be assumed as about 720 mm. The actual cover will depend on the number of rows of tensioned *HTS* wires. The area of tensioned steel with effective depth as 720 mm is:

$$A_p = \frac{1.15 M_c}{jdf_c} = \frac{1.15(432)^6}{0.81(720)(1500)} = 568 \text{ mm}^2$$

Provide 15 numbers of 7 mm wires in three rows. Area of steel provided is 577 mm^2. The minimum admissible spacing of 7 mm wires is 25 mm for consideration of anchoring. Provide 8 wires in bottom row spaced at 30 mm. The side cover available to the centre of the 7 mm wire is (300-210)/2 = 45 mm. The minimum clear cover required in the exposed environment is 30 mm. Arrange the balance 7 wires in the second row from the bottom. The centroid of the two rows of the wires can be obtained by taking the moment of the area of the wires about the bottom row. The centroid distance of the wires from the centre of the bottom row is = (7*30/15) = 14 mm. The minimum cover required to the wires in the lowest row of wires is 30 mm. provide a cover of 40 mm to the bottom row of wires. The actual effective depth available now is equal to the over all depth – cover of 40 mm – diameter of wires (7 mm) and the distance of the

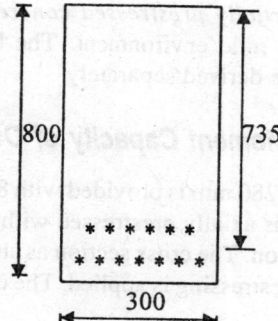

Fig. 5.5 Cross Section; Example 5.2.

centroid of the wires from the bottom row. The effective depth available is: The effective depth = 800 – 40 - 7 – 14 = 739 mm.

$$\text{Use } d = 735 \text{ mm}$$

This value is more than the assumed.

The moment capacity of the section can be based on the area of the tension steel, as the section is under-reinforced. The moment capacity of the section is then given by:

$$M_r = A_p f_p jd / 1.15 = 577*1500*0.81*735/1.15(10^6) = 448 \text{ kNm}$$

The enhanced capacity of the section is 448 kNm in place of the required value of 432 kNm, a marginal increase.

The design for shear, serviceability, anchorage and detailing is to be done to make the design complete.

5.3 TYPES OF PRESTRESSED CONCRETE STRUCTURES BASED ON SERVICEABILITY

Prestressed concrete structures may be classified into four groups based on the serviceability criteria. However the classification doesn't influence the design of limit state of strength. The four groups are:

- *Type 1 Structures*: Concrete shouldn't be subjected to tensile stress in type one structures. Such structures are considered to be important or exposed to aggressive environment. The exposed condition can be considered as very severe or extreme class. Important road bridges, railway bridges, some nuclear related structures come under this class.
- *Type 2 Structures*: Tensile stress in concrete is admissible but no cracking is allowed. Concrete is capable of resisting small tensile stress, and it is only a fraction of the allowed compressive stress. The codes of practice specify the allowable tensile stress. Important structure such as bridge girders; liquid retaining tanks come under this class.
- *Type 3 Structures*: Tension cracking in concrete is allowed with a limitation on pseudo tensile stress. Structures exposed to mild or even moderate environment come under this classification. High tensile steel is protected. All possible precautions must be taken against corrosion.
- *Type 4 Structures*: Some structures may be provided non-tensioned high yield deformed bars to resist part of the tension in addition to the high tensile tensioned steel. Such structures with tensioned and

non-tensioned steel are called *partially prestressed concrete structures*. These structures are not common but could be used in the mild environment. The limit state strength of the partially prestressed concrete sections has to be derived separately.

Example 5.3: Determination of Moment Capacity of Over-reinforced Section

A prestressed concrete section 100 by 280 mm is provided with 8 numbers of 5 mm pretensioned wires on each face of the section. The section is axially prestressed with an effective prestress of 200 kN. Determine the moment capacity of the section. The cross section as shown in Fig. 5.6 is subjected to reversible bending moment therefore an axial prestressing is applied. The concrete is grade $M50$ and the proof stress of HTS is 1600 N/mm^2.

Solution

The basic data of the problem is size $= b = 100$ mm, $h = 280$ mm, area of tension steel on one face $= A_{st} = 8*19.6 = 156.8$ mm^2, $f_{ck} = 50$ N/mm2, $f_p = 1600$ N/mm^2, and estimated $d = 220$ mm. The design property coefficients taken from table 5.1 are $K = 0.122$, $j = 0.83$, $k_u = 0.40$, $p_0 = 0.17$.

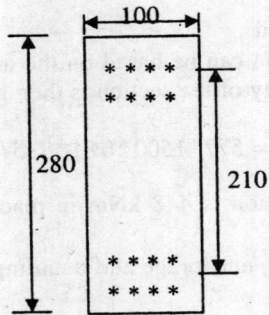

Fig. 5.6 Section of Example 5.3.

The balanced design compressive strength of concrete section and the tensile strength of steel provided are:

$$C = 0.36bk_u df_{ck} = 0.36*0.1*0.44*0.22*50 = 0.1742 \text{ MN} = 174.2 \text{ kN}$$

$$T = A_{st}f_p/1.15 = (156.8*1600/1.15)/1000 = 218.2 \text{ kN}$$

The modulus of elasticity of the concrete is: $E_c = 5000\sqrt{50} = 35355$ N/mm^2

The design compressive capacity of the concrete is less than the design capacity of the pretensioned steel on one face. Therefore the design is controlled by the primary compression failure of the concrete. The design moment capacity of the section is that of the concrete balanced section.

The design strength of the section is =

$$M_r = M_{rb} = Kbd^2f_{ck} = 0.123*100*220^2*50/10^6 = 29.8 \text{ kNm}$$

The compressive strain in concrete due to the effective prestressing is:

$$\varepsilon_c = \frac{P_e}{A_c E_c} = \frac{200000}{100*280*35355} = 0.0002$$

The pre-compressive strain in concrete is 0.0002 when compared to 0.0035 of the crushing strain. This will not have any influence on the capacity of the section.

Example 5.4: Determination of Moment Capacity of Under-reinforced Section

A prestressed concrete beam section of 100 by 280 mm is provided with 5 numbers of 5 mm pretensioned wires on each face of the section. The section is subjected to reversible bending moment therefore axial prestressing is applied with an effective prestress of 188 kN. Determine the moment capacity of the section. The cross section is shown in Fig. 5.6. The grade of concrete is $M50$ and the proof stress of HTS is 1600 N/mm^2.

Solution

The basic data of the problem is: size = b = 100 mm, h = 280 mm, f_{ck} = 50 N/mm^2, estimated d = 220 mm.

Area of tension steel on one face = A_{st} = 5*19.6 = 98mm^2, f_p = 1600 N/mm^2, and
The design property coefficients taken from table 5.1 are K = 0.123, j = 0.83, k_u = 0.40, p_0 = 0.17.

The design compressive strength of concrete and the tensile strength of steel are:

$$C = 0.36bk_u df_{ck} = 0.36*0.1*0.44*0.22*50 = 0.1742 \text{ MN} = 174.2 \text{ kN}$$

$$T = A_{st}f_p/1.15 = (98*1600/1.15)/1000 = 136.3 \text{ kN}$$

The design compressive capacity of the concrete is more than the design capacity of the pretensioned steel at one face. Therefore the design is controlled by the tension failure of the steel. Design moment capacity of the section is obtained from the steel capacity and it is:

The design strength of the section is = M_r = Tjd = 136.3*0.83*220/1000 = 24.9 kNm

As the steel yields, member deforms rapidly leading to larger curvature. The strain in concrete increases because of the larger curvature and finally crushing of the concrete takes place. The actual collapse of the section by secondary compression of concrete.

5.4. DESIGN FOR TRANSVERSE SHEAR

The force acting perpendicular to the axis of a member is called *transverse shear*. The shear force causes shear stress on a section of a beam. Figure 5.7(a) illustrates a rectangular cross section. The shear stress distribution of an uncracked cross section is parabolic along the depth of the section, however for the

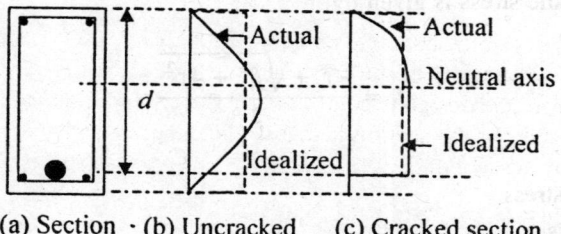

(a) Section · (b) Uncracked (c) Cracked section

Fig. 5.7 Shear Stress Distribution.

purpose of design, an average shear stress is considered as shown in Fig. 5.7(b) by broken lines. In case of cracked section, the variation of the shear stress up to the neutral axis is parabolic with zero stress at the top and then it is uniform till the level of the tension reinforcement. This variation is shown in Fig. 5.7(c). The figure also indicates the average shear stress that is used for design purpose. Shear stress produces diagonal tensile stress, and when the diagonal tensile stress reaches the limiting tension in concrete, the section is considered to have failed in shear.

Consider an infinitesimal element in the shear zone of Fig. 5.8(a). Fig. 5.8(b) illustrates the state of stress on the small element. The maximum principal stress is tensile stress on a plane at 45° with the horizontal for pure shear stress condition. The principal plane may be inclined at an angle less than 45° if the element is also subjected to compressive along with shear stress. The figure also indicates the plane on which the maximum principal stress acts. Fig. 5.8(c) shows the Mohr's circle of stresses. If the principal stress reaches the limiting tensile stress of concrete, the section cracks. The shear capacity of the section depends on whether the section is cracked or uncracked. In a simply supported beam, the maximum shear is at about the support. The critical shear force section is uncracked in simply supported beams. However in cantilever or continuous and frame beams, the maximum bending moment and the shear force occur at about the same location. In such cases the section is cracked.

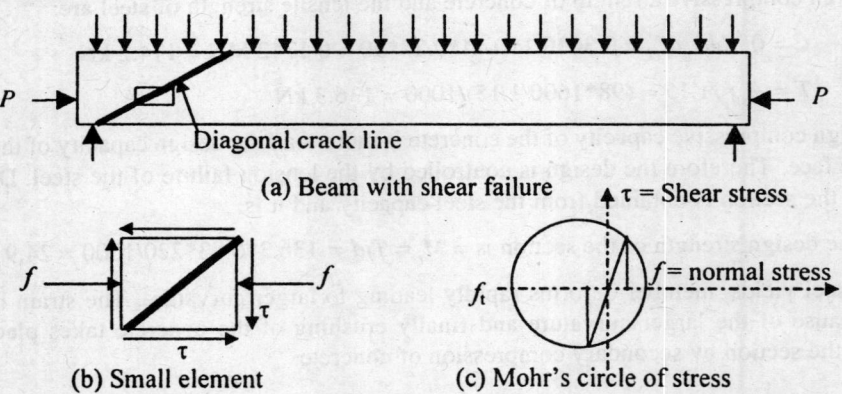

(a) Beam with shear failure

(b) Small element (c) Mohr's circle of stress

Fig. 5.8 Tensile Stress Failure Concept in Shear Design.

Shear Capacity of Uncracked Sections

The failure of the section is based on the principal tensile stress reaching the tensile capacity of the section. The principal tensile stress is given by:

$$f_1 = \frac{-f_c + \sqrt{f_1^2 + 4\tau^2}}{2} \tag{5.11}$$

where
  f_1 = principal tensile stress,
  f_c = compressive stress,
  τ = shear stress
  The rearrangement of the above equation results

$$\tau = \sqrt{f_1^2 + f_1 f_c} \tag{5.12}$$

The section is considered to have failed if the principal stress reaches the tensile capacity of the concrete. The corresponding shear stress of the uncracked section at failure is obtained by substituting the design tensile strength of the concrete in place of the principal stress. That is $f_1 = \overset{?}{f_t}$

where $\qquad\qquad f_t =$ design tensile strength of the concrete $= 0.24\sqrt{f_{ck}}$ $\tag{5.13}$

Further the code recommends that compressive stress at the section is taken as $f_c = 0.8 f_{cp}$. This is because that the compressive stress across the depth vary, so an approximate value is chosen for the purpose of design.

Where

$$f_{cp} = P/A \tag{5.14}$$

The shear failure criterion expression is the obtained from Eq. 5.14 as:

$$\tau_c = \sqrt{f_t^2 + 0.8 f_1 f_{cp}} \tag{5.15}$$

Multiplying the gross area of cross the section by the shear strength and then dividing it by the partial safety factor gives the design shear capacity of the section.

$$V_r = \frac{\tau_c bh}{\gamma_c} = 0.67bh\sqrt{f_1^2 + 0.8 f_{cp} f_t} \tag{5.16}$$

If the factored shear force exceeds the shear capacity, the shear reinforcement is provided. The vertical component of the prestressing cable compensates part of the external shear force. The effective shear force for which the section is to be designed is:

$$V_e = V_u - P_e \sin\theta \tag{5.17}$$

Where

V_u = factored static shear force at the section,
P_e = effective prestress force at the section,
θ = slope of the effective prestressing cable

The prestressing cable profile in beams is usually shallow, so the sine of the slope is approximately equal to the slope of the cable in radians. The effective shear force on the section can be rewritten as:

$$V_e = V_u - P_e \theta \tag{5.18}$$

The actual design of shear reinforcement is explained later.

Shear Capacity of Cracked Section

The shear strength of a cracked section that is subjected to axial force and bending moment is given by

$$V_r = \left[1 - \frac{0.55 f_{pe}}{f_p}\right] bd\tau_c + \frac{M_o V}{M} \tag{5.19}$$

and this value is subject to a minimum of :

$$V_{rmin} = 0.1 bd \sqrt{f_{ck}} \qquad (5.20)$$

Where

f_{pe} = effective prestress,

M_o = bending moment that can produce zero tensile stress in the section = $0.8 f_{pt} \, I/y_d$ (5.21)

y_d = distance of steel from the CG of the section = $d - x_u$ (5.22)

f_{pt} = compressive stress caused by the prestress at depth d

I = moment of inertia of the section,

M = bending moment on the section,

V = shear force at the section = V_e

τ_c = design shear strength of the concrete.

Tension capacity of concrete is also moderated by the amount of tensile reinforcement. Allowable shear capacity of the concrete is a function of the tensile reinforcement and these values are given in table 5.2. The shear reinforcement is called transverse reinforcement or stirrup steel. Further, the shear capacity of concrete even with shear reinforcement is limited to a maximum value. The absolute maximum admissible shear strength of concrete is given in table 5.2. The design criterion for shear can be stated in four cases:

Case 1. No shear reinforcement is needed

No shear reinforcement is required if the effective shear force is equal to or less than the half of the shear strength of the concrete section. Short roof floor slabs, lintels, railway sleepers, even culvert slab may have shear stress less than the half of the capacity.

$$V_e < 0.5 \, V_r \qquad (5.23)$$

Case 2. Nominal shear reinforcement required

Nominal shear reinforcement is required even if the shear force is less than the shear strength of the section, but more than half of the shear strength.

$$0.5 \, V_r \le V_e \le V_r \qquad (5.24)$$

Provide nominal reinforcement subject to the minimum of the nominal reinforcement spacing given below.

$$s_{max} = s_v = \frac{0.87 A_{sv} f_{yv}}{0.4 b} \qquad (5.25)$$

$$< 0.7 d_i \qquad (5.26)$$

$$< 4 b \qquad (5.27)$$

where

A_{sv} = area of cross section of the stirrup steel,

f_{yv} = proof strength of stirrup reinforcement,

d_i = effective depth of the beam for shear

    = h for uncracked section

    = d for cracked section

This expression implies that the reinforcement provided resists a shear stress of 0.4 N/mm².

Table 5.2 Design Shear Strength (τ_c) and the Absolute Maximum Admissible (τ_{cmax}) on concrete (N/mm²)

% steel	M30	M35	M40 + above
	Design shear strength of concrete without shear reinforcement		
0.25	0.37	0.37	
0.50	0.50	0.50	0.38
0.75	0.59	0.59	0.51
1.00	0.66	0.67	0.60
1.25	0.71	0.73	0.68
1.50	0.76	0.78	0.74
2.00	0.84	0.86	0.79
2.50	0.91	0.93	0.88
			0.95
	Absolute Maximum shear strength admissible with shear reinforcement		
$\tau_{cmax.}$	3.5	3.7	4.0 to 4.8

Case 3. In case, the shear force is greater than the shear capacity of the concrete, yet less than the absolute maximum admissible shear capacity with reinforcement; that is:

$$\text{If } V_r \leq V_e \leq V_{rmax} \tag{5.28}$$

Where

V_{cmax} = maximum shear strength of concrete with shear reinforcement = $bd_i\,\tau_{cmax}$
τ_{cmax} = shear strength of concrete with stirrup reinforcement and given in table 5.2.

The shear reinforcement has to be provided to resist the shear force over and above the shear capacity of the concrete section that was already defined earlier. The spacing of the stirrup reinforcement is given by:

$$s_v \leq \frac{0.87 A_{sv} f_{yv} d_i}{V_e - V_r} \tag{5.29}$$

This spacing of the stirrup steel is subject to a maximum as given below

$$s_{max} < 0.75\, d_i, \text{ for } V_e < 1.8 V_r \tag{5.30}$$

$$< 0.5\, d_i, \text{ for } V_e > 1.8 V_r \tag{5.31}$$

$$< 300 \text{ mm}$$

Case 4. The shear force exceeding the capacity of the concrete with stirrup reinforcement

$$\text{If } V_e > V_{cmax} \tag{5.32}$$

Table 5.2 gives the absolute maximum shear strength of concrete with shear reinforcement. If the effective shear force on the cross section exceeds that of the absolute maximum with shear reinforcement, the size of the section or the quality of concrete must to be upgraded.

Example 5.5: Simply Supported Beam-balanced Section-design for Shear

A simply supported beam of effective span 8 m in moderate environment is subjected to a superimposed uniformly distributed load of 30 kN/m over the entire span. The beam is pretensioned. The beam was designed for bending moment in example 5.1. The centroid of the three rows of the wires is at 26 mm from the bottom row. The effective depth $= d = 500$ mm. The profile of the centroid of the wires is parabolic with a sag of $(d - h/2) = 215$ mm. The cross section is shown in Fig. 5.4

Solution

The beam was designed for bending in example 5.1 and the data listed in the problem is summarized here.

$b = 300$ mm, $h = 570$ mm, sag of the cable $= g = 215$ mm, $f_{ck} = 45$ N/mm^2, $f_p = 1500$ N/mm^2,

$P_e = A_p(0.6f_p) = 808(0.6)(1500) = 727,200$ N $= 727$ kN,

The maximum shear force occurs at a distance d from the support. The self-weight of the beam is 4.3 kN/m, but a value of 5 kN/m is used for the purpose of design. The design shear force at a distance d from the support along with a partial safety factor of 1.5 is:

$$V_u = 1.5(w_g + w_l)(L/2 - d) = 1.5(5 + 30)(4 - 0.5) = 183.75 \text{ kN}$$

The component of the prestress force in the vertical direction reduces the effective shear on the concrete section and the effective shear force is:

$$V_e = V_u - P_e \sin\theta \cong V_u - P_e \theta$$

Where the slope of the cable in radians near the support is:

$$\theta = 4g/L = 4(215)/8000 = 0.1075 \text{ radians}$$
$$V_e = V_u - P_e \theta = 183.75 - 727(0.1075) = 105.6 \text{ kN}$$

The section is an uncracked one as it is near the support. The shear capacity of the section is given by:

$$V_r = \frac{\tau_c bh}{\gamma_c} = 0.67bh\sqrt{f_t^2 + 0.8f_{cp}f_t}$$

where

$$f_t = 0.24\sqrt{f_{ck}} = 0.24\sqrt{45} = 1.6 \text{ N/mm}^2 \text{ and}$$

$$f_{cp} = \frac{P_e}{A_c} = \frac{727000}{300 * 570} = 4.25 \text{ N/mm}^2$$

$$V_r = 0.67bh\sqrt{f_t^2 + 0.8f_{cp}f_t} = 0.67(0.3)(0.57)\sqrt{1.6^2 + 0.8(4.25)1.6)}$$

$$V_r = 0.324 \text{ MN} = 324 \text{ kN} > V_e = 105.6 \text{ kN}$$

The shear capacity of the section is about three times that of the effective shear force on the section, so no shear reinforcement is required. Nominal stirrups of 8 mm bars at a spacing of 0.75 times the effective

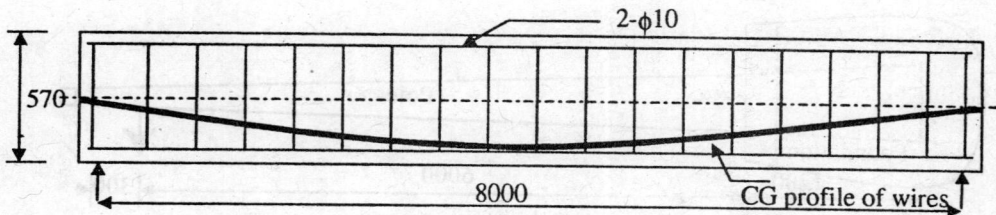

Fig. 5.9 LS of PSC Beam; Example 5.5.

depth are provided. Two legged 8 mm bars at 375 mm spacing. Use one 10 mm *HYSD* hanger bar at each of the corners of the cross section of the beam.

Each wire is set in parabolic profile. The centroid of all the wires is also parabolic. The wires are anchored at the end faces of the beam such that the centroid of the wires is at the centroid of the cross section at the end span. The combined eccentricity of the wires at the mid-span of the beam is 215 mm, which results in an effective depth of 500 mm. The sag of the centroid of the wires at mid span should be 215 mm. The sag in this case is also the eccentricity of the wires. The longitudinal profile of the centroid of the wires is shown in Fig. 5.9. The nominal shear reinforcement along with hanger bars is also shown in the figure. The cross sections at the mid-span and also at the end span are illustrated in Fig. 5.10.

Example 5.6: Design of Electric Pole

The total height of an electric pole is 7.5 m with 1.2 m length embedded into the ground. The electric cables were attached at 300 mm from the top of the pole. The wind force acting on the electric wires is 1800 N. Design a pretensioned concrete pole with *M45* concrete. The pole is in the mild environment

Solution

The pole that carries the electric wires acts as a cantilever beam with the wind load on the cables acting at 300 mm from top of the pole. 1.2 m length of the pole is embedded into the foundation so the cantilever length is 7.5 – 1.2 – 0.3 = 6 m. The pole is built with constant width and tapering depth from top to the bottom. The least depth can be at the top and it must be able to accommodate the wires. The effective span is taken equal to the cantilever plus about depth of the beam. That is the fixidity of the pole is at a depth of the effective depth of the beam into the foundation.

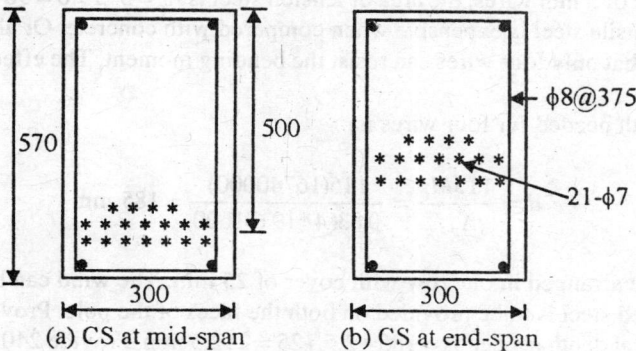

Fig. 5.10 Cross Sections of Beam; Example 5.5.

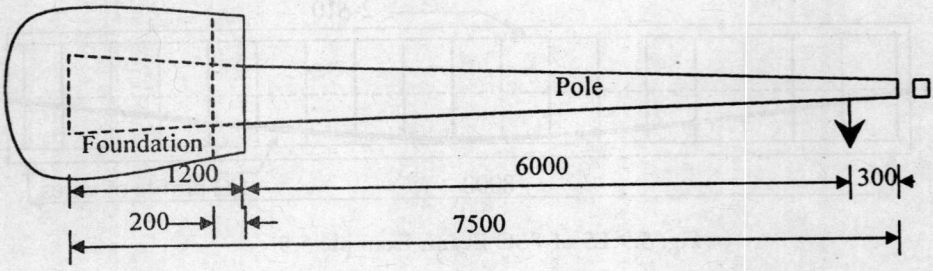

Fig. 5.11 Cantilever Electric Pole; Example 5.6.

The bending moment is critical at a depth of about half to one the effective depth from the top of the foundation. Assume the following specifications and data for the design.

Effective depth = 200 mm, 5 mm *HTS* wires with proof strength of 1600 MPa, Width of the pole as 100 mm. The partial load factor applied to wind load is = 1.5. The pole is upright so the self-weight cause only nominal compression and no bending moment. The total bending and shear forces are due to the wind load only.

The moment coefficient and the lever arm distance taken from the table 5.1 are: $K = 0.123, j = 0.83$

Design for Bending Moment

The bending moment span = $7.5 - 1.2 - 0.3 + 0.2 = 6.2$ m
The limit bending moment = $M_c = 1.5(1800)(6.2) = 16740$ Nm

The effective depth of the section can be obtained by equating the moment capacity to the limit moment and it is:

$$M_r = Kbd^2 f_{ck} = 0.123(120)(45)d^2 = 167400(1000) = M_c$$

Area of tension steel required is:

$$A_p = \frac{1.15M_c}{jdf_p} = \frac{1.15(16740000)}{0.83(175)(1600)} = 83 \text{ mm}^2$$

Provide 5 numbers of 5 mm wires, the area of tension steel is $A_p = 5*19.6 = 98$ mm^2. This is against 83 mm^2 needed. High tensile steel is expensive when compared with concrete. Or alternatively increase the effective depth such that only four wires can resist the bending moment. The effective depth needed with 4 wires is:

The effective depth needed for four wires is:

$$d = \frac{1.15M_c}{jA_p f_p} = \frac{1.15(16740000)}{0.83(4*19.6)1600} = 185 \text{ mm}$$

The 4 wires can be arranged in one row with cover of 25 mm. The wind can blow in either direction; therefore the tensioned steel is to be provided on both the faces of the pole. Provide a cover requirement of 25 mm, then the total depth needed is = 185 +2.5 +25 = 212.5 mm. Provide 240 mm depth at the bottom of the pole and 140 mm at the top of the pole. The beam depth at 1.5 m from the bottom of the pole is =

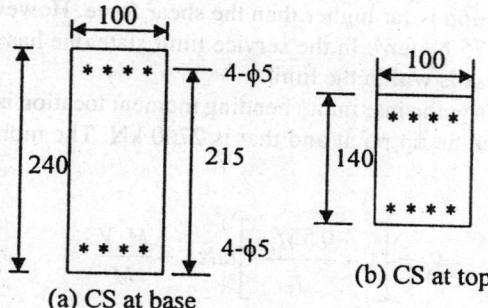

(a) CS at base

(b) CS at top

Fig. 5.12 Cross Sections of Example 5.6.

240 – (240-140)(1.5)/7.5 = 220 mm. The effective depth available is slightly more than that is needed and hence on the safer side. The section is also under-reinforced as the depth selected is more than that of the balanced section. The four wires are arranged in one row in 100 mm width. After deducting the cover of 25 mm, the spacing of the wires is = (100 – 50)/3 = 16.67 mm. This value is adequate to provide the bond strength for the 5 mm wires, however the special anchorages have to be designed to locate the wires at 16.67 mm spacing.

Design for Shear

The shear force on the pole is constant from top to the bottom of the pole. The size at the top of the pole is 140 by 100 mm with no bending moment. The size of the section at 300 mm below the top of the pole is taken same as that of the top. The section is uncracked and checked for shear capacity.

Shear Capacity of the Section at Top

The limit design shear force acting is = V_u = 1.5(1800) = 2700 N

There is no slope to the cable that will compensate the shear force so the effective shear to be resisted is same. The shear capacity of the section is:

$$V_r = \frac{\tau_c bh}{\gamma_c} = 0.67bh\sqrt{f_t^2 + 0.8f_{cp}f_t}$$

where

$$f_t = 0.24\sqrt{f_{ck}} = 0.24\sqrt{45} = 1.6\,\text{N/mm}^2 \text{ and}$$

$$f_{cp} = \frac{2P_e}{A_c} = \frac{2*4*19.6*0.6(1600)}{100*140} = 10.75\,\text{N/mm}^2$$

$$V_r = 0.67bh\sqrt{f_t^2 + 0.8f_{cp}f_t} = 0.67(0.1)(0.14)\sqrt{1.6^2 + 0.8(10.75)1.6}$$

$$V_r = 0.038\,\text{MN} = 38\,000\,\text{N} > V_e = 2700\,\text{N}$$

The shear capacity of the section is far higher than the shear force. However on should note that the axial stress on the section is 10.75 N/mm². In the service limit state one has to check for the allowable stresses. In this example the stress is within the limit.

The size of the cracked section at the maximum bending moment location is 100 by 215 mm. The limit shear force is same as that at the tie up point and that is 2700 kN. The moment capacity of a cracked section is:

$$V_r = \left[1 - \frac{0.55 f_{pe}}{f_p}\right] bd\tau_c + \frac{M_o V}{M}$$

and this value is subject to a minimum of :

$$V_{rmin} = 0.1 bd\sqrt{f_{ck}} \qquad (5.20)$$

Where

f_{pe} = effective prestress,
M_o = bending moment that can produce zero tensile stress in the section = $0.8 f_{pt}\, I/y_d$
y_d = distance of steel from the CG of the section = $d - x_u$
f_{pt} = compressive stress caused by the prestress at depth d
I = moment of inertia of the section,
M = bending moment on the section,
V = shear force at the section = V_e
τ_c = design shear strength of the concrete.

$$f_{pe} = \frac{2P_e}{A_c} = \frac{2(4*19.6)(0.6)1600}{100*215} = 7 \text{ N/mm}^2$$

$$y_d = d - x_u = 215 - 0.41(215) = 126.85 \text{ mm}$$

$$I = bh^3/12 = 100(215)^3/12 = 82819791 \text{ mm}^4$$

$$M_0 = \frac{0.8 f_{pt} I}{y_d} = \frac{0.8(7)(82819791)}{126.85} = 3656214\, Nmm = 3.65 \text{ kNm}$$

$$M = 6.2(1.8) = 11.16 \text{ kNm}$$

Percentage of the tension steel is = $4*19.6(100)/(215*100) = 0.36\%$
The allowable shear stress capacity from table 5.2 for 0.36 percentage of steel is: $\tau_c = 0.44$ N/mm²

$$V_r = \left[1 - \frac{0.55 f_{pe}}{f_p}\right] bd\tau_c + \frac{M_o V}{M} = (1 - 0.55*0.6)(100)215(0.44)/1000 + \frac{3.65*1.8}{11.16}$$

$$V_r = 6.33 + 0.59 = 6.92 \text{ kN} > V_u = 2.7 \text{ kN}$$

The shear capacity of the cracked section is far higher than the limit shear force acting on the section. *Note*: The design of the electric pole is not complete till the service limit state design is performed. The present example is only to illustrate the design for limit-state of strength only. The limit-state serviceability considers the allowable stresses, deflection and even the durability requirements. The bond strength and the development lengths must also be checked to make the design complete. The problem is solved in stages to give a better understanding of each step of limit-state design.

Example 5.7: Strength Limit State Design of 12 m Span Post-tensioned Prestressed Concrete Beam

A simply supported beam of effective span 12 m is subjected to dead load of 12 kN/m and live load of 28 kN/m. Design the beam with $M60$ concrete and 7 mm *HTS* wires cables. The exposure condition is moderate.

Solution

The 7 mm *HTS* wires have a proof strength of 1500 N/mm^2. The design coefficients from the table 5.1 are: $K = 0.128$, $j = 0.81$, $k_u = 0.44$, and other data is $L = 12$ m, $f_{ck} = 60$ N/mm^2, Let the width of the section be $= b = 500$ mm. The beam is exposed to the moderate environment and so the minimum cover to the reinforcement is 30 mm. The depth of the simply supported beam is assumed in the range of span/20 to span/12. In this case let the depth be $= h = 800$ mm and width $= b = 500$ mm. Then the self-weight is 10 kN/m.

Design load for limit-state of strength is:

$$w_u = 1.5(w_g + w_s + w_l) = 1.5(10 + 12 + 28) = 75 \text{ kN/m}$$

The maximum bending moment in simply supported beam is:

$$M_c = \frac{w_t L^2}{8} = \frac{75(12*12)}{8} = 1350 \text{ kNm}$$

Equating the balanced moment capacity to the limit moment, the effective depth of the section is:

$$d = \sqrt{\frac{M_c}{Kbf_{ck}}} = \sqrt{\frac{1350(10)^6}{0.128(500)60}} = 593 \text{ mm}$$

Assume 600 mm effective depth and the overall depth about 700 mm. The area of the tension reinforcement needed is:

$$A_p = \frac{1.15 M_c}{jdf_p} = \frac{1.15(1350)10^6}{0.81(1600(1500)} = 2156 \text{ mm}^2$$

Preliminary Detailing of Wires/Cables

Number of 7 mm wires needed is $= 2156/38.48 = 56.1$. Use 56 wires then the effective depth is proportionally increased. $d = 56.1(600)/56 = 601$ mm. Let the available cable contains eight wires of 7 mm wires. The number of cables required is seven to give 56 wires. The ducts are either metallic or plastic normally available is 50 mm diameter. The ducts have to placed with a cover that is equal to the diameter of the duct, further spacing of the ducts has be not less than 1.5 times the diameter of the duct, preferably twice the duct diameter. The width of the beam is 500 mm so only five ducts can be placed in a row. The total width required for five cables in a row is = 4*100 + 2*50 = 500 mm. That is 50 mm cover for each outer duct, and the five ducts are placed at 100 mm apart. The outer most row contains five ducts and inner row contains two ducts as shown in the Fig. 5.13.

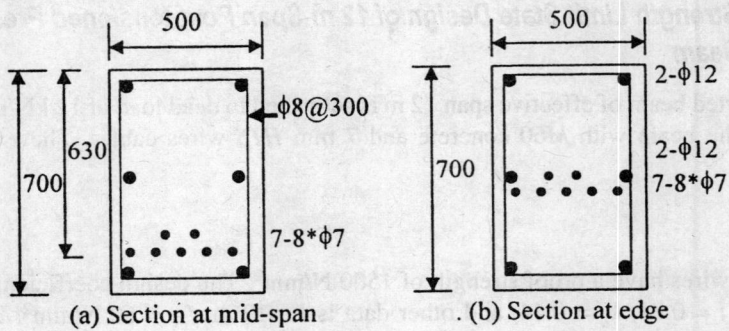

(a) Section at mid-span   (b) Section at edge

Fig. 5.13 Section of Example 5.7.

The centroid of the wires from the centre of the outer row of cable is = (2*75)/7 = 22 mm. The outer cable ducts are placed at 630 mm from the top fibre. This provides an effective depth of at least equal to 630 - 22 = 608 mm. This effective depth is more than that assumed. The cables are anchored at the ends such that the centroid of the cables should coincide with the centroid of the section. The profiles of the cables are set parabolic.

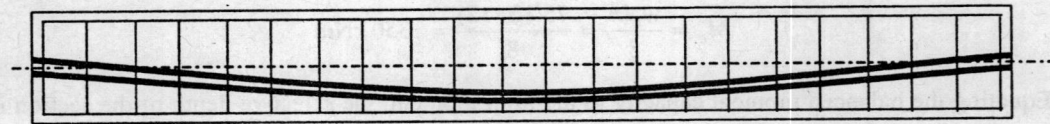

Fig. 5.14 Cable Profile and Stirrups; Example 5.7.

Design for Shear

The beam designed for bending has:

b = 500 mm, h = 700 mm, sag of the cable is = g = 608 – 700/2 = 258 mm, f_{ck} = 60 N/mm^2, f_p = 1500 N/mm^2,

$$P_e = A_p(0.6f_p) = 7*8*38.48(0.6)(1500) = 1939392 \text{ N} = 1939.4 \text{ kN}$$

The critical shear force occurs at a distance d from the support. The self-weight of the beam is 8.75 kN/m, but for the purpose of design a value of 9 kN/m can be used. The design shear force at a distance d from the support along with a partial safety factor of 1.5 is:

$$V_u = 1.5(w_g + w_s + w_l)(L/2 - d) = 1.5(9 + 12 + 28)(6 - 0.6) = 396.9 \text{ kN}$$

The effective shear on the concrete section is:

$$V_e = V_u - P_e \sin\theta = V_u - P_e \theta$$

Where the slope of the cable in radians near the support is:

$$\theta = 4g/L = 4(258)/12000 = 0.086 \text{ radians}$$

$$V_e = V_u - P_e \theta = 396.9 - 1939.4(0.086) = 230 \text{ kN}$$

The section is an uncracked as it is near the support. The shear capacity of the section is given by:

$$V_r = \frac{\tau_c bh}{\gamma_c} = 0.67bh\sqrt{f_t^2 + 0.8 f_{cp} f_t}$$

where

$$f_t = 0.24\sqrt{f_{ck}} = 0.24\sqrt{60} = 1.86 \text{ N/mm}^2 \text{ and}$$

$$f_{cp} = \frac{P_e}{A_c} = \frac{1939400}{500*700} = 5.54 \text{ N/mm}^2$$

$$V_r = 0.67bh\sqrt{f_t^2 + 0.8 f_{cp} f_t} = 0.67(0.5)(0.7)\sqrt{1.86^2 + 0.8(5.554)1.86}$$

$$V_r = 0.802 \text{ MN} = 802 \text{ kN} > V_e = 230 \text{ kN}$$

The shear capacity of the section is about three times that of the shear force therefore no shear reinforcement is required. However, it is desirable to provide nominal stirrups. Provide 8 mm bars at a spacing of 0.75 times the effective depth. Two legged 8 mm bars at 300 mm spacing are provided. Use 10 mm HYSD shear hanger bars at each of the corners of the cross section of the beam.

5.5 DESIGN OF FLANGE SECTIONED BEAMS

In prestressed concrete beam, compressive force is introduced at bottom to compensate the tensile force that is generated by the working load. The neutral axis distance from top is in the range of 0.45 times the effective depth of the section. A beam having flanges both at top and bottom is called I section or simply I beam. Such flanged sections are common in prestressed concrete construction.

Effective flange width

As a beam bends, the slab or flange of the beam bends along with the beam, however, only a portion of the slab width bends to the same curvature as that of the beam. The width of the flange that acts together with the beam is called *effective width*. The following controls the effective width of the flange that acts with the beam:

b_f = average of the widths of the slabs on either side of the beam,

$$= b_w + 6t + L_o/6$$

where

b_f = b_t = effective width of top flange,

b_w = width of the beam section, it is also called *web width*,

L_o = distance between two successive points of zero bending moments of the beam, = span in simply supported beam, or = twice the span of a cantilever,

t_t = t = thickness of the top slab,

And the other notations are:

b_b = effective width of bottom flange,

h = overall depth of the section,

t_b = thickness of the bottom slab,

A_p = area of tensioned steel.

Figure 5.15(a) illustrates *I* cross section notations. Fig. 5.15(b) indicates the stress distribution on the web portion and Fig. 5.15 (c) gives the stress distribution on the overhang portion of the flange at the limit-state of strength of the section. In most cases the entire flange area is subjected to compressive stress distribution of the rectangular shape intensity. The web portion is subjected to rectangular cum parabolic intensity of the stress. Idealization of the stress distribution of rectangular pattern on the total flange width is normally assumed. This is subject to the condition that the neutral axis falls beyond the flange thickness. If the neutral axis falls in the flange, the section can be considered as a rectangular one with the width of the section as equal to the width of the flange. The moment capacity expression of the rectangular section will then be extended to the *T* beam.

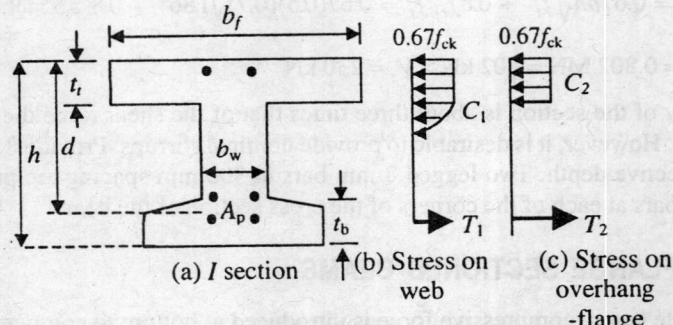

(a) *I* section   (b) Stress on web   (c) Stress on overhang -flange

Fig. 5.15 *I* Beam Section and Notations.

The maximum strains in concrete and steel at the time of failure are:

$$\varepsilon_{cu} = 0.0035$$
$$\varepsilon_y = 0.002 + f_y/1.15$$

Let:

x_u = neutral axis distance

d = distance of steel reinforcement from the extreme compression fibre,

f_p = proof stress in steel and 1.15 is the partial safety factor applied to steel

The neutral axis distance for simultaneous occurrence of the limiting strains is:

$$\frac{\varepsilon_{cu}}{\varepsilon_s} = \frac{\varepsilon_{cu}}{\varepsilon_y - \varepsilon_e} = \frac{x}{d - x_u} \qquad (5.21)$$

The rearrangement of the above equation gives:

$$x_u = \frac{\varepsilon_{cu}\, d}{\varepsilon_{cu} + \varepsilon_y - \varepsilon_e} = \frac{0.0035d}{0.0035 + 0.002 + f_p/1.15E_s - 0.04}$$

The expression after substitution of the modulus of elasticity of steel reduces to:

$$x_u = \frac{805d}{345 + f_p} = k_u d \tag{5.22}$$

Where

$$k_u = \frac{805}{345 + f_p}$$

The Eq. (5.22) gives the neutral axis distance in a balanced section. The value of the neutral axis coefficient is given in table 5.1.

The total compressive force on concrete in the flanged section is divided into two potions. The web width as rectangular and the second portion is on the overhang portions of the flange beyond the width of the web. The compressive force on the web is same as that of the rectangular is equal to:

$$C_1 = 0.36 k_u b_w d f_{ck} \tag{5.23}$$

The compressive force on the overhang portions of the flange with rectangular stress distribution is:

$$C_2 = (b_f - b_w)t(0.67 f_{ck})/\gamma_m = 0.45(b_f - b_w)t f_{ck} \tag{5.24}$$

Taking the moment of the above two forces about the centre of the reinforcement gives:

$$M_{rb} = Kbd^2 f_{ck} + 0.45(b_f - b_w)t f_{ck}(d - 0.5t) \tag{5.25}$$

Where
K = the value obtained for rectangular section in chapter 3

Equating the total compressive force to the tensile strength gives the area if tension steel.

$$A_p = \frac{C_1 + C_2}{f_p/\gamma_s} = \frac{1.15 p_0 b_w d f_{ck}}{f_p} + 0.518(b_f - b_w)t\frac{f_{ck}}{f_p} \tag{5.26}$$

Reasonable Idealization

For the case of neutral axis below the flange, the entire flange is assumed to be under compressive force with rectangular stress distribution and zero stress below the flange. This idealization is on the safer side and the estimate of the capacity of the section is slightly lesser than the real value. With this assumption, the compressive force on the flange and the moment capacity are:

$$C_d = 0.45 \, b_f t f_{ck} \tag{5.27}$$

$$M_r = 0.45 b_f t (d - 0.5t) f_{ck} \tag{5.28}$$

Effective depth of the section is obtained by equating the limit moment to the moment capacity of the section.

$$d = 0.5t + \frac{M_c}{0.45 b_f t f_{ck}} \tag{5.29}$$

And the corresponding area of tension reinforcement is:

$$A_p = 0.5t + \frac{1.15 M_c}{(d - 0.5t) f_p} \tag{5.30}$$

The above two expressions holds good only if the neutral axis falls in the web. In case the neutral axis lies in the flange, the sectional moment capacity is equal to that of rectangular section with width of the section equal to the width of the flange.

$$M_r = K b_f d2 f_{ck}, \; for \; x_u < t \tag{5.31}$$

$$Let \; x_u = k_u d > 1.15 \, t$$

For reinforcement *HTS*1500, the value of the neutral axis coefficient is 0.44; then the relation between the thickness and the effective depth is:

$$d > 2.6 \, t \tag{5.32}$$

This relation gives only a guideline about the possibility of neutral axis falling in the web in flanged sections.

Example 5.8: Moment Capacity of a Flange Section

A flange section shown in Fig.5.16 is made of *M*50 grade concrete, and the tensioned steel consists of ten cables, each cable consists of twelve 7 mm high-tension wires. The ultimate capacity of each cable is 690 kN. Determine the moment capacity of the section.

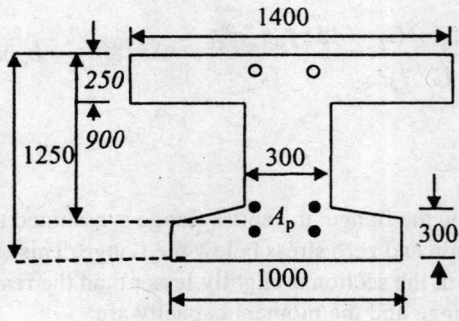

Fig. 5.16 Section Examples 5.8 & 5.9.

Solution

The details of the section are:

$b_f = 1400$ mm, $t = 250$ mm, $d = 900$ mm, $b_b = 1000$ mm, $t_b = 300$ mm, $A_p = 10*12*38.48 = 4617.6$ mm^2,

$$T_u = 6900 \text{ kN}, \quad f_{ck} = 50 \text{ N/mm}^2$$

The compression capacity of the top flange is:

$$C_d = 0.45\, b_f t f_{ck} = 0.45(1400)(250)50/1000 = 7875 \text{ kN}$$

The design tension capacity of the tensioned steel is:

$$T_d = T_u/1.15 = 6900/1.15 = 6000 \text{ kN}$$

The design strength of steel is far less than the compression capacity of the flange. Therefore the steel controls the moment capacity of the section. The moment capacity is:

$$M_r = T_d(d-t/2) = 6000(0.9 - 0.25/2) = 4650 \text{ kNm}.$$

Example 5.9: Design of Reinforcement for Balanced Section

Figure 5.16 gives details of a prestressed concrete cross section. Determine the maximum amount of tensioned steel the section can take without causing primary compression failure. Determine the maximum design moment capacity of the section and the approximate size of the bottom flange to withstand the pre-compression.

Solution

The details of the section are:

$b_f = 1400$ mm, $t = 250$ mm, $d = 900$ mm, $b_b = 1000$ mm, $t_b = 300$ mm, $f_{ck} = 50$ N/mm^2

The compression capacity of the top flange is:

$$C_d = 0.45\, b_f t f_{ck} = 0.45(1400)(250)50/1000 = 7875 \text{ kN}$$

The neutral axis distance for a balanced section for $f_p = 1500$ N/mm^2 can be:

$$k_u d = 0.44d = 396 \text{ mm} > b_t = 250 \text{ mm}$$

The neutral axis of the section fall below the top flange, therefore the total top flange can resist the compression. The corresponding tension steel can be obtained by equating the compression capacity to tension capacity.

$$A_p = C_d f_p/1.15 = 1.15(7875)/1.5 = 6037.5 \text{ mm}^2$$

Number of 7 mm wires required are:

$$N = 6037.5/38.28 = 156.9$$

Moment capacity of the section with 156 numbers of 7 mm wires is:

$$M_r = T_d(d-t/2) = 156*38.48*1.5(0.900 - 0.125)/1.15 = 7829.4(0.775) = 6069 \text{ kNm}$$

Assuming the web portion below the neutral axis along with the bottom flange will be under

Pre-compression, the area of concrete under compression is:

$$A_{cb} = (h - b_b - k_u d)b_w + b_b(t_b) = (1.25 - 0.3 - 0.396)*0.250 + 1(0.3) = 0.4385 \text{ m}^2$$

The design compressive capacity of the bottom portion of the section is:

$$C_{db} = 0.45A_{cb}f_{ck} = 0.45(0.4385)(50) = 9.866 \text{ MN} = 9866 \text{ kN} > T_d = 7829.8 \text{ kN}$$

The section is safe against crushing at design strength of the tensioned steel. But this check is not satisfactory. The check should be made for the ultimate capacity of the tensioned steel with a nominal partial safety factor. Alternatively one can calculate what is the most probable partial safety factor available against crushing of the concrete at the time of pre-tensioning. The partial safety factor available is:

$$\gamma_f = \frac{C_{db}}{T_u} = \frac{9866}{1.15(7829.8)} = 1.09$$

This value should not be less than 1 to ensure the safety at the time of tensioning the cables.

Example 5.10: Moment Capacity of I Section

A prestressed concrete beam section shown in Fig. 5.17 is made of *M50* grade concrete. Determine the maximum moment capacity of the section based on the concrete capacity. Also determine the pretension steel required matching with the compressive capacity of the section. The centroid of the steel is assumed to be at 920 mm from the top of the section.

Solution

Assume the high tensile steel is of 7 mm wires having an proof strength of 1500 N/mm². The neutral axis distance coefficient for such steel having a balanced section is 0.44 from table 5.1. The neutral axis for the section is at a distance from the top fibre is:

$$x_u = k_u d = 0.44d = 0.44*920 = 405 \text{ mm}$$

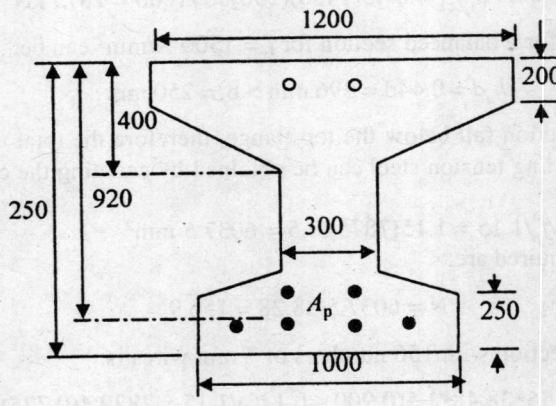

Fig. 5.17 Section Example 5.10.

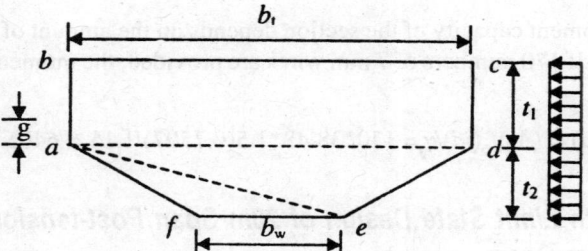

Fig. 5.18 Compression Area Notations.

This value is greater than the thickness of the top flange; therefore the top flange is under compression. However the neutral axis falls just outside the tapering thickness of the top flange, therefore the approximate expression must be taken with a pinch of salt. Since the section is tapering, the total compressive force on the flange has to be calculated based on the trapezoidal cross section. The flange portion of the section is shown in Fig. 5.18 for clarity of computation of the centroid of the flange etc. The section is divided into a rectangle *abcd*, and two triangles, *dea* and *aef*. Take area moments about line *ad* to obtain the centroid of the total section from line *ad*. The dimensional notations are marked on the figure. The distance of the centroid *g* from line *ad* is:

$$[b_t t_1 + 0.5(b_t + b_w)t_2]g = (b_t t_1) \, t_1/2 - [(b_t t_2/2) \, t_2/3 + (b_w t_2) \, t_2(2)t_2/3]$$

The above equation gives:

$$g = \frac{3b_t t_1^2 - (b_t + 2b_w)t_2^2}{3[2b_t t_1 + (b_t + b_w)t_2]}$$

The substitution of different quantities in the above equation gives

$$g = 30.7\text{mm}$$

The distance of the centroid of the compression force from top is $= x_c = 200 - 30.7 = 169.3$ mm

The lever arm distance is: $jd = d - x_c = 920 - 169.3 = 750.7$ mm

The compression area is $= A_c = [b_t t_1 + 0.5 (b_t + b_w)t_2] = 390,000$ mm^2 = 0.39 m^2.

The total compressive force on the section is:

$$C_d = A_c (0.45 f_{ck}) = 0.39(0.45)(50) = 8.775 \text{ MN} = 8775 \text{ kN}$$

The corresponding area of tension steel required to balance the compression is;

$$A_p = \frac{\gamma_f C_d}{f_p} = \frac{1.15(8775)1000}{1500} = 6727.5 \text{ mm}^2$$

This is equal to 174.8 number of 7 mm wires. It is desirable to design the section as an under-reinforced section. Therefore provide about 170 number of 7 mm wires.

The maximum moment capacity of the section based on the compression capacity is:

$$M_{rmax} = C_d (jd) = 8775(0.7507) = 6587.392 \text{ kNm}$$

The actual bending moment capacity of the section depends on the amount of steel provided if it is an under-reinforced section. If 170 numbers of 7 mm wires are provided, the moment capacity of the section is:

$$M_r = T_d(jd) = (A_p)f_p(jd)/\gamma_f = 170*38.48*1.5(0.7507)/1.15 = 6405.35 \text{ kNm}$$

Example 5.11: Strength Limit State Design of 20m Span Post-tensioned Prestressed Concrete Beam

A simply supported beam of effective span 20 m exposed to the moderate environment is supporting a 160 mm thick slab on the top. The spacing of the main beams is 4 m. The loads coming from the slab are: super-imposed dead load of 22 kN/m and live load of 28 kN/m. Design the beam with M60 concrete and 7 mm *HTS* wires cables.

Solution

The problem is solved in steps and each step is given a title.

(1) Data and Information

The 7 mm *HTS* wires have a proof strength of 1500 N/mm². The design coefficients are taken from the table 5.1 are: $k_u = 0.44$, and other data is $L = 20$ m, $t = t_t = 160$ mm, $f_{ck} = 60$ N/mm², Let the width of the web be $= b_w = 275$ mm. The beam is exposed to the moderate environment and so the minimum cover to the reinforcement has to be 30 mm. The area of web and the bottom flange are included in the self-weight of the beam. The top flange is already accounted in the superimposed load from the slab. Let the self-weight of the web and bottom flange of 300 mm by 1300 size. So let it be about 10 kN/m.

(2) Moment Capacity Computation

Design load for limit-state of strength is:

$$w_u = 1.5(w_g + w_s + w_l) = 1.5(10 + 22 + 28) = 90 \text{ kN/m}$$

The maximum bending moment in simply supported beam occurs at mid span and it is:

$$M_c = \frac{w_l L^2}{8} = \frac{90(20*20)}{8} = 4500 \text{ kNm}$$

(3) Design for Bending Moment

The effective width of the flange is subjected to the minimum of the following:

$$b_f = \text{centre to centre of the beams} = 4000 \text{ mm}$$
$$= b_w + 6t + L_o/6 = 275 + 6*160 + 20{,}000/6 = 4568 \text{ mm}$$

Select the effective width of the web as 4000 mm.

Equating the balanced moment capacity to the limit moment, one can get the effective depth of the section, Vide Eq. 5.29. This gives:

$$d = 0.5t + \frac{M_c}{0.45tb_f f_{ck}} = 80 + \frac{4500(10)^2}{0.45(160(4000)50} = 382.5 \text{ mm}$$

The effective depth for balanced section is only 312.5 mm. This value is rather small and may cause problems in serviceability limit state, so select at least span/20. Therefore assume effective depth of 800 mm and an over all depth of 1000 mm.

The area of tension reinforcement is:

$$A_p = \frac{1.15M_c}{(d - 0.5t)f_y} = \frac{1.15(4500)(10)^6}{(800 - 80)1500} = 4792 \text{ mm}^2$$

(4) Preliminary Detailing of Wires/Cables

Number of 7 mm wires needed is = 4792/38.48 = 125. Use 16 numbers of cables; each having eight wires of 7 mm. The ducts are normally available is 50 mm. The cover should be not less than the diameter of the duct, further spacing of the ducts be not less than 1.5 times the diameter of the duct, preferably twice the duct diameter. The width of the web is 275 that can accommodate two vertical rows of cables. Let six cables be placed in the bottom flange. The total width of the bottom flange required is two covers, five clear spacing of 75 mm and six ducts of 50 mm. It is equal to 2*50 + 5*75 + 6*50 = 775 mm. Fig 5.19 illustrates the detailing of the cable at the mid span section.

Six cables at a distance of 75 mm from bottom, next a pair of cables at a distance of 175 mm, next at 275 mm, next pair at 375 mm, next pair at 475 mm, next pair at 575 mim and the top most at 675 mm from the bottom fibre. The centroid of the cables from the bottom is computed by taking moment about the bottom fiber. And it is:

$$16(g) = 6(75) + 2[175 + 275 + 375 + 475 + 575]$$

From the above, the centroid of the cables from the bottom fibre is:

$$g = 4200/16 = 262.5 \text{ mm}$$

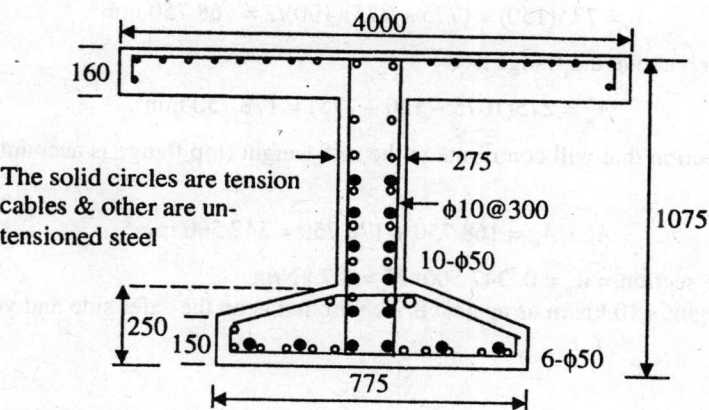

The solid circles are tension cables & other are un-tensioned steel

Fig. 5.19 CS of Example 5.11.

The minimum overall depth is $= 800$ mm $+ 262.5$ mm $= 1062.5$ mm
Use the overall depth of the beam as 1075 mm.

The cables are anchored at the ends such that the centroid of the cables coincides with the centroid of the section. The profiles of the cables are parabolic. At the moment only area of the cross section and the centroid are determined. The other properties is required for the design of service load condition.

The total area is divided into four parts, the total web area (top to bottom), overhangs of the top flange, bottom flange overhang triangle portion and the bottom flange overhang rectangle portion. The centroid of the section is computed through a tabulation of the three segments in table 5.3.

Tables 5.3 Calculation of Centroid of Section Example 5.11.

No. Segment	Area	First moment area
1. Web	275*1075= 295 625	158 898 437
2 Top flange	2*160(4000-275) =1192 000	1192 000 000
3 Bottom flange Triangle	2*250*100 =50 000	9166 500
4 Bottom flange Rectangle	2*250*150 = 75 000	5625 000
	1612625	1365689937

The Area column also shows: 295 625*1075/2= ; 1192 000*(1075-75)= ; 50 000(150+33.3)= ; 75 000(75)=

Centroid of the section = cumulative first moment of area/ area of the section = 1365689937/1612625= 846.9 mm

The centroid of the section is 847 mm from the bottom fibre or it is same as $1075 - 847 = 228$ mm from top fibre.

The centroid of the section from the top fibre is: $\qquad x_c = 228$ mm
The centroid of the cables is at 800mm from the top fibre.
The sag of the centroid of the cable is $= g = d - x_c = 800 - 228 = 572$ mm

(5) Check for the Original Assumptions and Make Corrections if Needed

The self-weight of the section is assumed in the computations of bending moment. The shear design is to be finalized. The area of cross section of the bottom flange is:

$$A_b = 775(150) + (775 + 275)(100)/2 = 168\ 750 \text{ mm}^2$$

The area of the web above the flange is:

$$A_w = 275(1075 - 150 - 275) = 178\ 750 \text{ mm}^2.$$

The area of the section that will contribute to the self-weight (top flange is accounted to be superimposed load) is:

$$A_b + A_w = 168\ 750 + 178\ 750 = 347\ 500 \text{ mm}^2$$

The weight of the section $= w_g = 0.\ 347\ 500*25 = 8.7$ kN/m
The assumed weight is 10 kN/m as against 8.7 kN/m that is on the safer side and yet quite close.

Design for Shear

The effective prestressing force is taken at 0.6 times the cable capacity and it is:

$$P_e = A_p(0.6 f_p) = 16*8*38.48(0.6)(1500) = 4432\ 896 \text{ N} = 4432.9 \text{ kN},$$

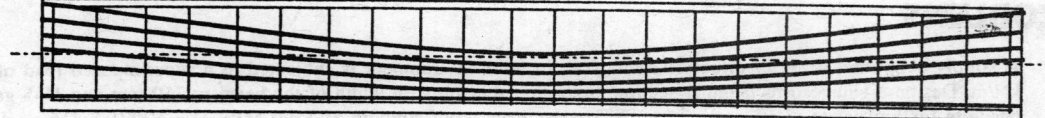

Fig. 5.20 Cable Profile and Shear Stirrups; Example 5.11.

The ultimate cable force $= T_u = A_p(f_p) = 7388$ kN

The critical section is at a distance d from the support. The shear with a partial safety factor of 1.5 is:

$$V_u = 1.5(w_g + w_s + w_l)(L/2 - d) = 1.5(10 + 22 + 28)(10 - 0.8) = 828 \text{ kN}$$

The effective shear on the concrete section is:

$$V_e = V_u - P_e \sin\theta = V_u - P_e\theta$$

Where the slope of the cable in radians near the support is:

$$\theta = 4g/L = 4(572)/20000 = 0.114 \text{ radians}$$

$$V_e = V_u - P_e\theta = 828 - 4432.9(0.114) = 321 \text{ kN}$$

The section is an uncracked one as it is near the support. The shear capacity of the section is given by:

$$V_r = \frac{\tau_c bh}{\gamma_c} = 0.67 b_w h \sqrt{f_t^2 + 0.8 f_{cp} f_t}$$

where

$$f_t = 0.24\sqrt{f_{ck}} = 0.24\sqrt{60} = 1.86 \text{ N/mm}^2 \text{ and}$$

$$f_{cp} = \frac{P_e}{A_c} = \frac{4432900}{1612625} = 2.75 \text{ N/mm}^2$$

$$V_r = 0.67 bh \sqrt{f_t^2 + 0.8 f_{cp} f_t} = 0.67(0.275)(0.7)\sqrt{1.86^2 + 0.8(2.75)1.86)}$$

$$V_r = 0.544 \text{ MN} = 544 \text{ kN} > V_e = 321 \text{ kN}$$

The shear capacity of the section is more than the effective shear force on the section. However, it is desirable to provide nominal stirrups in the section. Provide 8 mm stirrups at a spacing of 0.75 times the effective depth or at 300 mm spacing. Two leg 10 mm bars at 300 mm spacing. Use 12 mm *HYSD* shear hanger bars at each of the corners of the cross section of the beam, and at two intermediate level on either side. A minimum non-tensioned reinforcement of 0.15% is provided in each direction. This reinforcement is meant to resist secondary stresses, shrinkage, and secondary stresses that might arise during tensioning of the cables. Open circles in the cross section Fig. 5.19 indicate non-tensioned reinforcement.

PROBLEMS

5.1. A simply supported beam of effective span 10 m in mild environment is subjected to a superimposed load of 30 kN/m. Design a balanced rectangular section pretensioned beam with width of the beam as 350 mm and *M*45 grade concrete for limit state of strength. 7 mm *HTS* wires have proof strength of 1500 MPa. Use *HYSD-Fe*415 steel for stirrups.

5.2. A simply supported beam of effective span 10 m in mild environment is subjected to a superimposed load of 40 kN/m over the entire span. The size of the beam is prefixed as 300 by 700 mm. Design a pretensioned beam by limit state of strength with 7 mm *HTS* wires having a proof strength of 1500 MPa. The grade of the concrete is *M*45. Use *HYSD-Fe*415 steel for stirrups.

5.3 A simply supported beam of effective span 12 m in severe environment is subjected to a superimposed load of 40 kN/m over the entire span. The beam is post-tensioned and the size is 330 by 900 mm. Design the section for limit state of strength with 7 mm *HTS* wires having a proof strength of 1500 MPa. The grade of the concrete is *M*50. Use *HYSD-Fe*415 steel for stirrups.

5.4 A simply supported beam of effective span 12 m is in severe environment subjected to a superimposed load of 40 kN/m over the entire span. The beam is post-tensioned. The top flange is 200 by 800 mm and the bottom flange is 300 by 600 mm. The overall depth of the beam is 750 mm. Design the section for limit state of strength with 7 mm *HTS* wires having a proof strength of 1500 MPa. The grade of the concrete is *M*60. Use *HYSD-Fe*415 steel for stirrups.

5.5 A cantilever beam of 6 m span is subjected to a superimposed load of 40 kN/m over the entire span. The beam is exposed to moderate environment. The section is 300 mm by 700 mm. Design a post-tensioned beam with *M*50 concrete and 7 mm *HTS* wires having a proof strength of 1500 MPa. Design for limit state of strength. Use *HYSD-Fe*415 steel for stirrups.

5.6 A cantilever beam of span 8 m is subjected to a superimposed load of 40 kN/m over the entire span. The beam is post-tensioned and exposed to severe environment. The beam is flanged section with top flange 200 mm by 500 mm, the bottom flange is 300 by 500 mm. The overall depth is 850 mm. Design the section for limit state of strength with 7 mm *HTS* wires having a proof strength of 1500 MPa. The grade of the concrete is *M*60. Use *HYSD-Fe*415 steel for stirrups.

Serviceability Limit State Design of Beams

6.1 INTRODUCTION

A structure must be safe against collapse and give a satisfactory service during the designed life of the structure. The deflections of the structure or the elements must be within certain satisfactory limits such that the people within the building shouldn't feel them. Similarly, any visual damage or cracks shouldn't be observed. The method of design that ensures a satisfactory performance of the structure is called limit-state design of serviceability. The parameters considered for the serviceability of reinforced concrete structures are:

1. Deflection,
2. Crack width and control of crack and,
3. Stresses in some cases.

The deflection and crack width are the two criteria applied in the serviceability design. The third aspect, namely the stress limits is not being imposed in reinforced concrete structures. However it is a requirement in prestressed concrete structures. Only limits of deflection and crack width are considered in this chapter.

The environmental forces cause deterioration of the structure in course of time. Environmental forces such as normal wind, rain, wetting and drying, wear and tear, aggressive actions time dependent effects. Durability design is through implementation of specifications on materials, workmanship, tolerances etc. For example, the specifications on minimum strength and maximum water cement ration of the concrete, cover to the reinforcement etc. are specified. The durability specifications take care of:

1. Exposure conditions,
2. Aggressive exposure such as chemical nature saline, sulphate actions,
3. Wear exposure-surface exposed to movable objects, or traffic, or the nature of traffic.
4. Heat exposure-extreme changes in the temperature including heat generated by fire and accidents.
5. Repeated loads and fatigue.

6.2 DEFLECTION LIMIT STATE DESIGN

Deflections of concrete structures are usually small in comparison with those in metal ones as the moment of inertia of cross-section of the concrete member is more than that of an equivalent strong metal one. The deflections may not control the design in many cases. However, there are some cases such as cantilever beams and long span simply supported beams, the deflections are likely to control. Limits on deflections are set for the following reasons:

- Large deflections induce larger strains even without causing excessive stress. Such large deflections may be due to high slenderness, creep, shrinkage, etc.
- Large relative deflections cause psychological fear to the user during moving load, gust wind conditions,
- Large deflections may also cause damage on architectural members, large glass windows and doors, pipe and plumb lines, partition walls. Some deflections may not be large for the structural system or for occupants, but they may cause damage to the fixtures such ceilings, partition walls or to the service lines such as air conditioning ducts and water mains, etc.

Codes of practice specify limits on the deflections of the members or the structure. The design criterion is: the deflection caused by the external loads must be less than that specified by the codes of practice. The design criterion is expressed as:

$$v_i < v_{ai} \tag{6.1}$$

where

v_i = actual deflection in the ith load condition,
v_{ai} = allowable deflection in the ith load condition.

Table 6.1 Maximum Allowable (permissible) Deflections.

No.	Type of structure	Deflection
1	Beams, slabs, roofs in ordinary buildings: Total deflection including creep, shrinkage and temperature effects are measured from as cast level.	$L/250$
2	Beams and slabs: Deflection due lo live loads	$L/350$ Or 20 mm
3	Beams and slabs deflection under guest wind or sever earthquake load condition	$L/250$
4	Hand operated crane girders	$L/500$
5	Electrically operated crane girders	$L/1000$
6	Lateral drift(deflection) of tall building	$L/400$

Where L = span of the beam or slab or the height of the building

The duration of gust wind load may be for a short period of two to five minutes when compared to the duration of live load which act for much longer period. That is why allowable deflections in different load conditions are different. Table 6.1 gives the allowable values of the deflections.

Deformation is a more generalized term indicating the displacements and rotations of members or joints. Each point in space has six degrees of freedom, three displacements and three rotations. In plane structural members are only two degrees of freedom. Two displacements and one rotation are the degrees of freedom in plane structures. Major factors that produce deformations are:

1. External loads such as dead loads, superimposed loads, live loads, wind load, earthquake load etc.
2. Shrinkage of concrete,
3. Creep in concrete,
4. Temperature variations,
5. Settlement of foundations.

The deflection caused by loads in beams is normally expressed as:

$$v_f = \frac{c_v W L^3}{E_c I_e}$$

(6.2)

where,

W = total load on the beam,

L = effective span,

E_c = short term modulus of elasticity of the material,

I_e = effective moment of the section,

v_f = short term deflection,

c_v = coefficient which depends on the boundary conditions of the beam and load distributions.

The deflection due to bending moment acting on the beam can also be expressed as:

$$v_f = \frac{c_v M L^2}{E_c I_e}$$

(6.3)

Where

M = the applied bending moment on the beam.

Concrete forms the bulk of a member; the percentage of the reinforcement is in the range of 0.4 percent to about 3 percent. The effect of reinforcement is neglected in determination of the young's modulus of elasticity. Effect of reinforcement is neglected in computation of moment of inertia of uncracked section and accounted in the cracked section. The moment of inertia along the length of the beam changes as some part of the beam is cracked. It more complicated in framed buildings. In beam-slab construction, the beam near the supports is subjected to negative bending moment and that at mid span is subjected to positive bending moment. Therefore, it is rectangular section at negative bending moment zone and a T-beam near mid-span zone. In most structural analysis, the sectional properties of the beam are taken as a rectangular section. In statically determinate structure the error is negligible but in large indeterminate systems the error could be from zero to fifty percent.

Deflection coefficients for cantilever and simply supported beams for typical cases are listed in table 6.2. The deflections of statically indeterminate beams and structures have to be obtained by structural analysis. The deflections are analyzed by computer programs for complex structures such as trusses, frames, multistory-framed buildings, and space frames.

The Young's modulus of elasticity of concrete is idealized as a function of the characteristic strength of concrete. And it is repeated here for quick reference.

$$E_c = 5000 \sqrt{f_{ck}}$$

(6.4)

6.3 EFFECTIVE MOMENT OF INERTIA

The effective moment of inertia of a concrete section of isolated simple beams is idealized as:

$$I_e = \frac{I_{cr}}{1.2 - \dfrac{M_{cr}(j)}{M}\left[1 - \dfrac{x_u}{d}\right]\dfrac{b_w}{b_f}}$$

(6.5)

Table 6.2 Deflection Coefficients for Beams.

Type of beam	Type of loading	Deflection	
		Location	Coefficient
Cantilever beam,	moment at free end	Free end	1/2
Cantilever beam,	Concentrated load at free end	Free end	1/3
Cantilever beam,	Uniformly distributed load	Free end	1/8
Cantilever beam,	Triangular load max. at free end	Free end	11/60
Cantilever beam,	Triangular load max. at fixed end	Free end	1/15
Simply supported	beam, Moment at support	mid-span	1/15.6
Simply supported beam,	Concentrated mid-span load	mid-span	1/8
Simply supported beam,	Uniformly distributed load	mid-span	5/384
Simply supported beam,	Triangular load, max at mid-span	mid-span	1/60
Simply supported beam,	2 concentrated loads at 1/3 points	mid-span	23/1296
Two ends fixed beam,	Mid-span concentrated load	mid-span	1/192
Two ends fixed beam,	Uniformly distributed load	mid-span	1/384

Where

I_r = moment of inertia of the cracked section

M_{cr} = cracking moment of the section, and it is equal to

$$= \frac{f_{cr} I_g}{y_t}$$

M = maximum bending moment under service load

j = ratio of the lever arm to effective depth,

d = effective depth of steel,

b_w = width of web,

b_f = width of compression flange

f_{cr} = modulus of rupture of concrete,

I_g = moment of inertia of gross section about the centroidal axis,

y_t = distance of the centroid of the gross section to the extreme tension fibre.

The gross moment of inertia is computed neglecting the effect of reinforcement of the uncracked section. The effective moment of inertia is subject to the following limitations.

$$I_{cr} \le I_e \le I_g \tag{6.6}$$

Even in rectangular sections, the reinforcement at mid section may not be same as that at the support section. The modified sectional properties are:

$$x_e = k_1 \left[\frac{x_1 + x_2}{2} \right] + (1 - k_1) x_0 \tag{6.7}$$

where,

x = value of I_r or I_g or M_r,

x_e = modified value of X,

x_1, x_2 = values of X at supports (both may be same in most cases)

x_0 = value of X at mid-span,

k_1 = coefficient given in table 6.3 for different values of moments

The expression for T beams is long unlike that of rectangular section. For simplicity, different properties for rectangular sections are given here.

The moment of inertia of cracked rectangular section is:

$$I_{cr} = \frac{bx_u^3}{3} + mA_{st}(d - x_u)^2$$

(6.8)

Table 6.3 Coefficients for Computation of Modified Properties of Inertia in Continuous Beams.

k_2	0.5 or less	0.6	0.7	0.8	0.9	1.0	1.1	1.2	1.3	1.4
k_1	0.0	0.03	0.08	0.16	0.30	0.50	0.73	0.91	0.97	01.0

where $k_2 = \dfrac{M_1 + M_2}{M_{f1} + M_{f2}}$

where M_1 and M_2 are moments at supports, The beam sign convention for moments M_{f1} and M_{f2} are fixed end moments at supports

where

b = width of the section,
d = effective depth,
A_{st} = area of tension reinforcement,
m = modular ratio = $E_s/E_c = 40/\sqrt{f_{ck}}$,
x_u = neutral axis distance from top compression fibre.

The moment of inertia of the uncracked rectangular section is:

$$I_g = \frac{bh^3}{12}$$

(6.9)

Where h = overall depth of the section

The moment at cracking of the section is:

$$M_{cr} = \frac{I_g f_{cr}}{y_t} = \frac{bh^2 \sqrt{f_{ck}}}{8.6}$$

(6.10)

Example 6.1: Sectional Properties and Deflection due to Live Load

A 5 m simply supported beam of rectangular section, 250 mm by 470 mm is provided with a reinforcement of 3 numbers of 20 mm bars and one of 12 mm bar at a depth of 430 mm from top fibre. Determine the effective moment of inertia of the section. The beam is made of $M20$ grade concrete and the live load is 24 kN/m. Service load bending moment at mid span is 84.37 kNm. Compute the deflection caused by the dead and live loads.

Solution

The given data is: $L = 5$ m, $f_{ck} = 20$ N/mm^2, $f_y = 415$ N/mm^2, $w_l = 24$ kN/m, $b = 250$ mm, $d = 430$ mm, $h = 470$ mm,. $A_{st} = 1055$ mm^2, $M = 84.37$ kNm. The design coefficients for the present data can be obtained from table 3.1 and they are: $j = 0.80$, $k_u = 0.479$; $x_u = 0.479d = 206$ mm.

Modular ratio $= E_s/E_c = 200\,000/\,5000\,\sqrt{40} = 40/\sqrt{20} = 9.0$,
The gross moment of inertia and the moment of inertia of the cracked section are:

$$I_g = \frac{bh^3}{12} = \frac{0.250(0.470)^3}{12} = 0.00216\ \text{m}^4$$

$$I_{cr} = \frac{bx_u^3}{3} + mA_{st}(d - x_u)^2 = 0.25(0.26)^3/3 + 9(0.001055)(0.43 - 0.206)^2$$

$$= 0.000724 + 0.000476 = 0.00120\ \text{m}^4$$

The bending moment at fracture is:

$$M_{cr} = \frac{I_g f_{cr}}{y_t} = \frac{bh^2\sqrt{f_{ck}}}{8.6} = 250(470)^2(\sqrt{20})/8.6(1000\,000) = 28.72\ \text{kNm}$$

$$I_e = \frac{I}{1.2 - \dfrac{M_{cr}(j)}{M}\left[1 - \dfrac{x_u}{d}\right]\dfrac{b_w}{b_f}}$$

$$I_e = \frac{0.0012}{1.2 - \dfrac{28.72(0.8)}{84.37}\left(1 - \dfrac{0.206}{0.430}\right)\dfrac{0.25}{0.25}} = \frac{0.0012}{1.0581} = 0.001134\ \text{m}^4$$

This value is less than the moment of inertia of the cracked section so the cracked sectional values have to be used.

$$I_e = 0.0012\ \text{m}^4$$

The maximum deflection occurs at mid span due to uniformly distributed load and it is:

$$v = \frac{5wL^4}{384E_cI_e}$$

Where w is the UDL.
The Young's modulus of elasticity of the concrete is:

$$E_c = 5000\sqrt{20} = 22360\ \text{N/mm}^2 = 22.36(10^6)\ \text{kN/m}^2$$

The deflections due to dead and live loads are computed separately. The dead load is that due to the selfweight and it is 3 kN/m. The corresponding deflection is:

$$v_d = \frac{5(3)5^4}{384(22.36)(10^6)(0.0012)} = 0.0009\text{m} = 0.9\ \text{mm}$$

The corresponding deflection due to live load of 24 kN/m is:

$$v_1^* = 0.0072 \text{ m} = 7.2 \text{ mm}$$

The admissible deflection due to live load alone is:

$$v_a = L/350 = 5000/350 = 14.3 \text{ mm}.$$

The admissible deflection is 14.3 mm whereas that is caused by the live load is 7.2 mm only. The deflections due to shrinkage and creep are covered in the next section.

Example 6.2: Effective Moment of Inertia of Rectangular Section Continuous Beam

A three span continuous beam of rectangular section, 250 mm by 470 mm, each 5 m span is provided with a 3 numbers of 20 mm bars and one of 12 mm bar at a depth of 430 mm from top fibre at mid span; and Four 20 mm bars are at top fibre at intermediate supports. The beam is made of $M20$ grade concrete and subjected to a live load of 24 kN/m. Compute the effective moment of inertia of the beam.

Solution

The given data is: $L = 5$ m, $f_{ck} = 20$ N/mm^2, $f_y = 415$ N/mm^2, $w_l = 24$ kN/m, $b = 250$ mm, $d = 430$ mm, $h = 470$ mm. $A_{st} = 1055$ mm^2, $M_l = 84.37$ kNm. The design coefficients for the present data can be obtained from table 3.1 and they are: $j = 0.80$, $k_u = 0.479$; $x_u = 0.479d = 206$ mm.
Modular ratio $= E_s/E_c = 200\,000/(5000 \sqrt{20}) = 40/\sqrt{20} = 9.0$,
The gross moment of inertia section at mid span and the end span is the same and it is:

$$I_g = \frac{bh^3}{12} = \frac{0.250(0.470)^3}{12} = 0.00216 \text{ m}^4$$

The fracture bending moment of the sections at mid-span and at support are same and it is:

$$M_{cr} = \frac{I_g f_{cr}}{y_t} = \frac{bh^2 \sqrt{f_{ck}}}{8.6} = 250(470)^2(\sqrt{20})/8.6(1000\,000) = 28.72 \text{ kNm}$$

The moment of inertia of the cracked section at mid-span is:

$$I_{cr0} = \frac{bx_u^3}{3} + mA_{st}(d - x_u)^2 = 0.25(0.206)^3/3 + 9(0.001256)(0.43 - 0.206)^2$$

$$= 0.000724 + 0.000476 = 0.00120 \text{ m}^4$$

The sections at the intermediate supports are same. The amount of reinforcement at support is four of 20 mm bars at top ($A_{st} = 1256$ mm^2), the moment of inertia of the cracked section is:

$$I_{cr1} = \frac{bx_u^3}{3} + mA_{st}(d - x_u)^2 = 0.25(0.206)^3/3 + 9(0.001256)(0.43 - 0.206)^2$$

$$= 0.000728 + 0.000567 = 0.001396 \text{ m}^4$$

The total load acting on the span is dead plus live load, and it is $= 3 + 24 = 27$ kN/m. The fixed end bending moment and the final bending moment at the interior supports are (please note only magnitude is given here):

$$M_{f1} = M_{f2} = WL^2/12 = 56.25 \text{ kNm and}$$

$$M_1 = 0; M_2 = 72.3 \text{ kNm}$$

The k_2 factor from the table 6.3 is:

$$k_2 = \frac{M_1 + M_2}{Mf_1 + M_{f2}} = \frac{0 + 72.3}{2*56.25} = 0.64$$

The value of k_1 from the table is: 0.05
The effective modified sectional property can be obtained from Eq.6.6 and it is:

$$x_e = k_1 \left[\frac{x_1 + x_2}{2} \right] + (1 - k_1)x_0$$

in which
$x_0 = I_{cr0} = 0.0012 \text{ m}^4$
$x_1 = x_2 = I_{cr1} = 0.001396 \text{ m}^4$, and $k_1 = 0.05$.

The effective moment of inertia of the cracked section is:

$$I_e = x_e = k_1 \left[\frac{x_1 + x_2}{2} \right] + (1 - k_1)x_0 = 0.05(0.001396) + 0.95(0.0012) = 0.00121 \text{ m}^4$$

The gross moment of inertia of the section is more than that of the cracked section, so a lower value is selected.

6.4   DEFLECTIONS DUE TO SHRINKAGE AND CREEP

Deflection due to Shrinkage

The concrete shrinks as it gets hardened and dries. The shrinkage strain leads to deflection in members and the structure. The shrinkage deflection is a function of shrinkage strain, percentage of tension and compression reinforcements and boundary conditions of the member. The shrinkage of the concrete is reduced by increase in the percentage of reinforcement. A minimum reinforcement must be provided in all the reinforced concrete members to minimize or eliminate cracking due to shrinkage. In most cases 0.15 percent of the sectional area is provided as minimum reinforcement to avoid shrinkage cracking. This reinforcement is to be distributed almost equitably in the cross section. In case the structural reinforcement is more than the minimum, then no extra steel need be added, except that some reinforcement must be available on all faces of the member. The spacing of the reinforcement should also be reasonably spaced to avoid micro-cracking. The equitable distribution of the reinforcement on all the faces minimizes shortening or even eliminates the shrinkage deflection. Indian code of practice on plain and reinforced concrete suggests the following expression for shrinkage deflection.

$$v_{sh} = k_3 k_4 \varepsilon_{cs} \frac{L^2}{h} \tag{6.11}$$

Where

k_3 = boundary condition coefficient and it is:

= 0.5 for cantilever beams; 0.125 for simply supported beams,

= 0.086 for one end continuous and the other end discontinuous,

= 0.063 for both ends continuous

k_4 = the coefficient that depends on the percentage of reinforcement in the member and it is:

$$= \frac{0.72(p_t - p_c)}{\sqrt{p_t}} \le 1.0, \ for: 0.25 \le p_t - p_c < 1.0 \tag{6.12}$$

$$= \frac{0.65(p_t - p_c)}{\sqrt{p_t}} \le 1.0, \ for; \ p_t - p_c > 1.0 \tag{6.13}$$

p_t = percentage of tension reinforcement with respect to effective area ($=100A_{st}/bd$),

p_c = percentage of compression reinforcement with respect to effective area ($=100A_{sc}/bd$),

h = total depth of the section,

L = span of the member,

ε_{cs} = shrinkage strain of concrete, this is normally taken as 0.0003

The differential shrinkage problem is different from total shrinkage strain.

Deflection due to Creep

Creep is the increase in strain under constant stress. Fixed loads, dead loads cause creep strain. The deflection due to fixed loads increase due to creep. The deflection due to creep can be obtained by multiplying the elastic strain by the effective creep coefficient. The deflection due to creep alone is obtained as:

$$v_c = (c_c - 1)v_e \tag{6.14}$$

Where

v_c = deflection due to creep only,

c_c = creep coefficient,

v_e = elastic deflection

The values of creep coefficients are given the code. The deflection due to live load is normally un-effected by the creep coefficient.

Example 6.3: Deflection due to Shrinkage and Creep

A 5 m simply supported beam of rectangular section, 250 mm by 470 mm is provided with a reinforcement of 3 numbers of 20 mm bars and one of 12 mm bar at 40 mm from bottom fibre. The top hanger bars are two numbers of 8 mm bars. The beam is made of $M20$ concrete. The beam is loaded with live load of 24 kN/m six months after the construction.

Solution

The given data is: $L = 5$ m, $f_{ck} = 20$ N/mm², $f_y = 415$ N/mm², $w_l = 24$ kN/m, $b = 250$ mm, $d = 430$ mm, $h = 470$

mm. $A_{st} = 1055$ mm^2. The design coefficients are $j = 0.80$, $k_u = 0.479$; and $x_u = 0.479d = 206$ mm.
Modular ratio = E_s/E_c = 200 000/ 5000 $\sqrt{20}$ = 40/$\sqrt{20}$ = 9.0,
The gross moment of inertia and the moment of inertia of the cracked section are:

$$I_g = \frac{bh^3}{12} = \frac{0.250(0.470)^3}{12} = 0.00216 \, \text{m}^4$$

$$I_{cr} = \frac{bx_u^3}{3} + mA_{st}(d - x_u)^2 = 0.25(0.206)^3 / 3 + 9(0.001055)(0.43 - 0.206)^2$$

$$= 0.000724 + 0.000476 = 0.00120 \, \text{m}^4$$

The bending moment at fracture is:

$$M_{cr} = \frac{I_g f_{cr}}{y_t} = \frac{bh^2 \sqrt{f_{ck}}}{8.6} = 250(470)^2(\sqrt{20})/(8.6(1000\ 000)) = 28.72 \, \text{kNm}$$

$$I_e = \frac{I_{cr}}{1.2 - \dfrac{M_{cr}(j)}{M}\left[1 - \dfrac{x_u}{d}\right]\dfrac{b_w}{b_f}}$$

$$I_e = \frac{0.0012}{1.2 - \dfrac{28.72(0.8)}{84.37}\left(1 - \dfrac{0.206}{0.430}\right)\dfrac{0.25}{0.25}} = \frac{0.0012}{1.0581} = 0.001134 \, \text{m}^4$$

This value is less than the moment of inertia of the cracked section so the cracked sectional value has to be used.

$$I_e = 0.0012 \, \text{m}^4$$

The maximum deflection occurs at mid span due to uniformly distributed load and it is given by:

$$v = \frac{5wL^4}{384 E_c I_e}$$

The Young's modulus of elasticity of the concrete is:

$$E_c = 5000\sqrt{20} = 22360 \, \text{N/mm}^2 = 22.36(10^6) \, \text{kN/m}^2$$

The deflections due to dead and live loads are computed separately. The dead load is self weight and it is 3 kN/m. The corresponding deflection is:

$$v_d = \frac{5(3)5^4}{384(22.36)(10^6)(0.0012)} = 0.0009 \text{m} = 0.9 \, \text{mm}$$

The deflection due to live load of 24 kN/m is:

$$v_l = 0.0072 \, \text{m} = 7.2 \, \text{mm}$$

The percentages of tension and compression reinforcements are:

$$p_t = \frac{100A_{st}}{bd} = \frac{100(1055)}{250(430)} = 0.98$$

$$p_c = \frac{100A_{sc}}{bd} = \frac{100(2*50)}{250(430)} = 0.09$$

$$p_t - p_c = 0.98 - 0.09 = 0.89 < 1.0$$

So select the value of k_4 as

$$k_4 = \frac{0.72(p_t - p_c)}{\sqrt{p_t}} = \frac{0.72(0.98 - 0.09)}{\sqrt{0.98}} = 0.65$$

The beam is simply supported so the value of the factor on boundary conditions is: $k_3 = 0.125$. The shrinkage deflection is given by:

$$v_{sh} = k_3 k_4 \varepsilon_{cs} \frac{L^2}{h}$$

The span is 5 m, total depth of the beam is 0.47m and the shrinkage strain is 0.0003. The substitution of different values in the shrinkage deflection is:

$$v_{sh} = k_3 k_4 \varepsilon_{cs} \frac{L^2}{h} = 0.125(0.65)(0.0003)(5000)^2 / 470 = 1.3 \, \text{mm}$$

The deflection due to creep is obtained by multiplying the elastic deflection caused by the permanent loads by the creep coefficient. The deflection due to dead load is 0.9 mm. The creep coefficient for loading after 6 months can be taken as 1.2. So the creep deflection is:

$$v_c = c_c v_d = 1.2(0.9) = 1.1 \, \text{mm}$$

The total deflection including creep and shrinkage is:

$$v_t = v_d + v_1 + v_c + v_{sh} = 0.9 + 7.2 + 1.1 + 1.3 = 10.5 \, \text{mm}$$

The admissible live load and total deflections are:

$$v_{al} = L/350 = 5000/350 = 14.3 \, \text{mm.}$$

$$v_{at} = L/250 = 5000/250 = 20 \, \text{mm}$$

The admissible deflections are more than the actual ones. The beam is safe in the deflection serviceability.

Example 6.4: Cantilever Beam Deflection

A 3 m clear span cantilever beam is subjected to a concentrated load of 16 kN at 500 mm from free end after 6 months of construction. The beam is exposed to open non-aggressive environment. The beam was designed for strength criterion in example 3.3. The design parameters and details are:

Concrete grade is $M25$ and reinforcement is $HYSD$-$Fe415$. The corresponding design coefficients are: $K = 0.138, j = 0.80, k_u = 0.479. f_{ck} = 25, f_y = 415$ N/mm^2. The collapse moment is 88.7 kNm. The effective cantilever span is $= L =$ clear span + effective depth of the beam $= 3.4$ m. 3 numbers of 20 mm bars are provided at top, ($A_{st} = 942$ mm^2). Adequate development length is provided. 2 numbers of 8 mm bars at bottom as hanger bars. The width of the beam is 250 mm and the total depth is 370 mm. The effective depth is 330 mm. The depth of beam in this example is assumed as constant for the simplicity of illustration. Compute approximate deflection of the beam and compare with the allowable values.

Solution

Modulus elasticity of the concrete is: $E_c = 5000\sqrt{25} = 25,000$ N/mm^2
Modular ratio $= E_s/E_c = 200\,000/(25\,000) = 8$
The neutral axis distance is $xu = 0.479(330) = 158$ mm

The collapse moment computed in the example 3.3 is 88.7 kNm. The service load moment on the beam can be obtained by dividing the collapse moment by 1.5 in this case.
The gross moment of inertia section is:

$$I_g = \frac{bh^3}{12} = \frac{0.250(0.370)^3}{12} = 0.00105\,\text{m}^4$$

The fracture bending moment of the section at mid-span and at support are same and it is:

$$M_{cr} = \frac{I_g f_{cr}}{y_t} = \frac{bh^2 \sqrt{f_{ck}}}{8.6} = 250(370)^2(\sqrt{25})/8.6(1000\,000) = 20\,\text{kNm}$$

The moment of inertia of the cracked section at mid-span is same as that of the previous example:

$$I_{cr} = \frac{bx_u^3}{3} + mA_{st}(d - x_u)^2 = 0.25(0.156)^3/3 + 8(0.000942)(0.33 - 0.158)^2$$

$$= 0.000328 + 0.000223 = 0.000551\,\text{m}^4$$

The effective moment of inertia of the cross section is:

$$I_e = \frac{I_{cr}}{1.2 - \dfrac{M_{cr}(j)}{M}\left[1 - \dfrac{x_u}{d}\right]\dfrac{b_w}{b_f}}$$

$$I_e = \frac{0.00051}{1.2 - \dfrac{20.(0.8)}{88.7}\left(1 - \dfrac{0.158}{0.330}\right)\dfrac{0.25}{0.25}} = \frac{0.000551}{1.106} = 0.000498\,\text{m}^4$$

The effective moment of inertia must be limited to lower bound value of the moment of inertia of the cracked section. So select the effective moment of inertia as that of the cracked section.

$$I_e = 0.000551\,\text{m}^4$$

The dead load is that of the self weight and it is: $w_g = 2.3$ kN/m. The deflection due to the dead load is:

$$v_d = \frac{w_d L^4}{8 E_c I_e} = \frac{2300(3.4)^4}{8(25000)(10)^6(0.000551)} = 0.0028 m = 2.8 \text{mm}$$

The live load deflection due to 16 kN placed at 0.5 m from free end is:

$$v_t = \frac{W_l L_0^3}{3 E_c I_e} = \frac{16000(3.4 - 0.5)^3}{3(25000)(10)^6(0.000551)} = 0.0094 \text{ m} = 9.4 \text{mm}$$

Shrinkage Deflection

The percentages of tension and compression reinforcements are:

$$p_t = 100(942)/250(370) = 1.02$$

$$p_c = 100(100)/250(370) = 0.11$$

$$p_t - p_c = 0.91, \text{ that is less than 1.}$$

So select the value of k_4 as

$$k_4 = \frac{0.72(p_t - p_c)}{\sqrt{p_t}} = \frac{0.72(1.02 - 0.11)}{\sqrt{1.02}} = 0.65$$

The beam is cantilever so the value of the k factor on boundary conditions is: $k_3 = 0.5$. The shrinkage Maximum shrinkage strain is = 0.0003
Shrinkage deflection is given by

$$v_{sh} = k_3 k_4 \varepsilon_{cs} \frac{L^2}{h}$$

The span is 3.4 m, total depth of the beam is 0.37 m and the shrinkage strain is 0.0003. The substitution of different values in the shrinkage deflection is:

$$v_{sh} = k_3 k_4 \varepsilon_{cs} \frac{L^2}{h} = 0.5(0.65)(0.003)(3400)^2/370 = 3.05 \text{mm}$$

The deflection due to creep is obtained by multiplying the elastic deflection caused by the permanent loads by the creep coefficient. The deflection due to dead load is 2.8 mm. The creep coefficient for loading after 6 months can be taken as 1.2. So the creep deflection is:

$$v_c = c_c v_d = 1.2(2.8) = 3.36 \text{ mm}$$

The total deflection including creep and shrinkage is:

$$v_t = v_d + v_l + v_c + v_{sh} = 2.8 + 9.4 + 3.36 + 3.05 = 18.61 \text{ mm}$$

The admissible live load and total deflections are:

$$v_{al} = L/350 = 3400/350 = 9.71 \text{ mm}$$

$$v_{at} = L/250 = 3400/250 = 13.6 \text{ mm}$$

The admissible deflection under live load is more than the actual one. However the total deflection is 18.61 mm as against the allowable value of 13.6. This is not admissible. The clear span of the cantilever is 3 m but the effective span is taken as 3.4 m to be on the safer side. The effective span depends on the rigidity of the support.

Example 6.5: Deflection of T-beam

A simply supported beam of 8.4 m effective span is supporting a slab of 120 mm thickness. The slab is carrying a uniformly distributed load of 4 kN/m², in addition to the weight of the slab. The beam is in a protected environment. The beams are spaced at 4 m apart. (The example was solved for the strength consideration in chapter 4.) Determine the deflection of the beam. The beam can be treated as *T* beam section.

Solution

The given data and the assume information for preliminary calculation are:

Effective span $= L = 8.4$ m, $t = 120$ mm, $w_l = 4$ kN/m², $f_{ck} = 20$ N/mm²
The superimposed load from the weight of the slab is: $w_s = 0.125*4*25 = 12.5$ kN/m
The superimposed live load from the slab is: $w_l = 4*4 = 16$ kN/m

The width of the flange is the minimum of the following:

b_f = Centre to centre distance of the beams = 4 m = 4000 mm, or
$b_f = L_o/6 + b_w + 6t = 2.32$ m

Assume the size of the web, excluding the thickness of the slab as 300 by 500 mm for self weight purpose.

The assumed self-weight $= w_g = 0.3(0.5)*25 = 3.75$ kN/m, use 4 kN/m as self-weight.
Design service load on the beam is $= w_t = (w_g + w_s + w_l) = 4 + 12.5 + 16 = 32.5$ kN/m,

The service load maximum bending moment is $= M = \dfrac{wL^2}{8} = 286.65$ kNm

The overall and the effective depths based on the strength limit state design are: $h = 600$ mm and $d = 560$ mm

The area of reinforcement provided is:5 numbers *of 25 mm bars at the bottom. And 2 numbers of 10 mm at top for stirrup support bars.*
The area of tension reinforcement provided is $A_{st} = 5*490 = 2450$ mm²

$$p_t = \frac{2450*100}{300*560} = 1.45\%$$

The percentage of compression reinforcement in the top flange is not given. The reinforcement of the slab should also be included in the compression steel. Assume the percentage of the compression steel as 0.2 percent. Fig 6.1 illustrates the *T*-beam details.

The section is divided into two segments, the total web portion (300*600) and the overhang flange portion (2000*120). The centroid and the moment of inertia of the gross section are computed using the two segments. The area of the cross section is:

$$A_c = 0.3*0.6 + 2.0*0.12 = 0.18 + 0.24 = 0.42 \text{ m}^2$$

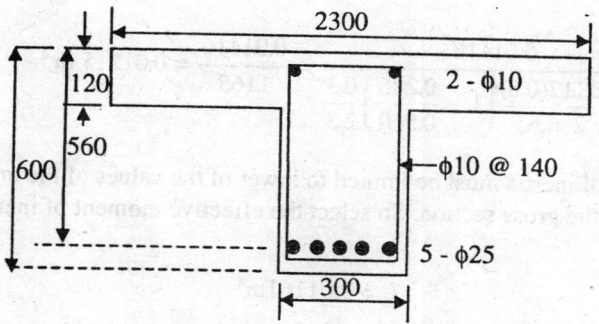

Fig. 6.1 T Section; Example 6.5.
(All Dimensions in mm)

Let y_t is the distance of the centroid of the section from top fibre is obtained by taking moment of the areas about the top fibre, then,

$$y_t = \frac{0.18*0.3 + 0.24*0.06}{0.42} = 0.163\,\text{m} = 163\,\text{mm}$$

$$I_g = \frac{0.3*0.6^3}{12} + 0.18(0.3 - 0.163)^2 + \frac{2*0.12^3}{12} + 0.24(0.163 - 0.06)^2 = 0.01161\,\text{m}^4$$

The neutral axis distance of the section is:

$$x_u = 0.479\,d = 0.479(0.56) = 0.268\,\text{mm}$$

The cracked moment of inertia of the section is computed taking the total flange width and its moment of inertia about the neutral axis. The effect of the small potion of the web below the flange is negligible so not included.

$$I_{cr} = \frac{b_f t^3}{12} + b_f t (x_u - t/2)^2 + m A_{st}(d - x_u)^2$$

$$= \frac{2.3(0.12)^3}{12} + 2.3(0.1)(0.268 - 0.06)^2 + 9(0.002450(0.56 - 0.268)^2 = 0.01416\,\text{m}^4$$

The cracked moment of inertia of the section is more than the gross moment of inertia. Therefore the lower of the values has to be taken.

The fracture bending moment of the sections at mid-span and at support are same and it is:

$$M_{cr} = \frac{I_g f_{cr}}{y_b} = \frac{0.01161(10)^{12}(0.7)\sqrt{20}}{(600 - 163)(10)^6} = 83.17\,\text{kNm}$$

The effective moment of inertia of the cross section is:

$$I_e = \frac{I_{cr}}{1.2 - \frac{M_{cr}(j)}{M}\left[1 - \frac{x_u}{d}\right]\frac{b_w}{b_f}}$$

$$I_e = \frac{0.01416}{1.2 - \dfrac{83.17(0.8)}{286.65}\left(1 - \dfrac{0.286}{0.560}\right)\dfrac{0.3}{2.3}} = \frac{0.01416}{1.165} = 0.01215 \text{ m}^4$$

The effective moment of inertia must be limited to lower of the values of the moment of inertia of the cracked section or that of the gross section. So select the effective moment of inertia as that of the gross section.

$$I_e = 0.01161 \text{m}^4$$

The dead load is that of the self-weight and it is: $w_d = w_g + w_s = 16.5$ kN/m.

The deflections are computed treating the beam slab construction idealized into a T beam. The two dimensional space action is not included in the analysis. The deflections computed on this basis are on the very much safe side.

The deflection due to the dead load is (all units in metres and Newtons):

$$v_d = \frac{5w_d L^4}{384 E_c I_e} = \frac{5(16500)(8.4)^4}{384(5000)\sqrt{20}(10)^6(0.01161)} = 0.0041m = 4.1 \text{ mm}$$

The deflection due to live load of 16 kN UDL is:

$$v_l = \frac{5w_l L_0^4}{384 E_c I_e} = \frac{5(16000)(8.4)^4}{384(5000)\sqrt{20}(10)^6(0.01161)} = 0.004m = 4 \text{ mm}$$

Shrinkage deflection

The percentage of tension reinforcement is computed with respect to web area. The compression reinforcement is considered to be negligible to be on the safer side.

$$p_t = 1.45, p_c = 0$$

$$p_t - p_c = 1.45, \text{ that is more than } 1.$$

So select the value of k_4 as

$$k_4 = \frac{0.65(p_t - p_c)}{\sqrt{p_t}} \le 1.0$$

$$k_4 = \frac{0.65(p_t - p_c)}{\sqrt{p_t}} = \frac{0.65(1.45)}{\sqrt{1.45}} = 0.54$$

The beam is simply supported so the value of the k factor for the boundary conditions is $k_3 = 0.125$. This factor could be much lesser than this value considering the beam slab integrated action.

Maximum shrinkage strain is $= 0.0003$

Shrinkage deflection is given by

$$v_{sh} = k_3 k_4 \varepsilon_{cs} \frac{L^2}{h}$$

The span is 8.4 m, total depth of the beam is 0.56 m and the shrinkage strain is 0.0003. The substitution of different values in the shrinkage deflection is:

$$v_{sh} = k_3 k_4 \varepsilon_{cs} \frac{L^2}{h} = 0.125(0.54)(0.0003)(8400)^2 / 560 = 2.55 \, mm$$

The deflection due to creep is obtained by multiplying the elastic deflection caused by the permanent loads by the creep coefficient. The deflection due to dead load is 4.1 mm. The creep coefficient for loading after 6 months can be taken as 1.2. So the creep deflection is:

$$v_c = c_c v_d = 1.2(4.1) = 4.9 \, mm$$

The total deflection including creep and shrinkage is:

$$v_t = v_d + v_l + v_c + v_{sh} = 4.1 + .4 + 4.9 + 2.55 = 15.55 \, mm$$

The admissible live load and total deflections are:

$$v_{al} = L/350 = 8400/350 = 24 \, mm.$$

$$v_{at} = L/250 = 3400/250 = 33.6 \, mm$$

The admissible deflections are more than the actual ones. The member sizes in beam slab constructions are normally large. Therefore the deflections of the beams in reinforced concrete beam slab constructions are usually small.

6.5 DESIGN FOR STRENGTH AND DEFLECTION

The design was carried out by piecemeal so far to give a student for better understanding and to enable him to do some exercises. A beginner is able to understand better and appreciate the subject with one section at a time. The total design of a reinforced concrete beam consists of the following main stages:

1. Identification of exposure condition,
2. Selection of materials and specifications,
3. Estimation of preliminary sections,
4. Selection of loads such as self weight, live load, seismic or wind coefficients etc,
5. Structural analysis and determination of limit forces such as moments, shear, torsion etc,
6. Design for strength to withstand collapse moment, size, reinforcement details,
7. Revise the structural analysis and strength design if the final required section is different from that of the assumed one,
8. Design for shear strength,
9. Design for torsional effect if present,
10. Analysis for service load deflections,
11. Check for deflection serviceability,
12. Check for crack width design criterion,
13. Detailing of the reinforcement

A typical example involving some of the steps is illustrated.

Example 6.6: Complete Design of Reinforced Beam

A simply supported beam of 6m effective span in mild environment is subjected to a live load of 24 kN/

m and it is to be designed with $M25$ concrete and $HYSD415$ bars. The depth is 450 mm for architectural consideration.

Solution

The given data is: $L = 6$ m, $h = 450$ mm, $f_{ck} = 25$ N/mm^2, $f_y = 415$ N/mm^2, $w_l = 24$ kN/m, the design coefficients from table 3.1 are: $K = 0.138$, $j = 0.80$, $k_u = 0.479$

Assume the width of the beam as $b = 300$ mm,

Self weight of the beam = $w_g = 0.30(0.45)25 = 3.375$ kN/m,

Total design load = $w_t = 1.5(3.375 + 24) = 41.0625$ kN/m, use 42 kN/m

Maximum bending moment at collapse = $M_c = \dfrac{w_t L^2}{8} = \dfrac{42(36)}{8} = 189$ kNm

At the moment, it is not known whether the given size is under-reinforced or over- reinforced. So it is necessary to find out the balanced moment capacity of the section. The effective depth of the section is equal to

$$d = 450-35 = 415 \text{ mm}$$

$$M_{rb} = Kbd^2 f_{ck} = 0.138(300)(415^2)(25)/10^6 = 178.25 < 189 \text{ kNm}$$

The balanced moment capacity of the section is less than the collapse moment; therefore the beam is designed as doubly reinforced concrete section.

The area of the tension reinforcement needed for balanced section portion is:

$$A_{sb} = \frac{1.15 M_{rb}}{jdf_y} = \frac{1.15(178.25)10^6}{0.8(415)415} = 1488 \text{ mm}^2$$

The area of compression and tension reinforcement for the remaining portion of the bending moment is;

$$A_{sc} = \frac{A_{st2} = 1.15(M_c - M_{rb})}{(d.dc)f_y} = \frac{1.15(189 - 178.25)10^6}{(415 - 35)415} = 78 \text{ mm}^2$$

The total area of tension reinforcement provided is:

$$A_{st} = A_{sb} + A_{st2} = 1488 + 78 = 1566 \text{ mm}^2$$

Provide 5 numbers of 20 mm bars at bottom.

A_{st} (provided) = 5*314 = 1570 mm^2; Actually required is 1566 mm^2, marginally more than that provided.

Provide 2 numbers of 10 mm to bars at 35mm below the top fibre. These bars will be used as hanger bars for shear reinforcement.

Design for Shear

Total design load at strength state is = $w_t = 42$ kN/m

The critical section for shear failure is at a distance d from the support. The critical shear failure occurs at a distance d from the support in simply supported beams. The effective depth of the section is $= d = 415$ mm.

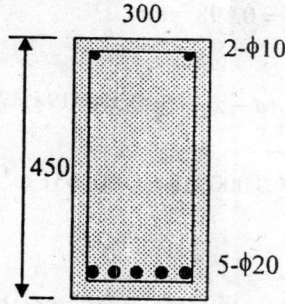

Fig. 6.2 Example 6.6.

The design shear force at the section is:

$$V_u = \frac{w_t(L - d)}{2} = 42(2.75 - 0.415) = 98.07 \text{ kN}$$

The nominal shear stress is =

$$\tau_v = \frac{V_u}{bd} = \frac{98070}{300(415)} = 0.79 \text{ N / mm}^2$$

Assuming only three 20 mm tension bars are available at the support,
The percentage of steel is = 3*314*100/(300*415) = 0.75%
The shear strength of concrete for 0.75 percent of reinforcement and the absolute shear strength of concrete with stirrup steel are taken from table 3.4. And they are:

$$\tau_c = 0.57 \text{ N/mm}^2 \text{ and } \tau_{cmax} = 3.1 \text{ N/mm}^2$$

The shear strength of concrete is 0.57 N/mm², which is less than the nominal shear stress. Select two legged 8 mm bar as shear stirrup, the area of two legs of the stirrup is: 100 mm². The spacing of the stirrup reinforcement is:

$$s_v = \frac{0.87 A_{sv} f_{yv}}{b(\tau_v - \tau_c)} = \frac{0.87(100(415)}{300(0.79 - 0.57)} = 547 \text{ mm}$$

The maximum spacing of the stirrups is controlled by:

$$S_{max} \le \frac{A_{sv} f_{yv}}{0.4b} = \frac{100(415)}{0.4(300)} = 345 \text{ mm}$$

$s_{max} = d = 435$ mm or 300 mm, which ever least. Provide two leg 8 mm stirrups at 300 mm spacing.

Modular ratio = E_s/E_c = 200 000/ 5000 $\sqrt{25}$ = 40/5 = 8.0,

The gross moment of inertia and the moment of inertia of the cracked section are:

$$I_g = \frac{bh^3}{12} = \frac{0.300(0.450)^3}{12} = 0.00227 \text{ m}^4$$

$$x_u = 0.479(0.415) = 0.198 \text{ mm}$$

$$I_{cr} = \frac{bx_u^3}{3} + mA_{st}(d - x_u)^2 = 0.30(0.198)^3/3 + 8(0.00227)(0.415 - 0.198)^2$$

$$= 0.000776 + 0.000855 = 0.001631 \text{ m}^4$$

The bending moment at fracture is:

$$M_{cr} = \frac{I_g f_{cr}}{y_t} = \frac{bh^2 \sqrt{f_{ck}}}{8.6} = 300(450)^2(\sqrt{25})/8.6(1000\,000) = 35.32 \text{ kNm}$$

The bending moment at service load condition is:

$$M = 189/1.5 = 126 \text{ kNm}$$

$$I_e = \frac{I_{cr}}{1.2 - \frac{M_{cr}(j)}{M}\left[1 - \frac{x_u}{d}\right]\frac{b_w}{b_f}}$$

$$I_e = \frac{0.001631}{1.2 - \frac{35.32(0.8)}{126}\left(1 - \frac{1.98}{0.415}\right)\frac{0.30}{0.30}} = \frac{0.001631}{1.082} = 0.0015 \text{ m}^4$$

This value is less than the moment of inertia of the cracked section so the cracked sectional value is used.

$$I_e = 0.0015 \text{ m}^4$$

The maximum deflection occurs at the mid span due to uniformly distributed load. The maximum deflection is:

$$v = \frac{5wL^4}{384E_c I_e}$$

Where w is the *UDL*.

The Young's modulus of elasticity of the concrete is:

$$E_c = 5000\sqrt{25} = 25000 \text{ N/mm}^2 = 25.(10^6) \text{ kN/m}^2$$

The deflections due to dead and live loads are computed separately. The dead load is that due to the self weight and it is 3.4 kN/m. The corresponding deflection is:

$$v_d = \frac{5(3.4)5.5^4}{384(25)(10^6)(0.0015)} = 0.0011 \text{ m} = 1.1 \text{ mm}$$

The corresponding deflection due to live load of 24 kN/m is:

$$v_l = 1.1(24)/3.4 = 7.7 \text{ mm}$$

The admissible deflection due to live load alone is:

$$v_a = L/350 = 5500/350 = 15.7 \text{ mm.}$$

The admissible deflection is 15.7 mm whereas that is caused by the live load is 7.7 mm only. The percentages of tension and compression reinforcements are:

$$p_t = \frac{100 A_{st}}{bd} = \frac{100(1570)}{300(415)} = 1.26$$

$$p_c = \frac{100 A_{sc}}{bd} = \frac{100(2*78)}{300(415)} = 0.13$$

$$p_t - p_c = 1.26 - 0.13 = 1.13 > 1.0$$

So select the value of k_4 as

$$k_4 = \frac{0.65(p_t - p_c)}{\sqrt{p_t}} = \frac{0.65(0.126 - 0.13)}{\sqrt{1.26}} = 0.65$$

The beam is simply supported so the value of the factor on boundary conditions is: $k_3 = 0.125$. The shrinkage deflection is given by:

$$v_{sh} = k_3 k_4 \varepsilon_{cs} \frac{L^2}{h}$$

The span is 5 m, total depth of the beam is 0.47 m and the shrinkage strain is 0.0003. The substitution of different values in the shrinkage deflection is:

$$v_{sh} = k_3 k_4 \varepsilon_{cs} \frac{L^2}{h} = 0.125(0.65)(0.0003)(5500)^2 / 450 = 1.6 \text{ mm}$$

The deflection due to creep is obtained by multiplying the elastic deflection caused by the permanent loads by the creep coefficient. The deflection due to dead load is 1.1 mm. The creep coefficient for loading after 6 months can be taken as 1.2. So the creep deflection is:

$$v_c = c_c v_d = 1.2(1.1) = 1.3 \text{ mm}$$

The total deflection including creep and shrinkage is:

$$v_t = v_d + v_l + v_c + v_{sh} = 1.1 + 7.7 + 1.3 + 1.6 = 11.7 \text{ mm}$$

The admissible live load and total deflections are:

$$v_{al} = L/350 = 5500/350 = 15.7 \text{ mm.}$$

$$v_{at} = L/250 = 5500/250 = 22 \text{ mm}$$

The admissible deflections are more than the actual ones. The beam is safe in the deflection serviceability.

6.6 SOME COMMENTS ON DEFLECTION COMPUTATIONS

Deflection of a beam depends very much on the moment of inertia of the member. Most beams are design on cracked section theory. This implies that all the tensile force is resisted by reinforcement and the tensile strength of concrete is neglected. The cross section is cracked at some locations and uncracked at many other locations. The moment of inertia of the member is variable along the length of the member. Effective moment of inertia of the member is an idealization and is applied to all types of beams such as simply supported, cantilevers and framed ones and subjected to variety of forces. The idealization is an over simplification. The result the deflection computed is not necessarily an exact value. The deflection caused by the shrinkage is usually small when compared with that caused by the loads or creep. Indian code of practice suggests a set of span to depth ratios and if one uses them, the deflection requirement is automatically satisfied. The desirables span to depth ratios to satisfy the limits of vertical deflections are recommended by the Indian code of practice, they are listed Table 6.4. The serviceability check for deflection need not be carried out provided that the spans to depth ratios are within the values specified by the code.

Table 6.4 Span to Depth Ratio (for Span < 10 m).

	Type of structure	L/h
1.	Cantilever beams	7
2.	Simply supported	20
3.	Continuous beams	26

Note: The limits of the spans to depth ratios have to be modified in case of the spans more than 10 m. The values of the ratios given in Table 6.4 are to be multiplied by 10/span (in metres) except for cantilever beams to find the admissible values. The deflection computations have to be made in case of cantilever beams of spans more than 10 m. The values given in Table 6.4 are further subject to modification based on the percentage of reinforcement and the actual stress allowed in the reinforcement during the service load condition. The code suggests a set of multiplying factors depending on the percentage of tension and compression reinforcements through a set of graphs. The range of percentage tension reinforcement in most beams is 0.8 to 2 for which the multiplication factors are:

1 to 0.8, for deformed bars and
1.5 to 1.0, for plain bars.

The recommended effective width of the flange in T beams is primarily for strength considerations. The effective width that comes into play for deflection consideration is normally less than that considered for strength. The moment of inertia of flanged sections needs reduction to be on the safe side in service load condition. The code recommends some reduction factors. The reduction factor recommended for flange beams is in the range of 0.8 to 1.0 depending on the ratio of web width to flange width. In the opinion and experience of the author, the design basis of the code for deflections is conservative especially for beam slab construction. Deflections of isolated T beams are different from those of T-beams coming from beam and slab constructions. The two dimensional continuity is ignored to be on the safer side. Hence the deflections of beams in beam slab constructions are not normally critical. The moment of inertia taken for deflection computations is close to the moment of inertia of the cracked sections. Therefore, the actual deflection of a beam will be smaller than that computed.

In addition to the deflection limit, the slenderness of the beam should also be restricted if the top flange is not supported laterally. The lateral stability of the beam is ensured only if the lateral supports are provided subject to the limits given below. The clear distance between the two lateral supports should be less than $60b$ or less than $250\ b^2/d$, whichever is less. This may be expressed as:

$$L_e < 60\ b \text{ and}$$

$$L_e < 250\ b^2/d. \tag{6.15}$$

In case of cantilever beams, the distance between the free end and the lateral support should confirm to the following.

$$L_e < 25\ b \text{ and}$$

$$L_e < 100\ b^2/d. \tag{6.16}$$

Moment of Inertia

The accuracy of the deflection computation depends on the accuracy in the assumption of computing the effective moment of inertia and the Young's modulus of elasticity. The question that often encountered is when does the gross moment of inertia dominate that of the cracked section. The computation of the moment of inertia of the cracked section is also made approximate. The moment of inertia should be computed about the centroid of the cracked section and not really about the neutral axis. The depth of cracking during the service load condition is different from that of the strength state. For convenience of computation, the moment of inertia of the cracked section is computed about the neutral axis of the cracked at the limit state and it is quite possible that the cracked section at service load has higher moment of inertia. To have an idea of the dominance of the gross moment of inertia, a balanced section is considered. The gross moment of inertia is:

$$I_g = \frac{bh^3}{12} \tag{6.17}$$

The moment of inertia of the cracked section is:

$$I_{cr} = \frac{bx_u^3}{3} + mA_{st}(d - x_u)^2 \tag{6.18}$$

The above equation is expressed in percentage of reinforcement and the neutral axis distance and it is:

$$I_{cr} = bd^3\left[\frac{k_u^3}{3} + mp(1 - k_u)^2\right]$$

For a simple comparison, a section with $M20$ concrete and $Fe415$ steel is considered. In that case some of the values of the above equation are: $p = 0.0095$, $m = 13$, and $k_u = 0.479$. Assume the overall depth of the section as $1.1d$. The cracked moment of inertia works out to be:

$$I_{cr} = (0.0366 + 0.0335)bd^3 = 0.0527bh^3$$

The ratio of the gross moment of inertia to the cracked one is:

$$\frac{I_g}{I_{cr}} = \frac{1}{12 * 0.0527} = 1.58$$

For balanced rectangular cross sections, the cracked moment of inertia is about sixty three percent of that of the gross section. If the reinforcement is about two percent, then there is a good chance that the cracked moment of inertia may exceed that of the gross section. The cracked moment of inertia of flanged section is likely to dominate in many cases.

6.7 INTRODUCTION TO CRACKS IN REINFORCED CONCRETE CONSTRUCTION

Concrete develops cracks for several reasons and each crack has its own characteristic. The cracking may be due to non-structural and structural reasons. The non-structural reasons are:

1. Micro-cracks are developed if heat of hydration is not absorbed properly by curing in the initial stage of hardening.
2. Shrinkage of concrete will cause micro-cracks,
3. Some atmospheric agents and pollution reacts with free lime resulting in carbonation of the concrete. The carbonation of concrete increases the volume that results in cracking,
4. Alkaline aggregate reactions result in the increase in the volume of the aggregate, and consequent cracking,
5. Concrete when exposed to about 100^0 C heat, the moisture in the concrete become steam. The steam will burst out of the concrete resulting in cracking or spalling,
6. Rusting (corrosion) of embedded reinforcement will cause increase in the volume of steel, then increase in the diameter of the bar which results in cracking of concrete,

The structural cracking can be due to the following reasons. It is primarily due to the stress caused in concrete by the loads exceeds the tensile strength of the concrete.

Every crack in concrete has is detrimental to the durability and life of the structure. The structural cracking is due to stress exceeding the tensile strength of the concrete and this is listed as number one item in the list. The adverse effects of cracks concrete are:

1. Psychological effects on the occupants, and owners of the buildings,
2. Ugly and unsightly appearance,
3. Increased permeability that results in further carbonation, chemical reactions,
4. Corrosion of reinforcement,
5. Decrease in sectional area,
6. Loss of durability and reduced life of the structure.

Cracks in concrete must be avoided at all costs. Small discussion on the reasons for cracking and possible precautions to avoid cracking is suggested here. The reader is also advised study the section on durability for additional information.

Micro-cracks

Micro cracking occurs randomly at many places due to heat of hydration not dissipated by proper curing or could be due to chemical reactions. Micro cracks due to heat of hydration are primarily from shrinkage effects. On the other hand the chemical reactions such as carbonation or alkaline aggregate reaction cause increase in volume. The increase in volume of the new compounds such as carbonates or alkali oxides cause expansion and consequent cracking in the concrete. A large number of small cracks develop randomly. Such cracks may not be seen on both faces at the same locations. Further the direction of the crack is oriented in any specific direction. The micro cracks can be prevented by proper curing of concrete and proper choice of the basic ingredients of the concrete.

Shrinkage Cracks

Three types of shrinkages take place: Plastic shrinkage, Drying shrinkage and Thermal shrinkage. When boundary and external conditions restrain shrinkage, tensile strains are produced in the concrete. As the tensile stress in concrete exceeds the tensile strength of the concrete, crack are developed. The shrinkage cracks can be controlled by shrinkage reinforcement. The shrinkage cracks are normally found in beams of frames. The columns restrain the free shrinkage of the beams, so tensile strains across the section along the axis of the beam are developed. Such shrinkage cracks are dominant in plain concrete but could be controlled by providing nominal reinforcement uniformly distributed across the section. Large sections say in the range of 1000 mm or more in depth have a higher tendency of developing shrinkage cracks. The cracks are perpendicular to the axis of the member and spaced about at a distance of 1000 mm to 2000 mm depending on the length of the member and the percentage of the reinforcement. The cracks appear on both the vertical faces of the beam. The crack width at the middle height of the beam is more and decreases towards the bottom and top faces. This implies less reinforcement at the middle height of the beam. Reinforcement is provided at the bottom and top faces of the beam for structural purpose. The middle height of the beam normally doesn't need the structural reinforcement unless there is torsional reinforcement is provided. The nominal longitudinal reinforcement can be placed at 500 to 600 mm height on each vertical faces to prevent shrinkage. The percentage of such nominal reinforcement on vertical face can be about 0.05 percent of the cross section. It is advisable to place smaller diameter bars at closer spacing rather than bigger diameter bars.

Shrinkage cracks also appear on the walls and in the slabs if the spacing of the bars is more than 300 mm. The maximum spacing of the bars in slabs and in walls is about 300 mm irrespective of the thickness of the member. Thickness of raft slabs is more than 400 mm and may even go upto 5000 mm for large bases. The differential shrinkage in thick raft slabs can cause internal stresses and cracking. It is therefore recommended that nominal layer reinforcement for every depth of 1000 mm be provided. Top and bottom reinforcement in raft is relatively high even then, middle layers of reinforcement is desirable when the depth of the raft exceeds 1500 mm. Further the slab behaves like a reinforced concrete construction if the spacing of the reinforcement is at reasonable spacing. The shrinkage is in all directions, but its effect in the longitudinal direction is more dominant because of the length. Differential shrinkage is another problem area. Continuous laying or poring of concrete is recommended any given member to minimize the differential shrinkage and to avoid construction joints. Each section at construction joint is expected to have the same strength as that of any monolithic section. The differential shrinkage sets interface stresses. Usually these stresses are self-balancing type. In other words, there will be tensile and compressive stresses distributed such that the net effect is self-balancing. Closely spaced small diameter bars can be placed in the differential shrinkage zone.

Figure 6.3 illustrates shrinkage cracks in a beam of a reinforced concrete frame. It is assumed that the reinforcement is at the top and bottom faces of the beam, further the size of the beam is more than 500 mm by 650 mm or above.

Structural Cracks

The cracking can be due to:
a) Bending moment,
b) Shear force, or combined moment and shear,
c) Tension force,
d) Torsional force.

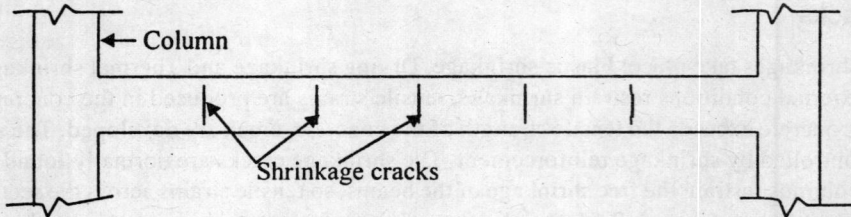

Fig. 6.3 Shrinkage Cracks in Beam, Spaced about the Same Distance Apart.

The bending moment cracks developed at the critical bending moment zone on the tension face and expand to the vertical faces in the tension zone. The cracks are normally vertical or near vertical in most cases and appear on both the vertical faces of the beam. The critical zones of cracking in beams are: 1 Middle span zone at the bottom fibre in simply supported beams, 2 at top fibres near fixed support in cantilever beams, 3 at top fibres near the column supports in framed and continuous beams. The cracks near the supports may also be the combined effect of the shear and bending moment. Fig. 6.4 illustrates the bending cracks in a simply supported beam. The cracks are wide at bottom and then close towards the neutral axis. This cracking occurs when the tension reinforcement is not adequate to withstand the loads.

The shear cracks in a beam occur near the support and it is about 40 degrees to the axis of the beam. Shear cracks occur if the diagonal tension developed at the maximum shear zone exceeds the tensile strain of the concrete. Typical shear cracks are indicated in Fig. 6.5. The failure by shear is more sudden when compared with the bending tension failure. Provide adequate shear reinforcement.

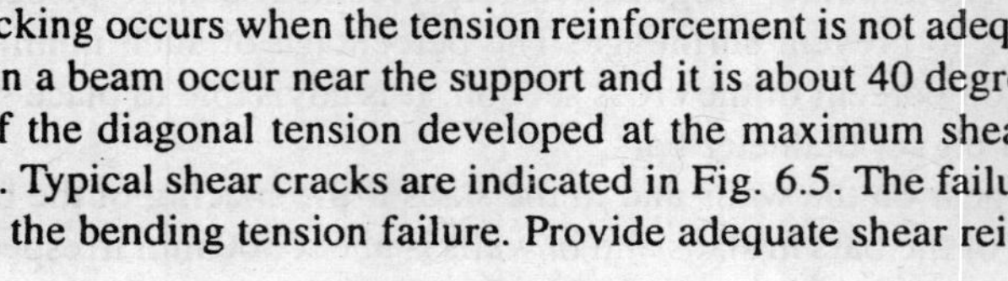

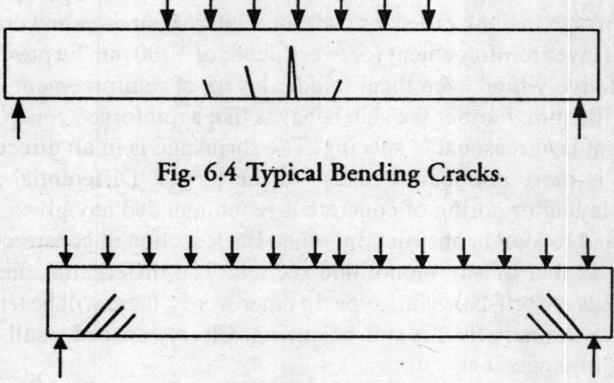

Fig. 6.4 Typical Bending Cracks.

Fig. 6.5 Typical Shear Cracks.

In continuous and framed beams, the maximum bending moment and shear forces occur near the support. The cracking in such cases is the combined effect of the bending and shear, occurs near the support. The inclination of the cracks is about 30 to 40 degrees to the axis of the beam.

6.8 CRACK WIDTH IN REINFORCED CONCRETE

Micro cracks are invisible to naked eye. Reinforced concrete beams crack in tension zone even in the service load condition. All tensile force must be resisted by the reinforcement. In water retaining structures, the tensile stress in concrete is limited to an admissible value, and yet the reinforcement must be designed to resist the total tension. The durability of the structure depends on the crack width. A crack width of 0.3 mm is allowed in structures in protected environment. A crack width of 0.2 mm is allowed in structures exposed to moderate environment. A maximum crack width of 0.1 mm is admissible in severe, very severe and extreme exposure conditions. A crack width of 0.1 mm is considered to be a no crack condition and permeability of the concrete is considered to be nil. The crack width at the surface in a member is also controlled by factors such as the diameter of the bar, the cover to the reinforcement and bonding of the reinforcement with concrete. The spacing and the layers of the reinforcement also has some marginal influence. A simple derivation about the spacing of the cracks in bending is illustrated here. Consider a typical beam in a cracked state as shown in Fig. 6.6(a). The cracks are shown to an exaggerated scale for illustration. The spacing of the cracks in the critical section is 2a. Fig. 6.6(b) illustrates a segment of the cracked section in which two consecutive cracks are illustrated along with notations.

Let 2a = spacing of cracks, and 2w = width of the crack

Consider Fig. 6.6(b) in which two sections of the segment are selected. Section A is at the cracked location and the stress distribution at this section is illustrated in Fig. 6.6(c). Let section B is the middle section between two successive cracks. The stress distribution on the section is shown in Fig. 6.6(d). The stress distribution in the uncracked location is similar to the elastic condition and that at the cracked location is similar to the strength state. There may be some minor variation of the stress distribution from that shown. The stress in steel reinforcement at the uncracked section is compatible with the strain in the concrete. The stress in the reinforcement at the cracked location is higher than that at the uncracked section. The bond stress between the concrete and the reinforcement over the segment of length a, is balanced by the difference of the stresses of bar of the two sections. The spacing of the cracks depends on the bond strength of the concrete and the diameter of the bar. Section A is cracked and the stress distribution is idealized as the limit state of strength. An approximate expression for the spacing of the cracks and crack width is derived in the simplest format. There are a number of empirical formulae exist in the literature. The one recommended by the Indian standard will be discussed later. The stress in steel reinforcement is:

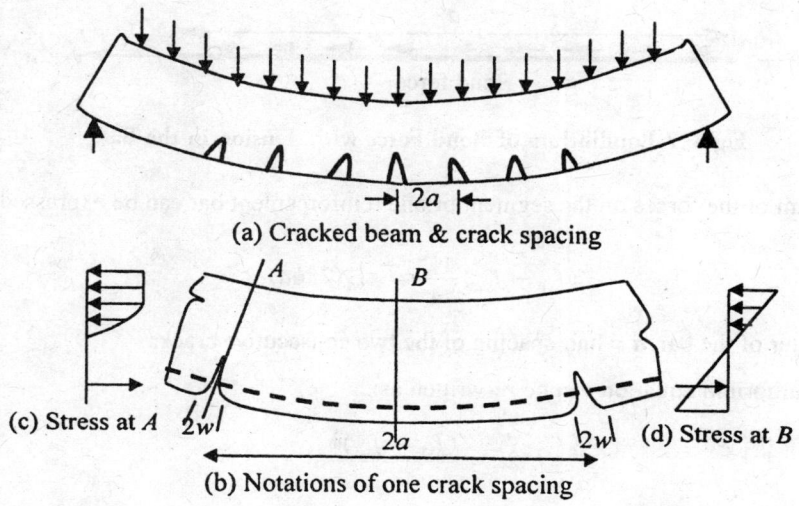

(a) Cracked beam & crack spacing

(c) Stress at A

(b) Notations of one crack spacing

(d) Stress at B

Fig. 6.6 Typical Cracked Section and Crack Spacing Notations.

$$f_{s1} = \frac{M}{jdA_{st}} \qquad (6.19)$$

where f_{s1} = stress in steel at the cracked section, jd = lever arm distance, M = bending moment.

The lever arm distance at service limit state will be different from that at the strength state. However the difference is marginal so the lever arm of the strength limit state is used.

The stress in steel at the uncracked location is controlled by the compatibility of strains in concrete and steel. The stress in concrete at the bottom fibre is:

$$f_{cr} = \frac{M}{Z_t} = \frac{6M}{bh^2(1 + 3mp)} \qquad (6.20)$$

where f_{cr} = fracture stress in concrete at the extreme fibre, Z_t = modulus of the section preferably that of a transformed section, p = the ratio of the reinforcement with respect to the sectional area, m = modular ratio, b and h are the width and overall depth of the section

The corresponding stress in steel, f_{s2} is:

$$f_{s2} = mf_{cr}\frac{d - x}{h - x} \qquad (6.21)$$

where m = modular ratio, d = effective depth of reinforcement, and x = neutral axis distance.

The neutral axis distance in the service load situation is different from that at the strength limit state.

Some approximations and idealizations of the quantities are made to get an idea of the crack spacing. The effective depth is assumed equal to 0.9h and similarly $(d-x)/(h-x)$ is assumed as 0.9. The crack develops if the bond of the steel fails to match the difference in the stress levels between the cracked and uncracked sections. The difference of the forces on steel bar between the cracked and uncracked sections must be equal to the bond force on the bar between the sections. This is illustrated in Fig. 6.7.

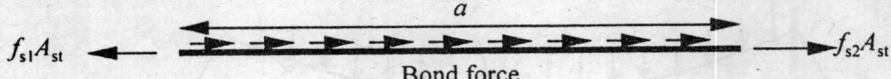

Bond force

Fig. 6.7 Equilibrium of Bond Force with Tension in the Bar.

The equilibrium of the forces on the segment of one reinforcement bar can be expressed as:

$$(f_{s1} - f_{s2})\frac{\pi\phi^2}{4} = \tau_{bd}(2\pi\phi a)$$

where ϕ = diameter of the bar, a = half spacing of the two consecutive cracks

The above equilibrium equation can be re-written as:

$$2a = \frac{(f_{s1} - f_{s2})\phi}{4\tau_{bd}} \qquad (6.22)$$

The difference in the stresses in steel bar can be approximated as:

$$f_{s1} - f_{s2} = M\left[\frac{1}{jdA_{st}} - \frac{(0.9)m}{Z_t}\right] = \frac{M}{Z_t}\left[\frac{(1+3mp)h}{6jdp} - 0.9m\right] = f_{cr}\left[\frac{(1+3mp)h}{6jdp} - 0.9m\right] \quad (6.23)$$

where the lever arm ratio is taken as 0.8, d/h as 0.9 and $p = 0.01$, is the percentage of the reinforcement. The spacing of the cracks can be approximated as:

$$2a = \frac{(f_{s1} - f_{s2})\phi}{4\tau_{bd}} = \frac{f_{cr}\phi}{4\tau_{bd}}\left[\frac{(1+3mp)}{54jp} - 0.9m\right] \quad (6.24)$$

The above equation can be expressed in terms of the bending moment as:

$$2a = \frac{M\phi}{Z_t 4\tau_{bd}}\left[\frac{(1+3mp)}{5.4jp} - 0.9m\right] \quad (6.25)$$

The Eq. (6.24) indicates the potential of the crack spacing while Eq. (6.25) gives the crack spacing for a given bending moment. To get an idea of the crack spacing, it is computed for balanced section of $M20$ concrete with $Fe415$ steel. The value of bond strength is taken as 1.9 N/mm^2 and the balanced reinforcement ratio is 0.01 and modular ratio as 9. The crack spacing is:

$$2a = 9\phi \quad (6.26)$$

The bending moment cracks at the maximum bending moment zone occur at a spacing of 9 times the diameter of the bar.

Crack width in the service load condition can be calculated in the similar lines. The total strain within the crack spacing can be lumped at the cracked location. The stress in steel is maximum at cracked location and minimum at the uncracked location. An average strain across the length can be taken as the average stress in the zone divided by the Young's modulus of elasticity multiplied by the length between the cracks. The total elongation can be taken as the width of the crack. The crack width can be expressed as a total elongation at steel level multiplied by a depth factor:

$$2w = \left(\frac{f_{s1} + f_{s2}}{2E_s}\right)\frac{h}{d}2a \quad (6.27)$$

The substitution of values of the stresses results:

$$2w = f_{cr}\left[\frac{1+3mp}{5.4pj} + 0.9\,m\right]\frac{a}{E_s}\frac{h}{d} \quad (6.28)$$

For $M20$ concrete, the crack width works out to be:

Crack width for $M20$ concrete $= 2.9\phi/1000$

The expressions of crack spacing or width given above are independent of the bending moment. The expression assumes that the bending moment causes flexural fracture at the extreme fibre. Therefore instead of fracture stress, one can use the actual bending moment divided by the section modulus. The crack width in terms of bending moment is:

$$2w = \frac{M}{Z_t}\left[\frac{1+3mp}{5.4pj} + 0.9\,m\right]\frac{a}{E_s}\frac{h}{d} \quad (6.29)$$

The crack width expression recommended by the code IS:456-2000 is:

$$2w = \frac{3a_{cr}\varepsilon_m}{1 + \dfrac{2(a_{cr} - C_{min})}{h - x}} \tag{6.30}$$

where

a_{cr} = distance from the point considered to the surface of the nearest longitudinal bar,
C_{min} = minimum cover to the longitudinal bar,
e_m = average strain in steel at the level considered,
h = overall depth of the beam,
x = depth of the neutral axis.

The average strain in steel can be computed assuming the plane section remains plane even after bending. The stresses in the concrete and steel are calculated from the moment. It may be that one can assume the maximum allowable stress in steel and the compute the strain in the steel.

The code suggests an empirical expression to compute the average strain in steel. The strain can be obtained as:

$$\varepsilon_m = \varepsilon_1 - \frac{b(h - x)(a - x)}{3E_s A_{st}(d - x)} \tag{6.31}$$

where

ε_1 = strain at the level under consideration,
A_{st} = area of the tension reinforcement,
a = distance from compression face to the point where the crack width is computed.

It is seen that the numerical value 3 in the expression has some dimension. The normally expected strain in steel is in the range of 0.00115 for Fe415 steel. The crack width expression given by the code also has another disadvantage. The diameter of the reinforcement is not reflected in it. There are a number of empirical expressions available in literature. The reliability and accuracy of the expressions is not as certain as the strength expression. The saving grace in the confusion is that the requirement is assumed to be satisfied if the detailing is carried out as per the code specification.

Example 6.7: Properties of Transformed Cross Section

A rectangular section 250 by 470 mm is provided with 3 numbers of 20 mm bars and one of 12 mm bar in a row at a 40 mm from bottom. Compute the transformed moment of inertia of the section and also the stresses in the $M20$ concrete and $HYSD$ steel for a service bending moment of 84 kNm. Determine the crack spacing and crack width. Vide Fig. 6.8 for the sectional details.

Solution

Area of the tension reinforcement is:

$$A_{st} = 3*314 + 113 = 1055 \text{ mm}^2.$$

The percentage of steel is = p = 1055*100/250*470 = 0.9%
The modular ratio for $M20$ concrete is = m = 200000/5000$\sqrt{20}$ = 9

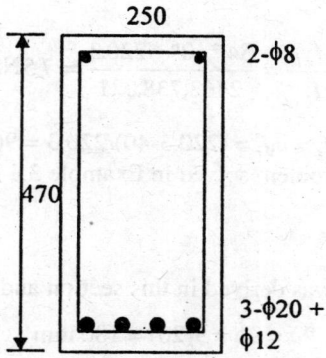

Fig. 6.8 Example 6.7.

The transformed area of the cross section is $= A_t = A_c + (m-1)A_{st} = 250*470 + 1055*8 = 125940$ mm^2.

The distance of the centroid of the transformed section from the middle height of the section can be obtained by taking the moment of the areas about centroid of the concrete section.

$$y_c = 1055*9(430 - 470/2)/125940 = 14.7 \text{ mm}$$

The second moment of the transformed area is obtained from the sum of the individual moment of inertia about their axes plus the sum of the products of the area multiplied by the square of the distance between the centroid of the corresponding area to the centroid of the total section. It is equal to:

$$I_t = \Sigma I_i + \Sigma A_i (y_c - y_{ci})^2$$

The moment of inertia of the transformed section, (neglecting the moment of inertia of the steel about its axis) is:

$$I_t = \frac{250*470**3}{12} + 250*470*(14.7)^2 + 8*1055*(430 - 470/2 - 14.7)^2$$

$$= 2162,979,166 + 25,390,575 + 274,368,279 = 2462,738,021 \text{ mm}^4$$

The gross sectional moment of inertia is:

$$I_g = 2162,979,166 \text{ mm}^4$$

The ratio of the transformed moment of inertia to the gross sectional inertia is: 2462,738,021/ 2162,979,166 =1.14

For a 0.9 percent of the reinforcement, the transformed moment of inertia is about 14 percent higher. It can be observed that the effect of the moment of inertia of the combined section taken about the centroid of the concrete section is generally close to that computed more accurately.

The stresses in the concrete section can be computed from the simple beam formula. The top and bottom fibre distance of the transformed section are: $y_t = 470/2 + 14.7 = 249.7$ and $y_b = 470/2 - 14.7 = 220.3$.

stresses in the section are:

$$\text{Stress at top fibre} = f_c = \frac{My_t}{I_t} = \frac{84*10^6*249.7}{2462,738,021} = 8.5 \text{ N/mm}^2$$

$$\text{Stress at bottom fibre} = f_c = \frac{My_b}{I_t} = \frac{84*10^6*220.3}{2462,738,021} = 7.5 \text{N/mm}^2$$

The compatible stress in steel is $= f_s = mf_t = (220.3-40)/220.3 = 9(7.5)*180.3/220.3 = 55.2 \text{ N/mm}^2$

Note: This example refers to the problem solved in Example 3.1 by limit state of strength.

Crack Spacing and Crack Width

The crack spacing for M20 concrete was derived in this section and the crack spacing is:

$$2a = 9\phi = 9(20) = 180 \text{ mm}$$

The crack width based on the simple derivation is:

$$\text{Crack width} = 2.9\phi/1000 = 2.9*(20)/1000 = 0.058 \text{ mm}$$

The crack width as per IS:456-2000 is given by:

$$2w = \frac{3a_{cr}\varepsilon_m}{1 + \dfrac{2(a_{cr} - C_{min})}{h - x}}$$

Various values given in the equation are:

a_{cr} = distance of the point under consideration from the surface of the nearest bar $= 40-10 = 30$ mm

in which the centre of reinforcement is 40mm from the outer fibre. It is same as $C_{min} = 40 - 10 = 30$ mm

The neutral axis is = 249.3mm   and $h = 470$ mm
The strain in steel $= f_s/E_s = 55.2/200\,000 = 0.00028$

The substitution of various quantities in the crack width calculation gives

$$2w = 3a_{cr}\, e_m = 3\,(30)(0.00028) = 0.025 \text{ mm}.$$

The crack width computed based on the Indian code of practice is 25 micron and that computed based on the approximate derivation is 58 microns. The derivation assumed d/h as 0.9 whereas the actual value is $=430/470 = 0.91$.

The denominator in the expression given by the code is likely to be more than 1. Therefore the maximum crack width is equal to $3\,a_{cr}\,e_m$. The allowable stress in steel in such exposure condition can be 150 N/mm^2 and the corresponding strain is 0.00075. The compatibility stress in steel is likely to be much less than this value. Assuming a maximum cover distance of 60mm in extreme exposure condition and the average strain in reinforcement as 0.00075, the maximum crack width as per the code is 0.135 mm. On the other hand if the cover provided is only 30 mm, the crack width is 0.078 mm. The crack width limitation in very severe and extreme exposure conditions is going to be important consideration. This is because of the large cover requirement in such exposure conditions. The durability requirement suggests more cover for more severe conditions. More cover distance provision results in more crack width. There is a dichotomy in the specifications of cover and the crack width in severe exposure conditions. Another drawback in the crack width expression is lack of term associated with diameter of bar. In the opinion of the author, diameter of reinforcement bar plays an important role in the crack width. Smaller diameter bar decreases the crack spacing and consequently the crack width.

PROBLEMS

(Use IS:456 – 2000)

6.1 A 5 m span cantilever beam in mild environment is subjected to a superimposed dead load of 10 kN/m and live load of 15 kN/m. The beam is made of M20 concrete and HYSD-Fe415 Bars. Design the beam by limit state design with a width equal to 300 mm and compute the deflection due to live load.

6.2 An eight metre simply supported beam is subjected to superimposed dead load of 10 kN/m and live load of 25 kN/m. The size of the beam is 250 by 700 mm. The concrete is M25 grade and six bars of 20 mm reinforcement is HYSD-Fe415 at an effective depth of 650 mm. Determine the deflections due to dead and live loads. The creep coefficient for the beam is 2.1. Compute the deflections caused by shrinkage and creep.

6.3 A five-metre span cantilever beam with a rectangular cross section of 300 by 500 mm is made of M20 concrete and 4 numbers of 20 mm HYSD-Fe415 bars at an effective depth of 40 mm at top and 2 numbers of 10 mm bars at the bottom. Estimate deflection caused by shrinkage for a shrinkage strain of 0.0003.

6.4 One-way 100 mm thick slab of effective span 4 m is made of M25 concrete with 12 mm HYSD-Fe415 bars spaced at 150 mm at an effective depth of 70 mm. Calculate the deflection caused by shrinkage for a shrinkage strain is 0.0003.

6.5 One-way 100 mm thick slab of effective span 4 m is made of M25 concrete with 12 mm HYSD-Fe415 bars spaced at 150 mm at an effective depth of 70 mm. The slab is subjected to a superimposed dead and live loads of 2kN/m² and 4 kN/m² respectively. Calculate the deflection caused by dead and total load. Check whether the deflections are with in the limits as per the cod. The shrinkage strain is 0.0003 and creep coefficient is 2.1.

6.6 A 200 mm thick cantilever slab of span 2.5 m is made of M20 concrete and has 12 mm HYSD-Fe415 bars at 150 mm at 35 mm from the top face. The slab carries dead load of 2 kN/m² and live load of 1.5 kN/m². Shrinkage is assumed as 0.0003. Determine the deflections caused by loads, shrinkage and creep for a creep coefficient of 1.6. Compute the total deflections.

6.7 A six metre effective span simply supported T-beam has the following cross sectional details b = 1.2 m, t = 120 mm, web size 250 by 450 mm and made of M20 concrete. Five numbers of 20 mm HYSD-Fe415 bars placed at 520 mm from top. The concrete shrinkage strain is given as 0.0003 determine the deflection caused by the shrinkage.

6.8 A simply supported with an effective span of 8.5 m is subjected to a superimposed dead and live loads of 10 and 20 kN/m respectively. The grade of the concrete is M30 and size is 350 by 650 mm. Six numbers of 20 mm bars are placed at bottom 45 mm from bottom face and three 20 mm bars at 45 mm from top face. The reinforcement is HYSD-Fe500 bars. Determine the deflections due to dead and live loads, shrinkage and creep. The shrinkage strain is 0.0003 and creep coefficient is 1.6.

6.9 A 250 by 450 mm beam is provided with five numbers of 20 mm HYSD-Fe415 bars at bottom with a clear cover of 30 mm. The grade of concrete is M20. Determine the fracture moment and spacing of 0.02 mm wide micro cracks at that bending moment.

6.10 A beam made of M30 concrete is 300 by 650 mm and is provided with six of 20 mm HYSD-Fe415 reinforcement bars at 50 mm from bottom and two of 20 mm bars at 50 mm from top. Determine the fracture moment and spacing of 0.03 mm wide micro cracks at the fracture moment.

6.11 A T-beam of M25 grade concrete has the following dimensions: flange width = 1200 mm and flange thickness 125 mm. The web size is 300 by 450 mm. Five numbers of 25 mm HYSD-Fe415 bars are placed 530 mm from top. Determine the fracture moment capacity of the section and spacing of 0.02 mm wide micro cracks.

Design of Staircases, Curved and Tie Beams

7.1 INTRODUCTION TO STAIRCASES

The architectural design, fixtures and finishes play important role in building of public utilities. Besides having a functional importance, there is the architectural challenge in making the stair to fit into the scheme of the facility. This chapter deals with the structural design. The staircases in the public utilities are subjected to overcrowding in peak hours at regular frequency. The width of the staircase is chosen based on the traffic. Some terminology of the staircase is illustrated in Fig. 7.1. The rise is the height of a step. The tread is the length of the step along the length of the staircase. The going is the horizontal distance between the first rise to the nose of the landing. Thickness of the staircase slab is the shortest distance measured perpendicular inclination of the staircase.

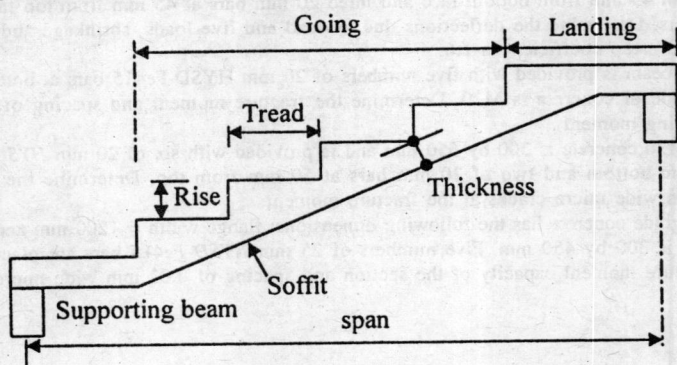

Fig. 7.1 Notation in Typical Staircase.

The width, rise and tread of a staircase depend on the functional needs. Some guidelines are given in Table 7.1.

The tread has both upper and lower bounds to provide a comfortable step movement of the users.

7.2 LOADS ON STAIRCASES

The live load on the staircase is for the horizontal projection of the staircase. If the landing is supported in one direction then the load on the landing is taken same as that of the live load. If the landing is

Table 7.1 Guidelines for Typical Dimensions of Staircases (Dimensions in mm).

Staircase	Range of dimensions in mm			
	Width	Rise	Tread	Landing
1. Service & maintenance	700-1000	110-170	225-300	700-1000
2. Residences	800-1200	110-160	225-300	800-1200
3. Apartments, offices,	1000-1500	110-160	250-300	1000-1400
4. Public houses & places	1200-3000	100-150	300-350	1200-1500

supported in two directions the live load in distributed equally between the two directions. The Indian standard code (IS: 456-2000) recommends effective spans and the load distribution for some cases. In view of variety of staircases and supporting conditions, the designer must make a rational choice of the effective spans in the analysis. Table 7.2 suggests the recommended live loads on the staircases. The weight of nominal bar railing is included in the live load.

Table 7.2 Live Load on Staircases and Balconies.

	Locality	Load in kN/m^2
1	Service stairs for maintenance, catwalks	1.5
2.	Residential buildings	3.0
3.	Office, public and industrial buildings	5.0
4.	Isolated steps concentrated load per step	1.5(kN)

Earthquake Load

Staircase is the vital link of communication and should have higher safety factor when compared with the other structural elements. The staircases are subjected to maximum loads during the emergency conditions. The entire country is divided into a number of earthquake zones. There are four Indian codes of practice that deal with design and detailing of reinforcement in earthquake zones. The reader must refer to these codes when designing for earthquake forces. The structural analysis must be performed as per the specifications of the earthquake resisting design code *IS*: 1893. The staircases are usually analysed as isolated structural elements and not integrated with the frames. The design and detailing of staircases in earthquake zones four and five be detailed as ductile members. An important factor of 1.5 must be applied to the staircases in such designs.

Fire Protection

The fire protection rating for the staircase should be atleast 30 minutes more than that assigned to the building. The cover to the reinforcement shouldn't be less than 25 mm in any building. Similarly the minimum thickness of the slab should be 110 mm. The thickness of the isolated step can be 75 mm for a fire rating of one hour and for higher rating the thickness has to be more. More important aspect is that the fixtures and railings must be fire proof. It is desirable to avoid plastic covers to the railings and steps. Fire resisting fibreglass covers are desirable.

7.3 DESIGN OF STAIRCASE

Commonly adopted staircases are:

1. Independent steps cantilevered from wall,
2. Slab staircase supported at each landing,
3. U shaped stairs with supports at landings,
4. Folded staircase supported at floor levels,
5. Saw tooth staircase supported at each landing,
6. Spiral staircase around a column or a well,
7. Helicoidal staircase supported at floor levels.

There are other special types based on the creativity of the architect.

Example 7.1: Cantilevered Steps

Independent steps 1.2 m cantilevering from the face of the wall are used for staircase in a residence. The width of step is 300 mm. Design the staircase with $M20$ concrete and $HYSD$-$Fe415$ reinforcement by limit state design.

Solution

Cantilever span = L = 1.2 m. The service load on each step is = W_1 = 1.5 kN at 150 mm from the free end, f_{ck} = 20 N/mm^2, f_y = 415 N/mm^2. The limit state design coefficients taken from chapter are K = 0.138, j = 0.8.

The bending moment increases to maximum at the support, so the thickness of the step can be nominal at the free end and maximum at the support. In the present case the overall thickness is taken as 100 mm at free end. For the purpose of the self-weight calculations, the average thickness is assumed as 150 mm.

$$\text{The self-weight of the step is} = W_g = bt\gamma_c(L) = 0.3(0.15)(25)(1.2) = 1.35 \text{ kN}$$

The collapse bending moment on the step with a partial safety factor of 1.5 to both the loads is:

$$M_c = 1.5(W_g(L)/2 + W_1(L - 0.15)) = 1.5(1.35(1.2)/2 + 1.5(1.2 - 0.15)) = 3.5775 \text{ kNm}$$

The bending moment is negative and causes tension at the top face.
Equating the collapse bending moment to the moment capacity of the section, we have:

$$Kbd^2 f_{ck} = M_c$$

The effective depth of the section is = $d = \dfrac{M_c}{Kbf_{ck}} = \dfrac{3.5775(10^6)}{0.138(300)(20)} = 66 \text{ mm}$

Use d = 80 mm with a cover of 25 mm, then the minimum thickness at the support is = t = 80 + 25 + 5 = 110 mm.

The depth recommended by the code for cantilever beams base on the deflection criterion is = $L/7$ = 170 mm. The author however recommends span/8 for such cantilevers. Assume the overall thickness at support as 160 mm and the effective depth as = d = 125 mm.

The area of reinforcement required is:

$$A_{st} = \frac{1.15 M_c}{jdf_y} = \frac{1.15(3.5775)(10^6)}{0.8(125)415} = 99 \text{ mm}^2$$

Provide 2 numbers of 8mm bars at top face of the step. The actual area of the reinforcement provided is

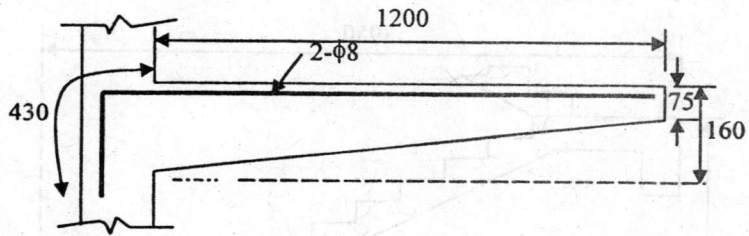

Fig. 7.2 Reinforcement details of Cantilever Step; Example 7.1

100 mm². Each of the bars must be anchored into the wall with the required development length as: 54f $\approx$ 430 mm with an L bent. Provide distribution reinforcement of three 6 mm bars one at each end and one at the middle of the step.

Note: The cantilever members are statically determinate and likely to be over loaded at free ends. Main reinforcement must be placed at the top face of the member with adequate development length. The workers have a tendency to trample down the top reinforcement during concreting. Therefore, it is necessary to provide proper chairs to the main reinforcement to ensure that the bars are at the top face. Another important aspect is the stability and strength of the supporting wall. The cantilever puts in a moment in the supporting wall. The reinforcement must be designed to resist the extra moment coming from the cantilever.

Example 7.2: Design of Slab Staircase

A two-flight staircase in a residential house connects two floors, with floor height of 3000 mm. There are ten steps in each flight having a tread of 250 mm and a rise of 150 mm. The landing is 700 mm clear length. The landings are supported by 300 by 450 mm beam. The width of the staircase is 1000 mm. Design the staircase using *M*25 concrete and *Fe*415 reinforcement. Fig. 7.3 illustrates the profile of the staircase. The exposure condition is mild.

Solution

Number of steps in one flight = 10, tread = 250 mm, rise = 150 mm
Length of going = 10 * 250 = 2500 mm
Effective span = L = going + two landings – width of the landing beam = 2500 + 1400 – 300 = 3600 mm.
Let the thickness of the slab for the purpose of self-weight = $L/$ 20 = 180 mm.
The slope of the staircase slab = θ = tan^{-1}(150/300 = 0.5) = 26.6°
The self-weight of the slab and the steps is computed for one metre of horizontal projection. The staircase is provided with nominal railing and its weight can be included in the live load. Nominal load can be added for floor finish etc.
The weight of the Staircase slab of 1m width is = (1) 0.18(25)/sinθ = 10.0 kN/m.
The weight of the step per metre length (4 steps) = 0.5(0.15*0.3)*4(25) = 2.25 kN/m
Weight of floor finish   = 0.75 kN/m
Total dead weight = w_g = 13.0 kN/m
The live load = w_l = 3 kN/m

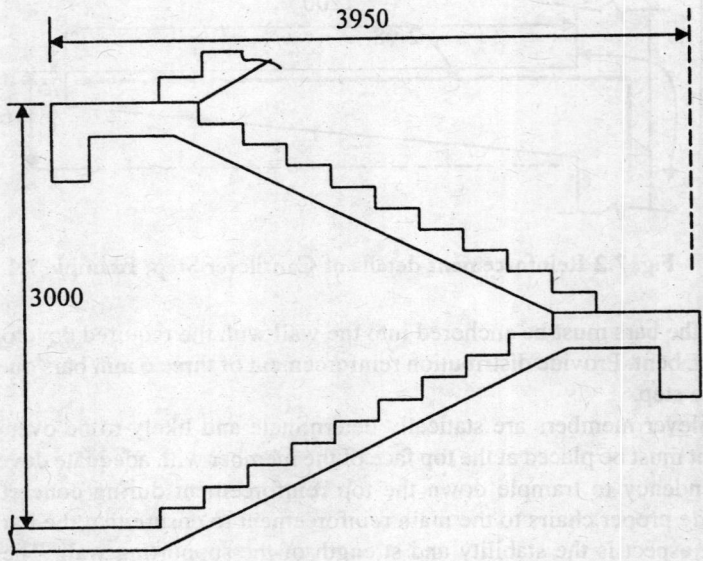

Fig. 7.3 Sectional Elevation of Staircase; Example 7.2.

Each flight is supported on the end beams, therefore the staircase slab can be considered as simply supported at the centre of the beam. The beam may offer a nominal restraint so nominal negative reinforcement must be provided in the landing slabs. The factored maximum bending moment occurs at the mid span:

Equating the collapse moment to the moment capacity, the effective depth of the slab is:

$$d = \sqrt{\frac{M_c}{Kbf_{ck}}} = \sqrt{\frac{38.88(10^6)}{0.138(1000)25}} = 106.1 \text{ mm}$$

The cover to the reinforcement is 25 mm, so the total depth needed is = 106 + 25 + 5 = 136 mm
The slab can be taken equal to 150 mm and the effective depth of the section as = d = 125 mm. The assumed self-weight of the slab is close to and on safer side. The area of tension reinforcement is:

$$A_{st} = \frac{1.15M_c}{jdf_y} = \frac{1.15(38.88)10^6}{0.8(125)(415)} = 1078 \text{ mm}^2$$

Select nine numbers of 12 mm bars, then the area of reinforcement provided is 10*113 = 1130 mm².

The Reinforcement Details

The main reinforcement of 10 of 12 mm bars of which five are curtailed at the face of the landing slabs and the remaining five bars are continued to the end. These bars are placed at the bottom face of the inclined slab. The bars are normally given a bent at the bottom landing to follow the profile of the slab. Five of 12 mm bars can be continued with a bent following the profile of the slab at bottom landing. However at the top-landing slab if the bars are bend down following the slab, then the radial component of the tension

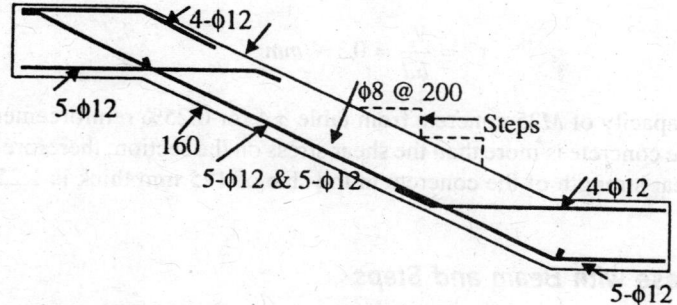

Fig. 7.4 Reinforcement Details of Slab Staircase; Example 7.2.

force of the bars will act away from bottom cover of the slab at the bent. This radial force has to be resisted by the cover concrete as the both the segments of the bars are on the cover face. Please see Fig. 7.5 that illustrates this radial force effect. To avoid the splitting out of the cover concrete, the bar shouldn't be bent but taken straight to the top face and then bent down. This detailing is illustrated in the Fig. 7.4. Since the bars are taken straight to the top face, extra reinforcement bars whose area of cross section is equal to half of the main are provided at the bottom face extending into the waist of the slab. Unsymmetrical loading on the flights and the landing slabs will cause. Distribution reinforcement has to be provided along the width of the slab. The minimum percentage of the shrinkage reinforcement in the slab is 0.15%. The reinforcement is also called nominal or even secondary reinforcement. The shrinkage reinforcement is:

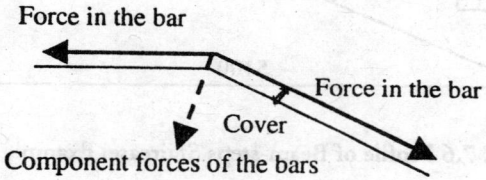

Fig. 7.5 Action of the Forces in the Bars.

$$A_s = 0.15(160)1000/100 = 240 \, mm^2/1000 \, mm.$$

Provide 8 mm bars at 200 mm spacing.

Check for Shear Strength

The maximum shear force occurs at the support. The critical shear force is at a distance of effective depth from the support. The total design load including dead load is: $13 + 3 = 16 \, kN/m$. The critical factored shear force is:

$$V_u = 1.5(w_t)\left(\frac{L}{2} - d\right) = 1.5(16)(1.8 - 0.134) = 40 \, kN$$

The nominal shear stress is:

$$\tau_v = \frac{V_u}{bd} = 0.3 \, N/mm^2$$

The shear strength capacity of $M25$ concrete from table 3.4 for 0.25% reinforcement is 0.36 N/mm². The shear strength of the concrete is more than the shear stress on the section; therefore the section is safe in shear. The actual shear strength of the concrete in the slab of 175 mm thick is 1.25 times that of the beams.

Example 7.3: Staircase with Beam and Steps

A staircase consists of 10 steps, 300 mm tread and 160 mm rise plus two landings. The width of the staircase is 1250 mm. The length of the landing is 1200 mm. The steps are fixed on a central beam as isolated steps cantilevering symmetrically on either side. The elevation of the part staircase is shown in Fig. 7.6 and the projection of the step is shown in Fig. 7.7. The staircase beam is supported by the end cross beams of width 300 mm that form the part of the frame structure. Design the staircase for moderate environment.

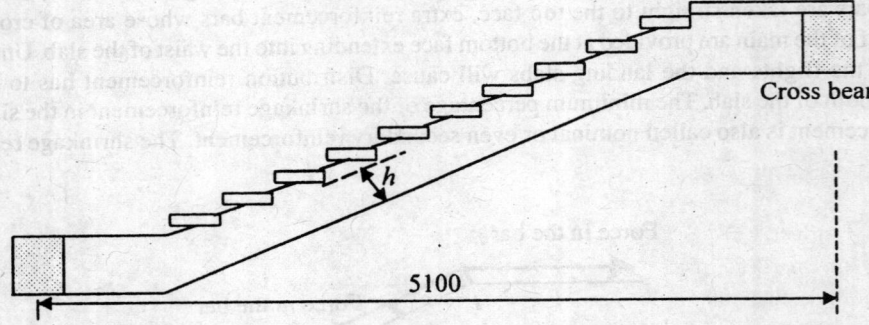

Fig. 7.6 Profile of Beam-steps Staircase; Example 7.3.

Solution

The design consists of 1250 mm wide step resting on the central beam. The second part is design of the landing slab. The design of the beam resting on the crossbeams and supporting the 10 steps and two-landing slabs. The live load on the staircase in commercial complex is 5 kN/m². Each independent step is designed to carry concentrated load of 1.5kN placed at about 150 mm from the free end of the step. The exposure is moderate so select $M25$ concrete and $HYSD$-Fe415 reinforcement. The cover to the reinforcement is taken as 25 mm for slabs and 30 mm for beams.

Design of Step

The top of the step is open so bending moment is maximum at the middle of the support .The effective cantilever of the step is 1250/2 = 625 mm. Assume the concentrated load is placed at 125 mm from the free end. The self-weight of the step including the railing is likely to be marginal so add extra 20 percent to the live load bending moment towards the self-weight bending moment.

The collapse moment on the step with partial safety factor of 1.5 applied to the loads is:

$$M_c = 1.5(1.2)(0.625 - 0.125)(1.5) = 1.35 \text{ kNm}$$

Equating the collapse moment to the moment capacity of the section gives

$$d = \sqrt{\frac{M_c}{Kbf_{ck}}} = \sqrt{\frac{1.35(10)^6}{0.138(300)25}} = 37 \text{ mm}$$

Provide an overall depth of 75 mm and an effective depth equal to $75 - 20 - 5 = 50$ mm. The area of reinforcement required is:

$$A_{st} = \frac{1.15(1.35)(10^6)}{0.8(50)415} = 94 \text{ mm}^2$$

Use two numbers 8 mm bars at the top. And provide 6 mm bars @250 mm as distribution reinforcement. Figure 7.7 illustrates the reinforcement details of the step.

Design of Beam for Bending Moment

The length of the going is =10*300 = 3000 mm
Total length of landings   = 2400 mm
Total end to end distance   = 5400 mm
Less the width of the cross beam   = 300 mm
Effective length of the beam   = 5100 mm
Weight of steps/m = 1.25*0.075*25   = 2.4 kN/m
Let self weight = 0.25*0.5*25   = 3.1 kN/m
Add extra for finishes etc   = 1.0 kN/m
    Total dead load   = 6.5 kN/m
    Live load   = 5.0 kN/m
    Total factored load = $w_t = 1.5(6.5 + 5) = 17.25$ kN/m

Collapse bending moment $= M_c = \dfrac{w_t L^2}{8} = \dfrac{17.25(5.1)^2}{8} = 56$ kNm

The effective depth of the section is obtained by equating the moment capacity to the collapse moment.

$$d = \sqrt{\frac{M_c}{Kbf_{ck}}} = \sqrt{\frac{56(10^6)}{0.138(250)25}} = 256 \text{ mm}$$

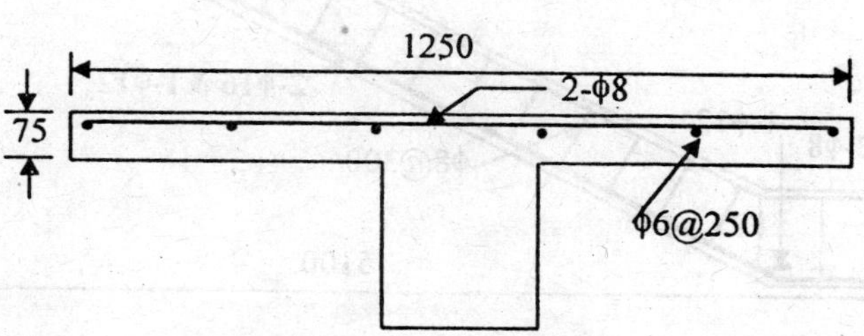

Fig. 7.7 Reinforcement in Step, Example 7.3.

Use overall depth of the beam as 450 mm and the effective depth as 410 mm, then the area of reinforcement is:

$$A_{st} = \frac{1.15 M_c}{jdf_y} = \frac{1.15(56)10^6}{0.8(410)415} = 474 \, \text{mm}^2$$

Provide two numbers of 16 mm bars and one of 12 mm bar. The 12 mm bar can be curtailed at the beginning of the landings. The areas of steel provided at the critical moment and shear zones are 515 and 402 mm² respectively.

Design for Shear

The critical shear force occurs at a distance effective depth from the support and it is:

$$V_u = 1.5(w_t)\left(\frac{L}{2} - d\right) = 17.25\left(\frac{5.1}{2} - 0.41\right) = 36.915$$

The nominal shear stress on the beam is:

$$\tau_v = \frac{V_u}{bd} = \frac{36915}{250 * 410} = 0.36 \, \text{N/mm}^2$$

The percentage of reinforcement provided at the critical shear zone is 402(100)/250*410 = 0.4% and the allowable shear strength from table 3.4 is about 0.44 N/mm². The nominal shear stress is less than the admissible shear strength so only nominal shear reinforcement needs to be provided. Select two legged 8 mm bars, then the spacing is

$$s_v = \frac{A_{sv}f_y}{0.4b} = \frac{100(415)}{0.4(250)} = 415 \, \text{mm}$$

However the maximum spacing admissible is 300 mm. Provide two legged 8 mm bars at 300 mm spacing. Fig. 7.8 illustrates the reinforcement details. The landing slab has the same cantilever span as that of the steps, therefore provide 8 mm bars at 250 mm spacing at the top face of the landing slabs. The

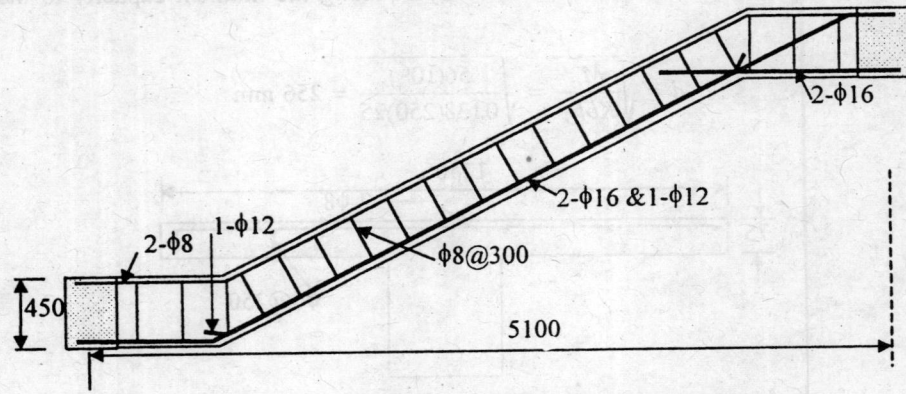

Fig. 7.8 Reinforcement in Beam of Example 7.3.

distribution reinforcement is 8 mm bars at 300 mm spacing. The thickness of the landing slab is 75 mm same as that of the steps.

Example 7.4 Design of Staircase in Severe Earthquake Zone

Redesign problem in example 7.2 for a most severe earthquake zone. A two-flight staircase in a residential house connects two floors, having a floor height of 3000 mm. There are ten steps in each flight having a tread of 250 mm and a rise of 150 mm. The landing is 700 mm clear length. The landings are supported by cross beams of 300 by 450 mm. The width of the staircase is 1000 mm. Design the staircase using M25 concrete and Fe415 reinforcement. Fig. 7.3 illustrates the profile of the staircase. The exposure condition is mild.

Solution

The basic live and dead loads don't change but the loads are to be increased by the appropriate accelerating force. The accelerating coefficient in the zone five is less than 0.1 of the gravitational acceleration. The partial load factors for the earthquake load condition are 1.2, 1.2 and 1.2 for dead, live and earthquake loads. The partial load factors for the live load condition are 1.5 and 1.5 for live and dead loads. However, the staircase must be designed as a ductile member and serviceable even in the earthquake load condition. It is therefore necessary to place reinforcement on the top face of the slab. Atleast 0.15 percent reinforcement must be placed at top face of the staircase. Landings are likely to be subjected to negative bending moment. The reinforcement detail of the staircase slab is shown in Fig. 7.9. This is very similar to that of example 7.2 (Fig. 7.4) except that additional nominal reinforcement is provided on the top face of the slab.

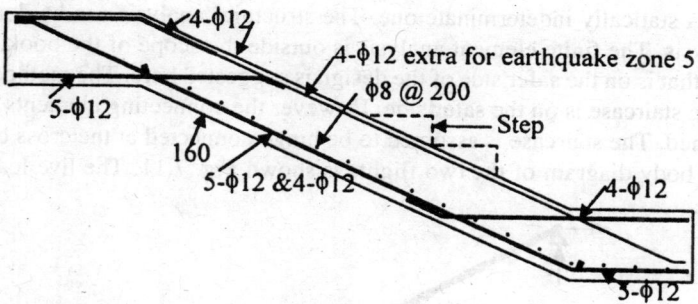

Fig. 7.9 Reinforcement Details of Slab Staircase; Example 7.4.

Example 7.5: Design of Folded Slab Staircase

A staircase of one metre width has two flights for each floor. The middle landing is not supported on wall or beam but to be treated as fold of a folded plate. There are 12 steps of each 250 mm tread and 150 mm rise. The lengths of floor and mid landings are 1.2 m each. The live load is 3 kN/m². The staircase is in protected environment. Fig. 7.10 gives the dimensions.

Solution

The staircase is designed with M25 concrete and HYSD-Fe415 reinforcement. M20 concrete is adequate

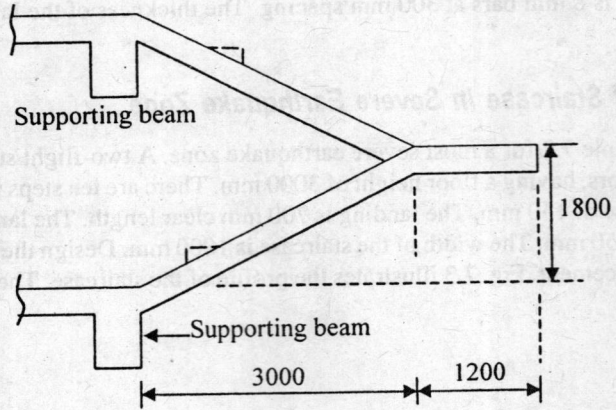

Supporting beam

Supporting beam

3000   1200

1800

Fig. 7.10 Folded Slab Staircase Dimensions; Example 7.5.

but the frames and the slabs are designed with M25 concrete, therefore M25 grade is chosen. Cover to the reinforcement is chosen as 25 mm.

Number of steps = 12
Length of going = 12(0.25) = 3 m
Length from the supporting beam = Going + 1.2 m of landing = 3 + 1.2 = 4.2 m
Gradient of the staircase is = tan θ = 150/250 = 0.6, Or θ = 31°
Half flight height = H = 12(0.15) = 1.88 m

The structure is a statically indeterminate one. The structural analysis can be done by plate element finite element analysis. The finite element analysis is outside the scope of the book. However, a simple method of analysis that is on the safer side of the design is suggested here. The method is based on simple statics; therefore the staircase is on the safer side. However the connecting elements have to be properly modeled and designed. The staircase is assumed to be hinge connected at the cross beams at each of the floor levels. A free body diagram of the two flights is shown Fig. 7.11. The live load is occupying two consecutive flights.

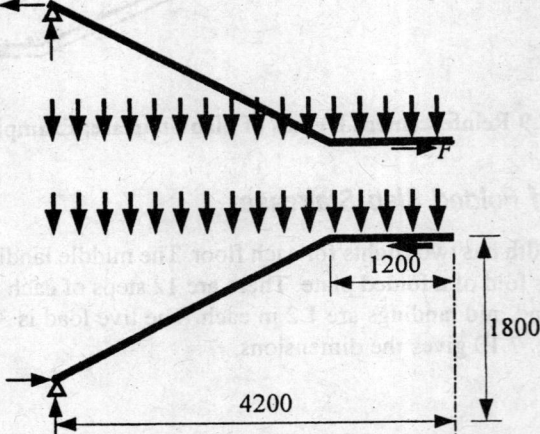

F

1200

1800

4200

Fig. 7.11 Idealized Free Body Diagram of Folded Slab Staircase; Example 7.5.

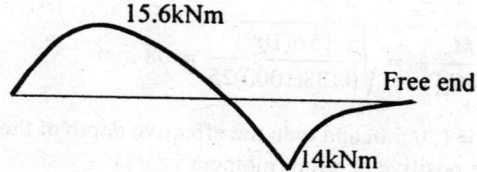

Fig. 7.12 Bending Moment Diagram on the Lower Flight; Example 7.5.

Let the thickness of the slab is assumed equal to 180 mm. The self-weight is

$$w_g = 0.18*25/\sin\theta = 9 \text{ kN/m},$$

Use the same intensity for the landing area and add extra 1 kN/m for other fixtures and finishes. The total factored load on the staircase is:

$$w_t = 1.5(9 + 1 + 3) = 19.5 \text{ kN/m}$$

From the free body diagram, it can be seen that there is an in plane shear force at the interface of the mid-landings. This force is indicated as equal to F. The interacting force is computed by taking moments about the idealized support of any one flight. The moment equilibrium about the support of the lower flight gives:

$$F(1.8) = w_t(4.2)(4.2)/2 = 172, \text{ or}$$

$$F = 95.55 \text{ kN}$$

The bending moment at a distance x from the free end of the mid-landing applicable beyound landing is:

$$M_x = F(x - 1.2)\tan\theta - w_t \frac{x^2}{2} = 95.55(x - 1.2)0.6 - 9.75x^2$$

This equation is applicable for value of x greater than 1.2 m. The maximum bending moment occurs when the derivative of the moment is equal to zero. This leads to:

$$95.55(0.6) - 19.5x = 0$$

$$\text{or } x = 2.94 \text{ m}$$

The second derivative of the moment is −19.5 As the second derivative is negative, the bending moment is maximum. The bending moment is zero at the free end of the mid-landing. The maximum bending moment is at a distance 2.94 m from the free end. The substitution of 2.94 m for x, the maximum bending moment is:

$$M_{max} = 95.55(1.74)0.6 - 9.75(2.94)^2 = 15.6 \text{ kNm}$$

The extreme negative bending moment occurs at the beginning of the mid landing, that is for x equal to 1.2 m. And this negative bending moment is:

$$M_n = 9.75(1.2)^2 = 14 \text{ kNm}$$

The bending moment profile on the lower flight of the staircase is given in Fig.7.13.

The effective depth of the slab is obtained by equating the maximum factored bending moment to the moment capacity of the section.

$$d = \sqrt{\frac{M_{max}}{Kbf_{ck}}} = \sqrt{\frac{15.6(10^6)}{0.138(1000)25}} = 68 \text{ mm}$$

Let the total depth be selected as $h = 120$ mm and then the effective depth of the section is 90 mm. The area of the reinforcement to resist the positive bending moment is:

$$A_{st} = \frac{1.15(15.6)10^6}{0.8(90)415} = 523 \text{ mm}^2$$

Provide five numbers of 12 mm bars at the bottom face, and then the area of tension reinforcement is 565 mm².

Similarly the area of tension reinforcement required for the negative bending moment is 467 mm². Provide five 12 mm bars at the top face. The distribution reinforcement is 0.15 percent and it is 180 mm². Provide 8 mm bars at 250 mm. The assumed thickness for dead load purpose was 160 mm while the actual thickness works out to be 120 mm.

Note

The analysis was carried out for the load on the two flights. However when the load is only on the upper flight, the bending moment profile will change and the bottom flight is subjected to positive bending moment due to live load. Similarly the bending moment on the upper flight is similar to that of the lower flight. The reinforcement detail has to be suitably made. The landing should have reinforcement both at top and bottom faces. The reinforcement details are shown in Fig. 7.13.

Design for Transverse Shear

The total load of the two flights of the staircase is equally distributed to the two cross-beams at the floor levels. The vertical reaction at the lower cross-beam is

$$R_u = 4.2(19.5) = 81.9 \text{ kN}$$

The staircase slab is inclined so the transverse shear force on the slab at a distance d from the support is:

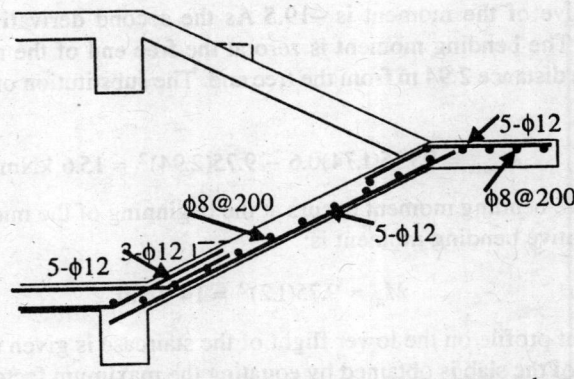

Fig. 7.13 Reinforcement Details at Lower Flight; Example 7.5.

$$V_u = (R_u - 19.5*0.09)(\cos\theta) = 80.1(0.857) = 68.7 \text{ kN}$$

The nominal shear stress on the slab is

$$\tau_v = \frac{V_u}{bd} = \frac{68700}{1000(90)} = 0.76 \text{ N/mm}^2$$

The percentage of reinforcement near the support is

$$p = 5*113(100)/120(1000) = 0.47\%$$

The shear strength with 0.47 % reinforcement is 0.47 N/mm^2 from table 3.4.

The admissible value for slab of thickness is $= 1.3(0.47) = 0.61$ N/mm^2.

The nominal shear stress is more the admissible strength. There are two options in this case. Increase the thickness of the slab or provide shear reinforcement. Since the folded slab is subjected to nominal torsion, it is better provide closed loop stirrups near the support. Provide only nominal shear reinforcement of 6 mm bars at 100 mm spacing upto 1000 mm from the support.

Note about Support

There will be a component of force along the inclination of the slab; therefore the support is to be properly designed. If the staircase slab is built continuous with the floor landing, the landing slab must be provided with reinforcement both at top and bottom. The continuity of the reinforcement from the staircase slab to the landing slab reduces the absolute maximum bending moment but introduce negative bending moment in to the slab. The reinforcement detailing has to take of such local moments.

7.4 DESIGN OF HELICOIDAL AND FOLDED PLATE STAIRCASES

Helicoidal staircase is in three-dimensional space requiring three-dimensional structural analysis. The staircase can be idealized as a curve beam in a plane, in such a case the design is on the conservative side. A more rational idealization is to treat it as a helical girder in space. The analysis of such girders is not in the scope of this book; however, results of forces on such girders are available in the literature. The three critical stress resultants acting on a girder are the bending moments about two principal planes and torsional moment; traverse shear and axial thrust. The axial thrust is absent in case it is treated as a curved girder otherwise it exists in helical girder. The effect of the axial thrust is not significant for the purpose of design.

The helicoidal staircase slab is subjected to the following forces

- M_r = Radial bending moment, considered to be positive if causing compression on the top fibre
- M_t = torsional moment
- V = shear force
- P = axial compressive force

The notations follow the beam conventions. The other notations are:
- R = radius to the centre line of the curve,
- 2θ = subtended angle in plan,
- α = slope of the helicoidal slab, $\tan(\alpha)$ = rise/tread for equal size steps.

For the sake of simplicity, the resultant forces are taken from the curved beam theory. The vertical load acting on the curved beam is multiplied by cosine of the slope of the slab. This load is considered to be

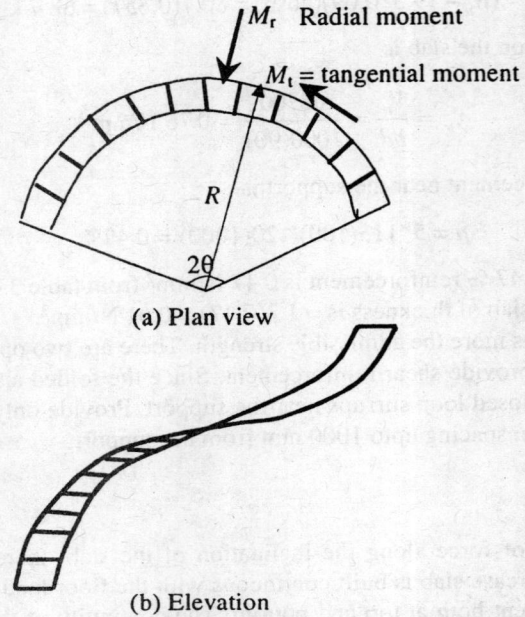

(a) Plan view

(b) Elevation

Fig. 7.15 Helicoidal Staircase Notations.

that acting normal to the surface of the slab. The effect of the axial thrust is neglected.

The radial and torsional moments on the slab for fixed support boundary conditions are given by:

$$M_r = wR_c^2(c\cos\phi - 1) \tag{7.1}$$

$$M_t = wR_c^2(c\sin\phi - \phi) \tag{7.2}$$

where R_c = radius to the centroidal axis of the slab = $R + e = R + b^2/12R$
  e = eccentricity of the load action = $b^2/12R$, $\qquad(7.3)$
  b = width of the slab,
  ϕ = angle measured from the middle point of the curve of the slab,

$$c = \frac{2(g+1)\sin\theta - 2g\theta\cos\theta}{(8+1)\theta - (g-1)\sin\theta\cos\theta} \tag{7.4}$$

$$g = \frac{EI}{GJ} \tag{7.5}$$

  EI = flexural rigidity,
  GJ = torsional rigidity,

The value of g can be approximated for concrete slab by assuming the following values
  $E/G = 2$
  I = moment of inertia = $bh^3/12$,
  J = torsional constant = $bh^3/3.5$,

h = thickness of the slab taken less than the width of the slab.

Therefore the value of g for rectangular section works out to be

$$g = 3.5/12 = 0.58$$

The helicoid is idealized as an in plane curved beam with a modification to the load on the beam. The boundary conditions are fixed ended. The connecting slabs at the floor levels must be designed to provide the fixity. Absolute fixity of the support is not possible, however the connecting slab reinforcement must provide adequate reinforcement to match the fixed moment conditions. The method of design is illustrated by an example.

Example 7.6: Design of Helicoidal Staircase

A helicoidal staircase is provided in a building where the floor height is 3.3 m. The width of the staircase is 1.2 m and the tread and rise are 300 mm and 150 mm respectively. The tread is measured at the centre line of the curve, it may be considered as the average of that actually available across the width of the staircase. The included angle of the helicoidal stair is 180 degrees. There is a mid-landing of 1.2 m length. The building is used for commercial purpose with mild exposure conditions.

Solution

The height of the floor is = 3.3 m, the rise of the step is 150 mm.

The number of rises = 3300/150 = 22,

Total length of the tread at the centre line = 22*300 =6600 mm

Total going of the staircase including mid landing is = 6600 +1200= 7800 mm

Since the included angle is 180 degrees, the radius of the helicoid is:

πR = length of the centre line = 7800 mm, or

$R = 7800/3.1415 = 2483$ mm

The gradient of the staircase neglecting the effect of the mid-landing is:

$$\alpha = \tan^{-1}\frac{150}{300} = \tan^{-1} 0.5 = 26.6°$$

Select M25 concrete with HYSD-Fe415 steel reinforcement. For the purpose of self-weight, the thickness of the slab is taken as about going/20, say 350 mm. The self-weight of the slab is:

$$w_g = 0.35*25*1.2 = 10.5 \text{ kN/m}$$
$$w_1 = 5*1.2 = 6.0$$
$$\text{Total load} = 16.5 \text{ kN/m}$$

Factored load normal to the surface of the slab is

$$w_t = 1.5*16.5 \cos \alpha = 22.2 \text{ kN/m}$$

The centroidal distance from the centre line is

$$e = \frac{b^2}{12R} = 0.05 \text{m}$$

The centroidal line radius is

$$R_c = 2.483 + 0.05 = 2.533 \text{m}$$

The ratio of the flexural rigidity to the torsional rigidity for rectangular section is $= g = 0.58$, the value of c for included half angle 90 degrees is

$$c = \frac{2(g+1)\sin\theta - 2g\theta\cos\theta}{(g+1)\theta - (g-1)\sin\theta\cos\theta} = \frac{2(1.58)}{1.58(1.57)} = 1.27$$

The bending and twisting moments at the mid-span for value of $\phi = 0$ are

$$M_r = wR_c^2(c\cos\phi - 1) = 22.2(2.53)^2(1.27 - 1) = 142.1(0.27) = 38.4 kNm$$

$$M_t = wR_c^2(c\sin\phi - \phi) = 0$$

The bending and twisting moments at the support for value of $\phi = 90°$ are

$$M_r = wR_c^2(c\cos\phi - 1) = 142.1(0 - 1) = -142.1 \text{ kNm}$$

$$M_t = wR_c^2(c\sin\phi - \phi) = 142.1(1.27 - 1.57) = -42.63 \text{ kNm}$$

Design for Critical Moments

The bending and the twisting moments are critical at the support. The size of the section is chosen based on the maximum design values. The combined effect of the bending and twisting into an equivalent bending moment is:

$$'M_e = M_r + M_t \left[\frac{1 + h/b}{1.7} \right] \tag{7.6}$$

The substitution of the quantities in the above equation gives

$$M_e = 142.1 + 42.63 \left[\frac{1 + 350/1200}{1.7} \right] = 174.5 \text{ kNm}$$

The effective depth of the section is obtained by equating the equivalent moment to the moment capacity.

$$d = \sqrt{\frac{M_e}{0.138 b f_{ck}}} = \sqrt{\frac{174.5(10)^6}{0.138(1200)25}} = 206 \text{ mm}$$

Let the total depth of the section be selected as 300 mm and the effective depth of the reinforcement as 260 mm then the tension reinforcement that is needed at the top face of the support is:

$$A_{st} = \frac{1.15 M_c}{j d f_y} = \frac{1.15(174.5)10^6}{0.8(260)(415)} = 2325 \text{mm}^2$$

Provide 12 numbers of 16 mm bars. The actual area of reinforcement provided is $12*201 = 2412 \text{ mm}^2$. This is the reinforcement at supports to resist the negative bending moment. It should be provided at the top face of the slab for distance of span/3 that is for a length of 7800/3 = 2600 mm. The positive bending moment is only 22 percent of that of the negative bending moment. The reinforcement required is about 500 mm². The minimum reinforcement required is 0.15 percent, and it is

$$A_{smin} = 0.15*1200*300/100 = 5400 \text{ mm}^2.$$

Provide six numbers of 12 mm bars at the bottom from end to end.

Design for Shear Force

The transverse shear force acting at a distance 260 mm from the support is:

$$V_u = 22.2(7.8/2 - 0.26) = 80.8 \text{ kN}$$

In addition to the shear force at the section, there is twisting moment. The combined effect of the two forces as equivalent transverse shear is:

$$V_e = V_u + \frac{1.6T}{b} \tag{7.8}$$

The substitution of the values in the above equation gives

$$V_e = 80.8 + \frac{1.6*42.63}{1.2} = 137.64 \text{ kN}$$

The nominal shear stress on the slab is:

$$\tau_v = \frac{V_u}{bd} = \frac{13764}{1200(260)} = 0.44 \text{ N/mm}^2$$

The percentage of bottom reinforcement is nominal in the order of 0.15 percent. The allowable shear strength from table 3.4 is $\tau_c = 0.29$ N/mm². Shear reinforcement is to be provided. As there is torsion on the slab, closed loops, as stirrups are provided.

Area of shear reinforcement required is given by

$$A_{sv} \geq \left[\frac{T}{b_1 d_1} + \frac{V_u}{2.5 d_1} \right] \frac{1.15 s_v}{f_y} \tag{7.9}$$

where T is the torsion, b_1 is horizontal distances between the vertical legs of the stirrup and d_1 is the vertical distance between the horizontals of the stirrup. These values are:

$$d_1 = 260 - 10 = 250 \text{ mm, and } b_1 = 1200 - 2*25 - 10 = 1140 \text{ mm}$$

The substitution of different quantities in the above expression gives:

$$A_{sv} \geq \left[\frac{42630000}{1140*250} + \frac{80800}{2.5*250} \right] \frac{1.15 s_v}{415} = 0.773 s_v$$

Select two legged 10 mm stirrups then the spacing of the stirrups from the above expression is:

$$s_v \leq \frac{2*78.5}{0.773} = 203 \text{ mm}$$

The minimum shear reinforcement or the maximum spacing of the stirrups is given by

$$s_{vmax} = \frac{A_{sv} f_y}{1.15b(\tau_v - \tau_c)} \qquad (7.10)$$

The substitution of the quantities in the above equation gives

$$s_{vmax} = \frac{2*78.5(415)}{1.15(0.44 - 0.29)(1200)} = 315 \text{ mm}$$

Provide two legged 10 mm bars at 200 mm spacing. Since the stirrups have to have hanger bars, two extreme to reinforcement must be provided end to end.

The sectional reinforcement at support and at mid span are shown in Fig. 7.16.

Note: The boundary conditions of the staircase are fixed. The bending and the twisting moments are substantial at the supports when compared with those at the mid-span. It is necessary to provide such boundary with the landing slabs and beams. The landing slabs must be capable of resisting the support moments and the reinforcement detail must match the assumption. It is often desirable to provide reinforcement at top and bottom for the landing slabs.

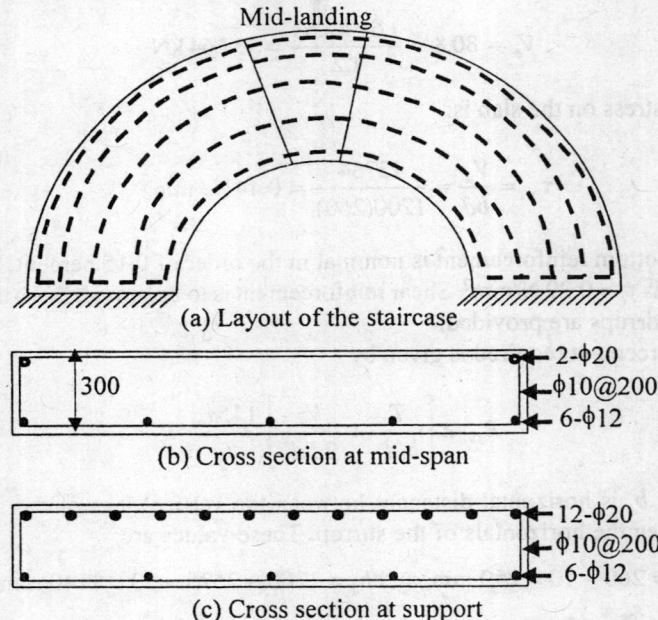

Mid-landing

(a) Layout of the staircase

300

2-φ20
φ10@200
6-φ12

(b) Cross section at mid-span

12-φ20
φ10@200
6-φ12

(c) Cross section at support

Fig. 7.16 Reinforcement Details; Example 7.6.

Example 7.7: Folded Plate Staircases

A folded staircase folded illustrated in Fig. 7.17 has 10 rises. Each step has 160 mm rise and 250 mm tread. Each landing is supported by cross beam. The width of the staircase is 1000 mm. Design the staircase in mild exposure condition with live load of 3 kN/m².

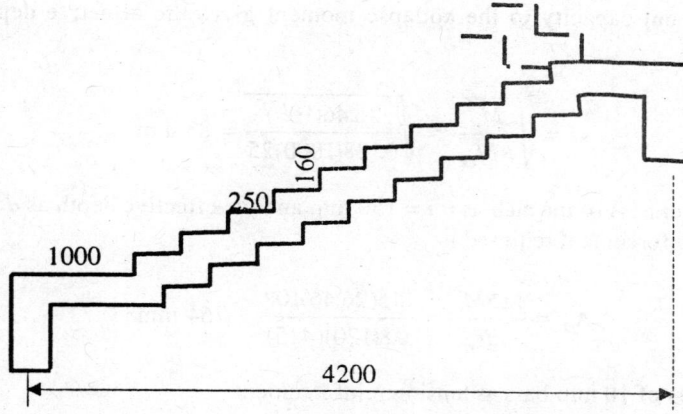

160

250

1000

4200

Fig. 7.17 Typical Folded Plate Staircase Flight; Example 7.7.

Solution

Select *M*25 concrete and *HYSD-Fe*415 reinforcement, which is same as the concrete that used in the other parts of the building. The going of the staircase is:

$$\text{Going} = 10*250 = 2500 \text{ mm}$$

The total horizontal length of the staircase is = 2500 + 2*1000 = 4500 mm.

Assuming the width of the crossbeam as 300 mm, the effective span of the one flight of the folded staircase is:

$$L = 4500 - 300 = 4200 \text{ mm} = 4.2 \text{ m}$$

Assume 150 mm thick plate for the purpose of self-weight.

Weight of one unit of step = (0.16 + 0.25)*1*0.15*25 = 1.5375 kN
The weight of 10 steps is = 10*1.5375 = 15.375, use 15.5 kN

Let the thickness of the landing is 150 mm, then the weight of the each landing is = 0.15*0.7*25*2 = 5.25 kN

The total weight of the one flight is:

$$W_g = 15.5 + 5.25 = 20.75, \text{ use } 21 \text{ kN}$$

The total live load on one flight is:

$$W_l = 4.2*3 = 12.6 \text{ kN}$$

The factored total load on a flight is:

$$W_t = 1.5(21 + 12.6) = 50.4 \text{ kN}$$

The collapse bending moment of the flight is:

$$M_c = \frac{W_t L}{8} = \frac{50.4(4.2)}{8} = 26.46 \text{ kNm}$$

Equating the moment capacity to the collapse moment gives the effective depth for a balanced section:

$$d = \sqrt{\frac{M_c}{Kbf_{ck}}} = \sqrt{\frac{26.46(10^6)}{0.138(1000)25}} = 88 \text{ mm}$$

Select the gross thickness of the slab as $= t = 150$ mm and the effective depth as $d = 120$ mm
Area of tension reinforcement required is:

$$A_{st} = \frac{1.15M_c}{jdf_y} = \frac{1.15(26.46)10^6}{0.8(120)(415)} = 764 \text{ mm}^2$$

Provide 10 numbers of 10 mm bars as tension reinforcement.

Note on Reinforcement Detail

There are too many bends in the folded slab so the bending of the bar along the bends is not only messy but not acceptable at the re-entrant corners. The tension bar needs to be parallel to the compression face as per the derivation but it is zigzagging here. Bars with re-entrant corners produces resultant tension that causes splitting of the corner. See Fig. 7.18 for the notations and typical reinforcement detail for a fold. For this reason, the bar has to be taken beyond the corner into the other end and then bent. For the same reason, the tension reinforcement of the tread has to be different from that of the riser and bending has to avoid the re-entrant corner. So two independent bars are provided for one fold. The bars can't be curtailed in the next immediate step as the required development length is not available. The reinforcement in the riser and the tread is parallel to the compression faces of the corresponding elements. The lower corner is not a serious problem because the compression face is not changing direction. But at the upper corner of the step, the compression face is changing the direction by 90 degrees. The tip of the corner is ineffective in providing the concrete compressive strength.

At the re-entrant corner, compression plane is diagonal and the reinforcement is not perpendicular to the plane. The tread's bar is horizontal and the riser bar is vertical. The components of these bars normal to the diagonal plane must be adequate to resist the moment. The total reinforcement is doubled here, the components of the reinforcement is in general adequate. The total diagonal depth is not available for bending plane.

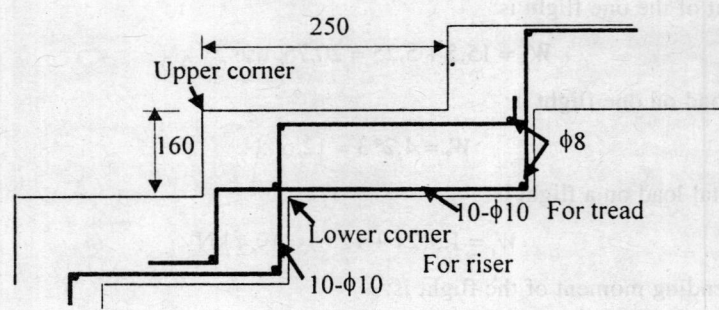

Fig. 7.18 Typical Reinforcement for Unit of Fold; Example 7.7.

Design for Shear

The maximum possible shear force at the crossbeam support is:

$$V_u = W_u/2 = 25.2 \text{ kN}$$

The nominal shear stress is:

$$\tau = \frac{V_u}{bd} = \frac{25400}{1000*120} = 0.21 \text{ N/mm}^2$$

The allowable shear strength of $M25$ concrete for 0.15 percent reinforcement is 0.29 N/mm². The nominal shear stress is with in the allowable strength of concrete.

7.5 CURVED BEAMS IN PLAN

A beam curved in plan is called curved beam. Curved beams are used to support circular shafts and slabs, circular porticos, water tanks, Cylindrical, and Intze tanks. A closed loop beam is called ring beam. The curved beam is subjected to bending and torsional moments. The moment is a function of the number of supports and the boundary conditions. Polar coordinate system is convenient to deal in the circular beams. The critical radial and tangential bending moments in a ring beam supported by equally spaced columns can be expressed as

$$M_p = c_p w R^2 \qquad (7.11)$$

$$M_n = c_n w R^2 \qquad (7.12)$$

$$M_t = c_t w R^2 \qquad (7.13)$$

where

M_p = maximum positive bending moment that occurs mid point of two successive supports,
M_n = maximum negative bending moment that occurs at the support,
M_t = maximum twisting moment which occurs at quarter points from the support,
c_p, c_n, and c_t are the corresponding moment coefficients listed in table 7.3.

Table 7.3 Moment Coefficients in Ring Beams.

Coefficient	Number of supports					
	4	6	8	9	10	12
c_p	0.110	0.047	0.023	0.019	0.015	0.009.
c_n	0.215	0.093	0.052	0.042	0.034	0.024
c_t	0.033	0.010	0.004	0.003	0.002	0.001

The maximum twisting moment occurs at a location where the bending moment is zero. At this location the shear force is not critical.

Example 7.8: Design of a Ring Beam

Four columns spaced at equal intervals support a circular beam having a radius of 3 m to the centre line of the columns. A circular roof slab of 150 mm thick is resting on the beam. The floor finish including the waterproof treatment is 90 mm thick. The structure is exposed to sever environment and the live load of 5 kN/m². No parapet wall is over the slab. Design the ring beam.

Solution

The structure is exposed to severe environment so $M30$ concrete and reinforcement of *HYSD-Fe*415 steel is selected. The required minimum cover to the reinforcement in beams is 45 mm and that in slab is 25 mm. Let the diameter of the column and the width of the beam be 300 mm and the slab extends 600 mm beyond the beam face to compute the weight of the slab etc.

The radius of the slab is = a = 3.0 + 0.15 + 0.6 = 3.75 m
Let the unit weight of waterproof treatment is = 22 kN/m².
Weight of the slab = $\pi a^2(0.15)(25)$ = 166 kN
Weight of waterproofing = $\pi a^2(0.09)(22)$ = 88 kN
Live load on the roof slab = $\pi a^2(3)$ = 133 kN
Total dead and live load from the roof = 387 kN
Superimposed load for unit length of beam = $387/2\pi R = 387/2\pi(3)$ = 20.6 kN/m
Let the self-weight of the beam is = $0.3*0.4*25$ = 3.0 kN/m
Total load per unit length is = w = 23.6 kN/m, use 24 kN/m
Factored limit load = $w_t = 1.5*24$ = 36 kN/m

The positive and negative bending moments and twisting moment coefficients taken from table 7.3 are: 0.110, 0.215 and 0.033 respectively. The moments are:

$$M_p = 0.11*36*9 = 0.11*324 = 35.64 \text{ kNm}$$
$$M_n = 0.215*36*9 = 0.215*324 = 70 \text{ kNm}$$
$$M_t = 0.033*36*9 = 0.033*324 = 11 \text{ kNm}$$

Torsion is zero at the maximum negative or positive bending moment locations. Therefore, the effect of the twisting moment is not cumulative. The magnitude of the negative bending moment is more than that of the positive moment, so the size of the beam can be selected based on the negative bending moment. The beam is a *T*-beam but at the negative bending moment location it is a rectangular section. Let the thickness of the beam be 300 mm. Equating the limit negative bending moment to the moment capacity of the section gives:

$$d = \sqrt{\frac{M_c}{Kbf_{ck}}} = \sqrt{\frac{70(10^6)}{0.138(300)(30)}} = 238 \text{ mm}$$

Fig. 7.19 illustrates the factored forces on one span of the ring beam.

Select the overall depth of the beam as = h = 360 mm and the effective depth as = d = 300 mm. A cover to the steel reinforcement in severe environment is 45 mm for the beam. The cover to the top as well as bottom bars must be 45mm. And add extra 15 mm to the centre of the reinforcement to obtain the effective depth. The actually available may be more. Then the area of negative reinforcement required is:

$$A_{st} = \frac{1.15(70)10^6}{0.8(300)(415)} = 809 \text{ mm}^2$$

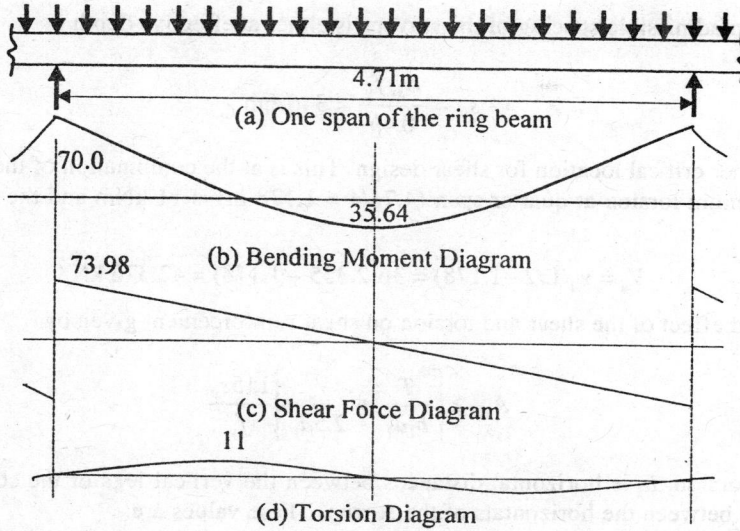

Fig. 7.19 One Span of a Ring Beam with Forces; Example 7.8.
(All Forces are in kN or kNm)

Provide 4 numbers of 16 mm bars at the top face over the supports. Two bars continued and two curtailed at one-third span. The actual area of reinforcement provided is 804 mm². All the four bars can be arranged in a row so the actual effective depth of the beam is $360 - 45 - 16/2 = 307$ mm.

The reinforcement required for an effective depth of the beam of 307 mm is = 791 mm².

The positive bending moment is about half of the negative bending moment. So provide 2 number of 16 mm bars at the bottom face throughout. The percentage of the positive reinforcement is

$$p = 402(100)/300(300) = 0.44\%$$

Design for Shear

L = distance between the two successive columns is = $2\pi R/4 = 4.71$ m

The maximum shear force at a distance d from the effective depth from the support is:

$$V_u = w_t (L/2 - d) = 36(2.355 - 0.3) = 73.98 \text{ kN}$$

The nominal shear stress is:

$$\tau_v = \frac{V_u}{bd} = \frac{73980}{300(300)} = 0.822 \text{ N/mm}^2$$

The admissible shear strength for 0.44% reinforcement is 0.41 N/mm². This value is less than the nominal shear strength so shear reinforcement must be provided.

Select two legged 8 mm bars for preliminary design, then the spacing of the stirrups required is:

$$s_v = \frac{A_{sv} f_y}{1.15 b_w (\tau_v - \tau_c)} = \frac{100 * 415}{1.15(300)(0.822 - 0.41)} = 292 \text{ mm}$$

The maximum admissible spacing of the stirrups is either at effective depth or

$$s_v = \frac{A_{sv}f_y}{0.4b} = 346\,\text{mm}$$

There is another critical location for shear design. This is at the combination of the torsion and shear force. The maximum torsion at quarter span (4.71/4 = 1.178 m) is 11 kNm and the shear force at that section is:

$$V_u = w_t(L/2 - 1.178) = 36(2.355 - 1.178) = 42.372\,\text{kN}$$

The combined effect of the shear and torsion on shear reinforcement given by

$$A_{sv} \geq \left[\frac{T}{b_1 d_1} + \frac{V_u}{2.5 d_1} \right] \frac{1.15 s_v}{f_y}$$

where T is the torsion, b_1 is horizontal distances between the vertical legs of the stirrup and d_1 is the vertical distance between the horizontals of the stirrup. These values are:
$d_1 = 300 - 10 - 45 = 245$ mm, and $b_1 = 300 - 2*45 - 10 = 200$ mm, the substitution of different quantities in the above expression gives

$$A_{sv} \geq \left[\frac{11000000}{200*245} + \frac{42372}{2.5*290} \right] \frac{1.15 s_v}{415} = 0.814 s_v$$

Select two legged 10 mm stirrups, then the spacing of the stirrups from the above expression is

$$s_v \leq \frac{2*78.5}{0.814} = 192\,\text{mm}$$

Provide two legged 10 mm bars at 190 mm spacing throughout the beam length. The reinforcement details of the longitudinal and cross-sections of one span segment are illustrated in Fig. 7.20.

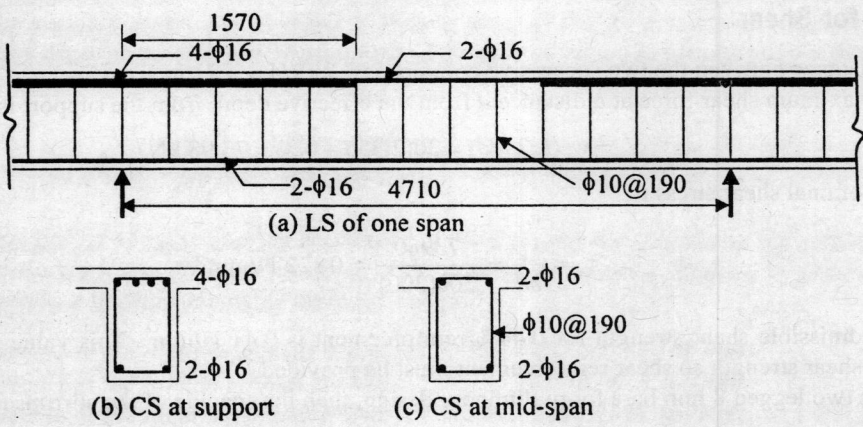

Fig. 7.20 Reinforcement Details; Example 7.8.

Example 7.9: Design of a Ring Beam

A 5 m radius ring beam is supported on six columns. The 450 mm diameter columns spaced at equal distance. The dead and live loads on the beam are 15 kN/m and 20 kN/m respectively. The beam is under *very severe exposure* environment. Design the ring beam with *HYSD-Fe*500 steel reinforcement.

Solution

The structure is exposed to very severe environment so select *M*35 concrete with a clear cover requirement of 50 mm. The beam is subjected to torsion in addition to the bending and shear forces. Select the width of the beam as 350 mm.

Total dead and live load the beam	= 15 + 20 =	= 35 kN/m
Let the self-weight of the beam is = 0.35*0.6*25		≅5.25 kN/m
Total load per unit length is = w		= 40.25 kN/m, use 40 kN/m
Factored limit load = w_t		= 1.5*40 = 60 kN/m

The critical positive and negative bending moments and twisting moment coefficients taken from table 7.3 are: 0.047, 0.093 and 0.010 respectively. The corresponding moments are:

$$M_p = 0.047*60*25 = 0.047*1500 = 70.5 \text{ kNm}$$
$$M_n = 0.093*60*25 = 139.5 \text{ kNm}$$
$$M_t = 0.01*60*25 = 15 \text{ kNm}$$

The twisting moment is zero at mid span and at supports. Therefore effect of the twisting moment is not cumulative with bending moment. The effective depth of the beam is obtained by equating the limit negative bending moment to the moment capacity of the section,

$$d = \sqrt{\frac{M_n}{Kbf_{ck}}} = \sqrt{\frac{139.5(10^6)}{0.138(350)(35)}} = 239 \text{ mm}$$

Fig. 7.21 illustrates the factored forces on one span of the ring beam. Select the overall depth of the beam = h = 500 mm and the effective depth = d = 500- 50 –10 = 440 mm. A 50 mm cover to the reinforcement should be available at top and bottom. Even the hanger bars and the stirrup steel must have the cover of 50 mm.

The area of negative reinforcement required is:

$$A_{st} = \frac{1.15(139.5)10^6}{0.8(440)(500)} = 793 \text{ mm}^2$$

Provide 4 numbers of 16 mm bars at the top face over the supports. Two bars are continued throughout and two are curtailed at one-third span. The actual area of reinforcement provided is 804 mm². All the four bars can be arranged in a row so the actual effective depth of the beam is 500 – 50 – 16/2 = 442 mm. The positive bending moment is about half of the negative bending moment.

2 of 16 mm bars at the bottom face are provided. The percentage of the positive reinforcement is:

$$p = 402(100)/350(440) = 0.26\%$$

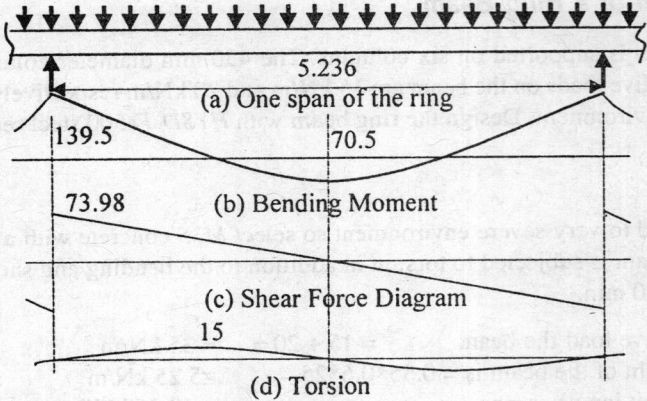

Fig. 7.21 One Span of a Ring Beam with forces; Example 7.9.
(All forces are in kN or kNm).

Design for Shear

The effective span = distance between the two successive columns is $= 2\pi R/6 = 5.236$ m
The maximum shear force at a distance d from the effective depth from the support is

$$V_u = w_t\,(L/2 - d) = 60(2.618 - 0.44) = 130.68 \text{ kN}$$

The nominal shear stress is:

$$\tau_v = \frac{V_u}{bd} = \frac{130680}{350(440)} = 0.85 \text{ N/mm}^2$$

The admissible shear strength for 0.26% reinforcement from table 3.4 is 0.0.38 N/mm². Shear reinforcement must be provided. Select two legged 8 mm bar as a preliminary design, then the spacing of the stirrups required is:

$$s_v = \frac{A_{sv}f_y}{1.15b(\tau_v - \tau_c)} = \frac{100 * 500}{1.15(350)(0.85 - 0.38)} = 265 \text{ mm}$$

The maximum torsion of 15 kNm occurs at quarter span (5.236/4 = 1.309 m) and the shear force at that section is:

$$V_u = w_t(L/2 - 1.309) = 60(2.618 - 1.309) = 78.54 \text{ kN}$$

The combined effect of the shear and torsion on shear reinforcement given by:

$$A_{sv} \geq \left[\frac{T}{b_1 d_1} + \frac{V_u}{2.5 d_1}\right] \frac{1.15 s_v}{f_y}$$

Where T is the torsion, b_1 is horizontal distances between the vertical legs of the stirrup and d_1 is the vertical distance between the horizontals of the stirrup. These values are:
$d_1 = 500 - 10 - 2*50 = 390$ mm, and $b_1 = 350 - 2*50 - 10 = 240$ mm, the substitution of different quantities in the above expression gives:

$$A_{sv} \geq \left[\frac{15000000}{240*390} + \frac{78540}{2.5*390}\right]\frac{1.15s_v}{415} = 0.667s_v$$

Select two legged 10 mm stirrups, and then the spacing of the stirrups from the above expression is:

$$s_v \leq \frac{2*78.5}{0.667} = 235 \text{ mm}$$

Provide two legged 10 mm bars at 220 mm spacing throughout the beam length. The first choice of 8 mm stirrups is not adequate. The reinforcement details of the longitudinal and cross-sections of one span segment are illustrated in Fig. 7.22.

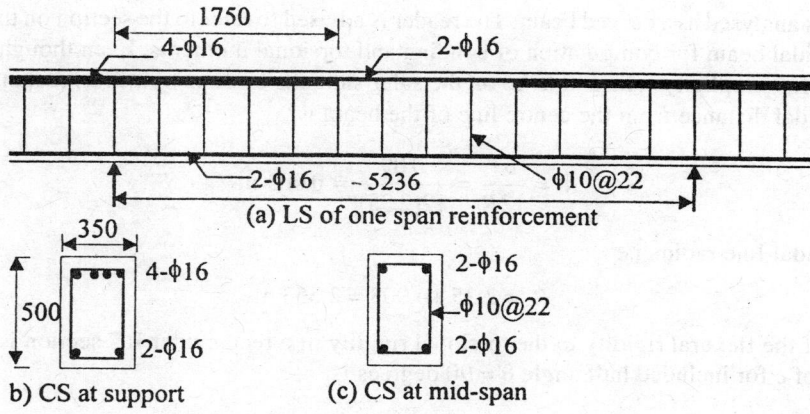

(a) LS of one span reinforcement

b) CS at support

(c) CS at mid-span

Fig. 7.22 Reinforcement Details, Examples 7.9.

Example 7.10: Beam Semi-circular in Plan

Two ends of a semi-circular beam are fixed to a large frame as a cantilever beam. The radius of the outer face of the beam is 2.5 m. The beam supports a balcony roof slab of 120 mm thick and waterproof treatment of 120 mm thick. The structure is exposed to moderate environment. Design the beam.

Solution

The structure is exposed to moderate environment so select *M*25 concrete, and *HYSD-Fe*415 reinforcement steel that is commonly available in the market. Since the roof slab is for balcony, actually recommended live load for such roofs is 1.5 kN/m², but it being a balcony, most balconies are excessively loaded and often misused during functions and processions if they are on the main roads. Some balconies are used for supports of scaffolding in the construction of the upper floors. The author recommends 3 kN/ m² roof balconies.

Dead load of the roof slab is $\quad$ = 0.12*25 = 3 kN/m²
Dead load of the water-proofing is $\quad$ = 0.12*22 =2.7, say = 3 kN/m²
Live load $\qquad$ = 3 kN/m².
Total superimposed load= 3 + 3 + 3 $\quad$ = 9 kN/m².

A parapet of 120 mm thick and 300 mm height is also assumed at the outer face of the slab.

Outer radius of the beam and slab is $= R_0 = 2.5$ m
Let the width of the beam is $= b = 300$ mm
Let the radius of the curved beam $= R = 2.5 - 0.15 = 2.35$ m.
The total load from the roof slab is $= W = \pi R_0^2 * 9/2 = 3.1415(2.5)^2 * 9/2 = 88.5$ kN
Load from the slab per metre of the beam is $= w = W/\pi R$ $= 12$ kN/m
Let the self weight of the beam $= 0.3*0.5*25$ $= 3.75$ kN/m
Then the weight of the parapet $= 0.12(0.3)(22)$ $= 0.8$ kN/m
The factored load on the beam is $= 1.5(12 + 3.75 + 0.8)$ $= 24.8$ kN/m
Use the total load as $= w_t$ $= 25$ kN/m

The beam is analysed as a curved beam. The reader is advised to refer to the section on the curved beam and the helicoidal beam for computation of bending and torsional moments. Even though the beam acts as an L beam, for simplicity sake and to be on the safer side, the beam is treated as rectangular one.
The centroidal distance from the centre line of the beam is:

$$e = \frac{b^2}{12R} = \frac{0.09}{12(2.35)} = 0.003 \text{ m}$$

The centroidal line radius is:

$$R_c = 2.35 + 0.003 = 2.353 \text{ m}$$

The ratio of the flexural rigidity to the torsional rigidity of a rectangular RC section is $= g = 0.58$,
The value of c for included half angle $\theta = 90$ degrees is:

$$c = \frac{2(g + 1)\sin\theta - 2g\theta\cos\theta}{(g + 1)\theta - (g - 1)\sin\theta\cos\theta} = \frac{2(1.58)}{1.58(1.57)} = 1.27$$

The bending and twisting moments at the mid-span for value of $\phi = 0$ are:

$$M_r = w_t R_c^2 (c\cos\phi - 1) = 25(2.353)^2 (1.27 - 1) = 37.37 \text{ kNm}$$

$$M_t = w_t R_c^2 (c\sin\phi - \phi) = 0$$

The bending and twisting moments at the support for value of $\phi = 90^0$ are

$$M_r = w_t R_c^2 (c\cos\phi - 1) = -138.4 \text{ kNm}$$

$$M_r = w_t R_c^2 (c\sin\phi - \phi) = 138.4(1.27 - 1.57) - 41.5 \text{ kNm}$$

Design for Critical Moments

The bending and twisting moments at the support are critical. The combined effect of the bending and twisting into an equivalent bending moment is:

$$M_e = M_r + M_t \left[\frac{1 + h/b}{1.7}\right]$$

The substitution of the quantities in the above equation gives:

$$M_e = 138.4 + 41.5\left[\frac{1 + 500/300}{1.7}\right] = 203.5 \text{ kNm}$$

The effective depth of the section is obtained by equating the equivalent moment to the moment capacity.

$$d = \sqrt{\frac{M_e}{0.138 bf_{ck}}} = \sqrt{\frac{203.5(10^6)}{0.138(300)25}} = 445 \text{ mm}$$

Let the total depth of the section be selected as 500 mm and the effective depth of the reinforcement as 465 mm, then the tension reinforcement that is needed at the top face of the support is:

$$A_{st} = \frac{1.15 M_e}{jdf_y} = \frac{1.15(203.4)10^6}{0.8(465)(415)} = 1318 \text{ mm}^2$$

Provide 3 numbers of 20 mm bars and 2 numbers of 16 mm bars at the top face at the support. The actual area of reinforcement provided is 1344 mm². Two of the 20 mm bars are continuous and the remaining three bars can be curtailed at one-third span. The total length of the beam is 7.39 m.. The positive bending moment is only 22 percent of that of the negative bending moment. The reinforcement required is about 300 mm². The minimum reinforcement required is 0.15 percent, and it is:

$$A_{smin} = 0.15*300*500/100 = 225 \text{ mm}^2.$$

Provide two numbers of 16 mm bars at the bottom from end to end.

Design for Shear Force

The transverse shear force acting at a distance 465mm from the support is:

$$V_u = 25(7.39/2 - 0.465) = 80.8 \text{ kN}$$

The combined effect of shear and twisting moment as equivalent transverse shear is:

$$V_e = V_u + \frac{1.6T}{b}$$

The substitution of the values in the above equation gives:

$$V_e = 80.8 + \frac{1.6*41.5}{0.3} = 302 \text{ kN}$$

The nominal shear stress on the slab is:

$$\tau_v = \frac{V_u}{bd} = \frac{302000}{300(465)} = 2.16 \text{ N/mm}^2$$

The percentage of bottom reinforcement is nominal of the order of 0.26 percent. The allowable shear strength from table 3.4 is $\tau_c = 0.38$ N/mm². This value is less than the allowable strength. As torsion exists on the slab, closed loop stirrups must be provided.

Area of shear reinforcement required is:

$$A_{sv} \geq \left[\frac{T}{b_1 d_1} + \frac{V_u}{2.5 d_1} \right] \frac{1.15 s_v}{f_y}$$

where T is the torsion, b_1 is horizontal distances between the vertical legs of the stirrup and d_1 is the vertical distance between the horizontals of the stirrup. These values are:
$d_1 = 500 - 2*30 - 10 = 430$ mm, and $b_1 = 300 - 2*30 - 10 = 230$ mm, the substitution of different quantities in the above expression gives:

$$A_{sv} \geq \left[\frac{41540000}{230 * 430} + \frac{80800}{2.5 * 430} \right] \frac{1.15 s_v}{415} = 1.37 s_v$$

Select two legged 10 mm stirrups, then the spacing of the stirrups from the above expression is:

$$s_v \leq \frac{2 * 78.5}{1.37} = 115 \text{ mm}$$

The minimum shear reinforcement or the maximum spacing of the stirrups is given by:

$$s_{v\,max} = \frac{A_{sv} f_y}{1.15 b (\tau_v - \tau_c)}$$

The substitution of the quantities in the above equation gives:

$$s_{v\,max} = \frac{2 * 78.5(415)}{1.15(2.16 - 0.38)(300)} = 106 \text{ mm}$$

Provide two legged 10 mm bars at 100 mm spacing.
The sectional reinforcement at support and at mid span is shown in Fig. 7.23.
Note: The bending and the twisting moments are substantial at the supports when compared with those at the mid-span. It is necessary to provide such fixed boundary conditions through the landing slabs and beams. The landing slabs must be capable of resisting the support moments and the reinforcement detail must match the assumption. It is desirable to provide reinforcement at top and bottom for the landing slabs. Provide two nominal bars of 8 mm diameter at mid height to avoid any torsional cracking.

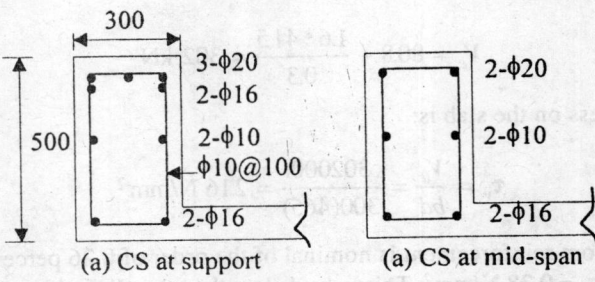

Fig. 7.23 Cross Section Reinforcement, Semi-circular Beam; Example 7.10.

7.6 DESIGN OF TIES OR TENSION MEMBERS

The straight tension members are called ties. Cylindrical pipes, water tanks, suspended roofs etc are under direct tension. Concrete is weak in tension when compared with compression. All tension on a member must be resisted by the reinforcement. Concrete tension members are classified into two groups based on the crack width limitation. A member in which, the crack width is less than 0.2 mm is often considered as uncracked section. Crack width is inter-linked with tensile stress on the transformed cross section and the stress in the reinforcement. The tensile stresses in reinforcement and in the concrete are limited to control the crack width indirectly. Recommended maximum crack width for different exposure conditions are:

Mild or protected environment   = 0.3 mm
Moderate exposure   = 0.2 mm
Severe or very severe exposure   = 0.1 mm

The serviceability stress governs the design of tension members that are exposed to moderate to severe exposure conditions. The strains in steel reinforcement under mild and moderate exposure conditions are normally limited to 0.00115 and 0.00075 respectively. Applying the compatibility of the strains between steel and concrete, the tensile strains in the concrete at the allowable stress levels in *HYSD* bars are 0.00115 and 0.00075 in mild and moderate exposure conditions. In both the cases, the strains are larger than the concrete cracking strains. However, the stress in the transformed cross section is kept below the fracture stress so that the concrete does not crack.

The criteria of design of tension members can be stated as:
- The total tension force must be resisted by the reinforcement alone,
- The stress in reinforcement steel must be less than the allowable stress in the serviceability design,
- The allowable stress in concrete of the transformed section in the limit state of serviceability must be less than the allowable tension in concrete.

The above three statements can be expressed as:

$$A_{st} \geq \frac{1.15 P_u}{f_y} \tag{7.14}$$

$$\frac{P}{A_{st}} \leq f_{ast} \tag{7.15}$$

and

$$\frac{P}{A_c + A_{st}(m-1)} \leq f_{act} \tag{7.16}$$

where
P = axial tensile force on the section,
P_u = factored axial tensile force on the section,
A_{st} = area of the reinforcement,
A_c = area of the concrete section,
f_y = proof stress of the reinforcement,
f_{ast} = allowable tensile stress in steel (also called as permissible or admissible stress) and
f_{act} = allowable direct tensile stress in concrete

The allowable stresses are specified by the code of practice. Indian concrete code on limit state design

(IS:456-2000) is clear on bending and compression elements and almost silent on tension members. The code that deals with liquid retaining structures is still in working stress design. In the absence of code recommendations, the author suggests allowable stress values very similar to the working stress design values. The design of tension members is dominated by the serviceability criterion. Suggested allowable tensile stresses in steel under different exposure conditions are listed in table 7.4.

Table 7.4 Recommended Allowable Tensile Stresses in Steel Reinforcement (N/mm²).

Exposure condition	Fe415	Fe500
Mild	230	275
Moderate	150	150
Severe/very severe	125	125

Similarly recommended allowable tensile stresses in concrete are listed in table 7.5.

Table 7.5 Recommended Allowable Tensile Stresses in Concrete (N/mm²).

Exposure condition	M20		M25		M30		M35	
	Direct	Bending	Direct	Bending	Direct	Bending	Direct	Bending
Mild	1.8	2.5	2.0	2.7	2.2	3.0	2.4	3.3
Moderate	-	-	1.3	1.8	1.5	2.0	1.6	2.2
Severe/very severe*	-	-	-	-	1.2	1.7	1.3	2.0

Example 7.11: Design of Tie in a Truss

A reinforced concrete truss is built in a protected environment. A tie of the truss is subjected to a tensile force of 600 kN. This force includes dead and live load effects. Design the tie with $HYSD-Fe415$ reinforcement steel.

Solution

The truss is designed with $M25$ concrete. The admissible stresses in the mild environment for $Fe415$ steel and $M25$ concrete are:

$$f_{ast} = 230 \text{ N/mm}^2, f_{act} = 2.0 \text{ N/mm}^2 \text{ and modular ratio is} = m = 200000/5000\sqrt{25} = 8.$$

The factored tension force is:

$$P_u = \gamma_f * P = 1.5(600) = 900 \text{ kN}$$

The area of tension reinforcement required is:

$$A_{st} = \frac{1.15 P_u}{f_y} = \frac{1.15(900000)}{415} = 2494 \text{ mm}^2$$

The area of the tension reinforcement that is needed based on the serviceability condition is:

$$A_{st} = \frac{P_u}{f_{ast}} = \frac{600000}{230} = 2609 \text{ mm}^2$$

The area of reinforcement required by the serviceability criterion is more than that of the strength state. Select six numbers of 20 mm bars and four of 16 mm bars. The total area of reinforcement is 2688 mm^2.

The area of concrete section required is given by Eq. 7.15 is:

$$A_c = \frac{P}{f_{act}} - (m - 1)A_{st} = \frac{600000}{2} - 7*2688 = 281184 \text{ mm}^2$$

Select 475 mm by 600 mm section. The area of concrete available is 285000 mm^2.

Distribute the main reinforcement uniformly across the cross section. Provide 3 number of 20 mm bars at top and bottom faces with a clear cover of 30 mm, and 4 number of 16 mm bars on the two side faces in the middle level.

The minimum percentage of the transverse reinforcement needed is 0.15 percent. And it should be provided in loops. The transverse reinforcement needed is:

$$A_{sv} = 0.0015(475)(1000) = 712.5 \text{ mm}^2 \text{ per metre length}$$

Select 10 mm diameter ties. The area of two legs of the tie is 157 mm^2. The spacing of the ties is:

$$s_v = 157(1000)/A_{sv} = 157(1000)/712.5 = 220 \text{ mm}.$$

The maximum ties spacing required is 0.75 times the least dimension of the section, and it is 356.25 mm. Provide 10 mm ties at 220 mm spacing. Fig. 7.24 illustrates the cross sectional details.

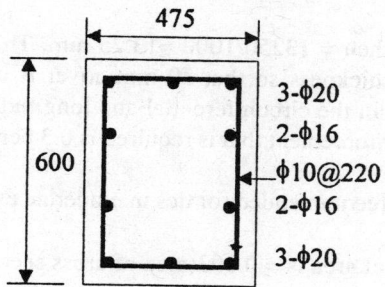

Fig. 7.24 Cross Section of Tie; Example 7.11.

Example 7.12: Design of Water Pipe

A reinforced concrete pipeline of 500 mm radius is carrying water at a pressure of 6 m head of water. Design the pipeline with *HYSD-Fe*415 steel reinforcement.

Solution

The pipeline that carries water can be considered as exposed to moderate environment. Minimum grade of concrete for moderate exposure is *M*25. The allowable stresses in reinforcement and in concrete are 150 and 1.3 N/mm^2 respectively.

Water pressure in the pipe = p = 6 m head = 60 kN/m
Hoop tension in the pipe is = P = pR = 60(0.5) = 30 kN/m

(The self-weight of the pipe is neglected)

The pipe is designed for 1 m length. The collapse load or the factored hoop tension is:

$$P_u = 1.5P = 45\ 000 \text{ N/m}$$

The area of tension reinforcement required is:

$$A_{st} = \frac{1.15P_u}{f_y} = \frac{1.15(45000)}{415} = 124 \text{ mm}^2/\text{m}$$

The area of the tension reinforcement based on the serviceability condition is:

$$A_{st} = \frac{P}{f_{ast}} = \frac{30000}{150} = 200 \text{ mm}^2/\text{m}$$

The serviceability condition controls the design.

Provide 8 mm bars at 200 mm spacing. The total area of reinforcement provided per metre length is 250 mm^2.

The area of concrete section required from Eq. 7.15 is:

$$A_c = \frac{P}{f_{act}} - (m-1)A_{st} = \frac{30000}{2} - 7*250 = 13250 \text{ mm}^2/\text{m}$$

The thickness of the pipe is then = 13250/1000 =13.25 mm. This thickness is too small to provide adequate cover. Select 75 mm thickness so that 30 mm cover is available to the reinforcement. The minimum reinforcement needed in the circumferential and longitudinal directions is 0.3 percent.

Minimum circumferential reinforcement that is required is 0.3 percent & it is = 0.003*75*1000 = 225 mm^2/m.

The minimum reinforcement recommended for ties in moderate exposure condition is 0.30 percent for pipes.

The longitudinal reinforcement area is = 0.003(area of cross section) = (0.003)2π(500 + 37.5)(75) = 760 mm^2.

Provide 10 number of 10 mm bars uniformly distributed along the circumference of the pipe. Figure 7.25 illustrates the sectional detail.

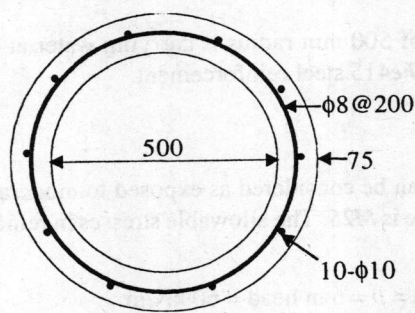

Fig. 7.25 Cross Section of Pipe; Example 7.12.

7.7 TIE BEAMS

A member subjected to primary tension and some bending moment is called a tie beam. The member that connects the two springings of an arch, the horizontal girder in a bowstring girder bridge, and tie beam in industrial foundations, etc are some of the common examples. The tie beam can be designed as cracked or uncracked section depending on the exposure condition. Structures exposed to mild environment can be designed as cracked sections in the sense that the crack width is of the order 0.3 mm. Two possible cases of failure exist in tie beams. If the tension force is very large when compared with the bending moment, the member will fail by fracture of the tension reinforcement. If the bending moment is very large and tension force is nominal, the failure is by crushing of the concrete. The design of tie beam is carried out by interaction of the forces. The design for shear is very similar to that of the beams. The tension force on a member decreases the shear capacity of the section.

Consider a rectangular section as shown in Fig. 7.26 that is subjected to tension and bending moment. Let P is the axial force and M is the bending moment acting on the section.

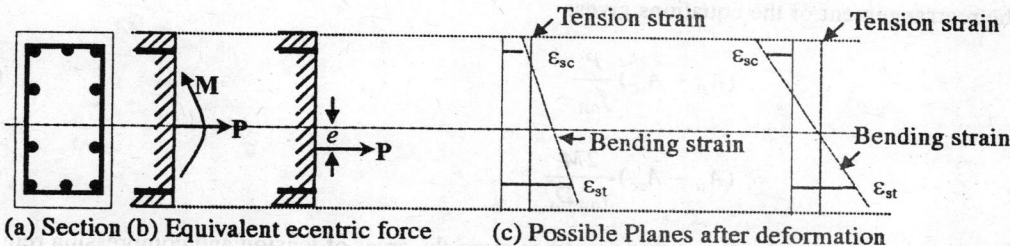

(a) Section (b) Equivalent ecentric force   (c) Possible Planes after deformation

Fig. 7.26 Tie-Beam Section under Tension & Bending.

The two forces can be combined into an equivalent eccentric force P acting at an eccentricity of:

$$e = \frac{M}{P} \tag{7.17}$$

The above eccentricity if less than the kern distance, that is one sixth of the depth in rectangular section, the total section is in tension. This case in tension dominated.

Figure 7.26 (a) indicates the cross-section with reinforcement. It also indicates the axial force and the bending moment. Fig. 7.27(c) indicates the deformed plane, a case in which the tension force dominates and the neutral axis is outside the section. Fig.7.27(c) indicates plane of deformed section a case in which the bending moment dominates and the neutral axis is within the section.

Case 1 Tension dominates

The limit state of strength is not governing but the serviceability of cracking controls the design. Once the section is cracked, all the forces are resisted by the reinforcement only. The Equilibrium of forces controls the design. Equating the external axial force to the resisting forces from the reinforcement gives

$$A_{st} f_{s1} + A_{sc} f_{s2} = P \tag{7.18}$$

Equilibrium of the forces about the centroid of the section gives:

$$(A_{st} f_{s1} - A_{sc} f_{s2}) \frac{D_s}{2} = M \tag{7.19}$$

where

A_{st} = area of the tension face (bottom) reinforcement

A_{sc} = area of the pseudo compression face (top) reinforcement, the steel is in real tension,

f_{s1} = tension stress in the tension face reinforcement,

f_{s2} = tension stress in the compression face reinforcement,

D_s = distance between the top and bottom reinforcement.

In the limit, when the section cracked, one can assume the stress in the top and bottom reinforcement as equal to the limit stress f_{ast} both at top and bottom. The equations of equilibrium reduce to:

$$(A_{st} + A_{sc})f_{ast} = P \qquad (7.20)$$

$$(A_{st} - A_{sc})\frac{f_{ast}D_s}{2} = M \qquad (7.21)$$

The rearrangement of the equations gives:

$$(A_{st} + A_{sc})\frac{P}{f_{ast}} \qquad (7.22)$$

$$(A_{st} - A_{sc})\frac{2M}{f_{ast}D_s} \qquad (7.23)$$

The rearrangement of the above two equations gives the areas of tension and compression reinforcements.

$$A_{st} = \left(\frac{P}{2} + \frac{M}{D_s}\right)\frac{1}{f_{ast}} \qquad (7.24)$$

$$A_{sc} = \left(\frac{P}{2} + \frac{M}{D_s}\right)\frac{1}{f_{ast}} \qquad (7.25)$$

The allowable tensile stress in reinforcement is based on the allowable crack width. The size of the cross section is obtained by limiting the tensile stress in the concrete with in the allowable value. The code specifies the allowable values for direct tension or for bending tension. The desirable to tests for direct tension stress and bending tension are:

$$\frac{P}{A_t} \leq f_{act} \qquad (7.26)$$

$$\frac{P}{A_t} + \frac{My}{I_t} \leq f_{cbt} \qquad (7.27)$$

where

A_t = transformed cross sectional area,

I_t = transformed sectional moment of inertial,

The design of tie beams subjected to dominate tension force is illustrated by an example.

Example 7.13: Tie Beam under Dominant Tension Force

A tie beam of length 6 m in moderate environment is subjected to a direct tensile force of 100 kN and a bending moment of 5 kNm. Design the tie beam.

Solution

The tie beam is exposed to moderate environment so the lowest strength of concrete that is permitted is *M*25. Select *M*25 grade concrete and reinforcement of *HYSD-Fe*415. The permissible tensile stress in reinforcement is 150 N/mm^2, and the allowable direct tensile and the bending tensile stresses from table 7.5 are: 1.3 and 1.8 N/mm^2 respectively. The data of the problem is:

$$P = 100 \text{ kN}, M = 5 \text{ kNm}, f_{ck} = 25 \text{ N/mm}^2, f_{ast} = 150 \text{ N/mm}^2, f_{act} = 1.3 \text{ N/mm}^2, f_{bct} = 1.8 \text{ N/mm}^2, m = 8$$

The effective eccentricity of the forces is:

$$e = M/P = 5/100 = 0.05 \text{ m}$$

The length of the member is 6 metres so the depth can be taken as 400 mm. The eccentricity ration is 0.05/0.4 = 0.125. This is less than 0.17 so the tension force dominates. Assume the size of the section as 250 by 400 mm for preliminary purpose. Let the distance between the top and bottom reinforcement as 300 mm. The areas of bottom and top reinforcements are:

$$A_{st} = \left(\frac{P}{2} + \frac{M}{D_s}\right)\frac{1}{f_{ast}} = \left(\frac{100000}{2} + \frac{5000000}{300}\right)\frac{1}{150} = 445 \text{ mm}^2$$

$$A_{sc} = \left(\frac{P}{2} - \frac{M}{D_s}\right)\frac{1}{f_{ast}} = \left(\frac{100000}{2} - \frac{5000000}{300}\right)\frac{1}{150} = 223 \text{ mm}^2$$

Select four numbers of 12 mm bars at bottom and two numbers of 12 mm bars at top. The areas of reinforcement are 452 and 226 mm^2. The transformed cross sectional area is:

$$A_t = A_c + (m-1)(A_{st} + A_{sc}) = bD + 7(452 + 226) = bD + 4746 \text{ mm}^2$$

Equating the concrete sectional capacity to the axial force gives

$$(bD + 4746)f_{act} = 100\,000\text{N}$$

$$bD = 100000/1.3 - 4746 = 71177 \text{ mm}^2.$$

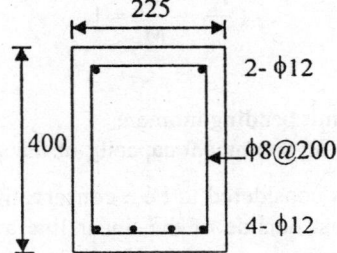

Fig. 7.27 Reinforcement Details of Tie Beam; Example 7.13.

Select 225 by 400 mm size. With a cover of 30 mm to top and bottom reinforcement, the distance between the two reinforcements is $= D_s = 400 - 60 - 12 = 328$ mm. The transformed sectional properties are:

$$A_t = 225*400 + 4746 = 94746 \text{ mm}^2.$$

Centroid of the transformed section is obtained by taking moments of areas about the centre of the concrete cross section.

CG distance $= y_c = m(A_{st} - A_{sc})*D_s/2A_t = 7(452 - 226)(164)/ 94746 = 2.8$ mm.

The moment of inertia of the transformed section is obtained by taking the 2nd moment of areas about the CG of the section. And it is:

$$I_t = \frac{bD^3}{12} + bD*y_c^2 + mA_{st}(D_s/2 - 2.8)^2 + mA_{sc}(D_s/2 + 2.8)^2$$

$$I_t = 1200000000 + 14400 + 85514567 + 42342166.72 = 1327871133 \text{ mm}^4$$

The eccentricity of the CG if the section from the centre of the cross section is small. For all practical purposes, one can use the centre of the section as the centroid of the section. The combined tensile stress in the section that occurs at the bottom fibre of the section is given by:

$$f_t = \frac{P}{A_t} + \frac{My}{I_t} = \frac{100000}{94746} + \frac{5(10)^6(200)}{1292524120} = 1.05 + 0.77 = 1.82 \text{ N/mm}^2$$

The allowable combined bending tensile stress is 1.80 N/mm^2 and it is about the same as the actual value. The design is safe.

7.8 DESIGN OF TIE BEAM BY INTERACTION FORMULA

In case the member is under large bending moment and small tension, one extreme fibre may be subjected to compression and the other to tension. One can solve the problem by limit-state of strength interaction. However the serviceability criterion is likely to control in the severe and extreme exposure cases. In such cases, the serviceability design or the working stress design is more appropriate. This section presents strength limit state design by interactive formula. A number of interactive formulae are available. Of which a simpler method is suggested here. A linear interactive formula that is considered to be on the safer side by static consideration is may be expressed as:

$$\frac{P_u}{P_r} + \frac{M_u}{M_r} = 1 \tag{7.28}$$

Where
$\quad P_u$ = limt axial force, M_u = limit bending moment
$\quad P_r$ = axial force capacity and M_r = moment capacity of the section.

The linear interaction diagram is considered to be a conservative one for design point of view. The experimental data normally falls just outside of the linear line and it is shown by dotted line in the Fig. 7.28. The linear interaction criterion is acceptable for limit-state of strength consideration. However this can't be extended to limit-state of serviceability crack width consideration. As it is difficult to

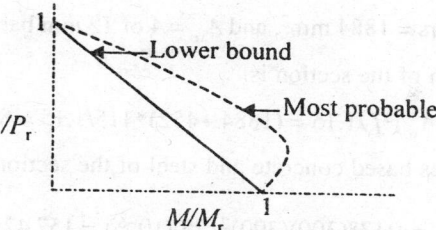

Fig. 7.28 Interaction Diagram.

calculate a reliable crack width under combined action of tension and bending one can increase the partial safety factors applied to the loads to a safe level. Partial safety factors applied to loads at serviceability are 1.0. The same can be achieved by selecting partial safety factors in limit state of strength in the range of 2.8, 3.2, 3.5 and 3.5 respectively for moderate, severe, very severe and extreme exposure conditions. The allowable stress in steel in moderate exposure condition is 150 N/mm². This gives a partial safety factor equal to 415/150 = 2.77. The procedure is illustrated by examples.

Example 7.14: Tie Beam with Large Moment

A tie beam embedded in foundations of moderate exposure condition is subjected to a tension of 100kN and a bending moment of 50 kNm. Design the tie beam.

Solution

The exposure condition is moderate so the minimum grade of concrete is *M25*. *HYSD-Fe*415 reinforcement is used along with the concrete. In a problem like this one has to start with a section and then check its safety with the interaction formula. Make a guess of the section based on the tension consideration and increase the size suitably based on the bending moment. The allowable direct and bending tensile stresses for *M25* concrete are 1.3 and 1.8 N/mm². The ratio of the moment to the tension force is equal to 50/100 = 0.5 m. Area of required concrete section neglecting the reinforcement area is:

$$Approx, \ A_c = \frac{P}{f_{ct}} = \frac{100000}{1.3} = 76923 \ mm^2$$

But as there exists bending moment in addition to *P*, the section selected is 300 by 450 mm. The area of tension reinforcement required to resist the tension force is:

$$Approx, \ A_{st} = \frac{P}{f_s} = \frac{100000}{150} = 667 \ mm^2$$

The approximate bending tension reinforcement required for an effective depth of 390 mm is:

$$A_{st} = \frac{M_u}{jdf_{st}} = \frac{1.5(50)10^6}{0.8(390)(150)} = 1603 \ mm^2$$

Try the solution with the following initial section
       Size = 300 by 450 mm

A_{st} = 6 of 20 mm bars = 1884 mm², and A_{sc} = 4 of 12 mm bars = 452 mm².

The limit of strength in tension of the section is:

$$P_r = (A_{st} + A_{sc})*f_y/1.15 = (1884 + 452)*415/1.15 = 843 \text{ kN}$$

The bending moment capacities based concrete and steel of the section are:

$$M_r = Kbd^2 f_{ck} = 0.138(300)(390)^2(25)(10^{-6}) = 157.4235 \text{ kNm or}$$
$$M_r = A_{st} jdf_y /1.15 = 1884(0.8)(390)(415)/1.15(10)^{-6} = 212 \text{ kNm}$$

The lower of the two controls the limit. So select the lower value.

$$M_r = 157 \text{ kNm}$$

The limiting factored forces on the member are:

$$P_u = 2.8P = 2.8(100) = 280 \text{kN and}$$
$$M_u = 2.8M = 2.8*50 = 140 \text{ kNm.}$$

The interaction expression gives

$$\frac{P_u}{P_r} + \frac{M_u}{M_r} = \frac{280}{843} + \frac{140}{157} = 0.332 + 0.892 = 1.224 > 1.0$$

The section is not acceptable. The bending moment dominates the situation. Further the compression failure in compression controls the moment capacity. Therefore increase the depth of the section to 510 mm. The effective depth of the beam is 450 mm, and the moment capacity of the section is:

$$M_r = Kbd^2 f_{ck} = 0.138(300)(450)^2(25)(10^{-6}) = 209.6 \text{ kNm or}$$
$$M_r = A_{st} jdf_y /1.15 = 1884(0.8)(440)(415)/1.15(10)^{-6} = 239 \text{ kNm}$$

The interaction expression for the revised is:

$$\frac{P_u}{P_r} + \frac{M_u}{M_r} = \frac{280}{843} + \frac{140}{209.6} = 0.332 + 0.668 = 1.0$$

The section 300 by 510 mm with 6 of 20 mm at bottom and 4 of 12 mm is adequate. However the section is to be checked for allowable stress at the Serviceability State. The allowable stresses in concrete in axial tension and in bending are 1.3 and 1.8 N/mm² respectively. The properties of the transformed section are computed here.

The transformed cross sectional area is:

$$A_t = 300*510 + 7(1884 + 452) = 169352 \text{ mm}^2.$$

The distance between the centres of steel is = 510 – 2*60 = 390 mm

By taking moments of areas about the centre of the concrete cross, the centroid of the transformed section is obtained.

$$\text{CG distance} = y_c = (A_{st} - A_{sc})*D_s/2A_t = (1884 - 452)(195)/169352 = 1.6 \text{ mm.}$$

The moment of inertia of the transformed section about the CG of the section is:

$$I_t = \frac{bD^3}{12} + bD * y_c^2 + mA_{st}(1.6 + D_s/2)^2 + mA_{sc}(D_s/2 - 1.6)^2$$

$$I_t = 3952804930 \text{ mm}^4$$

The direct tensile and bending tensile stresses in the transformed section are:

$$f_{ct} = \frac{P}{A_t} = \frac{100000}{169352} = 0.59 \text{ N/mm}^2$$

$$f_{bt} = \frac{MD}{2I_t} = \frac{50(10)^6 510}{2 * 3952804930} = 3.22 \text{ N/mm}^2$$

The tensile stress is too high so the section is a cracked one. The moment of inertia of the section has to be doubled so as to make the stress about half of the present level. Increase the depth of the section to 760 mm from 510 mm. By this way the section modulus is likely to double. The effective depth of the section is $d = 700$ mm and the distance between the compression and tension reinforcements is $D_s = 640$ mm. The properties of the section are:

The transformed cross sectional area is:

$$A_t = 300*760 +7(1884 + 452) = 244352 \text{ mm}^2.$$

The distance between the centres of steel is $D_s = 760 - 2*60 = 640$ mm

By taking moments of areas about the centre of the concrete cross, the centroid of the transformed section is:

$$\text{CG distance} = y_c = m(A_{st} - A_{sc})*D_s/2A_t = 7(1884 - 452)(320)/244352 = 13 \text{ mm}.$$

The moment of inertia of the transformed section about the CG of the section is:

$$I_t = \frac{bD^3}{12} + bD * y_c^2 + mA_{st}(D_s/2 - 13)^2 + mA_{sc}(D_s/2 + 136)^2$$

$$I_t = 10974400000 + 38532000 + 1242955812 + 350852796 = 2924664728 \text{ mm}^4$$

The direct tensile and bending tensile stresses in the transformed section are:

$$f_{ct} = \frac{P}{A_t} = \frac{100000}{244352} = 0.41 \text{ N/mm}^2$$

$$f_{bt} = \frac{MD}{2I_t} = \frac{50(10)^6 760}{2 * 12606740608} = 1.5 \text{ N/mm}^2$$

The interaction formula gives:

$$0.31 + 0.83 = 1.14 > 1.0$$

The section is not acceptable.

Note: The Indian Standard handbook SP-16 gives design aid curves for the design of beam columns.

These interaction curves are based on the limit state of strength and suitable for limited application. This example illustrates how the service load condition controls the design. All structures exposed to environments other than the mild exposure are controlled by the service state limit. The crack width or the tensile stress in concrete controls the design.

PROBLEMS

7.1 A staircase slab of 1.25 m wide is simply supported on cross beams. The tread and the rise are 250 mm and 150 mm respectively. There are twelve treads in each flight and a landing of 1.2 m length at each end. The staircase exposed to mild environment carries a live load of 5 kN/m². Design the staircase slab by limit state method using $M25$ concrete and $HYSD$-$Fe415$ N/mm² reinforcement.

7.2 A staircase of 1.6 m wide consists of a central beam on which independent steps are fixed overhanging symmetrically on either side of the beam. The tread and rise of the stair are 300 mm and 150 mm. The length of the going is 3600 mm and length of the landing at each end is 1250 mm. The ends of the beam are simply supported on rigid supports. Each step 300 mm wide is to be designed to carry a live load of 1.5 kN at 100 mm from its free end. The 300 mm wide beam is to carries a live load of 5 kN/m². Design the step and the beam using $M25$ concrete and $HYSD$-$Fe415$ reinforcement for moderate exposure condition for limit state of strength.

7.3 A three metre wide staircase of an open air stadium has two longitudinal beams placed at 1600 mm apart and the steps that rest on the two beams overhanging by 750 mm on each side. Each step is a double overhang beam. The tread and rise of the staircase are 350 mm and 160 mm respectively. The main beams can be treated as simply supported over a span of 6.5 m, This includes 1.25 m landings and 3.5 m going. The step must be capable of carrying 1.5 kN concentrated load at any point. The staircase has to be designed for a live load of 5 kN/m². Design the staircase beams using $M30$ concrete and $HYSD$-$Fe415$ steel.

7.4 Cantilever steps of 1.2 m span are built into a vertical concrete wall. The rise and tread of the step is 160 mm and 300 mm respectively. The steps are in a protected environment and subjected to a live load of 3 kN/m². Each step should withstand a concentrated load of 1.5 kN placed at 100 mm from the free end of the step. Design the staircase step using $M20$ concrete and $HYSD$-$Fe415$ bars.

7.5 Two floors are separated by 4 m height are connected by a helicoidal staircase semi-circular in plan with a 35⁰ slope. The staircase slab of width 1.5 m is to carry a live load of 3 kN/m². The radial and torsional moments on the slab for fixed support boundary conditions are given by:

$$M_t = wR_c^2(c \sin \phi - \phi)$$

Where R_c = radius to the centroidal axis of the slab = $R + e = R + b^2/12R$
   e = eccentricity of the load action = $b^2/12R$, b = width of the slab,
   ϕ = angle measured from the middle point of the curve of the slab,

$$c = \frac{2(g + 1)\sin \theta - 2g\theta \cos \theta}{(g + 1)\theta - (g - 1)\sin \theta \cos \theta} = \text{and } g = \frac{EI}{GJ}$$

   EI = flexural rigidity, GJ = torsional rigidity, $E/G = 2$
   I = moment of inertia = $bh^3/12$, J = torsional constant = $bh^3/3.5$,
   h = thickness of the slab taken less than the width of the slab

7.6 A helicoidal staircase consists of independent steps placed on a rectangular beam of 400 mm wide. The subtended angle is 240 degrees and the gradient of the staircase is 1.6. Each step is 300 by 1400 mm and overhangs symmetrically over the helicoidal beam. Design the steps and the helicoidal beam with $M20$ grade concrete and $HYSD$-$Fe415$ reinforcement bars. The live load is 3 kN/m². The exposure condition is mild and each step should also carry 1 kN placed at the free end. The radial and torsional moments on the beam for fixed support boundary conditions are given by:

$$M_r = wR_c^2(c \cos \phi - 1)$$

$$M_t = wR_c^2(c \sin \phi - \phi)$$

Where R_c = radius to the centroidal axis of the slab = $R + e = R + b^2/12R$
   e = eccentricity of the load action = $b^2/12R$, b = width of the slab.
   ϕ = angle measured from the middle point of the curve of the slab,

$$c = \frac{2(g + 1)\sin \theta - 2g\theta \cos \theta}{(g + 1)\theta - (g - 1)\sin \theta \cos \theta} = \text{and } g = \frac{EI}{GJ}$$

EI = flexural rigidity, GJ = torsional rigidity, E/G = 2

I = moment of inertia = $bh^3/12$, J = torsional constant = $bh^3/3.5$,

h = thickness of the slab taken less than the width of the slab

7.7 One metre wide staircase connects two floors of clear height 3 m with two flights with slab staircase without external support at mid landing. The going of each flight is 2.7 m and mid landing is 1.0 m. The two flights are in the same direction making the total length of the staircase as 6.4 m. The live load on the staircase is 3 kN/m². Assume suitable rise and tread of the staircase. Use M25 concrete and HYSD-Fe415 reinforcement steel. The exposure condition is mild.

7.8 One metre wide staircase connects two floors of clear height 3 m with a folded slab with a fold at mid-landing. The going of each flight is 2.7 m and the landing length is 1.0 m. The second flight is in the opposite direction making the total length of the staircase as 3.7 m excluding the floor landing of 1 m length. The live load on the staircase is 3 kN/m². Assume suitable rise and tread of the staircase. Use M25 concrete and HYSD-Fe415 reinforcement steel. Illustrate the reinforcement detail. Make suitable assumptions wherever needed. The exposure condition is moderate.

7.9 Each step of a staircase is made of L bent RC slab and each flight has ten such folds. The tread and rise are 250 mm and 150 mm respectively. The length of mid landing is 1000 mm and it is supported at beginning of the landing. Similarly the support to the staircase at the floor landing is also at the start of the staircase. In other words the total length of the flight can be taken as 2500 mm. Design the folded plate type of staircase with M25 concrete and HYSD-Fe415 reinforcement steel. Sketch the reinforcement details.

7.10 A tie of 15 m length exposed to severe environment is subjected to an axial tension of 700 kN. Design the tie using M30 concrete and HYSD-Fe415 bars by limit state design. Sketch the cross sectional details.

7.11 A circular cylindrical pipe of 4 m diameter is subjected to an internal pressure of 100 m head of water. Design a reinforced concrete pipe by limit state design using M25 concrete and HYSD-Fe415 reinforcement bars. Assume the serviceable stress in tension in concrete as 1.6 N/mm² and that in steel as 150 N/mm².

7.12 Derive the governing equations for limit state serviceability design of a reinforced concrete tie beam subjected to an axial tension and bending moment. Area of reinforcement is equally divided on top and bottom faces. The design criterion is that the actual tensile stress in concrete is less than that specified for a pre-assigned crack width. The compression stress is not controlling. The transformed sectional properties should be considered.

7.13 A tie beam of 600 by 800 mm is subjected to an axial tension of 700 kN and bending moment of 70 kNm. The beam is made of M30 concrete and HYSD-Fe500 reinforcement bars with an area of steel 9800 mm² equally divided on tension and compression faces. The distance of the reinforcement from the outer fiber of concrete is 70 mm. Analyse the beam and determine the stresses in the reinforcement.

7.14 A reinforced concrete ring beam has a mean radius of 5 m and supported by 8 columns spaced at equal intervals. A load of 20 kN/m is acting on the girder. Design the girder by Limit state design using M30 concrete and HYSD-Fe500 steel bars. The width of the beam is restricted to 400 mm.

7.15 A five-metre outer radius circular slab is supported by a ring beam having mean radius of 4 m. The thickness of the slab is 150 mm and carries a live load of 4 kN/m². 6 columns spaced at equal intervals support the beam. Design the circular beam using M20 concrete and HYSD-Fe415 steel reinforcement bars. The width of the beam is 300 mm.

Design of Slabs by Limit State Design

8.1 INTRODUCTION

Reinforced concrete slab is the most commonly used floor in buildings. The thickness of the slab in most cases is considered small when compared with the length or width of the slab. Bending moment, shear force and torsional moment are the internal resisting forces in the slab and are functions of the x and y coordinates. The shape of the slab can be rectangular, circular or polygonal. Shorter span is normally termed as width and carries more load compared to that of the long span. A slab having length to width ratio more than two is called *one-way slab*. The recommended thickness of slabs in different constructions is given in Table 8.1. The lower value can be used for higher loads.

Table 8.1 Recommended RC Slab Span to Thickness Ratio.

	Type	L/t
1.	One-way slabs	12 to 20
2.	Cantilever slabs	6 to 10
3.	Two-way	25 to 35
4.	Flat slabs	20 to 30
5.	Continuous slabs	25 to 35

The effective span of a slab can be taken as:
Effective span = clear span + effective depth in simply supported and cantilever slabs,
= clear span in fixed edges.

Minimum reinforcement and cover. Restraints of the supports against free shrinkage or thermal expansion of a slab cause bending, even cracking. Minimum reinforcement must be provided in each direction in the slabs to minimize the shrinkage and temperature effects. The *minimum percentage of reinforcement* in slabs is 0.12% of the area of the section for high yield strength-deformed bars or welded fabric.

Minimum cover to reinforcement should be more than 20 mm or diameter of the bar; which ever is more in mild exposure. It is 30 mm in moderate exposure condition, 45 mm in severe exposure and 75 mm in extreme exposure. In the case of foundations, the absolute minimum cover is 50 mm. The minimum cover for fire rating of slabs vary 20 mm for one hour fire rating, 35 mm for 2 hour fire rating, and 45 mm for 3 hour fire rating. Table 8.2 gives some important limits in case of reinforced concrete slabs. *Minimum cover* to the reinforcement should be provided so as to make steel effective and avoid damage, which might be caused by the environmental effects such as humidity, moisture, temperature, fire etc.

Maximum area of reinforcement is limited to *four per* cent of the cross section. *Diameter of the reinforcement* bar should not be more than one-eighth of the thickness of the slab. *Spacing of the main reinforcement* should not be more than three times the effective depth of or 300 mm in case of structural reinforcement. *The spacing of the nominal reinforcement* bars should not be more than five times the effective diameter or 450 mm to withstand shrinkage and temperature effects.

Table 8.2 Some Limits in Reinforced Concrete Slabs.

Functionality	Limits
Minimum percentage of (*HYSD*) reinforcement	0.12
Maximum percentage of reinforcement	4.0
Maximum diameter of reinforcement	Thickness/8
Maximum spacing of structural reinforcement	3*d* or 300 mm
Maximum spacing of nominal reinforcement	5*d* or 450 mm
Minimum clear cover to reinforcement:	
Mild exposure	20 mm
Moderate exposure	30 mm
Severe exposure	45 mm
Very severe exposure	50 mm
Extreme exposure	75 mm
Minimum cover in foundation slabs	50 mm
Minimum cover for 1 hour fire rating	20 mm
Minimum cover for 2 hour fire rating	35 mm
Minimum cover for 3 hour fire rating	45 mm

8.2 CRITICAL BENDING MOMENTS IN SLABS

Reinforced concrete thin slabs are analyzed as elastic plates. However the bending moments and shear forces are idealized based on modest redistribution of the moments. The bending moment across width of the slab varies almost sinsoidal for the common load of uniformly distributed load. The spacing of the reinforcement is more or less kept constant across the width. In large slabs, the spacing may be varied in segments of middle strip and end strips of the slab. Design bending moment coefficients for uniformly distributed loads are recommended by code of practice, IS:456-2000. The design bending moment is given by:

$$M_{ix} = c_{ix}L^2 \tag{8.1}$$

$$M_{iy} = c_{iy}L^2 \tag{8.2}$$

Where

L = short span (say along the y–direction),

M_{ix} = bending moment along the x-axis, the subscript i refers to 'n' for negative and 'p' for positive

M_{iy} = bending moment along the y-axis, the subscript i refers to 'n' negative and 'p' for positive

c_{ix} and c_{iy} are the corresponding bending moments given in the table

Figure 8.1 illustrates the notations of the bending moments on slab, with reference to the moment coefficients. The bending moment on x-*plane* is called the bending moment M_x. The arrow marked refers to the axis. The plane is referred by the normal to the plane. The bending moment that causes compression on the load's face of the slab is denoted positive bending moment. In case of foundation slabs, the

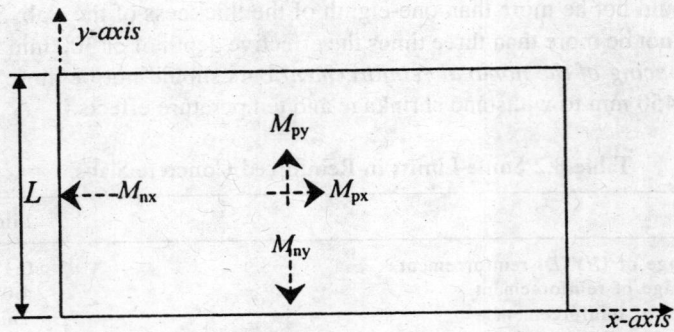

Fig. 8.1 Bending Moment Notation on Slab.

earth pressure face is considered as the loading face. Therefore, the bending moment that causes tension on the bottom face of the foundation slab is negative bending moment.

The corners of slabs have a tendency to lift up due to twisting moment at the corners. The corners are invariably loaded with walls or parapet walls that prevent uplift. Therefore, the corners of the slabs are designed for twisting moment. The twisting moment is usually small and a nominal top reinforcement at one-fifth span is be provided. Incase the corners are allowed to lift; there is no need for the corner top reinforcement. Table 8.3 gives the design bending moment coefficients for different boundary conditions of slabs. The first set for slabs, which are continuous over supports, and the corners of the slabs are restrained against warping. This requires placement of reinforcement at top face at the corners. Otherwise corner cracking is developed. The second portion of the table gives bending moment coefficients at mid span of simply supported slabs in which the corners are allowed to warp. These slabs don't need top reinforcement at corners. The bending moment on either side of any continuity support must be same, but bending moments that one computes depend on the aspect ratio of the panel and may differ if the boundary conditions of the adjacent panels are different. In the design of continuous slabs one has to select the higher of the two bending moments at the support and provide the same reinforcement on either side.

Minimum reinforcement be placed along the long span to resist the nominal bending moments and shrinkage effects. The bending and shear forces in one-way slab can be computed using the beam theory for one metre width.

8.3 MOMENT CAPACITY OF A SECTION AND DESIGN PROCEDURE

The moment capacity derived in chapter 3 for rectangular beam sections holds good for the slab sections in which the width of the section is taken as one unit. The unit width as one metre in this book. The balanced bending moment capacity of the slab section is given by:

$$M_{rb} = Kd^2 f_{ck} \tag{8.3}$$

where

M_{rb} = balanced bending moment capacity of a section of unit width,

K = coefficient that is equal to 0.138 for *HYSD-Fe*415 reinforcement bars and it is 0.133 for *Fe*500 steel,

d = effective depth and

f_{ck} = characteristic strength of the concrete.

Table 8.3 Bending Moment Coefficient for Rectangular Panels Supported on Four Sides.

No. Type of Panel	Short span								Long span
Aspect ratio	1.0	1.1	1.2	1.3	1.4	1.5	1.75	2.0	
(a) Panels continuous over supports with restrained corners (Top reinforcement at corners is needed)									
1. Interior panels									
Negative moment at cont. edge	0.032	0.037	0.043	0.047	0.051	0.053	0.060	0.065	0.032
Positive moment at mid-span	0.024	0.028	0.032	0.036	0.039	0.041	0.045	0.049	0.024
2. One short edge discontinuous									
Negative moment at cont. edge	0.037	0.043	0.048	0.051	0.055	0.057	0.064	0.068	0.037
Positive moment at mid-span	0.028	0.032	0.036	0.039	0.041	0.044	0.048	0.052	0.028
3. One long edge discontinuous									
Negative moment at cont. edge	0.037	0.044	0.052	0.057	0.063	0.067	0.077	0.085	0.037
Positive moment at mid-span	0.028	0.033	0.039	0.044	0.047	0.051	0.059	0.065	0.028
4. Two adjacent edges discontinuous									
Negative moment at cont. edge	0.047	0.053	0.060	0.065	0.071	0.075	0.084	0.091	0.047
Positive moment at mid-span	0.035	0.040	0.045	0.049	0.053	0.056	0.063	0.069	0.035
5. Two short edges discontinuous									
Negative moment at cont. edge	0.045	0.049	0.052	0.056	0.059	0.060	0.065	0.069	-
Positive moment at mid-span	0.035	0.037	0.040	0.043	0.044	0.045	0.049	0.052	0.035
6. Two long edges discontinuous									
Negative moment at cont. edge	-	-	-	-	-	-	-	-	0.045
Positive moment at mid-span	0.035	0.043	0.051	0.057	0.063	0.068	0.080	0.088	0.035
7. Three edges discontinuous (One long edge continuous)									
Negative moment at cont. edge	0.057	0.064	0.071	0.076	0.080	0.084	0.091	0.097	-
Positive moment at mid-span	0.043	0.048	0.053	0.057	0.060	0.064	0.069	0.073	0.053
8. Three edges discontinuous (One short edge discontinuous)									
Negative Moment at cont. edge	-	-	-	-	-	-	-	0.057	-
Positive moment at mid-span	0.043	0.051	0.059	0.065	0.071	0.076	0.087	0.096	0.043
9. Four edges discontinuous									
Positive moment at mid-span	0.056	0.064	0.072	0.079	0.085	0.089	0.100	0.107	0.056

(b) Bending moment coefficients in simply supported slabs with no torsion reinforcement at the corners.

Aspect ratio	1.0	1.1	1.2	1.3	1.4	1.5	1.75	2.0	3.0
At mid-span of short span	0.062	0.074	0.084	0.093	0.099	0.104	0.113	0.118	0.124
At mid-span of Long span	0.062	0.061	0.059	0.055	0.051	0.046	0.037	0.029	0.014

The area of tension reinforcement required can be obtained as:

$$A_{st} = \frac{1.15 M_c}{jdf_y}$$

$$(8.4)$$

where

M_c = collapse bending moment, and f_y = proof strength of the reinforcement,

j = lever arm distance of the reinforcement and it is 0.8 for Fe415 and Fe500 steels for balanced sections.

In case of under-reinforced sections, it is marginally different. However the same value is used even in case of under-reinforced sections.

The bending moment M_y at $x = 0.5L$ for $y = 0$ to L varies close to a sinusoidal variation for uniformly distributed load. Consequently, the theoretical spacing of the reinforcement along y varies with inverse relation of the sinusoidal. The slab in each direction may be divided into three strips to minimize the problem of detailing of the reinforcement. The middle $3/4^{th}$ portion is called the middle strip and the remaining two extreme edges of $0.125L$ are called edge strips. A uniform spacing of the reinforcement in each strip is practicable. In small slabs, the width of the edge strip is too small to be considered as a unit, so the spacing of the reinforcement can be same throughout the slab. The main reinforcement in the edge strips can be equal to half of that in the middle strip or the minimum reinforcement. However, in case of corners constrained slabs this limit is overruled by the minimum reinforcement required for torsion.

Design for shear: The shear force per unit width of the slab is normally computed knowing the boundary bending moment and the intensity of the load. Slabs are usually thin and providing shear reinforcement in building slabs is not convenient. Provision of shear reinforcement in slabs is messy when compared to that in beams. The thickness of the slab can be chosen so that the shear stress is less than the shear capacity of the section. The shear strength of solid slab is given by:

$$\tau_{cs} = k_s \tau_c \tag{8.5}$$

where

k_s = modified shear stress coefficient given in table 8.4,

τ_{cs} = shear strength of slab section

τ_c = shear strength of concrete given in chapter 3.

The shear failure occurs at a distance d from the face of the support in simply supported slabs, and at the face of the slab in cantilever slabs.

Table 8.4 Modified Shear Strength Coefficient in Solid Slab.

Slab thickness in mm >	150 or less	175	200	250	300
k_s	1.3	1.25	1.2	1.1	1.0

Torsion reinforcement: The transverse shear force when combined with twisting moment on the slabs gives rise to Kirchoffs shear force. The tendency of this shear force is to lift the corner of the slabs from the supports. If the corners of the slabs are prevented from lifting up then torsional moment is developed at each of the corners of the slab. The prevention of the lifting up of the corners can be due to either the walls over the slab or continuity of the slab over rigid or beam supports. Torsional reinforcement must be provided at the corners of the slab. The torsional moment depends on the aspect ratio, however, for convenience of design a reinforcement that is equal to three-fourths of the main reinforcement of the short span is recommended both at the top and bottom and in each direction at each of the corners. The torsional reinforcement must be placed in the edge for a distance equal to 1/5 of short span in both directions and both faces of the slab. Table 8.5 gives reinforcement areas for different spacing of the bars for convenience.

Table 8.5 Area of Reinforcement Bars in mm² for Different Bars per meter width.

Spacing of bars in mm	Diameter of the bars in mm							
	6	8	10	12	14	16	18	20
90	314	559	873	1257	1710	2234	2827	3491
95	298	529	827	1191	1620	2116	2679	3307
100	283	503	785	1131	1539	2011	2545	3142
105	269	479	748	1077	1466	1915	2424	2992
110	257	457	714	1028	1399	1828	2313	2732
115	246	437	683	983	1339	1748	2213	2732
120	236	419	654	942	1283	1676	2121	2618
125	226	402	628	905	1232	1608	2036	2513
130	217	387	604	870	1184	1547	1957	2417
135	209	372	582	838	1140	1489	1885	2327
140	202	359	561	808	1100	1436	1818	2244
145	195	347	542	780	1062	1387	1755	2167
150	188	335	524	754	1026	1340	1696	2094
155	182	324	507	730	993	1297	1642	2027
160	177	314	491	707	962	1257	1590	1963
165	171	305	476	685	933	1219	1542	1904
170	166	296	462	665	906	1183	1497	1848
175	162	287	449	646	880	1149	1454	1795
180	157	279	436	628	855	1117	1414	1745
185	153	272	425	611	832	1087	1376	1698
190	149	265	413	595	810	1058	1339	1653
195	145	258	403	580	789	1031	1305	1611
200	141	251	393	565	770	1005	1272	1571
205	138	245	383	552	751	981	1241	1532
210	135	239	374	539	733	957	1212	1496
215	132	234	365	526	716	935	1184	1461
220	129	228	357	514	700	914	1157	1428
225	126	223	349	503	684	894	1131	1396
230	123	219	341	492	669	874	1106	1366
235	120	214	334	481	655	856	1083	1337
240	118	209	327	471	641	838	1060	1309
245	115	205	321	462	628	821	1039	1282
250	113	201	314	452	616	804	1018	1257
255	111	197	308	444	604	788	998	1232
260	109	193	302	435	592	773	979	1208
265	107	190	296	427	581	759	960	1186
270	105	186	291	419	570	745	942	1164
275	103	183	286	411	560	731	925	1142
280	101	180	280	404	550	718	909	1122
285	099	176	276	397	540	705	893	1102

Bar Bending Details in Slab

The reinforcement is normally computed for the maximum bending moment and curtailed or bent at suitable locations. The following are some of the important detailing aspect of the slab reinforcement.

- The slab can be divided into middle and edge strips. The width of the middle strip is 0.75L and each edge strip is 0.125L. The amount of the reinforcement in the edge strip can be three-quarters of that of the middle strip. In many or most slabs, the width of the edge strip is less than one metre. Even though the code recommends the middle and edge strip division, it may not be practicable in slabs of span 6m or less. Therefore the spacing of the reinforcement of the middle strip is continued even in the edge strip. This is on the safer side.

- Atleast 50 percent of the positive bending moment reinforcement should be continued over the supports.
- All the positive bending moment reinforcement placed at the bottom must continue up to 0.25L at the continuous edge and 0.15L at the discontinuous edge, where L is the corresponding span.
- Fifty percent of the bottom reinforcement at mid span can be bent up to top face to resist the secondary bending moment that might arise at the discontinuous edge. As a simple guideline, fifty percent of the bottom reinforcement can be bent up to top at 0.15L from the support.
- The reinforcement placed at top face of the slab for the negative bending moment must continue up to 0.15L from the support. Fifty percent of the reinforcement placed at top face for the negative bending moment must continue up to 0.3L from the support.
- Torsion reinforcement needs to be provided at any corners at top and bottom. The area of the torsion reinforcement must be three-quarters of the mid-span bottom reinforcement, and it should extend 0.2L distance from the support. There are going to be four layers of reinforcement, two at top and two at bottom at the corners. This is code requirement but unfortunately, in most simply supported slabs half of the reinforcement is bent up consequently, three-quarters of the reinforcement may not be available. The reinforcement requirement of the edge strip is only three-quarters of the middle strip.
- The torsion reinforcement provided at corners of one continuous edge and other discontinuous edge can be three-eighths of the positive reinforcement. It is to be provided at top and bottom faces.
- Torsion reinforcement need not be provided at the corners at which two continuous edges meet.

8.4 DESIGN EXAMPLES OF SIMPLE SLABS

Example 8.1: Design of One-way Slab

A simply supported slab 4 by 10 m clear spans in mild environment carries a floor finish of marble and is used as a living room in a residential house. Design the slab using M20 concrete and HYSD-Fe415 reinforcement bars.

Solution

Effective spans and design load

Thickness of the slab for the purpose of self-weight is assumed as = short span/25 = 4000/25 = 160 mm
The effective span is equal to clear spans plus the effective depth.
Let the effective depth = $d = t - 25$ mm = 135 mm
The effective short span is = 4 + 0.135 = 4.135 mm; similarly the long span is = 10.135 m.
The slab is simply supported and with aspect ratio as = 10.135/4.135 = 2.51

The slab is assumed to have its corners free to lift and designed as a one-way slab. The maximum bending moment occurs at the mid-span of the short span. Either it can be taken equal to that of a simply supported beam of one unit width or it can be taken from the moment coefficients given in the table 8.3. Case (b) of table 8.3 gives the moment coefficient as 0.121 by interpolation. Bending moment coefficient corresponds to simply supported beam is of 0.125. (Total floor finish of marble with mortar varies 40 mm to 60 mm)

Self weight $= w_g = 0.16*25$ $= 4 \text{ kN/m}^2$
Floor finish $= w_s = 0.06*25$ $= 1.5$
Live load $= w_1 =$ $= 2.5$
Total working load $= w = 4 + 1.5 + 2.5 = 8 \text{ kN/m}^2$
The factored limit load at strength limit load is $= 1.5*8 = 12 \text{ kN/m}^2$

Design for Bending

The bending moment from simple beam theory at collapse is:

$$M_c = \frac{w_c L^2}{8} = \frac{12 * 4.135^2}{8} = 26.64 \text{ kNm/m}$$

Equate the collapse bending moment to the moment capacity of a section of one metre width, this gives:

$$d = \sqrt{\frac{M_c}{Kbf_{ck}}} = \sqrt{\frac{26.64(10^6)}{0.138(1000)20}} = 99.2 \text{ mm}$$

Let the overall depth of the slab be taken as 140 mm, and then the effective depth is 115 mm. The area of tension reinforcement is:

$$A_{st} = \frac{1.15 M_c}{jd f_y} = \frac{1.15(26.64)10^6}{0.8(115)415} = 803 \text{ mm}^2/\text{m}$$

Provide 12 mm bars at 140 mm spacing along the short span.
The minimum reinforcement required is 0.12 percent and it is:

$$A_{smin} = 0.12(140)(1000)/100 = 182 \text{ mm}^2/\text{m}$$

Provide 8 mm bars at 275mm spacing along the long span.

Design for Shear

The maximum shear force that occurs at a distance of effective depth from the face of the support, and it is:

$$V_c = w_c \left(\frac{L}{2} - d \right) = 12(2.0 - 0.115) = 22.62 \text{ kN/m}$$

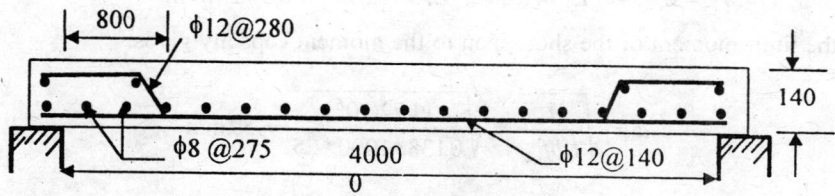

Fig. 8.2 Cross Section of Slab: Example 8.1; Dimensions in mm.

The nominal shear stress is

$$\tau_c = \frac{V_c}{bd} = \frac{22620}{1000(115)} = 0.19\,\text{N/mm}^2$$

The shear strength of concrete for 0.25 percentage of reinforcement is obtained from table 3.4 and it is 0.28 MPa. The nominal shear stress is far below the capacity, so no shear reinforcement is needed.

Alternate bars can be bent up at 0.2 times the span to provide some moment capacity in case of any negative bending moment coming at the edges. Fig. 8.2 illustrates the typical reinforcement details across the mid span of the slab. The corner reinforcement is not shown.

Example 8.2: Design of Two-way Slab with Simply Supported Edges and Unrestrained Corners

A simply supported roof slab with clear spans of 3.5 by 4.5 m is subjected to a live load of 1.5 kN/m^2. Design the slab.

Solution

The superimposed load due to waterproof treatment is 2 kN/m^2. The design live load on flat roof slab is 1.5 kN/m^2. The slab is assumed as exposed to moderate environment to protect it against any damage due to seepage of water. It is recommended that the grade of the concrete as $M25$ even though M20 grade concrete can be used with waterproof treatment. The cover to the reinforcement on the bottom face can be taken as 30 mm.

Let the thickness of the slab be = span/30, Use thickness   = 120 mm;

Design for bending moment

Self weight of the slab	= 0.12*25	= 3 kN/m^2,
Water-proof load	= 0.08*25	= 2 kN/m^2,
Live load	= 1.5 kN/m^2,	

Total factored load at limit of strength design is = w_c = 1.5(3+2+1.5) = 9.75 kN/m^2.
The effective spans of the slab are 3.5 + 0.095   = 3.595 m; and 4.595 m.
The aspect ratio of the slab is = 4.595/3.595   = 1.28

The corresponding bending moment coefficients for slab allowed to the corners lifted are taken from table 8.3, row (b). And they for the aspect ratio of 1.28 are 0.093 and 0.054 for short and long spans.

The limit factored bending moments in the short and long span directions are:

$$M_x = 0.093 w_c L^2 = 0.093 * 9.75 * 3.595^2 = 11.72 \text{ kNm /m}$$
$$M_y = 0.054 w_c L^2 = 0.054 * 9.75 * 3.595^2 = 6.8 \text{ kNm /m}$$

Equating the limit moment of the short span to the moment capacity gives:

$$d = \sqrt{\frac{M_x}{Kbf_{ck}}} = \sqrt{\frac{11.72(10^6)}{0.138 * 1000 * 25}} = 59\,\text{mm}$$

The effective depth required for a balanced section is only 59 mm. However because of practical considerations of waterproof requirement and to reduce the steel, the effective depth is taken as 95 mm as assumed. This means that the total thickness of the slab is 130 mm. The area of tension reinforcement in the short span direction is:

$$A_{st} = \frac{1.15M_x}{jdf_y} = \frac{1.15(11.92)10^6}{0.8(95)415} = 435 \text{ mm}^2/\text{m}$$

Use 10 mm bars at 180 mm spacing in the short span direction.

$$A_{st} \text{ (Provide)} = 78.5/0.180 = 436 \text{ mm}^2/\text{m}$$

Minimum reinforcement required for shrinkage etc. is 0.12 percent and it is:

$$A_{smin} = 0.12(120)(1000)/100 = 144 \text{ mm}^2/\text{m}$$

The bending moment on the long span direction is 6.8 kNm/m and the area of reinforcement in that direction is

$$A_{st} = \frac{1.15M_y}{jdf_y} = \frac{1.15(6.8)10^6}{0.8(85)415} = 278 \text{ mm}^2/\text{m}$$

Provide 8 mm bars at 175 mm spacing.

Design for Shear

In the absence of the coefficients for two way slabs, the load transfer to the supports can be take inversely proportional to the fourth power of spans of the slab. The load transferred to the short span is:

$$w_x = \frac{L_y^2}{L_x^2 + L_y^2} w_c = \frac{4.595^2(9.75)}{3.595^2 + 4.595^2} = 6.1 \text{ kN/m}$$

The maximum shear force that occurs at a distance of effective depth from the face of the support, and it is:

$$V_c = w_x \left(\frac{L_x}{2} - d \right) = 6.1(1.8 - 0.095) = 10.4 \text{ kN/m}$$

The nominal shear stress is

$$\tau_c = \frac{V_c}{bd} = \frac{10400}{1000(95)} = 0.11 \text{ N/mm}^2$$

The shear strength of concrete even for 0.25 percentage of reinforcement is higher than the nominal shear stress, so no shear reinforcement is needed.

Alternate bars are bent up at a distance of 0.2 times the span to resist any secondary bending moment coming at the supports. Fig. 8.3 illustrates the reinforcement details of the slab at mid-span of the long edge. Four extra nominal bars are shown in the section at edges and kinks for construction purpose.

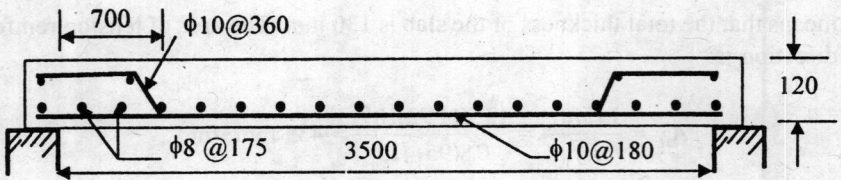

Fig. 8.3 Cross Section of Slab: Example 8.2; Dimensions in mm.

Example 8.3: Design of two-way slab with two Adjacent Sides Continuous

A rectangular panel slab is continuous over two adjacent edges and the other two edges are simply supported and restrained against lifting up of the corners. The slab is 4 by 5.5 m and subjected to a live load of 4 kN/m² and floor finish is 1 kN/m². The exposure condition of the slab is moderate. Design the slab.

Solution

The slab is exposed to moderate environment so $M25$ grade of concrete with 30 mm cover.

Let the thickness of the slab be = span/35, Use thickness = 120 mm;

The effective depth of reinforcement in the short span direction is = $d = 120 - 30 - 5 = 85$ mm

Design for bending moment

Self weight of the slab $= 0.12*25$ $= 3$ kN/m²,

Superimposed dead load $= 1$

Live load $= 4$

Total factored load at limit of strength design is = $w_c = 1.5(3 + 1 + 4) = 12$ kN/m².

The effective spans of the slab are 4m and 5.5 m.

The aspect ration of the slab is = 5.5/4 = 1.375

The bending moment coefficients taken from row 4, table 8.3 are 0.070 and 0.047 for negative bending moments in short and long spans respectively. The positive bending moment coefficients are 0.052 and 0.035 respectively.

The limit factored negative bending moments in the short and long span directions are:

$$M_{nx} = 0.07w_c L^2 = 0.07*12*4^2 = 13.44 \text{ kNm / m}$$
$$M_{ny} = 0.047w_c L^2 = 0.047*12*4^2 = 9.024 \text{ kNm / m}$$

Equating the limit negative bending moment of the short span to the moment capacity gives:

$$d = \sqrt{\frac{M_{nx}}{Kbf_{ck}}} = \sqrt{\frac{13.44(10^6)}{0.138*1000*25}} = 63 \text{ mm}$$

The effective depth required for a balanced section is only 63 mm. However, for practical considerations and to reduce the requirement, the effective depth is taken as 85 mm. This leads to thickness of the slab as 120 mm. The area of tension reinforcement in the short span direction is:

$$A_{st} = \frac{1.15M_{nx}}{jdf_y} = \frac{1.15(13.44)10^6}{0.8(85)415} = 548 \ mm^2/m$$

Use 10 mm bars at 140 mm spacing in the short span direction at top face at support. Alternate bars from the positive bending moment reinforcement can be bent up and some additional bars can be placed instead of 10 mm at 140 mm. The actual detail is finalized later.

Minimum reinforcement required for shrinkage is 0.12 percent:

$$A_{smin} = 0.12(120)(1000)/100 = 144 mm^2/m$$

The bending moment on the long span direction is 9.024 kNm/m and the area of reinforcement in that direction for the same effective depth is:

$$A_{st} = \frac{1.15M_{ny}}{jdf_y} = \frac{1.15(9.024)10^6}{0.8(85)415} = 320 \ mm^2/m$$

Provide 10 mm bars at 240 mm spacing.

Design for positive bending moment

The positive bending moment coefficients for short and long spans are: 0.052 and 0.035 taken from the table 8.3. The area of the reinforcement for the positive bending moment of the short span can be computed proportionately based on the bending moment coefficient, as the effective depth is same. The area of reinforcement is equal to:

$$A_{st} = (0.052)(548)/(0.07) = 408 \ mm^2/m$$

Select 10 mm bars. The spacing of the bars from table 8.5 is 190 mm.
The maximum positive bending moment for the long span is:

$$M_{py'} = 0.035w_c L^2 = 0.035 * 12 * 4^2 = 6.72 \ kNm/m$$

The bottom face reinforcement at mid-span is:

$$A_{st} = \frac{1.15M_{py}}{jdf_y} = \frac{1.15(6.72)10^6}{0.8(75)415} = 311 \ mm^2/m$$

Spacing of the reinforcement is 10 mm bars at 250 mm. Bend alternate bars at 0.2(5500) = 1100 mm

Detailing of the reinforcement

The reinforcement provided at mid-span at bottom is 10 mm bars at 190 mm spacing. Alternate bars are bent up at 1000 mm from the continuity edge and 800 mm from the discontinuous edge. Alternate bars are bent, so the area of the bent up bars at support is (10 mm at 380 mm) = 206 mm². The area of the top reinforcement required was computed earlier and it is 548 mm². Additional reinforcement to be provided at top is = 548 – 209 = 339 mm². From table 8.5, it can be seen that 10 mm bars can be placed at 230 mm spacing as extra give 341 mm². Similarly, the reinforcement in the long direction, bent up bars are 10 mm at 500 mm. The area of the bent up bars is 157 mm². The additional area of the reinforcement needed at to face in the long span direction is = 339 –157 = 182 mm². Provide 8mm bars at 250 mm spacing as additional reinforcement at top at the continuity edge.

Design for Shear

The load transferred to the short span is:

$$w_x = \frac{L_y^2}{L_x^2 + L_y^2} w_c = \frac{5.5^2(12)}{5.5^2 + 4^2} = 7.85 \text{ kN/m}$$

In continuous and discontinuous opposite edges, the reaction from the continuous support can be assumed to be about 10 percent more than the average reaction computed. The maximum reaction on short span is:

$$R = 1.1 * 7.85 \frac{L_x}{2} = 17.27 \text{ kN}$$

The maximum shear force that occurs at a distance of effective depth from the face of the support, and it is:

$$V_c = R_x - dw_x = 17.27 - 0.085 * 7.85 = 16.6 \text{ kN/m}$$

The nominal shear stress is:

$$\tau_v = \frac{V_c}{bd} = \frac{16600}{1000(85)} = 0.2 \text{ N/mm}^2$$

Alternate bars are bent up at a distance of 0.2 times the span to resist any secondary bending moment coming at the supports. Fig. 8.4 illustrates the reinforcement details of the example.

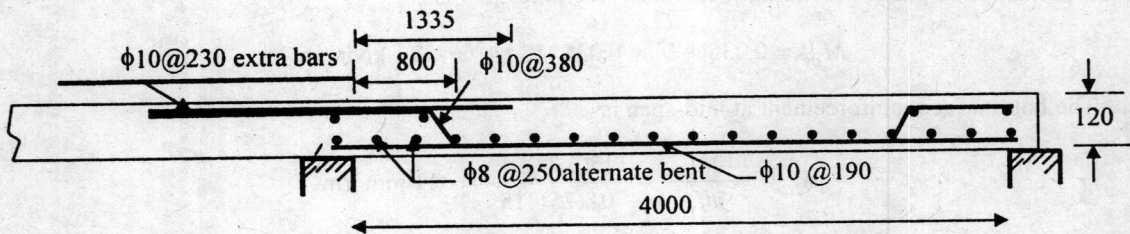

Fig. 8.4 Cross Section of Slab: Example 8.3; Dimensions in mm.

Example 8.4: Design of Cantilever Slab

A portico slab of 6 metres width has five-metre clear cantilever span. The portico is exposed to moderate environment. Design the slab with *HYSD-Fe*415 steel and *M*25 concrete.

Solution

Width of the portico $= B = 6$ m
Cantilever span $= L = 5$ m
Live load $= w_l = 3$ kN/m^2,

Note: code may recommend 1.5 kN/m^2 but it is advisable to design all the porticos for at least 3 kN/m^2 as it may be used as a dumping place or many people may climb in the event of procession etc.

Finish superimposed load (waterproofing) = w_{ds} = 2 kN/m^2

Design of the section for bending

The thickness of the slab at free end can be nominal of 150 mm. Let the average thickness of the slab be assumed as L/14 = 0.35 m for computing the self-weight.

Self weight   = w_g = 25(0.35)   = 8.75 kN/m^2
Total load   = w_t = w_g + w_{ds} + w_l = 8.75 + 2 + 3   = 13.75 kN/m^2
The factored design load of collapse is = w_c = 1.5 w_t = 20.625 kN/m^2, use 21 kN/m^2

The portico can be solid cantilever slab or beam and slab type. In the present case, a solid slab is chosen. The maximum bending moment occurs at a distance of effective depth inside of clear face of the slab.

The effective span can be chosen as 5 + 0.45 = 5.45 m
The maximum bending moment at collapse is:

$$M_c = \frac{w_c L^2}{2} = \frac{21(5.45^2)}{2} = 311.876 \text{ kNm / m}$$

Equating the moment capacity to the collapse bending moment, gives:

$$d = \sqrt{\frac{M_c}{Kbf_{ck}}} = \sqrt{\frac{311.876(10^6)}{0.138(1000)(25)}} = 301 \text{ mm}$$

The total depth of the cantilever slab has to be span/7 for deflection consideration. Therefore the depth of the slab at support is taken equal to 5.45/7 = 0.75 mm. Alternatively a small camber at free end is given to offset the deflection. This is desirable in case of portico slabs. Let the overall thickness at the support be 500 mm

The effective depth of the slab at support = d = 445 mm
Area of the tension reinforcement at top at the support is:

$$A_{st} = \frac{1.15 M_c}{jdf_y} = \frac{1.15 * 311.876(10^6)}{0.8(445)(415)} = 2428 \text{ mm}^2/\text{m}$$

Provide 20 mm bars at 125 mm spacing at top face of the slab. The actual area of tension reinforcement is 2512 mm^2. The cover to the reinforcement is 35 mm.

The percentage of the reinforcement is = p = 2512*100/500(1000) = 0.5 %

Check for shear capacity

The shear stress at the face of the section is:

$$\tau_v = \frac{V_u}{bjd} = \frac{21*5(1000)}{1000(0.8)(445)} = 0.3 \text{ N/mm}^2$$

The allowable shear strength of the $M25$ concrete for 0.5% reinforcement from table 3.4 is 0.47 MPa. Therefore the slab is safe in shear capacity.

Let the overall depth of the section at free edge is 150 mm. The average thickness of the slab works out to be $(500 + 150)/2 = 325$ mm. This checks closely to that assumed 350 mm.

The actual bending moment from the trapezoidal section of the slab will be slightly lass than that obtained from the average thickness. The actual bending moment can be calculated from the actual thickness of the slab.

Distribution reinforcement, Curtailment of bars and Development lengths

Distribution reinforcement of 0.12 percent must be provided in the other direction. The area of the distribution reinforcement for the average thickness of 325 mm is:

$$A_{smin} = 0.12*325(1000)/100 = 390 \text{ mm}^2.$$

Provide 100 mm bars at 150 mm spacing near the support and increase the spacing to 250 mm near the free edge. The distribution bars be placed below the main reinforcement for more effective depth of main bars.

The full development length for $M25$ concrete taken from table 3.3 is 47 times the diameter.

The development length is = $47\phi = 47*20 = 940$ mm

All the bars must be anchored into the support with an embedment length equal to 940 mm. In case of bends provided in the bar, suitable reduction can be included.

The bending moment is proportional to the square of the cantilever span. The bars can be curtailed at a distance equal to the development length beyond the point where they are not required bases on the bending moment. The curtailment span can be obtained by equating the ratio of the reinforcement to the ratio of the square of the cantilever spans and it can be expressed as:

$$\frac{1}{2} = \frac{L_1^2}{L^2} = \frac{L_1^2}{25}$$

where L_1^2 = curtailed length of the half reinforcement from the free edge. The curtailed cantilever span is:

$$L_1 = \sqrt{\frac{25}{2}} = 3.53 \text{ m}$$

The distance at which the bars are not required to resist the bending moment is = $5.0 - 3.53 = 1.47$ m

The actual curtailment is done at a distance beyond the development length.

The curtailment length is = $1470 + 940 = 2410$ mm from support.

The spacing of the bars after the curtailment is = $2*125 = 250$ mm.

This spacing is less than twice the depth of the slab at support. No further curtailment is to be done.

The cantilever support must be capable of resisting the overturning bending moment without rotation. If the support is a beam or a continuity slab, then suitable correction to the elastic rotation of the support should be applied through built-in upward camber. Figure 8.5 illustrates the reinforcement details of the slab.

Check for deflection: It is desirable to compute deflection in cantilever slabs in case the span to depth ratio is more than the specified by the code. Calculation of deflections is illustrated in chapter on serviceability design.

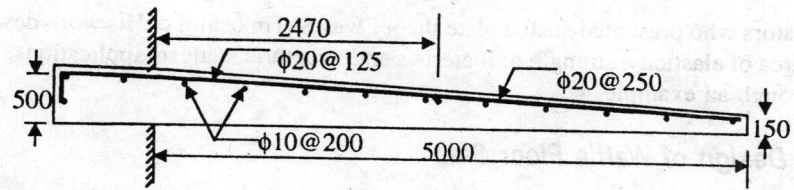

Fig. 8.5 Sectronal Details of Cantilever, Example 8.4 (Dimensions in mm).

8.5 WAFFLE SLABS (GRID FLOORS)

An assembly of closely spaced intersecting beams in two directions and integrated with slab is called *waffle floor*. The beam portion is referred as rib. The rib is shallower when compared with the beam in beam and slab construction. If the ribs intersect at right angle, then it is called orthogrid. In some cases the ribs are placed parallel to the diagonals of the floor. In such a case the grid is called a diagrid. Grid slabs are likely to be economical for spans of 8 to 25 m. The spacing of the ribs in each direction can be different depending on the architectural considerations. A minimum grid spacing of about 750 mm and a maximum spacing of two metres is desirable. Since tube lights are normally fixed in the pockets of the grid, a clear spacing of 750 mm is desirable. The waffle floor can be designed as an orthotropic plate. There are other theories that can also be applied; however, for convenience only the orthotropic plate theory is presented in this book. The bending moments per unit width of an orthotropic plate can be expressed as:

$$M_y = (c_y + vc_x)wL_y^2 \tag{8.6}$$

$$M_x = \frac{1}{r\sqrt{e}}(c_x + vc_y)wL_x^2 \tag{8.7}$$

where

L_x and L_y are the spans of the slab in x and y-directions respectively,
M_x and M_y are the bending moments on x and y-planes respectively,
c_x and c_y are the moment coefficients given in table 8.6,
v = Poisson's ratio, r = aspect ratio = L_x/L_y; L_y is the smaller span and usually set equal to L,

$$e = \sqrt{\frac{D_y}{D_x}}$$

D_x and D_y are the area moment of inertii of the plate about x and y-axes respectively

Table 8.6 Bending Moment Coefficients in Orthotropic Slab.

$r\sqrt{e}$ =	1	1.2	1.4	1.6	1.8	2.0	2.5	4
c_x	0.037	0.034	0.030	0.026	0.021	0.017	0.01	0.0020
c_y	0.037	0.052	0.067	0.079	0.088	0.096	0.11	0.0123

The bending moment coefficients for orthotropic plate theory are available in the literature. One of the earliest investigators who presented such a plate theory was S. Timoshinko. His work deserves a special mention in the area of elasticity, strength of materials and structural analysis applications. The procedure is illustrated through an example.

Example 8.5: Design of Waffle Floor Slab

A hall of 15 by 20 m is exposed to moderate environment and is covered by a waffle roof slab. Design the slab.

Solution

Roof slab is in moderate exposure, so $M25$ grade concrete is selected. Assuming access to roof exists, the loads are:

> Roof live load where access is provided = 1.5 kN/m^2,
>
> Waterproofing load = 2 kN/m^2,
>
> Aspect ratio of the slab = r = 20/15 = 1.333

Let the spacing of the grid be 1.25 m. The slab is divided into 12 grids in the short span direction and 16 in the long span direction.

> Grid spacing = a = 1.25 m

As the grid, spacing is small, only nominal and practicable slab thickness is selected. The outer face of the slab is exposed to moderate environment and the inside face to mild exposure. Select outer cover as 35 mm and inside cover as 20 mm. Therefore the minimum thickness of the slab is about 20 + 35 + 15 = 70 mm. Similarly, the depth of the web of the grid beam can be selected in the range of L/25 to L/20 depending on the load intensity on the floor.

Let the following dimensions be selected.

> Thickness of the slab = t = 90 mm
> Thickness of the web = b = 180 mm
> Depth of the rib (web) = h_w = 710 mm

The grid slab is considered as a T-beam in each direction with flange thickness same as the thickness of the slab and the width of the flange equal to the grid spacing but limited to $12t + b = 12*90 + 180 = 1200$ mm. The basic dimensions of grid beam are listed in Fig. 8.6.

As the properties of the grid beam in both the directions are same, the ratio of the flexural stiffness of the slab is equal to unity ($e = 1$). As the slab is designed as a cracked section, the Poisson's ratio is taken as 0.1

Design of the Section

The bending moment coefficients are to be selected from table 8.6. The value of $r\sqrt{e}$ *is equal to 1.333*. The moment coefficients taken from the table 8.6 are:

> c_y = 0.052 for r = 1.2 and 0.067 for r = 1.4, the interpolated value is equal to 0.0545,
>
> c_x = 0.034 for r = 1.2 and 0.030 for r = 1.4, the interpolated value is equal to 0.033,
> The weight of the top slab is = 0.09*25 = 2.25,

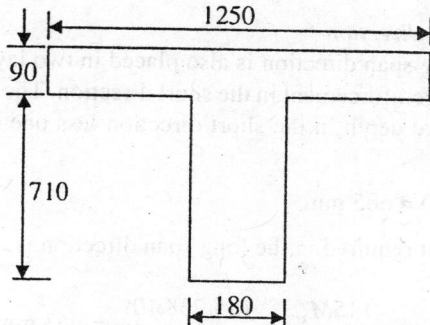

Fig. 8.6 Grid-beam Section, Example 8.5.

The weight of the rib is = 0.18*0.71*25 = 3.2 kN per one grid in one direction =3.2/1.25 = 2.6 kN/m^2,

The weight of the ribs in one square metre of the slab = 2*2.6 = 5.2 kN/m^2,
The self weight of the waffle slab is = w_g = 2.25 + 5.2   = 7.45 kN/m^2,
    Total load on the slab = 7.45 + 2 + 1.5   = 10.95 kN/m^2,
Use the total load on the slab = w = 11 kN/m^2,
The factored limit load = w_c = 1.5*11   =16.5 kN/m^2,

The bending moments at mid-span in short and long span directions are:

$$M_y = (c_y + vc_x)w_c L_y^2 = (0.0545 + 0.1*0.033)16.5(16^2) = 244.1 \text{ kNm/m}$$

$$M_x = \frac{1}{r\sqrt{e}}(c_x + vc_y)w_c L_x^2 = \left(\frac{1}{1.333}\right)(0.033 + 0.1*0.0545)16.5(20^2) = 190.4 \text{ kN/m}$$

The corresponding bending moments for one grid spacing are obtained by multiplying the above moments by the grid spacing, and the bending moments for a grid spacing are:

$$M_{yc} = 1.25*244.1 = 305.125 \text{ kNm}$$

$$M_{xc} = 1.25*190.4 = 238 \text{ kNm}$$

The reinforcement in the short span direction is placed bottom layer with a cover of 20 mm. The bars are placed in two rows in view of the smaller rib width. The section is treated as a T-beam so the lever arm for the short span direction is:

$d = h -$ cover – reinforcement thickness– 0.5 times the slab thickness = 800 – 20 – 30 – 45 = 705 mm
The area of the reinforcement in the short direction is:

$$A_{yst} = \frac{1.15M_{yc}}{df_y} = \frac{1.15(305.125)10^6}{705*415} = 1200 \text{ mm}^2$$

Provide 4 numbers of 20 mm bars in two layers. The actual area of reinforcement provided is 1256 mm^2. By this way, the reinforcement from the other direction can also be placed in two layers with one layer in between the two layers of the short span back up.

Reinforcement in the long span direction

The reinforcement in the long span direction is also placed in two layers. The bottom layer is placed in between the two layers of the reinforcement in the short direction. The effective depth of the reinforcement is the equal to the effective depth in the short direction less one diameter of the diameter of the bottom bars of the shorter span.

It is equal to: $d = 705 - 20 = 685$ mm

The area of the reinforcement required in the long span direction is:

$$A_{xst} = \frac{1.15 M_{xc}}{d f_y} = \frac{1.15(238)10^6}{685 * 415} = 963 \text{ mm}^2$$

Provide two numbers of 20 mm bars in the bottom layer and two numbers of 16 mm bars in the top layer.

The area of the reinforcement provided is: $\quad A_{xst} = 2(314 + 201) = 1030 \text{ mm}^2$

Check for Shear Stress

The distribution of load to spans is inversely proportional to the square of the spans. Maximum shear force is on the middle grid beam of the short span. The approximate intensity of the load on the middle grid beam of the short span is:

$$w_{xc} = \frac{L_x^2 w_c (spacing \ of \ grid)}{L_x^2 + L_y^2}$$

$$w_{xc} = \frac{20^2 (16.5)(1.25)}{20^2 + 16^2} = 12.57 \text{ kN/m}$$

The maximum shear force that occurs at a distance effective depth of the beam is:

$$V_c = w_{xc}(L_x/2 - d) = 12.57 * (8 - 0.705) = 91.7 \text{ kN}$$

The nominal shear stress in the beam is:

$$\tau_v = \frac{V_c}{b_w d} = \frac{91700}{180(705)} = 0.73 \text{ N/mm}^2$$

The percentage of reinforcement in the web of the short span grid beam is:

$$p = \frac{1256 * 100}{180 * 800} = 0.87\%$$

The corresponding shear strength of the M25 grade concrete from table 3.4 is: 0.6 N/mm². The shear strength is less than the nominal shear stress, therefore shear reinforcement must be provided. Select two legged 8 mm HYSD-Fe415 stirrups, then the area of the stirrup is $A_{sv} = 100 \text{ mm}^2$.

The spacing of the stirrups is:

$$s_v = \frac{A_{sv}f_y}{1.15(\tau_v - \tau_c)b_w} = \frac{100(415)}{1.15(0.73 - 0.6)(180)} = 1542 \text{ mm}$$

The maximum spacing of the stirrups is:

$$S_{v\max} = \frac{A_{sv}f_y}{0.4b_w} = \frac{100(415)}{0.4(180)} = 577 \text{ mm}^2$$

The maximum spacing is also subject to 300 mm, or 0.75 times the effective depth. So, provide two legged 8 mm stirrups at 300 mm spacing.

Design of the Top Slab

The slab over the grid beams continuous in both directions except at the corners of the floor. The intensity of the load is small. The span to depth of a unit is 13. The clear span is only $1.25 - 0.18 = 1.07$ m with a depth of 0.9 m. It is therefore provide only nominal reinforcement at the middle surface of the top slab. Provide 0.12% of reinforcement in each direction. The spacing of the reinforcement is subject to a maximum spacing of five times the thickness of the slab. Provide 8 mm *HYSD* bars at 300 mm spacing in the middle level of the slab in each direction. Figure 8.7 illustrates the typical reinforcement details of the grid slab.

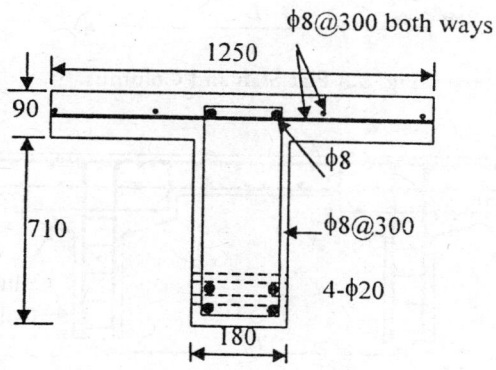

Fig. 8.7 Reinforcement Grid Slab, Example 8.5.

8.6 INTRODUCTION TO FLAT SLABS

The floor slabs are invariably of reinforced concrete built integral with the beams. Slabs supported by the beams are called beam-slab construction. Normal spacing of the beams in varies from four metres to seven metres. The thickness of the slabs in such spans is in the order of 100 mm to 180 mm. If the span of the slab exceeds six metres, intermediate beams that may not be part of the frame action are provided to keep slab thickness to reasonable. A construction in which the slabs rest on columns without beams is called *flat slab construction*. In such situations, the thickness of the slabs is larger than that in the beam-slab construction but the net clear ceiling height available is more. Warehouses, godowns, large office, and public halls need unobstructed space. Flat slabs are invariably two-way slabs and rest on several columns as shown in Figure 8.8. Sometimes the top of the column is widened so as to provide wider base to support the slab and reduce the punching shear. Such widened portions are called *column heads*. There is a limit up to which one can treat the widened portion as a part of the column. The width must be limited to the portion within 90 degrees of the segment as shown in Figs. 8.9 and 8.10. The projection beyond the column head should be really treated as a part of the slab rather than that of the column. In other words, it should be treated as thickening of the slab. Such a thickened portion of the slab is called drop. The drops are sometimes known as capital of the columns.

A portion that is enclosed in two directions by four adjacent columns is called a panel. The slab in each panel is divided into column strip and middle strip in each direction. The width of the column strip is equal to half of the column spacing and it is placed half on either side of the column line. In case of unequal spans, it can be taken equal to half of the average columns spacing. In addition, it should also be restricted to 0.5 times the column spacing in any direction. The middle strip is the one that is bounded by the column strips and its width is equal to the spacing of the columns minus the width of the column strip. The width is usually equal to or greater than half of the spacing of the columns.

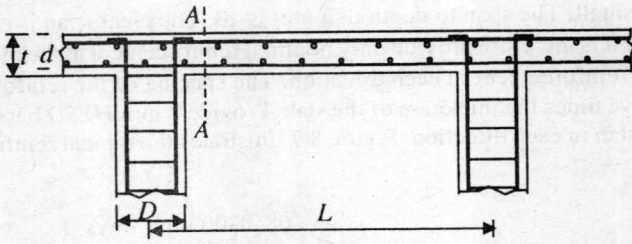

Fig. 8.8 Flat Slab and Columns.

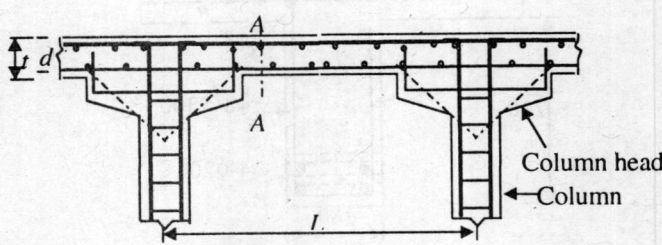

Fig. 8.9 Flat Slab and Column Heads.

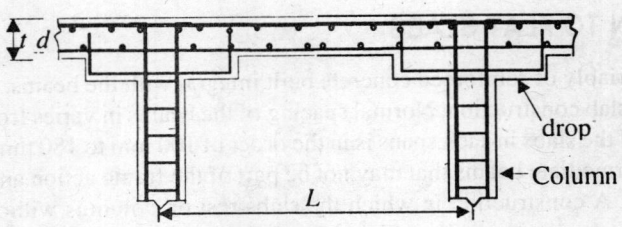

Fig. 8.10 Flat Slab with Drops

Let b_{ci} = width of column strip in ith column row

b_{mi} = width of the middle strip

L_{xi} = spacing of the columns in the x-direction in the ith panel

L_{yi} = spacing of the columns in the y-direction in the ith panel

Then widths of the column and middle strips in x direction are given by

$$b_{ci} = 0.25(L_{yi-1} + L_{yi}) < 0.25 (L_{yi} + L_{yi+1}) \tag{8.8}$$

$$b_{mi} = L_{yi} - 0.25 (L_{yi} + L_{yi+1}) \tag{8.9}$$

Similarly, the widths of the strip in the y-direction are calculated. The slab and columns can be analysed as equivalent frames having idealized columns to continuous wall supports along the column lines. The stiffness of the column is divided by the panel width and considered as the stiffness of the vertical element per unit width of the frame. The analysis in both directions is to be carried out independently as two sets of independent frames. Such an idealization introduces higher bending moments in the middle strip. Therefore, the moment and shear force computed by this method must be proportioned with higher weightage to the column strip when compared with that of the middle strip. The analysis is to be carried by loading full dead load and only three fourths of the total live load in each panel. However, in case of mat foundation slabs, full load coming from the column should be taken as the load and also no reduction should be given to liquid loads. The frames should be analysed for two load combinations, namely, all panels loaded and alternative panels loaded. The critical section for design of moment is the section at the face of the column or column head or at the face of the drop. The critical section for shear force design is peripheral line around the column at a distance $0.5d$ from the face of the column or the face of the drop. Where d is the effective depth of the slab. The punch shear is likely to dominate the failure of the slab. Usually the thickness of the slab or the size of the column or the size of the drop is taken large enough to eliminate the shear failure of the section around the periphery of the column. The design bending moments for each of the strips are obtained from the analysed bending moment by distributing the total bending moment between the two strips.

Table 8.7 Relative Values of Positive and Negative Bending Moment in a Strip.

Bending moment and location	Distribution ratio
1. Interior spans	
(a) Negative bending moment (M_n)	0.65
(b) Positive bending moment (M_p)	0.35
2. End span	
(a) Exterior negative bending moment (M_n)	$0.65/\alpha$
(b) Interior negative bending moment (M_n)	$0.75 - 0.1/\alpha$
(c) Positive bending moment (M_p)	$0.63 - 0.28/\alpha$

Where $\alpha = 1 + 1/\alpha_c$, K_{c1} and K_{c2} are the stiffness of the columns, and K_s is the stiffness of the slab $\alpha_c = \dfrac{K_{c1} + K_{c2}}{K_s}$

The sum of the magnitudes of the positive and the negative bending moments computed based on the coefficients of the table is more than M_o. The fixed end bending moment in beam fixed both sides and subjected to uniformly distributed load is: $WL/12 = 0.0833 \, WL$. And the corresponding positive bending moment at mid span for the beam is $WL/24 = 0.0416 \, WL$. Since there is some redistribution at the column supports, the negative bending moment coefficient at the interior support is take in the range of 0.75

minus some value. However, the positive bending moment is chosen on the higher side. A more accurate method of analysis is to treat the slab as an elastic plate supported on columns and analyse by finite element analysis.

8.7 APPROXIMATE DIRECT METHOD OF DESIGN OF FLAT SLAB

Direct design moments of flat slabs satisfying the equilibrium criterion can be obtained from the bending moment between the column and middle strips satisfying the following limitations:

1. At least three continuous spans in each direction are available.
2. The panels must be rectangular and have two-way action. The aspect ratio in each panel is less than two.
3. The end span should not be larger than the interior spans.
4. The ratio of any two successive span lengths should be within 0.75 to 1.33.
5. A cantilever projection of about one third of the exterior span is permitted with appropriate static modification in the bending moment at the cantilever.
6. The design live load should not be more than three times the dead load. This when ignored effects the magnitude of the positive bending moment.

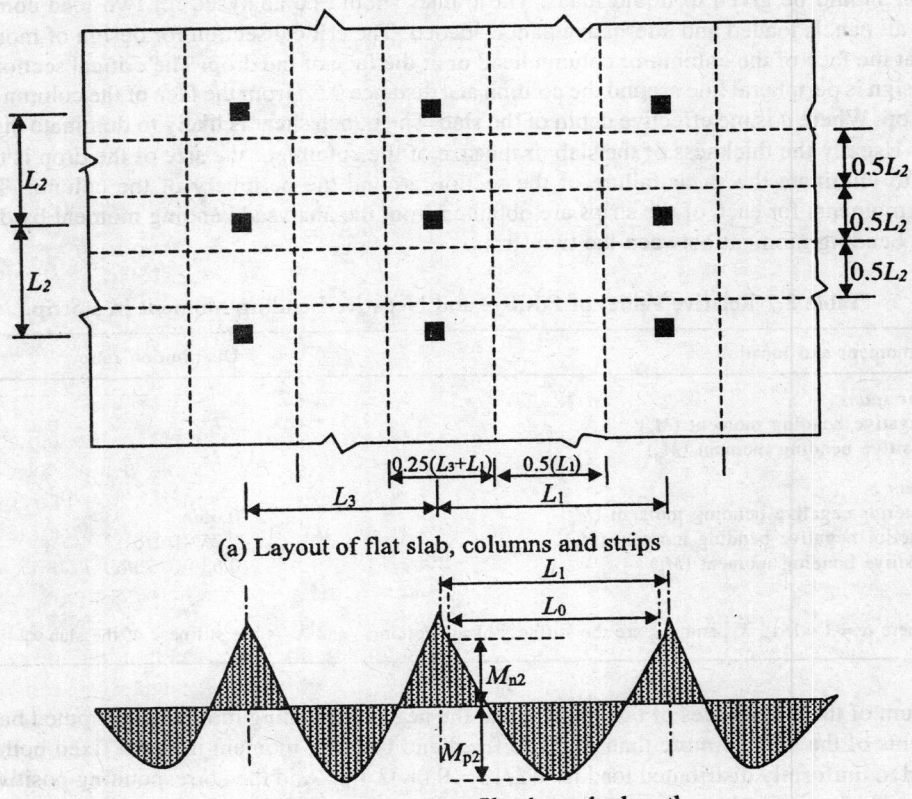

(a) Layout of flat slab, columns and strips

(b) Bending moment profile along the length

Fig. 8.11 Flat Slab with Columns: Notations.

The sum of the magnitudes of the positive and the negative bending moments in a panel is equal to that of the maximum bending moment of simply supported span under uniformly distributed load and it is give by:

$$M_0 = \frac{W_c L_n}{8} \tag{8.10}$$

where

L_n = clear span between columns or drop faces, and subject to a minimum of 0.65 times the spacing of the columns.

W_c = total of the uniformly distributed load on one clear panel,

M_0 = sum of the magnitudes of the positive and negative bending moments.

Figure 8.11(a) illustrates typical layout of a flat slab along with the notations. Figure 8.11(b) illustrates the bending moment profile in one direction.

Let

M_n = the magnitude of the negative bending moments at the face of the column,

M_p = positive bending moment at the mid-span,

Then the equilibrium of moments of the free body diagram of one panel gives

$$|M_n| + |M_p| = \frac{W_c L_n}{8} \tag{8.11}$$

Table 8.8 Redistribution of Bending Moments Across a Panel.

Strip and its boundary conditions	Percent of the total BM
1. Column strip	
(a) Negative BM at exterior support	100
(b) Negative BM at the interior support	75
(c) Positive BM	60
2. Middle strip	
The difference between the panel and column strip moments.	

The total magnitude of the combined bending moments in a panel is computed by the equation 8.11. Further the total bending moment in a panel is distributed into negative and the positive bending moments. The ratio of such distribution of the bending moment is given in table 8.7. These moments have to be distributed between the column and middle strips in the proportions given in table 8.8.

Design moments on columns: Columns are built integral with the slab and consequently the joints between the slab and the columns share the bending moment at the support. The bending moment on the column or the wall supporting the flat slabs is due to lack of symmetry in spans or due to unsymmetrical load on the spans or due to stiffness of joints between the column and slab. The bending moment on the column can be computed based on the relative stiffness of the column. The bending moment on the column can be approximated as:

$$M = 0.08 \frac{(w_d + 0.5 w_l) L_2 L_n^2 - w_d L_2 L_1^2}{1 + \dfrac{1}{\alpha_c}}$$

where

   w_d and w_l are the design dead and live load intensities respectively,

   L_{1n} is the shorter of the span in the direction of the moment.

L_1 is the center to center distance of the columns along the moment computing span

L_2 is the center to center distance of the columns transverse to the direction of the bending moment.

To avoid possible damages to columns due to lateral and unsymmetrical loads on the slabs, the columns are to be designed with at least the minimum relative stiffness that are given in table 8.9.

Table 8.9 Minimum Required Relative Stiffness of Column in Flat Slabs.

Ratio of Live Load to Dead Load *	L_2/L_1	α_c (minimum required)
0.5	0.5 to 2	0
1.0	0.5 to 1	0.7
	1.25	0.8
	2.0	1.2
2.0	0.5	1.3
	1.0	1.6
	1.25	1.9
	2.0	4.9
3.0	0.5	1.8
	1.0	2.3
	1.25	2.8
	2.0	13.0

There are a number of other conditions to be satisfied while designing the flat slabs. The reader is advised to refer to the IS: 456-2000 for more information.

Design for Shear

The critical section for shear is a peripheral plane at a distance $0.5d$ from the face of the column or the face of the column head or drop. In case there is an opening near this zone, appropriate deduction in the peripheral length should be made proportional to the distance of the critical section to that of the opening from the centre of the column. Figure 8.12 illustrates the critical shear stress section.

The shear force on the periphery of shear failure of the slab around the column support is:

$$V_c = w_c[L_1 L_2 - (a_1 + d)(a_2 + d)] \tag{8.12}$$

in which a_1 and a_2 are the sides of a rectangular column section, and d is the effective depth of the slab.

The periphery at a distance $0.5d$ from the face of the column is:

$$b_0 = 2(a_1 + a_2 + 2d)$$

The nominal shear stress on the periphery of failure is:

$$\tau_v = \frac{V_c}{2(a_1 + a_2 + 2d)d} \tag{8.13}$$

The admissible shear strength of the concrete without shear reinforcement is:

$$\tau_c = 0.25(0.5 + \beta_c)\sqrt{f_{ck}} \tag{8.14}$$

Where β_c = ratio of the short side to the long side of the column subject to a maximum of 0.5.

There are three cases of shear design and they are:

Case 1: The nominal shear stress is less than the shear strength of the concrete without shear reinforcement:

The section is safe and no need for shear reinforcement in the slab.

Case 2: The nominal shear stress exceeds the shear strength but less than 1.5 times the strength. Provide shear reinforcement in the slab to resist the shear stress beyond 0.5 times the shear strength of the concrete.

Case 3: The nominal shear stress is greater than 1.5 times the shear strength of the concrete. Increase the thickness of the slab such that the nominal shear stress is less than 1.5 times the shear strength.

Reinforcement Detailing

The following conditions must be met in case of solid flat slabs with or without drops or column heads.

The spacing of the reinforcement should not be more than the twice the thickness of the solid slab.

The minimum percentage of the reinforcement is same as that in solid slabs, that is 0.12 percent.

The thickness of the drop for determination of reinforcement is lesser of the following:

a) Thickness of the drop or
b) Thickness of the slab plus one quarter distance between the edge of the drop and that of the capital.

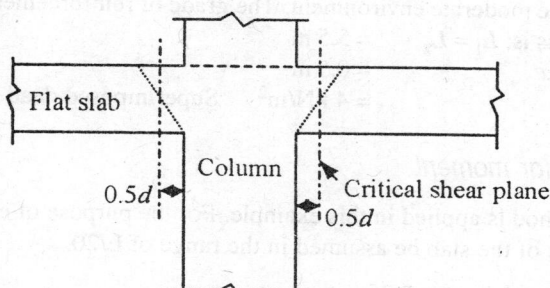

Fig. 8.12 Critical Shear Stress Periphery in Slab.

As some idealizations are made in the analysis, the minimum length of the reinforcement must be maintained. IS:456-2000 gives these minimum lengths for different cases. Some broad guide-lines are indicated here, but the reader is advised to refer to the code for more detailed specifications.

Detailing of the Reinforcement in Column Strip

1. At least half of the top bars must extend up to 0.3 times the span beyond the column support, and the balance of the other half of the bars must extend up to 0.2 times the span beyond the column support,
2. At least 50 percent of the bottom bars must be from support to support, the balance of the bars must extend up to 0.125 times the span from the support,
3. Fifty percent of the bottom bars can be bent up, and the distance of the bending up from the support can be 0.24 times the span,

Detailing of reinforcement in the middle strip
4. All top bars must extend at least 0.22 times the span from the support,
5. Fifty percent of the bottom bars must extend from support to support, and the balance of the bars can be curtailed at 0.15 times the span from the support. Or Not more than fifty percent of the bars can be bent up and the distance of the bending up from the support can be 0.2 times the span.
   Detailing of reinforcement is an important design aspect and should not left to the bar bender.

8.8   DESIGN EXAMPLES OF FLAT SLABS

The design examples illustrate different steps and criteria in the design of flat slab without and with drop panels.

Example 8.6: Design of Flat Floor Slab

A floor slab is supported on columns spaced 5.5 metres in both the directions. The column is 500 mm square and the live load on the floor is 4 kN/m². The floor finish load on the slab is 1 kN/m². Design a flat slab without drops or column heads. Height of each floor is 5 m. The floor slab is exposed to moderate environment.

Solution

Select *M25* grade concrete moderate environment. The grade of reinforcement is *HYSD-Fe415*.
Spacing of the columns is: $L_1 = L_2$ $= 5.5$ m
Size of the column $= a$ $= 0.5$ m,
Live load $= w_1$ $= 4$ kN/m²   Superimposed dead load $= w_{sd} = 1$ kN/m²

Design of the section for moment

The direct design method is applied in this example. For the purpose of estimating the self weight of the slab, let the thickness of the slab be assumed in the range of L/20.

Let the thickness of the slab $= t = 0.25$ m
Self weight $= 0.25(25)$ $= 6.25$ kN/m².
Total dead load $= w_d = 6.25 + 1.0$ $= 7.25$ kN/m²
Design factored load $= w_c = 1.5(7.25 + 4)$ $= 16.875$ kN/m²

Clear spacing between the columns is $= L_n = L_1 - a = 5.5 - 0.5 = 5.0$ m
The total design load in a panel is: $W = w_c L_n L_2 = 16.875 (5)(5.5) = 464.0625$ kN/one panel
Sum of the magnitudes of positive and negative bending moments in a panel is:

$$M_0 = \frac{WL_n}{8} = \frac{464.0625 * 5}{8} = 290.1 \text{ kNm / panel}$$

The negative bending moment acting on the column strip in the exterior column is likely to govern the depth of the slab. Therefore, the depth of the section is calculated first based on the bending moment and then shear force.
Magnitude of the negative BM at the face of the columns in the interior panels is:

$$M_{ni} = 0.65 M_0 = 188.5 \text{ kNm/5.5m}$$

The relative stiffness of the columns and the slab determine the distribution of the negative and positive bending moments in the exterior panel.

The building is not restrained against lateral sway so the effective column height can be taken equal to 1.2 times the clear height of the column. Hence the effective height of the column is:

$$L = H - t = 5 - 0.25 = 4.75 \text{ m}; \qquad\qquad L_e = 1.2(4.75) = 5.7 \text{ m}$$

The relative stiffness of the column is:

$$K_{c1} = \frac{I_c}{L_e} = \frac{a^4}{12 L_e} = \frac{0.5^4}{12 * 5.7} = 0.001$$

The relative stiffness of the slab panel is:

$$K_s = \frac{I_s}{L_{se}} = \frac{bt^3}{12 L_{se}} = \frac{5.5 * 0.25^3}{12 * 5.5} = 0.0013$$

Live load to dead load ratio = 4/7.25 = 0.55

The exterior columns must be designed to have a minimum relative stiffness so as to withstand the panel bending moment. Such a desired minimum relative stiffness is given in the table 8.9 for different aspect ratios and live load to dead load ratios. From table 8.9, it can be seen that the minimum required relative stiffness ratio is zero for the live load ratio of 0.55 and the aspect ratio of 1.0. The relative stiffness ratio in the present case is:

$$\alpha_c = \frac{K_{c1} + K_{c2}}{K_s} = \frac{2 * 0.001}{0.0013} = 1.55$$

The *alfa* factor that decides the relative distribution of bending moment between the negative and the positive bending moments is:

$$\alpha = 1 + 1/\alpha_c = 1 + 1/1.55 = 1.6$$

Table 8.7 gives the relative negative and the positive bending moments in the exterior panel, and these bending moment coefficients are:

> **2. End span**
> Exterior negative bending moment coeff. = $c_{n0} = 0.65/\alpha$ = 0.41
> Interior negative bending moment coeff. = $c_{ni} = 0.75 - 0.1/\alpha$ = 0.69
> Positive bending moment coeff. = $c_{p0} = 0.63 - 0.28/\alpha = 0.455$

The corresponding end panel bending moments, the negative at the outer support, negative at the inner support and the positive bending moment in the panel are:

$$M_{no} = c_{n0}M_0 = 0.41*290.1 = 118.9 \text{ kNm/panel}$$

$$M_{ni} = c_{ni}M_0 = 0.69*290.1 = 200.2 \text{ kNm/panel}$$

$$M_{po} = c_{p0}M_0 = 0.455*290.1 = 132 \text{ kNm/panel}$$

The negative bending moment corresponding to the interior panel support is:

$$M_{nii} = 0.65M_0 = 0.65*290.1 = 188.6 \text{ kNm/panel}$$

In which the subscript n refers to the negative bending moment, p refers to positive bending moment, and the subscript i refers interior support or interior panel and o refers to the outer panel.

The bending moment is distributed between the column and middle strips as per table 8.8. The subscript c refers to column strip and m refers to the middle strip. The corresponding bending moments in the outer panel strip are:

$$M_{cno} = (1)\,M_{no} = 118.9 \text{ kNm/2.75m}$$

$$M_{cni} = 0.75 M_{ni} = 0.75*200.2 = 150.15 \text{ kNm/2.75m}$$

$$M_{cpo} = 0.65 M_{po} = 0.65*132 = 85.8 \text{ kNm/2.75m}$$

The negative bending moment corresponding to the interior panel column strip is:

$$M_{nii} = 0.75 M_{cnii} = 0.75*188.6 = 141.45 \text{ kNm/2.75m}$$

The critical bending moments on the column strips are listed above. The difference remaining bending moments have to be assigned to the middle strip. It is advisable to determine the reinforcement in the column strip and the reinforcement spacing in the middle strip can be adjusted proportionally. The thickness of the slab is primarily controlled by the absolute maximum bending moment in the column strip. The most critical bending moment occurs at the interior support of the outer panel column strip and it is:

$$M_{cni} = 150.15 \text{ kNm/2.75m}$$

The effective depth of a balanced section to resist the bending moment is obtained by equating this bending moment to the moment capacity of a balanced section. The effective depth required is:

$$d = \sqrt{\frac{M_c}{Kbf_{ck}}} = \sqrt{\frac{150.15(10^6)}{0.138(2750)25}} = 125.8 \text{ mm}$$

Let the overall thickness of the slab be chosen as 250 mm and then the effective thickness of the slab in one direction for the lower layer of the reinforcement is = 250 – 35 – 6 = 209 mm and for the upper layer it is 197 mm. The diameter of the reinforcement is assumed as 12 mm in calculating the effective depth. Since the panel is a square panel, it is desirable to have spacing of the reinforcement in both the directions. Use effective depth of the slab as 197 mm. Area of the reinforcement in column strip at the support is:

$$A_{st} = \frac{1.15 M_c}{jdf_y} = \frac{1.15*150.15(10^6)}{0.8*197*415} = 2640 \text{ mm}^2 / 2.75\text{m}$$

Provide 12 mm bars at 115 mm spacing at top face of the slab over the columns in the column strip.

The actual reinforcement at different locations of the column strip can be obtained by proportioning with the corresponding bending moments obtained earlier.

Spacing at the outer support at top = 117.7*150.15/118.9 = 148 mm
Spacing at the middle span at bottom = 117.7*150.15/85.8 = 206 mm
Spacing at the inner support at top = 117.7*150.15/141.45 = 125 mm

The spacing of the reinforcement is rationalized later after the reinforcement in the middle strip is computed.

The bending moments in middle strip are the balance of the panel moments and the column strip, and they are:

$$M_{mno} = 0.$$
$$M_{mni} = 0.25M_{ni} = 0.25*200.2 = 50.05 \text{ kNm/2.75m}$$
$$M_{mpo} = 0.35M_{po} = 0.35*132 = 46.2 \text{ kNm/2.75m}$$

The negative bending moment corresponding to the interior panel middle strip is

$$M_{mii} = 0.25M_{cnii} = 0.25*188.6 = 47.15 \text{ kNm/2.75m}$$

The effective depth of the section in the middle strip is same as that of the column strip, therefore the reinforcement can be obtained by prorating with respect to the bending moment. However the minimum percentage of the reinforcement and the maximum spacing of the reinforcement of twice the depth of the slab must be maintained.

The minimum reinforcement in the slab is:

$$A_{smin} = 0.12*250*2750/100 = 825 \text{ mm}^2 \text{ per 2.75 m.}$$

The maximum spacing of the reinforcement is twice the depth of the slab and it is equal to 500 mm. One can use 12 mm bars consistently in both the strips. Spacing of the 12 mm bars for the minimum reinforcement is:

$$\text{Spacing of 12 mm bars for minimum steel} = 113*2750/825 = 375 \text{ mm.}$$

It is therefore restrict the maximum spacing of the 12 mm bar reinforcement to 375 mm.

The reinforcement in the middle strip can be prorated based on the moments. The moments in the strip are about one third of the maximum bending moment, so the required spacing of the reinforcement is about three times that of 117.7 mm. That is about 353 mm. Before finalizing the reinforcement details, it is desirable to check for the shear capacity of the section.

Design of Section for Shear

Let the spacing of the 12 mm bars is about 360 mm at the support, then the percentage of positive tension reinforcement in section is:

$$p = \frac{113*100}{0.36*250*1000} = 0.125\%$$

The critical shear plane is the peripheral plane at a distance $0.5d$ from the face of the column. The length of the critical section is:

$$b = 4(a + d) = 4(0.5 + 0.20) = 2.8 \text{ m}$$

The shear force on the plane is:

$$V_u = w_c [L*L - (a + d)(a + d)] = 16.875(5.5*5.5 - 0.6*0.6) = 505 \text{ kN}$$

The nominal shear stress is:

$$\tau_v = \frac{V_u}{bd} = \frac{505000}{2800*200} = 0.9 \text{ N/mm}^2$$

The shear strength of the concrete without shear reinforcement is given by:

$$\tau_c = 0.25(0.5 + \beta_c)\sqrt{f_{ck}}$$

The aspect ratio of the slab is one, so the maximum admissible value of b is only 0.5. Therefore the shear strength of the M25 concrete is:

$$\tau_c = 0.25(5) = 1.25 \text{ N/mm}^2.$$

The nominal shear stress is less than the shear capacity of the M25 concrete; therefore no need for transverse reinforcement or thickening of the slab.

Detailing of the Main Reinforcement

The main reinforcement consists of top bars at the column line and bottom bars at mid-span, in both the directions. There is a restriction of minimum curtailment and cranking of the bars. Further maximum spacing based on the minimum percentage of the reinforcement at a section is 375 mm. The maximum spacing of the 12 mm bars can be 500 mm if the reinforcement is available both at top and bottom faces of the slab. Cranking of the bars may not be very helpful because of restrictions and the development lengths. It is advisable to avoid the cranking of the bars for faster construction. The detailing of the bars is made in the column and middle strips separately. A uniform spacing in all the panels in each direction in any one strip is adopted. The positive reinforcement is the most important one to protect the slab from yield line failure, even though the reinforcement at top near the support is high. A rationalized reinforcement detail is presented here.

Column Strip

12 mm bars at 180 mm spacing at the bottom in mid-span and alternate bars are curtailed at 0.125 span (about 650 mm) from support. The length of the curtailed bottom bars is 5500 − 1300 = 4200 mm.

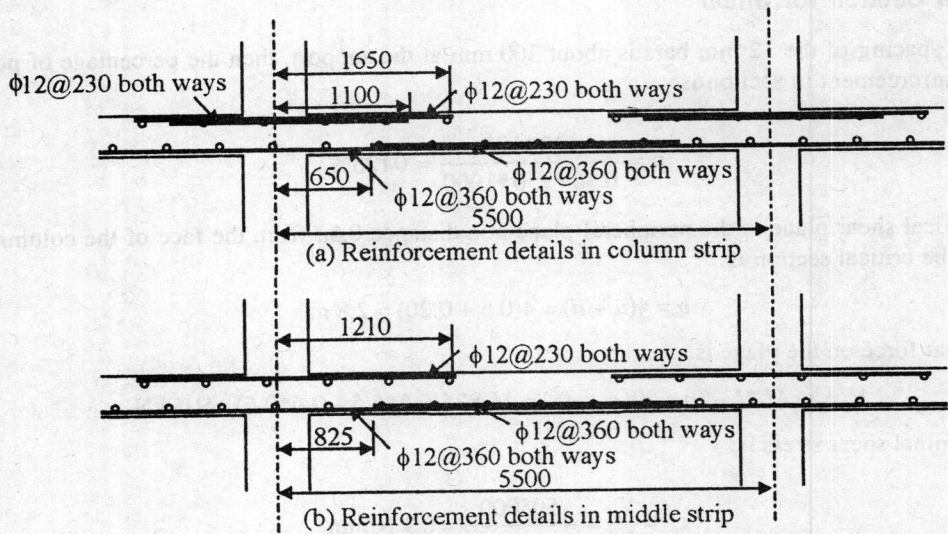

(a) Reinforcement details in column strip

(b) Reinforcement details in middle strip

Fig. 8.13 Reinforcement Details of Flat Slab, Example 8.6.

12mm bars at 115 mm at top face extending up to 0.3 times the span (about 1650 mm from support on each side). Alternate bars can be curtailed at 0.2 times the span (about 1100 mm) from the support

Middle Strip

12 mm bars at 180 mm spacing at the bottom in mid-span and alternate bars are curtailed at 0.15 times span (about 825 mm) from support. The length of the curtailed bottom bars is 5500 – 1650 = 3850 mm. 12 mm bars at 230 mm at top face extending to 0.22 times the span (about 1210 mm from support on each side).

Example 8.7: Flat Slab with Drop Panels

A large single storey flat roof slab of a warehouse is supported by 500 mm square columns spaced 6 m apart in both directions. The number of panels in one direction is six and the other direction it is five. The clear ceiling height is six metres. The slab extends by one metre all round the beyond the centre line of the outer most columns. The slab is exposed to sever environment. The live load on the roof is 3.0 kN/m^2.

Solution

The minimum cover to the reinforcement in severe exposure condition is 45 mm. Assume the weight of the waterproof treatment as 2.5 kN/m^2. The span of the slab in both directions is six metres so it is desirable to have drop panel flat slab construction. The size of the drop can be in the range of one-fourth of the span. M30 grade concrete is chosen for other and economic considerations.

The design data is consolidated here:
Span of the slab = $L_1 = L_2 = L$ = 6 m in both directions,
Size of the column (square) = a = 500 mm, clear height of column = H = 5 m;
Size of the drop panel = $D = L/4$ = 1.5 m,
Grade of concrete = M30, grade of reinforcement is = $HYSD-Fe$415;
Cover to the reinforcement is = 45 mm,
Live load = w_l =3. kN/m^2, Waterproof load = w_{sd} = 2.5 kN/m^2.

Structural Analysis

Let the thickness of the slab is assumed 300 mm in the range of L/20 for the purpose of self-weight including the effect of the drop.

Weight of the slab = w_d = 0.3*25   = 7.5 kN/m^2.
The total load on the slab is = w = 7.5 + 2.5 + 3   = 13 kN/m^2
Factored limit load = w_c = 1.5(13)   = 19.5 kN/m^2
Clear spacing between the drops = $L_e = L - D$ = 4.5 m > 0.65 L
Total design load between the drops in a panel is = $W_c = w_c*L*L_e$ = 19.5(6)(4.5) = 526.5 kN; use 530 kN.

The sum of the magnitudes of the positive and negative bending moments in a panel of slab is

The column is on isolated footing and fixed in the drop of the slab. The columns are not prevented against sway. The effective height of the column is taken as 1.2 times the clear height.
The effective height of the column is = 1.2*5 = 6 m.
The effective thickness of the slab in a panel is taken as 0.3 m for the purpose of relative stiffness.

The relative stiffness of the column and slab are:

$$K_c = \frac{L_c}{L_c} = \frac{0.5^4}{12(1.2)(5)} = 0.0008$$

$$K_s = \frac{I_s}{I_s} = \frac{6(0.3^3)}{12(4.5)} = 0.003$$

The relative stiffness of the column with slab is: $\alpha_c = 0.0008/0.003 = 0.27$

The minimum α_c value required is zero for an aspect ratio of the slab equal to one and the live load to dead load ratio less than 0.5. The size of the column is acceptable.

The value of $\alpha = 1 + 1/\alpha_c = 4.7$

Table 8.7 gives the relative negative and the positive bending moments coefficients in the exterior panel, and moment coefficients are:

>   *2. End span*
>   Exterior negative bending moment coeff. $= c_{n0} = 0.65/\alpha$ $= 0.14$
>   Interior negative bending moment coeff. $= c_{ni} = 0.75 - 0.1/\alpha$ $= 0.73$
>   Positive bending moment coeff. $= c_{p0} = 0.63 - 0.28/\alpha = 0.57$

The corresponding end panel bending moments, the negative at the outer support, negative at the inner support and the positive bending moment in the panel are:

$$M_{no} = c_{n0}M_0 = 0.14*298.125 = 41.7 \text{ kNm/panel}$$
$$M_{ni} = c_{ni}M_0 = 0.73*298.125 = 217.6 \text{ kNm/panel}$$
$$M_{po} = c_{p0}M_0 = 0.57*298.125 = 170 \text{ kNm/panel}$$

The negative bending moment corresponding to the interior panel support is:

$$M_{nii} = 0.65M_0 = 0.65*298.125 = 193.8 \text{ kNm/panel}$$

In which the subscript *n* refers to the negative bending moment, *p* refers to positive bending moment, and the subscript *i* refers interior support or interior panel and *o* refers to the outer panel. The bending moment is distributed between the column and middle strips as per table 8.8. The subscript *c* refers to column strip and *m* refers to the middle strip.

The corresponding bending moments in the outer (exterior) panel strips are:
$$M_{cno} = (1)\,M_{no} = 41.7 \text{ kNm/3m}$$
$$M_{cni} = 0.75M_{ni} = 0.75*217.6 = 163.2 \text{ kNm/3m}$$
$$M_{cpo} = 0.65M_{po} = 0.65*170 = 110.5 \text{ kNm/3m}$$

The negative bending moment corresponding to the interior panel column strip is:

$$M_{nii} = 0.75M_{cnii} = 0.75*193.8 = 145.35 \text{ kNm/3m}$$

The remaining bending moment has to be assigned to the middle strip. The most critical bending moment occurs at the interior support of the outer panel column strip and it is:

$$M_{cni} = 163.2 \text{ kNm/3m}$$

The effective depth of a balanced section to resist the bending moment is obtained by equating this bending moment to the moment capacity of the balanced section. The effective depth required is:

$$d = \sqrt{\frac{M_c}{Kbf_{ck}}} = \sqrt{\frac{163.2(10^6)}{0.138(3000)30}} = 115 \text{ mm}$$

Let the overall thickness of the slab be chosen as 240 mm and then the effective depth for the lower layer of the reinforcement is = 240 – 45 – 6 = 189 mm and for the upper layer is 177 mm for 12 mm bar. The spacing of the reinforcement in both the directions is kept same. Use effective depth of the slab as 177 mm. Area of the reinforcement in column strip at the support is:

$$A_{st} = \frac{1.15 M_c}{jdf_y} = \frac{1.15 * 163.2(10^6)}{0.8 * 189 * 415} = 2991 \text{ mm}^2 / 3m$$

The number of 12 mm bars required per one column strip at the support is:

$$N = 2991/113 = 26.5/3 \text{ m}$$

The spacing of the bars is = 3000/26.5 = 113 mm

The actual reinforcement at different locations of the column strip can be obtained by proportioning with the bending moments. The spacings required are:

Spacing at the outer support at top   = 113*163.2/41.7   = 442 mm
Spacing at the middle span at bottom = 113*150.15/110.5 = 167 mm
Spacing at the inner support at top   = 113*150.15/145.35 = 127 mm

The spacing of the reinforcement is rationalized after the reinforcement in the middle strip is also computed.

The bending moments in middle strip are the balance of the panel moments and the column strip, and they are:

$M_{mno} = 0.$
$M_{mni} = 0.25 M_{ni} = 0.25*217.6 = 54.4 \text{ kNm/3m}$
$M_{mpo} = 0.35 M_{po} = 0.35*170 = 59.5 \text{ kNm/3m}$

The negative bending moment corresponding to the interior panel middle strip is

$$M_{mii} = 0.25 M_{cnii} = 0.25*193.8 = 48.45 \text{ kNm/3m}$$

The effective depth of the section in the middle strip is same as that of the column strip, therefore the reinforcement can be obtained by prorating with respect to the bending moment. However the minimum percentage of the reinforcement and the maximum spacing of the reinforcement of twice the depth of the slab must be maintained.

The minimum reinforcement in the slab is:

$$A_{smin} = 0.12*240*3000/100 = 864 \text{ mm}^2 \text{ per 3 m}$$

The maximum spacing of the reinforcement is twice the depth of the slab and it is equal to 480 mm. One can use 12 mm bars consistently in both the strips. Spacing of the 12 mm bars for the minimum reinforcement is:

Spacing of 12 mm bars for minimum steel is = 113*3000/864 = 392 mm.

The reinforcement in the middle strip can be prorated based on the moments. The moments in the strip are about one third of the maximum bending moment, so the required spacing of the reinforcement will be about three times that of 113 mm.

Design of Section for Shear

The depth of the drop panel can be selected such that the shear stress in the drop is within the shear capacity of the concrete. Let the spacing of the 12 mm bars is 380 mm at the support, then the percentage of positive tension reinforcement is:

$$p = \frac{113*100}{0.38*240*1000} = 0.124\%$$

Let depth of the drop = 400 mm

The critical shear plane is the peripheral plane at a distance $0.5d_o$ from the face of the column, where the effective depth (d_o) of the drop slab at support is equal to 345 mm. The length of the critical section is:

$$b_o = 4(a + d_o) = 4(0.5 + 0.345) = 3.38 \text{ m}$$

The shear force on the plane is:

$$V_u = w_c [L*L - (a + d_o)(a + d_o)] = 19.5(6*6 - 0.845*0.845) = 688 \text{ kN}$$

The nominal shear stress is:

$$\tau_v = \frac{V_u}{bd} = \frac{688000}{3000*345} = 0.67 \text{ N/mm}^2$$

The aspect ratio of the column is one, so the maximum admissible value of b is only 0.5.
The shear strength of the concrete without shear reinforcement is:

$$\tau_c = 0.25(0.5 + \beta_c)\sqrt{f_{ck}} = 0.25\sqrt{(30)} = 1.37 \text{ N/mm}^2.$$

The nominal shear stress is less than the shear capacity; therefore, there is no need for transverse reinforcement. The main reinforcement consists of top bars at the support line and bottom bars at mid-span, in both the directions. Cranking of the bars may not be very helpful. The detailing of the bars is made in the column and middle strips separately. A rationalized reinforcement detail is presented here.

Column Strip

Detailing of the main reinforcement: 12 mm bars at 160 mm spacing at the bottom in mid-span and alternate bars are curtailed at 0.125 span (750 mm) from support.

12 mm bars at 110 mm at top face extending up to 0.3 times the span (1800 mm from support on each side). Alternate bars can be curtailed at 0.2 times the span (1200 mm) from the support.

Middle Strip

12 mm bars at 160 mm spacing at the bottom in mid-span and alternate bars are curtailed at 0.15 span (900 mm) from support.

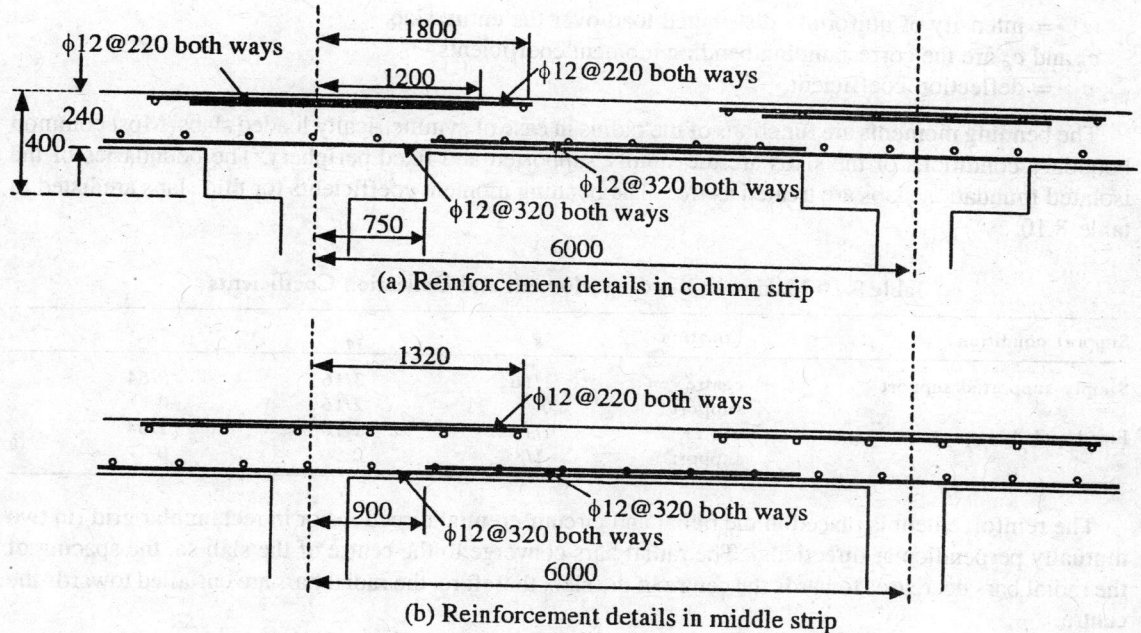

(a) Reinforcement details in column strip

(b) Reinforcement details in middle strip

Fig. 8.14 Reinforcement Details of Drops Panel Flat Slab, Example 8.7.

12 mm bars at 220 mm at top face extending up to 0.22 times the span (1320 mm) from support on each side.

8.9 CIRCULAR SLABS

Circular slabs are used in buildings that are circular in plan, circular tanks and well foundations. Some polygonal slab can also be approximated as circular slab to minimize the problem of the boundary conditions. The bending moments and deflection at critical points in circular slabs can be expressed as:

$$M_R = c_R w R^2 \tag{8.15}$$

$$M_\theta = c_\theta w R^2 \tag{8.16}$$

$$v = \frac{c_v w R^4}{EI} \tag{8.17}$$

where
M_R = maximum radial bending moment per unit width on the radial plane,
M_θ = maximum circumferential bending moment per unit width on the circumferential plane,
R = radius of the slab in plain,
I = moment of inertia of the slab = $t^3/12$ per unit width, E = Young's modulus of elasticity of material,
v = deflection,

w = intensity of uniformly distributed load over the entire slab,

c_R and c_θ are the corresponding bending moment coefficients,

c_v = deflection coefficient.

The bending moments are functions of the radius in case of symmetrically loaded slabs. Most common boundary conditions of the slabs are the simply supported and fixed periphery. The boundaries of the isolated foundation slabs are treated as free. The bending moment coefficients for thin slabs are listed in table 8.10.

Table 8.10 Maximum Bending Moment and Deflection Coefficients

Support conditions	Location	c_R	c_q	c_v
Simply supported support	centre	3/16	3/16	5/64
	support	0	2/16	0
Fixed ended support	centre	1/16	1/16	1/64
	support	-1/8	0	0

The reinforcement is placed in the radial and circumferential directions or in rectangular grid (in two mutually perpendicular directions). The radial bars converge to the centre of the slab so, the spacing of the radial bars decreases towards the centre of the slab, therefore, the radial bars are curtailed towards the centre.

Example 8.8: Simply supported circular slab

A circular slab in mild environment with simply supported boundary is subjected to a uniformly distributed load of 4 kN/m². The radius of the slab to the centre of the support is 3.5 m. Design the slab.

Solution

The slab is exposed to mild environment so the minimum cover to the reinforcement is 20 mm. Concrete grade $M20$ and reinforcement of $HYSD$-$Fe415$ can be used in the design.

Design of the Section for Bending Moment

Let $t = R/15 = 3.5/15 = 0.23$ m, for computing the self-weight. Use 250 mm thickness.

Self-weight $= w_g = 0.25(25) = 6.25$ kN/m².
Let Floor finish load $= 1$ kN/m²,
Live load $= w_1$ $= 4$ kN/m²
Factored total load on the slab $= w_c = 1.5(6.25 + 1 + 4) = 16.875$ kN/m²,

Maximum bending moment on the slab occurs at the mid-point and the moment coefficient is from Table 8.10. The bending moment is:

$$M_{Rc} = \frac{3 w_c R^2}{16} = \frac{3 * 16.875 * 3.5}{16} = 38.8 \text{ kNm/m}$$

Equating the collapse moment to the moment capacity of a section gives the effective depth required as:

$$d = \sqrt{\frac{M_{Rc}}{Kbf_{ck}}} = \sqrt{\frac{38.8(10^6)}{0.138*1000*20}} = 118.5 \text{ mm}$$

Select the total depth of the section as 200 mm. The effective depth in one direction is = 200 – 20 – 6 = 174 mm, and in the other direction it is 153 mm. The diameter of the bar is assumed as 12 mm in both the directions. The bending moment in the circumferential direction is same as that of the radial direction.

The area of reinforcement is:

$$A_{st} = \frac{1.15 M_c}{jdf_y} = \frac{1.15*38.8(10^6)}{0.8*153*415} = 879 \text{ mm}^2/\text{m}$$

The reinforcement is placed in any two mutually perpendicular directions. Provide 12 mm bars at 125 mm spacing in both the directions. The overall radius of the slab is chosen as the effective radius plus 150 mm. The diameter of the slab is 2(3.5 + 0.15) = 7.3 m. Fig. 8.15(a) illustrates the reinforcement layout and Fig. 8.15(b) illustrates a cross section.

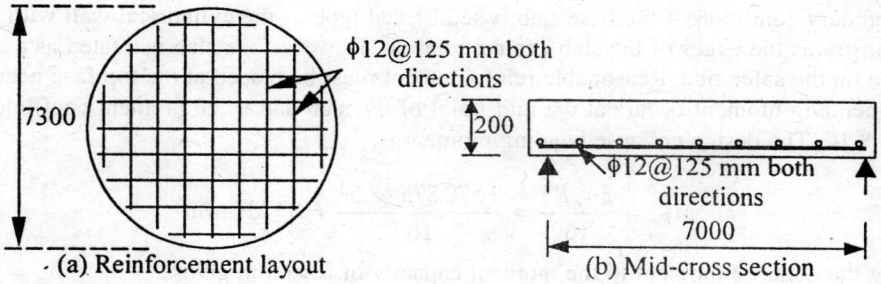

(a) Reinforcement layout

(b) Mid-cross section

Fig. 8.15 Reinforcement Details of Circular Slab, Example 8.8.

Design for Shear

The load is uniform so the reaction will also uniform. The total load divided by the perimeter of the support give the reaction per unit width. The critical section for shear is at a distance of effective depth from the support. The shear force at the critical section is:

$$V_u = w_c \left(\frac{R}{2} - d \right) = 16.875(1.75 - 0.153) = 27 \text{ kN}$$

The actual width of the slab at a distance d from the support is slightly less than a unit width at the support because of the radial coordinates. However the shear stress is checked at this location for a unit width as a first approximation. The nominal shear stress is:

$$\tau_v = \frac{V_u}{bd} = \frac{27000}{1000*153} = 0.18 \text{ N/mm}^2.$$

The nominal shear stress is less than the shear capacity of the section.

Example 8.9: Circular Slab of Water Tank

A circular cylindrical water tank stores 3.5 m head of water. Effective diameter of the cylindrical wall is 5 m. The cylindrical wall itself acts as a ring beam that supports the base slab. Design the base slab.

Solution

The slab can be considered as exposed to moderate environment so the minimum cover requirement is 35 mm. Concrete grade $M25$ and reinforcement of $HYSD$-$Fe415$ can be used in the design.

Design of the Section for Bending Moment

Assume the thickness of the slab around $R/12$ for the purpose of computing the self-weight.
Let $t = 5/12 = 0.42$ m. Use 450 mm thickness for the purpose of self-weight.

Self weight $= w_g = 0.45(25)$	$= 11.25$ kN/m².
Let Floor finish such as plaster finish etc. load	$= 1$ kN/m²,
Water load $= w_1 = 3.5*10$	$= 35$ kN/m²
Factored total load on the slab $= w_c = 1.5(11.25 + 1 + 35) = 70.875$ kN/m²,	

The boundary condition of the base slab is semi-fixed type as the cylindrical wall with hydrostatic pressure constrains the edges of the slab against rotation. However, the slab is treated as a simply supported to be on the safer side. Reasonable reinforcement must be placed at the top face near the edges. Maximum bending moment occurs at the mid-point of the slab and it the moment coefficient is taken from Table 8.10. The design collapse bending moment is:

$$M_{Rc} = \frac{3w_c R^2}{16} = \frac{3*70.875*2.5^2}{16} = 83 \text{ kNm/m}$$

Equating the collapse moment to the moment capacity of a section gives:

$$d = \sqrt{\frac{M_{Rc}}{Kbf_{ck}}} = \sqrt{\frac{83(10^6)}{0.138*1000*25}} = 155 \text{ mm}$$

The water tank has to be designed as an uncracked section. Select the total depth of the section as 450 mm. The effective depth for the section for one direction is $450 - 35 - 8 = 407$ mm, and in the other direction is 391 mm. The diameter of the reinforcement bar is assumed as 16 mm in both the directions. The bending moment in the circumferential direction is same as that of the radial direction.

The area of reinforcement required for smaller effective depth is:

$$A_{st} = \frac{1.15 M_c}{jdf_y} = \frac{1.15*83(10^6)}{0.8*391*415} = 639 \text{ mm}^2/\text{m}$$

The slab forms the base of a water tank, so it is designed as an uncracked section

The service load bending moment in the slab is $= M = M_c/1.5 = 55.3$ kNm/m. . The admissible crack width is 0.2 mm and the corresponding allowable stress in the concrete under bending tension is 1.8 N/mm². So the thickness of the slab required for serviceability is:

The thickness of the slab required $= t = \sqrt{\frac{M}{b\sigma_{cbt}}} = \sqrt{\frac{6*55.3(10^6)}{1000*1.8}} = 430 \text{ mm}$

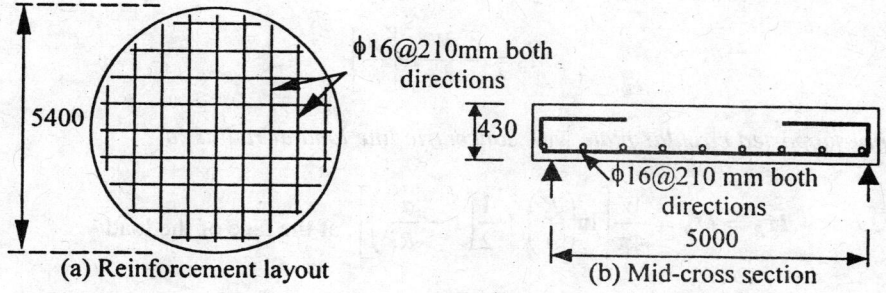

Fig. 8.16 Reinforcement Details of Circular Slab, Example 8.9.

The allowable stress in reinforcement on the face of the water up to a distance of 225 mm is 150 MPa. That beyond 225 mm is 190 MPa. The minimum percentage of reinforcement in thickness of slabs more than 425 mm is 0.16 percent otherwise it is 0.24 percent. The required area of the reinforcement is:

$$A_{st} = \frac{M}{jdf_{st}} = \frac{(55.3)10^6}{0.8*391*190} = 930 \text{ mm}^2/\text{m}$$

The minimum percentage of the reinforcement required is: $A_{smin} = 0.16*1000*430/100 = 688 \text{ mm}^2/\text{m}$ Provide 16 mm bars at 210 mm in each direction. Figure 8.16 illustrates the reinforcement details.

8.10 CIRCULAR SLAB WITH DIFFERENT LOAD AND BOUNDARY CONDITIONS

The analysis of circular slabs with different load and boundary conditions is beyond the scope of this book. However, the maximum bending moments of slabs can be taken from standard textbooks on theory of plates like Theory of Plates and Shells by Timoshenko and Woinowsky-Krieger. Some maximum bending moment coefficients are listed here for ready reference.

Case 1 Bending moments at the face of concentrated central load in simply supported slab:

$$M_R = \frac{W}{4\pi}\left[\ln\left(\frac{R}{a}\right) - \frac{1}{4}\left(1 - \frac{a^2}{R^2}\right)\right] \tag{8.18}$$

$$M_\theta = \frac{W}{4\pi}\left[\ln\left(\frac{R}{a}\right) + \frac{1}{4}\left(3 - \frac{a^2}{R^2}\right)\right] \tag{8.19}$$

where
  a = radius of the area over which the load is placed,
  R = radius of the circular slab,
  W = total load on the slab at centre.
  ln = natural logarithm

Case 2 Bending moments at the face of the load in fixed support slab with central load W:

$$M_R = -\frac{W}{4\pi}, \text{ at support} \tag{8.20}$$

$$M_\theta = -\frac{W}{4\pi}\ln\left[\frac{R}{a}\right] \tag{8.21}$$

Case 3 Simply supported circular plate with concentric line load at radius 'a'

$$M_R = M_\theta = \frac{W}{4\pi}\left[\ln\left(\frac{R}{a}\right) + \frac{1}{2}\left(1 - \frac{a^2}{R^2}\right)\right], \text{ at the face of the load} \tag{8.23}$$

Case 4 Circular footing with free edges and with central load:

The critical bending moments at the face of the column obtained from superposition of a simply supported slab with uniform load in the reverse direction, with simply supported slab with central load over an area of radius *a*.

$$M_r = \frac{W}{4\pi}\left[\ln\left(\frac{R}{a}\right) - \frac{1}{4} + \frac{a^2}{R^2}\right], \text{ at the face of the column} \tag{8.24}$$

$$M_\theta = \frac{W}{4\pi}\left[\ln\left(\frac{R}{a}\right)\right], \text{ at the face of the column} \tag{8.25}$$

The notations are illustrated in Fig. 8.17

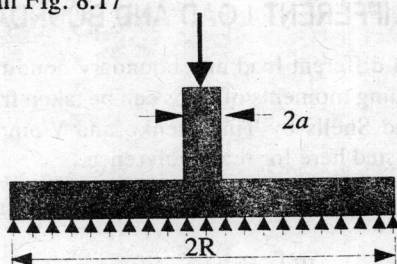

Fig. 8.17 Circular Footing with Column Load.

Example 8.10: Design of Well Cap

A ten metre internal diameter well supports a circular pier of 2 m diameter. The load on the pier is 1200 kN. Design the well cap. The well cap is in riverbed and subjected to wetting and drying by river water.

Solution

The well cap is in the riverbed so subjected to wetting and drying with ordinary water. The exposure condition is severe so M25 grade concrete to be used. The minimum cover to the reinforcement in well cap is 50 mm. Let *HYSD-Fe*415 grade reinforcement steel is used. The effective radius of the slab is equal to the inside radius plus the effective depth of the slab and let it be 5.0 + 0.6 = 5.6 m. The self-weight of the slab is assumed on the higher side of thickness of the slab as 0.65 m. The factored collapse loads on the slab are:

Factored self-weight $= w_c = 1.5*0.65*25 = 24.375$ kN/m^2,
Factored live load $= W_c = 1.5*1200 = 1800$ kN,

The maximum radial bending moment already mentioned is:

$$M_R = \frac{3w_c R^2}{16} + \frac{W_c}{4\pi}\left[\ln\left(\frac{R}{a}\right) - \frac{1}{4}\left(1 - \frac{a^2}{R^2}\right)\right] \qquad (8.26)$$

$$M_R = \frac{3*24.375*5.6^2}{16} + \frac{1800}{4\pi}\left[\ln\left(\frac{5.6}{1}\right) - \frac{1}{4}\left(1 - \frac{1}{5.6^2}\right)\right]$$

$$= 143.325 + 212.1 = 355.425 \text{ kNm/m}$$

The critical bending moment for uniformly distributed load under the central load is at the face of the pier. The maximum circumferential bending moment is at the edge of the pier support, that is a 1000 mm from the centre.

$$M_\theta = \frac{3w_c R^2}{16} + \frac{W_c}{4\pi}\left[\ln\left(\frac{R}{a}\right) + \frac{1}{4}\left(3 - \frac{a^2}{R^2}\right)\right] \qquad (8.27)$$

$$M_\theta = \frac{3*24.375*5.6^2}{16} + \frac{1800}{4\pi}\left[\ln\left(\frac{5.6}{1}\right) + \frac{1}{4}\left(3 - \frac{1}{5.6^2}\right)\right]$$

$$= 143.325 + 353 = 496.325 \text{ kNm/m}$$

The circumferential bending moment is more than the radial bending moment. The effective depth required for a balanced section is:

$$d = \sqrt{\frac{M_\theta}{Kbf_{ck}}} = \sqrt{\frac{496.325(10^6)}{0.138*1000*25}} = 380 \text{ mm}$$

Let the overall thickness of the slab be 715 mm then, the effective depth is equal to $715 - 50 - 10 = 655$ mm. This assumes the reinforcement bar diameter as 20 mm. The required circumferential reinforcement is:

$$A_{st} = \frac{1.15M_\theta}{df_y} = \frac{1.15*496.325(10^6)}{655*415} = 2100 \text{ mm}^2/\text{m}$$

Provide 20 mm bars at 145 mm spacing in the radial direction at the edge of the pier support, that is at a distance of 1000 mm from the centre. The actual area of the reinforcement provided is 2165 mm^2/m. The effective depth of the section for the radial reinforcement is $655 - 20 = 635$ mm and the required area of reinforcement is:

$$A_{st} = \frac{1.15M_R}{df_y} = \frac{1.15*355.425(10^6)}{635*415} = 1552 \text{ mm}^2/\text{m}$$

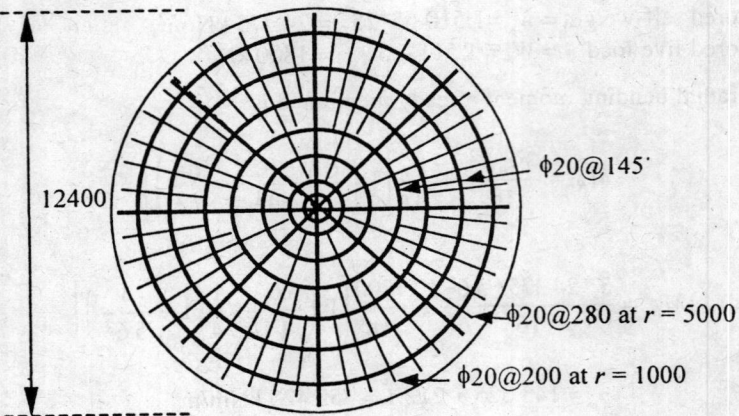

12400

φ20@145·

φ20@280 at r = 5000

φ20@200 at r = 1000

Fig. 8.18 Reinforcement Detail of Circular Well, Cap, Example 8.10.

Provide 20 mm bars at a spacing of 200 mm at 1000 mm from the centre. The spacing of the radial reinforcement changes along the radius. The slab will yield as a truncated cone with flat top at the column support. It is desirable to keep the spacing of the circumferential reinforcement at the same spacing. The minimum percentage of the reinforcement in such a case of slab is 0.16 percent. The minimum reinforcement area is:

$$A_{smin} = 0.16*700*1000/100 = 1120 \text{ mm}^2/\text{m}$$

Maximum spacing = 280 mm, or twice the thickness of the slab.

So provide 20 mm bars at a spacing less than 280 mm spacing.

Figure 8.18 illustrates reinforcement details of the circular well cap slab. The overall radius of the slab is chosen as 6200 mm. The clear radius is 5000 mm and the effective radius is 5600 mm.

Example 8.11: Design of Circular Footing Slab

A 600 mm diameter column transfers 1500 kN to a circular foundation. The safe net bearing capacity of the soil is 100 kN/m². The footing is exposed to medium exposure condition.

Solution

M25 grade concrete is selected for other site. The minimum cover to the foundation is 50 mm. The grade of reinforcement is chosen as HYSD-Fe415 bars. The depth of the foundation is taken to 1200 mm below the ground level as per the soil investigation report.

Safe bearing pressure on the soil   = p   = 100 kN/m²,

Superimposed column load   = W_s = 1500 kN

Let the difference in the load of the slab and the soil   = 150 kN

Total gross load on the foundation   = W   = 1650 kN

Required bearing area of the foundation   = W/p =1650/100 = 16.5 m²,

Minimum required radius of the footing = $\sqrt{\dfrac{16.5}{3.1416}}$ = 2.29 m

Select the radius of the foundation as $= R = 2.4$ m; the radius of the column is $= a = 300$ mm. The factored collapse effective loan on the footing for the purpose of limit state design is:

$$W_c = 1.5 * W_s = 1.5 * 1500 = 2250 \text{ kN}$$

The bending moments on a circular footing from Eqs. 8.24 and 8.25 are:

$$M_R = \frac{W_c}{4\pi}\left[\ln\left(\frac{R}{a}\right) - 1 + \frac{a^2}{R^2}\right] = \frac{2250}{4\pi}\left[\ln\left(\frac{2.4}{0.3}\right) - 1 + \frac{0.3^2}{2.4^2}\right] = 196 \text{ kNm/m}$$

$$M_\theta = \frac{W_c}{4\pi}\left[\ln\left(\frac{R}{a}\right)\right] = \frac{2250}{4\pi}\left[\ln\left(\frac{2.4}{0.3}\right)\right] = 372.4 \text{ kNm/m}$$

Equating the maximum bending moment to the moment capacity of a section gives:

$$d = \sqrt{\frac{M_c}{Kbf_{ck}}} = \sqrt{\frac{372.4(10)^6}{0.138 * 1000 * 25}} = 329 \text{ mm}$$

Select the overall thickness of the slab as 650 mm, and then the effective depth is equal to $650 - 50 - 10 = 590$ mm. The effective depth that is needed to resist shear force without stirrups is usually higher than that required for bending resistance. So it is recommended to select higher than that needed bending resistance.

Check for Shear Stress

The critical shear stress plane is at a distance $0.5d$ from the face of the column. The peripheral distance of the critical shear plane surface is:

$$b_s = \pi(2a + d) = 3.1416(0.6 + 0.59) = 3.74 \text{ m}$$

Shear force on this plane is:

$$V_c = W_c\left[1 - \left(\frac{a + d/2}{R}\right)^2\right] = 2250\left[1 - \left(\frac{0.3 + 0.295}{2.4}\right)^2\right] = 2112 \text{ kN}$$

The nominal shear stress at the critical surface is:

$$\tau_v = \frac{V_c}{b_s} = \frac{2112000}{3740 * 1000} = 0.56 \text{ N/mm}^2$$

The shear strength of the concrete without shear reinforcement is given by:

$$\tau_c = 0.25(0.5 + \beta_c)\sqrt{f_{ck}}$$

The maximum admissible value of β is only 0.5. Therefore the shear strength of the $M25$ concrete is:

$$\tau_c = 0.25(5) = 1.25 \text{ N/mm}^2.$$

The nominal shear stress is less than the shear capacity of the concrete; so no shear reinforcement is needed.

The circumferential reinforcement that is needed is:

$$A_{st} = \frac{1.15 M_{\theta c}}{d f_y} = \frac{1.15 * 372.4(10)^6}{590 * 415} = 1750 \text{ mm}^2/\text{m}$$

Provide 20 mm bars at 175 mm spacing in the circumferential direction.
Similarly the required reinforcement in the radial direction is:

$$A_{st} = \frac{1.15 M_{Rc}}{d f_y} = \frac{1.15 * 196(10)^6}{570 * 415} = 953 \text{ mm}^2/\text{m}$$

Provide 20 mm bars at 325 mm spacing as radial reinforcement and the details are illustrated in Fig. 8.19.

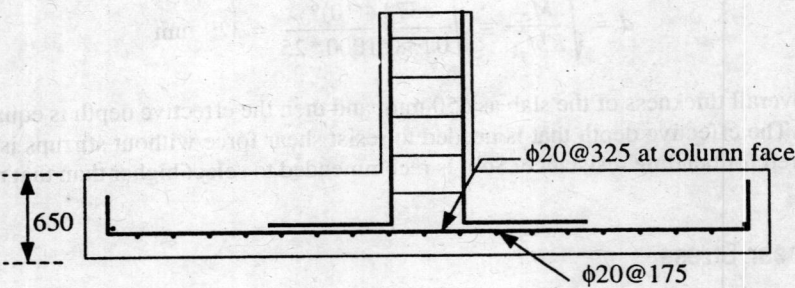

Fig. 8.19 Reinforcement Details of Footing, Example 8.10.

8.11 DESIGN OF DIFFERENT SHAPED SLABS

The bending and shear forces for any boundary condition can be determined by finite element analysis. Computer software for such purpose is available in the market. This section provides an approximate but reliable method of determination of bending moments in simple slabs such as polygonal and semicircular slabs. Similarly this section gives a method of detailing for slabs with openings. The procedure is illustrated through a set of examples.

Example 8.11: Design of Slab Hexagon in Plan

A slab hexagon in plan with 3 metres side is subjected to a live load of 3 kN/m^2 and floor finish of 1 kN/m^2. The slab is exposed to moderate environment. Design the slab with *HYSD-Fe*415 grade reinforcement.

Solution

*Select M*25 grade concrete with cover of 30 mm to the reinforcement. The hexagonal slab can be idealized into a rectangular shaped slab and the bending moment can be computed from the rectangular slab. The main principle associated with the idealization is to enclose the polygon by a square or a circle or a rectangle to the nearest possible accuracy and then determine the bending moment etc. for the idealized shape.

The method of idealization is illustrated. Figure 8.20 illustrates a polygonal shaped slab *ABCDEF* that is idealized by an enclosed rectangle *PQRS*. From simple geometry, the size of the rectangle can be determined.

$PA = AF \sin 30 = 1.5$ m,
$PF = AF \cos 30 = 3*0.866 = 2.6$ m
The side *PQ* of the rectangle *PQRS* is $= (2*1.5 + 3) = 6$ m
The side *PS* of the rectangle *PQRS* is $= 2*2.6 = 5.2$ m $= L$

The aspect ratio of the rectangle is $= 6/5.2 = 1.15$, and the corresponding bending moments of a rectangular slab are taken from table 8.3. The corresponding bending moment coefficients along the short and long span are 0.068 and 0.056 respectively.

Let the thickness of the slab be selected in the range of short span/25, that is 200mm for the purpose of calculation of self-weight.

Self-weight $= w_g = 0.2*25$ $= 5$ kN/m²,
Floor finish $= w_s$ $= 1$ kN/m²,
Live load $= w_1$ $= 3$ kN/m²,
The factored design load $= w_c = 1.5*(5 + 1 + 3) = 13.5$ kN/m²,

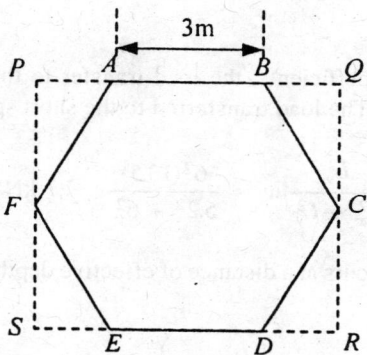

Fig. 8.20 Hexagonal Slab Idealized as Rectangular, Example 8.11.

The limit factored bending moments in the short and long span directions are:

$$M_x = 0.068 w_c L^2 = 0.068*13.5*5.2^2 = 24.8 \text{ kNm/m}$$
$$M_y = 0.056 w_c L^2 = 0.056*13.5*5.2^2 = 20.44 \text{ kNm/m}$$

Equating the limit moment of the short span to the moment capacity gives

$$d = \sqrt{\frac{M_x}{Kb f_{ck}}} = \sqrt{\frac{24.8(10)^6}{0.138*1000*25}} = 85 \text{ mm}$$

The effective depth is taken as 125 mm because of practical considerations of waterproof requirement and to reduce the steel. This means that the total thickness of the slab is 160 mm. The area of reinforcement in the short span direction is:

$$A_{st} = \frac{1.15M_c}{jdf_y} = \frac{1.15(24.8)10^6}{0.8(125)415} = 687 \text{ mm}^2/\text{m}$$

Use 12 mm bars at 160 mm spacing in the short span direction.

$$A_{st} \text{ (Provide)} = 113/0.160 = 706 \text{ mm}^2/\text{m}$$

Minimum reinforcement required for shrinkage etc. is 0.12 percent and it is:

$$A_{smin} = 0.12(160)(1000)/100 = 192 \text{ mm}^2/\text{m}$$

The bending moment in the long span direction is 20.44 kNm/m and the reinforcement in that direction is:

$$A_{st} = \frac{1.15M_y}{jdf_y} = \frac{1.15(20.44)10^6}{0.8(113)415} = 627 \text{ mm}^2/\text{m}^2$$

Provide 12 mm bars at 175 mm spacing.

Design for Shear

In the absence of the shear force coefficients, the load transfer to the supports can be take inversely proportional to the square of spans. The load transferred to the short span is:

$$w_x = \frac{L_y^2}{L_x^2 + L_y^2} \, w_c = \frac{6^2(13.5)}{5.2^2 + 6^2} = 7.7 \text{ kN/m}^2$$

The maximum shear force that occurs at a distance of effective depth from the face of the support, and it is:

$$V_c = w_x\left(\frac{L_x}{2} - d\right) = 7.7(2.6 - 0.125) = 19.1 \text{ kN/m}$$

The nominal shear stress is:

$$\tau_c = \frac{V_c}{bd} = \frac{19100}{1000(125)} = 0.16 \text{ N/mm}^2$$

The shear strength of concrete for 0.25% of reinforcement is higher, so no shear reinforcement is needed.

Alternate bars are bent up at a distance of 0.2 times the span to resist any secondary bending moment coming at the supports. Fig. 8.21 illustrates the reinforcement details of the slab at mid-span of the long edge. Four extra nominal bars are shown in the section at edges and kinks for construction purpose.

Example 8.12: Design of Slab Semi-circular in Plan

A semi-circular slab with 3 metres radius is subjected to a live load of 3 kN/m² and floor finish of 1 kN/m². The slab is exposed to moderate environment. Design the slab with *HYSD-Fe*415 grade reinforcement.

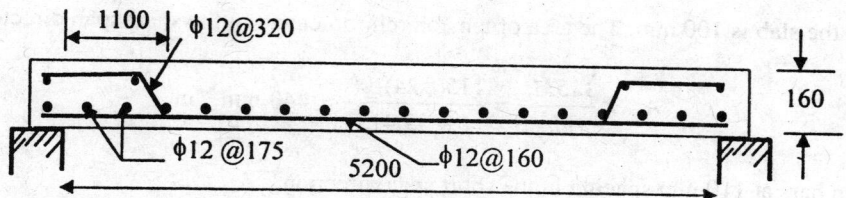

Fig. 8.21 Cross Section of Slab: Example 8.11; Dimensions in mm.

Solution

Figure 8.22 illustrates a semi-circular slab ABC that is idealized by an almost enclosed rectangle $ABDE$. From simple geometry, the size of the rectangle can be determined. The short side of the rectangle is selected as 0.866 of the radius of the circle.

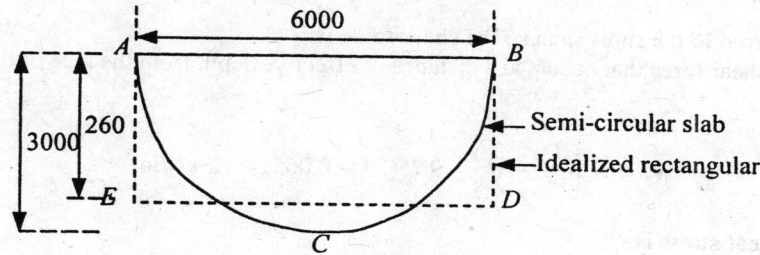

Fig. 8.22 Semicircular Slab Idealization, Example 8.12.

Select *M25* grade concrete with cover of 30 mm to the reinforcement..

$$BD = 0.866*3 = 2.6 \text{ m},$$

The rectangle $ABDE$ is therefore is chosen as 6 m by 2.6 m. This becomes a one-way slab and the maximum bending moment is given by:

Let the thickness of the slab be selected in the range of short span/25, that is 100 mm for the purpose of calculation of self-weight.

Self-weight	$= w_g = 0.1*25$	$= 2.5 \text{ kN/m}^2,$
Floor finish	$= w_s$	$= 1 \text{ kN/m}^2,$
Live load	$= w_l$	$= 3 \text{ kN/m}^2,$

The factored collapse design load $= w_c = 1.5*(2.5 + 1 + 3) = 9.75 \text{ kN/m}^2,$
The limit factored bending moments in the short span directions is:

$$M_x = 0.125 w_c L^2 = 0.125*9.75*2.6^2 = 8.24 \text{ kNm/m}$$

Equating the limit moment of the short span to the moment capacity gives:

$$d = \sqrt{\frac{M_x}{Kbf_{ck}}} = \sqrt{\frac{8.24(10^6)}{0.138*1000*25}} = 49 \text{ mm}$$

The effective depth is taken as 65 mm because of practical considerations. This means that the total thickness of the slab is 100 mm. The area of tension reinforcement in the short span direction is:

$$A_{st} = \frac{1.15M_c}{jdf_y} = \frac{1.15(8.24)10^6}{0.8(65)415} = 440 \text{ mm}^2/\text{ m}$$

Use 8 mm bars at 110 mm spacing in the short span direction.

$$A_{st} \text{ (Provide)} = 50/0.110 = 454 \text{ mm}^2/\text{m}$$

Minimum reinforcement required for shrinkage etc. is 0.12 percent and it is:

$$A_{smin} = 0.12(100)(1000)/100 = 120 \text{ mm}^2/\text{m}$$

Provide 8 mm bars at 300 mm spacing.

Design for Shear

The load is transferred to the short span so the shear force is:
    The maximum shear force that occurs at a distance of effective depth from the face of the support, and it is:

$$V_c = w_c\left(\frac{L_x}{2} - d\right) = 9.75(1.3 - 0.065) = 12 \text{ kN/m}$$

The nominal shear stress is:

$$\tau_c = \frac{V_c}{bd} = \frac{12000}{1000(65)} = 0.19 \text{ N/mm}^2$$

The shear strength of concrete even for 0.15 percentage of reinforcement is 0.29 MPa.

Slabs with Openings

The following three possible cases of openings in slabs can be considered:

Case 1. Very small hole: Hole size is smaller than three times the thickness of the slab.

Precaution must be taken to place the hole during construction and avoid punching the hole after hardening of the concrete. Avoid cutting of the reinforcement, and the reinforcement be suitably bent or shifted not to come in the way of the hole. Add nominal reinforcement around the hole if it is in a critical zone.

Case 2. Small opening compared with the size of the slab. The size of the opening is less than one-fourth the short span and not at the critical section.

The edges of the opening must be thickened and the thickening should be at least equal to the material lost in the opening. The reinforcement bars that were supposed to be in the opening zone can be shifted to the edges of the opening. Further extra reinforcement has to be placed in the diagonal direction. This amount of reinforcement in each of the four diagonal directions must be about half of the reinforcement placed at each edge. Cracks develop and propagate along the re-entrant corners, therefore additional reinforcement must be placed at such re-entrant corners. The Fig. 8.23 illustrates the reinforcement detail at small opening in slabs.

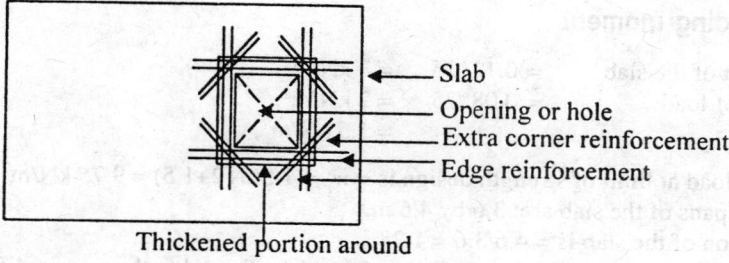

Fig. 8.23 Reinforcement Detail at Openings.

Case 3 Large openings or the openings at critical locations.

Openings more than one-fourth the size of the slab may be classified as large openings in slabs. In such a case, secondary beams must be placed at least on two edges of the opening. If the opening is large or it is at the critical location, four edge beams have to be placed, of which two will extend up to the supports and the edge beams in turn support the other two. The slab and the beams have to be designed to meet the requirements. A typical illustration of the secondary beams and the opening in the slab is shown in Fig. 8.24.

Example 8.13: Rectangular Slab with a Small Openings

A simply supported roof slab with effective spans of 3.6 and 4.6 m is subjected to a live load of 1.5 kN/m^2. The water proofing on the slab is 2 kN/m^2. An opening of 800 mm by 800 mm is placed at 1.5m by 1.5 m from the top right corner of the slab.

Solution

The size of the opening is 800 mm by 800 mm that is less than one-fourth the size of the span of the slab. The slab is designed as a solid slab and then suitable modifications are made in the reinforcement detail. The slab is roof slab so some waterproof treatment and sloping lean concrete will be laid to drain the rainwater. The design live load on flat roof slab is 1.5 kN/m^2. The slab has to be assumed as exposed to moderate environment to protect it against any damage due to seepage of water. Recommended grade of concrete is $M25$. The cover to the reinforcement on the bottom face can be taken as 20 mm.

Let the thickness of the slab be = span/30,

Use thickness = 120 mm;

The effective depth of reinforcement in the short span direction is = d = 120 – 20 -5 = 95 mm

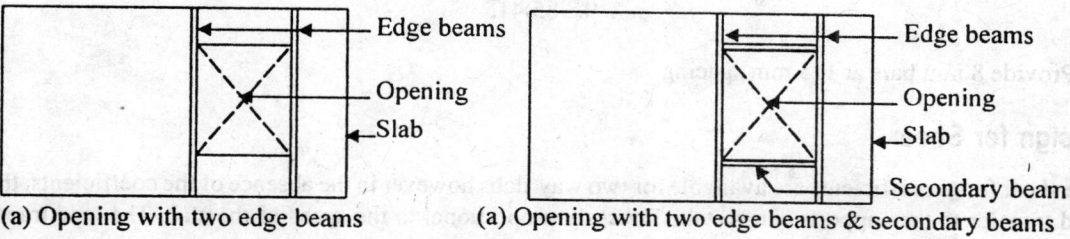

(a) Opening with two edge beams   (a) Opening with two edge beams & secondary beams

Fig. 8.24 Slabs with Large Openings and Edge beams.

Design for bending moment

Self weight of the slab	$= 0.12*25$	$= 3 \text{ kN/m}^2$,
Water-proof load	$= 0.08*25$	$= 2 \text{ kN/m}^2$,
Live load		$= 1.5 \text{ kN/m}^2$,

Total factored load at limit of strength design is $= w_c = 1.5(3+2+1.5) = 9.75 \text{ kN/m}^2$.
The effective spans of the slab are: 3.6 by 4.6 m.
The aspect ration of the slab is $= 4.6/3.6 = 1.28$
The corresponding bending moment coefficients for slab allowed to the corners lifted are taken from table 8.3, row (b). And they for the aspect ratio of 1.28 are 0.093 and 0.054 for short and long spans.
The limit factored bending moments in the short and long span directions are:

$$M_y = 0.054 w_c L^2 = 0.054 * 9.75 * 3.6^2 = 6.8 \text{ kNm/m}$$

Equating the limit moment of the short span to the moment capacity gives:

$$d = \sqrt{\frac{M_x}{Kbf_{ck}}} = \sqrt{\frac{11.8(10^6)}{0.138*1000*25}} = 59 \text{ mm}$$

The effective depth required for a balanced section is only 59 mm. However an effective depth of 95 mm is assumed because of practical considerations. The total thickness of the slab is 120 mm. The area of tension reinforcement in the short span direction is:

$$A_{st} = \frac{1.15 M_c}{j d f_y} = \frac{1.15(11.8)10^6}{0.8(95)415} = 435 \text{ mm}^2/\text{m}$$

Use 10 mm bars at 180 mm spacing in the short span direction.

$$A_{st} \text{ (Provide)} = 78.5/0.180 = 436 \text{ mm}^2/\text{m}$$

Minimum reinforcement required for shrinkage etc. is 0.12 percent and it is:

$$A_{smin} = 0.12(120)(1000)/100 = 144 \text{ mm}^2/\text{m}$$

The bending moment on the long span direction is 6.8 kNm/m and the area of reinforcement in that direction is:

$$A_{st} = \frac{1.15 M_y}{j d f_y} = \frac{1.15(6.8)10^6}{0.8(85)415} = 278 \text{ mm}^2/\text{m}$$

Provide 8 mm bars at 175 mm spacing.

Design for Shear

The shear force coefficients are available for two way slabs however in the absence of the coefficients, the load transfer to the supports can be take inversely proportional to the spans of the slab. The load transferred to the short span is:

$$w_x = \frac{L_y^2}{L_x^2 + L_y^2} \, w_c = \frac{4.6^2(9.75)}{3.6^2 + 4.595^2} = 6.1 \text{ kN/m}$$

The maximum shear force that occurs at a distance of effective depth from the face of the support, and it is:

$$V_c = w_x\left(\frac{L_x}{2} - d\right) = 6.1(1.8 - 0.095) = 10.4 \text{ kN/m}$$

The nominal shear stress is:

$$\tau_c = \frac{V_c}{bd} = \frac{10400}{1000(95)} = 0.11 \text{ N/mm}^2$$

The shear strength of concrete even for 0.25% of reinforcement is higher so no shear reinforcement is needed.

Alternate bars are bent up at a distance of 0.2 times the span to resist any secondary bending moment at the supports.

Correction to the opening

The size of the opening is 800 mm and the thickness of the slab is 120 mm. The opening is not at the critical zone of the slab. 200 mm strip all round the edges of the opening are thickened by 250 mm. The reinforcement at the critical location is 10 mm at 180 mm in the short span and 8 mm at 175 along the long span. The actual requirement of the reinforcement at the opening location is smaller that that is at the critical location. Provide the following extra reinforcement at the opening.

Provide two of 10 mm bars at each edge along the short span direction, two of 8 mm at each of the long span direction. Similarly provide two of 8mm bars at the four diagonal corners of the opening. Fig. 8.25 illustrates the reinforcement details of the slab.

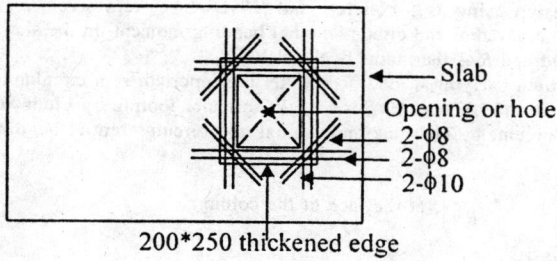

Fig. 8.25 Reinforcement Detail at Openings.

PROBLEMS (USE IS: 456 - 2000)

8.1 A simply supported rectangular slab with effective spans 3.5 m by 9 m and is exposed to mild environment. Design a reinforced concrete slab using $M20$ concrete and $HYSD\text{-}Fe415$ reinforcement bars by Limit State Design method to with stand a live load of 4 kN/m² and floor finish of 1 kN/m².

8.2 A floor slab of effective spans 3.5 m by 10 m is simply supported along the long edges and continuous over short edge supports. The slab is exposed to moderate environment and is subjected to a live load of 4 kN/m². Design a RC slab using $M25$ concrete and $HYSD\text{-}Fe415$ reinforcement bars by Limit State Design. Sketch the reinforcement details.

8.3 A verandah slab of 10 m long cantilevers 2 m from a wall. The load on the slab is 3 kN/m². Design a RC slab with $M25$ concrete and $HYSD\text{-}Fe500$ reinforcement bars by Limit State Design. Sketch the reinforcement details and indicate the minimum embedment length.

8.4 A slab is simply supported on 4 edges with effective span of 3.5 m by 5.0 m and subjected to a live load of 5 kN/m². Design a reinforced concrete slab assuming the corners of the slab are allowed to warp and using $M20$ concrete and $HYSD\text{-}Fe415$ reinforcement bars. Design the slab by Limit State Design giving reinforcement details.

8.5 A simply supported slab of effective spans 4.5 m by 5.5 m is subjected to a live load of 2 kN/m². Design a RC slab assuming the corners of the slab are fixed against warping. Design the slab with $M20$ concrete and $HYSD\text{-}Fe415$ reinforcement bars by Limit State Design. Sketch the reinforcement details.

8.6 A 4.5 m square slab simply supported at two adjacent edges and has continuity at the other two adjacent supports. Design a reinforced concrete slab by Limit State Design by using $M25$ concrete and $HYSD\text{-}Fe500$ reinforcement bars for a live load of 4 kN/m². Illustrate the reinforcement details.

8.7 A rectangular slab of effective spans 4 m by 5 m is simply supported on two opposite short edges and continuous on the two long edges and is subjected to a live load of 4 kN/m² in mild exposure condition. Design a reinforced concrete slab by Limit State Design using $M20$ concrete and $HYSD\text{-}Fe415$ reinforcement bars.

8.8 A flat slab is supported by a set of 450 mm square columns spaced at 5 m intervals in two directions. The effective height of the columns is 4 m. The slab is subjected to a live load of 5 kN/m². Design a flat slab by Limit State Design using $M25$ concrete and $HYSD\text{-}Fe415$ reinforcement bars.

8.9 A flat slab is supported by a series of columns spaced at 6 m in two orthogonal directions. The size of the column is 500 mm square and with a column head of 800 mm square. The slab is subjected to a live load of 5 kN/m². Design a flat slab by Limit State Design using $M25$ concrete and $HYSD\text{-}Fe415$ reinforcement bars.

8.10 A flat slab with one metre drop panels is supported by columns spaced at 7 m in two perpendicular directions. The size of the column is 600 mm square with effective height as 5 m. The slab is subjected to a live load of 6kN/m². Design a reinforced concrete flat slab by Limit State Design using $M30$ concrete and $HYSD\text{-}Fe415$ reinforcement bars. Sketch reinforcement details.

8.11 A floor slab simply supported on 4 edges having effective spans of 12 m by 16 m is subjected to a live load of 5 kN/m². The slab is designed as a waffle floor with grid spacing of 1.2 m by 1.6 m. Design the waffle floor by Limit State Design using $M25$ concrete and $HYSD\text{-}Fe415$ reinforcement bars.

8.12 A circular floor slab having a radius of 4.5 m is subjected to a live load of 4 kN/m². Design a reinforced concrete slab by Limit State Design using $M25$ concrete and $HYSD\text{-}Fe415$ reinforcement bars. Sketch the reinforcement details. Given the maximum radial and circumferential bending moment on the slab equal to $(3/16)\,wR^2$. where the w is intensity of the load and R is the radius of the slab.

8.13 A 600 mm circular column carrying a load of 800 kN is supported by a circular footing resting on a soil having a net safe bearing capacity of 100 kN/m². Design a RC circular footing by Limit State Design using $M25$ concrete and $HYSD\text{-}Fe415$ reinforcement. The maximum radial and circumferential bending moments are given by

$$M_r = \frac{W}{4\pi}\left[\ln\left(\frac{R}{a}\right) - 1 + \frac{a^2}{R^2}\right],\ \text{at the face of the column}$$

$$M_\theta = \frac{W}{4\pi}\left[\ln\left(\frac{R}{a}\right)\right],\ \text{at the face of the column}$$

where W = total superimposed load, R = outer radius of the slab, a = radius of circular column

Yield Line Theory of Slabs

9.1 INTRODUCTION AND ASSUMPTIONS

Figure. 9.1 illustrates typical simple slab in xy-plane subjected to vertical load. Stresses in concrete and reinforcement increase more or less proportionately with the load upto proof stress of the reinforcement as the load on a slab increases. Increase of load beyond proof strength of reinforcement causes excessive strain in the reinforcement and large deformation. The deflections of the slab will be elasto-plastic upto a level called collapse load. At *the collapse load*, the slab continues to deform without additional load and then collapses. Visible cracks developed during elasto-plastic state and progresses in the least resistance paths. The line along which the reinforcement reaches the yield point is called *the yield line*. A set of yield lines in the slab results into a mechanism leading a collapse. A typical load deflection curve of a slab is shown in Fig. 9.2. Redistribution of bending moments takes place after the initial yielding of the reinforcement. Some strain hardening of reinforcement takes place afterwards. In addition to the bending resistance of the slab, some membrane action of the plate comes in to play. The following assumptions apply in selecting the collapse yield line pattern.

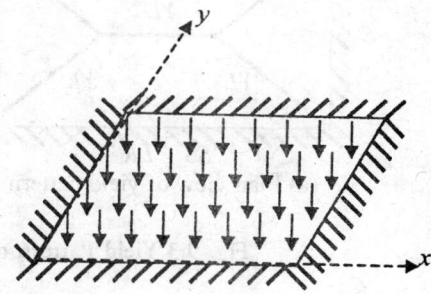

Fig. 9.1 Simple Slab with Load.

1. *The elastic deformations are negligible when compared with the plastic deformations,*
2. *Plastic hinge lines are formed along the yield lines and the plastic deformation is lumped at the yield lines.*
3. *The segments between the yield lines are treated as plane rigid elements,*
4. *Yield lines either terminate at the boundary of the slab or at the intersection of other yield lines.*

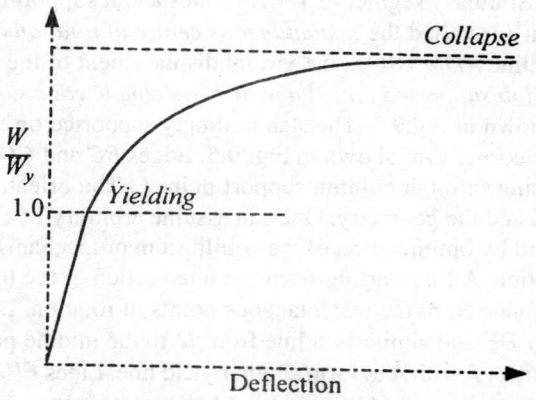

Fig. 9.2 Load Deflection Behaviour of Slab.

5. *Yield lines are idealized as straight.*
6. *Each segment tends to rotate as a rigid body and the axes of rotation lie either along the lines of supports.*
7. *The yield lines must pass through the intersection of axes of rotation of the adjacent segments.*

9.2 SIMPLE EXAMPLES OF YIELD LINE PATTERNS

Some examples of yield line patters are illustrated here.

Simply supported rectangular slab. Fig. 9.3 illustrates the yield line pattern of a simply supported rectangular slab. The maximum bending moment occurs at middle of the short span. As the load increases, the yield that got initiated at mid-span spreads sideways and then move towards the corners of the supports. The slab is now divided into four segments. Each segment rotates about the support line. The deformation of the slab is an inverted pyramid. Fig. 9.3(b) indicates the profile of the mechanism at mid-section of the long supports. There are five yield lines and the moment capacity of the section along the yield line is normally taken same.

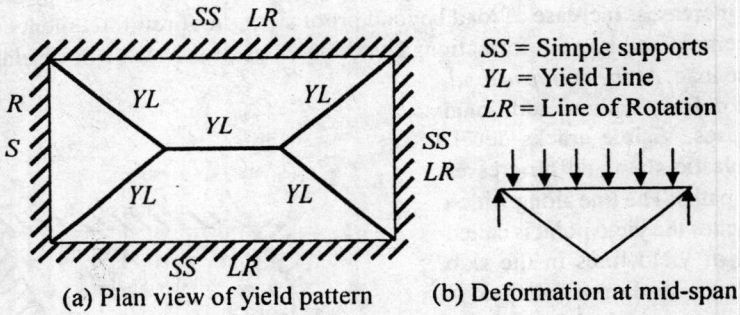

SS = Simple supports
YL = Yield Line
LR = Line of Rotation

(a) Plan view of yield pattern   (b) Deformation at mid-span

Fig. 9.3 Yield Pattern of Simply Supported Slab.

Trapezoidal slab. A trapezoidal slab in plan is shown in Fig. 9.4. *AB, BC, CD* and *DA* are the four simply supported edges. The yield lines are *MN, AM, BN, CN* and *DM*. Segment *ABNM* rotates about support line *AB*. Similarly segment *CDMN* rotates about support line *DC*. Lines *AB* and *DC* intersect at point *I*. The point *I* is called the *instantaneous centre of rotation*. The rotation of line *INM* controls the mechanism. The line *ADMNBC* is the virtual displacement of the compatible mechanism.

Slab supported on column and two simple edge supports. A slab supported on two edges and column is shown in Fig.9.5. The slab is simply supported on edges *AB* and *AD*, and supported by a column at *E*, near corner *C* as shown in Fig. 9.5. Edges *BC* and *CD* are free. The lines of rotation are *AB, AD* and line passing through column support point *E*. The orientation of line that passes through *E* depends on the load and the geometry. One can assume arbitrary line inclined with line *AB* at angle *q*. This angle can be found by optimization of the equilibrium or mechanism. Points *I1* and *I2* are the points of instantaneous rotation. A line starting from the intersection of the lines of rotation is one of the yield lines. Yield lines originate from the instantaneous points of rotation. Draw a line from *I1* connecting the middle point of span *DE* and similarly a line from *I2* to the middle point of span *BE*. Let these two lines intersect at *F*. Connect *F* with *A* to form another yield line. Lines *FH* and *FG* are the two yield lines. Slab segment *AFGB* rotates about line *AB*, segment *AFHD* rotates about *AD* and segment *EHFG* rotates about line *I1-I2*. The rotations of the segments have to be compatible.

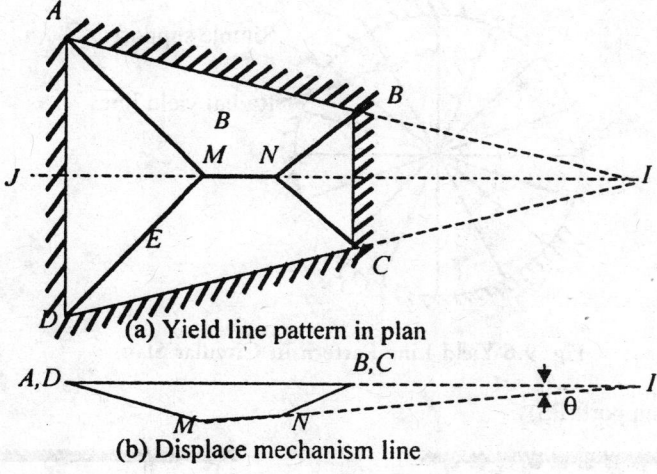

(a) Yield line pattern in plan

(b) Displace mechanism line

Fig. 9.4 Trapezoidal Slab & Yield Line Pattern.

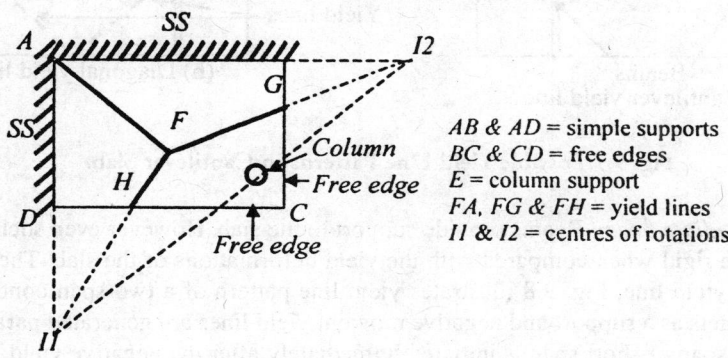

AB & AD = simple supports
BC & CD = free edges
E = column support
FA, FG & FH = yield lines
I1 & I2 = centres of rotations

Fig. 9.5 Slab with Corner Column.

Simply supported circular slab. A circular slab simply supported on the periphery and subjected to a uniformly distributed load is shown in Fig. 9.6. The segments of the yielded slab rotate about the support lines. The support is circular, so the rotation is about the tangents at points on the circumference. The radius is perpendicular to the tangent and the radial lines tilt towards the centre. The conical deformation of the slab is the yield shape. Fig. 9.6 illustrates the yield line pattern of the circular slab.

Cantilever slab with or without beams. The bending moment at the fixed end of the cantilever slab is called negative bending moment. The slab doesn't understand positive or negative signs but the negative bending moment means that the tension occurs on the load face. Fig. 9.7(a) illustrates a cantilever slab with beams and possible yield line. The yield line is next to and parallel to the support. It is a simple cantilever mechanism and the slab collapses along with beam as a whole. Fig. 9.7(b) illustrates another possible mechanism in which a part of the slab may collapse along with a portion of beam. The broken line indicates the negative bending moment. The critical failure depends on the relative strength properties of the slab in the two perpendicular directions. If the distribution reinforcement is nominal, the second mode of failure may occur.

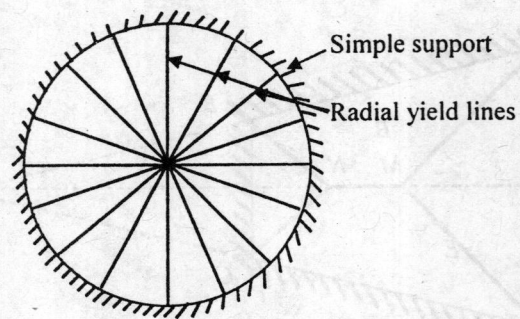

Fig. 9.6 Yield Line Pattern in Circular Slab.

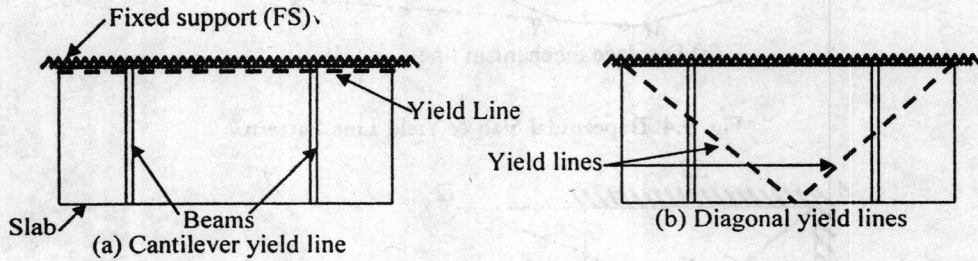

Fig. 9.7 Possible Yield Line Patterns in Cantilever Slab.

Slab with intermediate beam. Beams provide support to the slab. However even such elastic supports are considered to be rigid when compared with the yield deformations of the slab. The yield line at the beam is a negative yield line. Fig. 9.8 illustrates yield line pattern of a two-span continuous slab. The Intermediate beam acts as a support and negative moment yield lines are generated parallel to the beam. A yield line at mid span of short span is initiated immediately after the negative yield line at the beam. The collapse mechanism is very similar to that of a simply supported slab along with the negative yield line at the beam.

9.3. METHODS OF ANALYSIS IN YIELD LINE THEORY

Two basic methods of analysis of determination of the collapse load are:

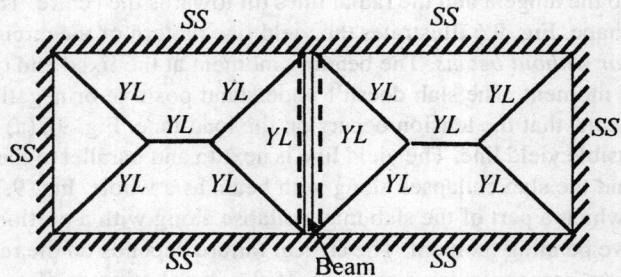

Fig. 9.8 Yield Line Pattern in Two Span Continuous Slab.

1. Equilibrium method, also called Statical method and
2. Mechanism method, also called virtual work

The equilibrium method is based on the equilibrium of forces. The moment at any section is equal or less than the capacity of the section. A method that satisfies the principles of equilibrium is called the statically admissible field. The method gives a lower bound collapse load. A lower bound predicts a load lower than or equal to that when compared with the actual collapse value. In a design problem, the collapse load is fixed and a reverse process is applied to find the moment capacity required resisting the load. Therefore, the lower bound analysis problem gives a safe design. Normally, any statical method gives a safe design. A load based on the elastic analysis can be considered as a particular solution of an equilibrium method. One selects any statically admissible field and maximizes the moments until a failure of a section takes place.

A mechanism is formed, if adequate numbers of hinges are available in the system that allows free rotation of the segments of the slab. The mechanism method assumes a set of yield lines that are kinematically admissible and then estimate the collapse load. The simple geometric laws govern the formation of yield lines and mechanisms. The virtual work concept is used in estimating the collapse load. The mechanism method is also called as the virtual work method and gives an upper bound solution. This means that the estimated collapse load is higher than or equal to actual collapse load. The upper bound solution may result into an unsafe design if the lowest mechanism is not chosen. However, selection of the most probable true mechanism in slabs is not difficult. The actual mechanism may be marginally different at some corner points. In case of cantilever or continuity supports, the yield lines are formed with tension at the top fibers. The bending moment capacity of the sections along the yield line is assumed to be same. This may not so in many cases. In the actual slab, there may be some minor constraints at the corners of the slab with the result the yield lines bifurcates into branches near the corners. Further the concrete is brittle so the yield lines wiggle and may not be straight lines. Fig. 9.9 (a) illustrates idealized pattern of the yield lines for a simply supported rectangular slab, while Fig. 9.9(b) illustrates the most probable yield line mechanism. The pattern in Fig. 9.9(b) avoids sharp bends that occur in the idealized pattern.

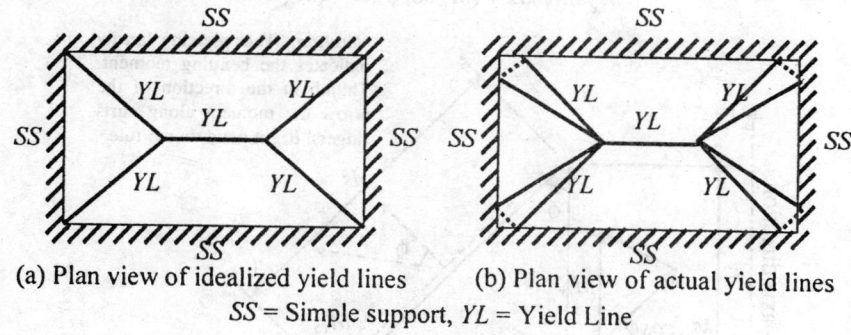

(a) Plan view of idealized yield lines   (b) Plan view of actual yield lines

SS = Simple support, YL = Yield Line

Fig. 9.9 Idealized and Actual Yield Lines.

9.4 MOMENT CAPACITY ALONG AN INCLINED LINE

Reinforcement bars are normally placed in two orthogonal directions in slabs. The direction of the yield line need not coincide with the direction of the bars. Typical yield line formations shown in earlier Figures indicates that the yield line can be at an angle with the reinforcement direction. It is, therefore, essential to find the resisting moment capacity along the yield line for any given orthogonal moment capacities. The moment capacity at an inclined line is derived using the notations indicated in Fig. 9.10.

Notations

M_{rx} = moment capacity of slab for unit width on x-plane. Positive if compression is caused on the top of the slab .

M_{ry} = moment capacity of a unit width slab on the y-plane.

$M_{r\phi}$ = moment capacity of a unit width slab about an axis making ϕ-degrees with the x-axis.

A piercing arrow shown in Fig. 9.10 indicates the bending moment by the right hand thumb rule. The curling of the fingers indicates the direction of the bending when the right hand thumb is directed towards the arrow.

Subscript notations are: r = resisting capacity, x = x-axis and y = y-axis

Consider an infinitesimal element of sides' dx and dy along x and y axes respectively as shown in Fig. 9.10. Let ds is the diagonal distance of the infinitesimal element. A plane under consideration is inclined at an angle of f with the x-axis. A plane is indicated by the direction of the normal to the plane. x-plane means that the plane is perpendicular to the x-axis. ϕ-plane means the normal to the plane is indicated by a simple arrow broken line. The bending moments acting on two perpendicular planes are known, then the bending moment acting on any arbitrary plane can be determined from the equilibrium of the moments on the infinitesimal element. The two moments M_x and M_y are indicated on the faces of the plans and the components of these moments in the f directions are also shown by triangle of moments. The components of the bending moments along the f direction for unit width are listed here.

$$\text{Component moment on } x\text{-plane} = M_x \sin\phi \tag{9.1}$$

$$\text{Component moment on } y\text{-plane} = M_y \cos\phi \tag{9.2}$$

The moment equilibrium about the ϕ-axis on the infinitesimal triangular element is:

$$(M_x \sin\phi)dx + (M_y \cos\phi)dy = (M_\phi)ds$$

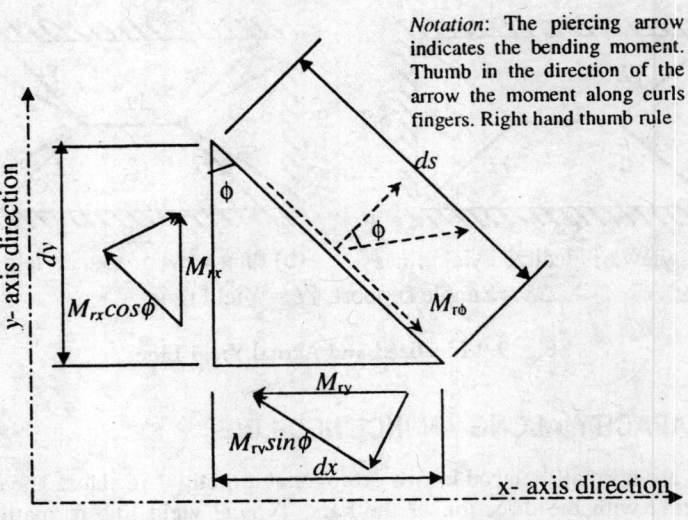

Notation: The piercing arrow indicates the bending moment. Thumb in the direction of the arrow the moment along curls fingers. Right hand thumb rule

Fig. 9.10 Moments on Three Phases-Equilibrium of Moments.

Dividing the above expression by ds gives:

$$(M_x \sin\phi)\frac{dx}{ds} + (M_y \cos\phi)\frac{dy}{ds} = M_\phi \tag{9.3}$$

The values of dx/ds and dy/ds are $\sin\phi$ and $\cos\phi$ respectively.
The above equation reduces to:

$$M_x \sin^2\phi + M_y \cos^2\phi = M_\phi \tag{9.4}$$

Let $M_x = c\, M_y$ then the above equation can be written as

$$M_\phi = M_y(c\,\sin^2\phi + \cos^2\phi) \tag{9.5}$$

In case the slab is isotropic the value of $c = 1$, and the moments relation reduces to

$$M_\phi = M_y = M_x \tag{9.6}$$

This is obviously the isotropic slab moment capacity relation
Given the moment capacities of a slab in orthogonal directions, the moment capacity of any plane section can be obtained by the Eq. 9.4 and it is

$$M_{r\phi} = M_{rx} \sin^2\phi + M_{ry} \cos^2\phi \tag{9.7}$$

Where

$M_{r\phi}$ = moment capacity of a section on plane inclined at ϕ with the x-axis
M_{rx} = moment capacity of the x-plane section
M_{ry} = moment capacity of the y-plane section

Example 9.1: Computation of Moment Capacity at 45 Degrees

A reinforced concrete rectangular slab has a moment capacity of M_r in the short span direction and $0.5M_r$ in the long span direction. Determine the moment capacity of the slab on a section whose normal makes an angle of 45 degrees with the long span axis.

Solution

The moment capacities of sections of x and y-axes are:

$$M_{rx} = M_r \text{ and } M_{ry} = 0.5M_r$$

The moment capacity of a section at 45 degrees with the short axis is

$$M_{ri} = M_{rx} \cos^2 45 + M_{ry} \sin^2 45 = M_r(1/2 + 0.5*(1/2)) = 0.75M_r$$

9.5 METHOD OF VIRTUAL WORK

Work done by a force is equal to force multiplied by the distance moved in the direction of the force. In most cases, the load or the force starts at zero level and reaches its full level. In that case the work done is integrated over the distance with the variable load. The total work done is in moving through the displacement caused by it is equal to half of the load multiplied by the displacement. Figure 9.11 illustrates work done by the load by moving through displacement caused by it.

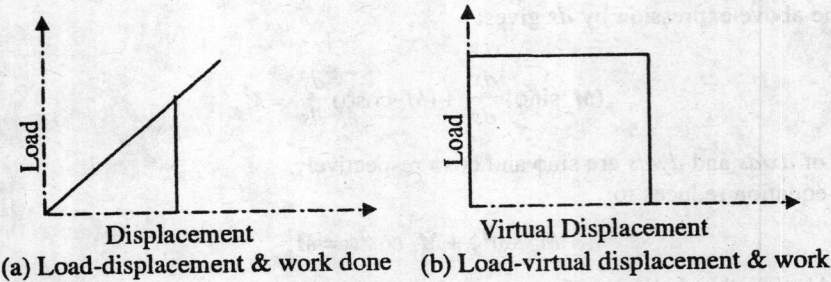

Fig. 9.11 Actual Work Done and Virtual Work Done Concepts.

At a pre-assigned load, if an additional displacement is given in the direction of the load, then the work done by the load is equal to the load multiplied by the displacement. The displacement is called virtual displacement and the work done is called virtual work.

$$Work\ done = \frac{Wv}{2} \qquad (9.8)$$

where $\qquad W$ = load, and v = displacement due to the load.

$$Virtual\ work\ done = V^* = \int (wv)\, dx\, dy \qquad (9.9)$$

Where v^* = virtual displacement

The virtual work concept is a powerful tool in solving the collapse load. A most probable yield line pattern is assumed and a virtual displacement compatible with the boundary conditions along with the yield line formation is assumed. The segments of the slab within the yield lines go through rigid body displacements with the collapse load acting on the structure. The virtual work done by the load is equal to the load multiplied by the virtual displacement and is illustrated in Fig. 9.11(b). The Eq.(9.9) integrated over the total area of the slab gives the total virtual work.

The moment capacity along a yield line is assumed to be constant, then the internal resisting work is equal to the product of the relative rotation and the corresponding resisting moment capacity of the section. The internal work done is:

$$Internal\ work = U = \sum_{i} \theta_i^* \, M_{ri} s_{ii} \qquad (9.10)$$

where
θ_i^* = relative virtual rotation of the surface at the ith yield.
M_{ri} = resisting moment capacity of the section at ith yield line,
s_i = total length of the ith yield line.

The principle of conservation of energy states that the variation in the total energy of the system is equal to zero if no external energy is supplied. In a conservative energy system the external virtual work done must be equal to the internal work done by the resisting forces. The virtual work is equated to the internal work done.

$$V^* = \int (wv^*)\, dx\, dy = U = \sum \theta_i M_{ri} s_s \qquad (9.11)$$

The same relation can be rewritten as:

$$\int (wv^*)\,dx\,dy = \sum \theta_i M_{ri} s_s \tag{9.12}$$

The relative rotations at the yield lines are obtained from the compatible virtual displacement.

9.6 MOMENT CAPACITY OF SLAB SECTION

Most slabs are singly reinforced and are under-reinforced. Raft slabs and pile caps are normally doubly reinforced. The moment capacity of slab section for one unit width of the section is given by

$$M_{rb} = Kd^2 f_{ck} \tag{9.13}$$

and the corresponding area of tension reinforcement is,

$$A_{st} = \frac{1.15\,M}{jdf_y} \tag{9.14}$$

The values of the constants K and j for commonly used reinforcement steels are listed in table 9.1.

Table 9.1 Moment Capacity Factors

f_y	K	j
415	0.138	0.80
500	0.133	0.81

The balance moment capacities for one metre width of slabs of different depths of slab are given in table 9.2 using the *effective depth of the slab* from 65 mm to 240 mm. Similarly the area of reinforcement for one metre width for a balanced section are also listed in the table 9.2.

Example 9.2: Selection of Slab Depth & Area of Reinforcement using Table 9.2

The ultimate moment on a concrete slab exposed to mild environment is 30 kNm/m. Determine the depth of the section and the reinforcement required.

Solution

The structure is exposed to mild environment so the minimum grade of concrete that can be used is $M20$. This minimum grade is selected. Move down the column 2 of table 9.2 till the moment capacity just exceeds limit value of 30,000 Nm/m. This happens at an effective depth of 105 mm where the moment capacity is 30429 Nm/m and the corresponding area of reinforcement is 997 mm²/m. The minimum cover in slabs for mild exposure is 20 mm so select 105 mm plus half diameter of bar and plus cover to the reinforcement. The total thickness is = 105 + 5 + 20 = 130 mm. Provide 12 mm bars at 110 mm spacing, then the actual area of reinforcement provided is 1027 mm²/m. More rational way of providing the reinforcement is to compute it from the bending moment and actual effective depth.

Example 9.3: Design of Slab for a Given Collapse Limit Moment

A reinforced concrete slab is in sever exposure condition and subjected to a collapse bending moment of 30 kNm/m. Select a suitable section and reinforcement.

Solution

The structure is exposed to sever environment so the minimum grade of concrete that can be used is $M30$. Move down the column 6 of table 9.2 till the moment capacity just exceeds limit value of 30,000 Nm/m. This happens at an effective depth of 95 mm where the moment capacity is 37365 Nm/m and the corresponding area of reinforcement is 1362 mm^2/m. The minimum cover for sever exposure is 45 mm so select 95 mm plus half diameter of bar and plus cover to the reinforcement. And it is equal to $95 + 5 + 45 = 145$ mm overall thickness of the slab. Provide 12 mm bars at 825 mm spacing.

Table 9.2 Moment Capacity and Area of Balanced Reinforcement (Per one metre width).

Effective depth	M20 concrete		M25 concrete		M30 concrete	
	M_{rb} (Nm/m)	A_{st} (mm^2)	M_{rb} (Nm/m)	A_{st} (mm^2)	M_{rb} (Nm/m)	A_{st} (mm^2)
65	11661	617	14576	776	17491	932
75	15525	712	19406	896	23287	1075
85	19941	807	24926	1015	29911	1218
95	24909	902	31136	1135	37363	1362
105	30429	997	38036	1254	45643	1505
115	36501	1092	45626	1374	54751	1649
125	43125	1187	53906	1493	64687	1792
135	50301	1282	62876	1613	75451	1935
145	58029	1377	72536	1732	87043	2079
155	66309	1472	82886	1852	99463	2222
165	75141	1567	93926	1971	112711	2366
175	84525	1662	105656	2091	126787	2509
185	94461	1757	118076	2210	141691	2652
195	104949	1852	131186	2330	157423	2796
205	115989	1947	144986	2449	173983	2939
215	127581	2042	159476	2569	191371	3083
225	139725	2137	174656	2688	209587	3226
235	152421	2232	190526	2808	228631	3369
245	165669	2327	207086	2927	248503	3513

9.7 EXAMPLES OF DETERMINATION OF COLLAPSE LOAD FROM MECHANISM METHOD

The design procedure of reinforced concrete slab or frame is not based on the collapse load analysis even though the load is often termed as collapse load. The reinforced concrete structures have limited ductility therefore total redistribution of the moments on the sections to collapse state is not achieved. The structural analysis of a structure in limit state design is linear-elastic with limited redistribution of the moments.

Example 9.4: Determination of Collapse Load on One-way Slab

A rectangular slab of short span L having a moment capacity of M_r per unit length in the short span direction (y-axis) and rM_r in the long span direction is simply supported on all the four edges. Determine the collapse load of the slab.

Figure 9.12(a) illustrates the plan view of the slab along with yield lines. The yield line that starts from corner A is assumed at 45 degrees to the edge support. It may not exactly be at 45 degrees with x-axis but a simple compatible mechanism is assumed. One can assume an arbitrary inclination of the yield line and obtain it from the principle of optimization. The slab is divided into four segments and the numbers of the segment are marked as 1 or 2 or 3 or 4 on the segment itself. The intersecting yield lines point C is at a distance $L/2$ from the edge AC. Fig. 9.12(b) illustrates the virtual deflection at the middle of the long span. AB, BC, DE, DF, and BD are the yield lines.

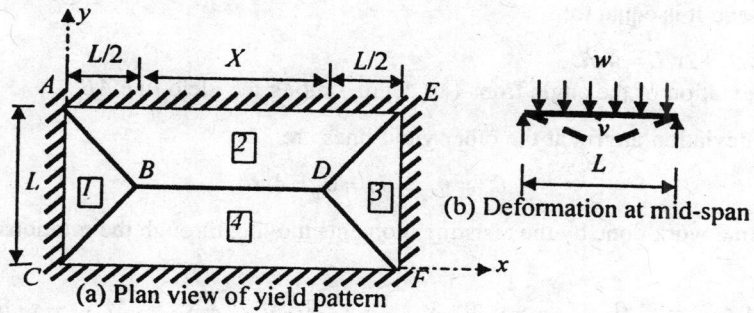

Fig. 9.12 Yield Pattern of Example 9.4.

Let v = virtual deflection of the points B and D

Let w_c = intensity of the collapse load

Virtual Work

The virtual work done by the loads on the segments moving through the virtual deflections are computed one by one. The virtual work done by the load on segment ABC (also called as segment 1) is equal to the total load on the triangle multiplied by the virtual displacement of the centroid of the load. The centroid of the load on the triangle is at the centroid of the triangle.

The corresponding displacement of the centroid of the load is $= v/3$

The work done by the load of the triangle ABC is $= \frac{1}{2}w_c(L)\left(\frac{L}{2}\right)\left(\frac{v}{3}\right) = \frac{w_cL^2v}{12}$

The trapezium $ABDE$ is divided into two triangles ABE and EBD and the work done by the loads on the two separate triangles is computed separately. It is easier to locate the centroid of a triangle.

The work done by the load of the triangle ABE is $= \frac{1}{2}w_c(L+X)\left(\frac{L}{2}\right)\left(\frac{v}{3}\right) = \frac{w_c(L+X)Lv}{12}$

The work done by the load of the triangle BDE is $= \frac{1}{2}w_c(X)\left(\frac{L}{2}\right)\left(\frac{2v}{3}\right) = \frac{w_cXLv}{6}$

There is a symmetry in the yield lines and the segments, so the total virtual work done by the 6 segments is

$$V = 2w_c Lv\left(\frac{L}{12} + \frac{L+X}{12} + \frac{X}{6}\right) = \frac{w_c Lv}{6}(L + L + X + 2X) = \frac{w_c Lv}{6}(2L + 3X) \qquad \text{(a)}$$

The rotation of the segment ABC is $= v/(L/2) = 2v/L$
Similarly the rotation of segment $ABDE$ is $= 2v/L$

The total angle change at the yield line AB is the sum of the angles of slope of the adjoining segments ABC and $ABDE$, and it is equal to:

$$\theta_{ab} = 2v/L + 2v/L = 4v/L$$

where $\quad \theta_{ab}$ = deviation of the angle from 180 degrees along the yield line AB.

Similarly the deviation angles at the other yield lines are:

$$\theta_{bc} = \theta_{bd} = \theta_{de} = \theta_{bf} = 4v/L$$

The total internal work done by the resisting moments moving through the rotations at the yield lines is

$$U = \Sigma M_i s_i \theta_i = s_{ab}\theta_{ab}M_{rab} + s_{ac}\theta_{ac}M_{rac} + s_{bd}\theta_{bd}M_{rbd} + s_{de}\theta_{de}M_{rde} + s_{df}\theta_{df}M_{rdf} \qquad \text{(b)}$$

where s_{ab} = length of the yield line AB, similarly the other notations

The length of the yield line AB is $= (L/2)\ddot{O}2 = L/\ddot{O}2$, similarly the other lengths can be calculated easily.
Alternatively one can compute the internal work done by considering each segment and its rotation about x and y axes separately and then multiplies the rotation by the corresponding moment capacity component of the slab about the respective axis. This approach is used in this book with the following notations.

Let $\quad x_i$ = projection of the yield line of segment i on x-axis, similarly
$\quad\quad y_i$ = projection of the yield line of segment i on y-axis

The internal work done can be expressed as

$$U = \Sigma x_i \theta_{ri} M_{rxi} + y_i \theta_{yi} M_{ryi} \qquad (9.15)$$

where $\quad \theta_{xi}$ = rotation of segment I about the x-axis, similarly the other notations,
$\quad\quad M_{rxi}$ = moment capacity of the segment i about x-axis

Segments 1 and 3 rotate about y-axis only and the rotation is equal to
$\quad\quad \theta_{y1} = \theta_{y3} = 2v/L$. (Only magnitude of rotation is given and not the sign)
And $\quad \theta_{x1} = \theta_{x3} = 0$; That is no rotation of the segment about x -axis.

The projection of the yield lines of segments 1 and 3 on y-axis is equal to $L_y = L$.
Similarly segments 2 and 4 rotate about x-axis only.

$\quad\quad \theta_{x2} = \theta_{x4} = 2v/L$. (Only magnitude of rotation is given and not the sign)
And $\quad \theta_{x2} = \theta_{y4} = 0$; That is no rotation of the segment about y -axis.

The projection of the yield lines of segments 2 and 4 on x-axis is equal to L_x.

$$\text{Let } M_{rx} = rM_{ry} = rM_r$$

The work done at the yield line is always positive because the rotation is always opposite to the direction of the resistance. The work done by the yield line of the four segments is equal to:

$$U = L_y(\theta_{y1} + \theta_{y3})M_{rx} + L_x(\theta_{x2} + \theta_{x4})M_{ry} = L\left(\frac{2v}{L} + \frac{2v}{L}\right)rM_r + (L + X)\left(\frac{2v}{L} + \frac{2v}{L}\right)M_r$$

$$U = 4v(rM_r + (1 + X/L)M_r) = 4v[r + (1 + X/L)]M_r \tag{c}$$

Equating the external virtual work to the internal work done from Eqs.(a) and (c) gives

$$V = \frac{w_c L v}{6}(2L + 3X) = U = 4v[r + (1 + X/L)]M_r$$

The virtual deflection gets cancelled from each side of the expression. The above equation reduces to:

$$w_c = \frac{24[r + (1 + X/L)]M_r}{L(2L + 3X)} = \frac{24[r + (1 + X/L)]M_r}{(2 + 3X/L)L^2} \tag{9.16}$$

For square slab the value of X is equal to zero and let the moment capacity in both the directions is same, $r = 1$. Then the above equation reduces to:

$$w_c = \frac{24 M_r}{L^2} \tag{9.17}$$

Consider a slab of length 3 times the short span, then the value of X is $2L$, and let the moment capacity along the long span be negligible small, say equal to zero. The substitution of the values $X = 2$ and $r = 1$ in the Eq.(9.16) give the collapse load as:

$$w_c = \frac{9 M_r}{L^2} \tag{9.18}$$

The collapse load of one-way slab that is computed above is the upper bound solution. The mechanism method gives an upper bound solution and the margin depends on how close the mechanism that is selected to the true one. In the present case the upper bound solution is too high. Normally accepted collapse load of one-way slab is $8M_r/L^2$ as against the solution of $9M_r/L^2$. This variation is primarily due to the assumption that the yield line at the corners makes 45 degrees with the edge.

Example 9.5: Collapse Load of Rectangular Slab

A rectangular slab simply supported on all four edges is subjected to uniformly distributed load. Determine the optimal collapse mechanism and the load.

Solution

The problem is very similar to the one solved earlier except that the yield lines point of intersection is given an arbitrary distance from the short support. The yield line pattern is illustrated in Fig. 9.13. The yield line that starts from corner A is assumed at an arbitrary inclination to the edge support. The actual inclination of the yield line is obtained from the principle of optimization. The slab is divided into four segments marked as 1 or 2 or 3 or 4. The intersecting yield lines point C is at a distance αL from the edge AC. Similarly the yield lines intersection point on the other end is assumed by symmetry. Fig. 9.13(b)

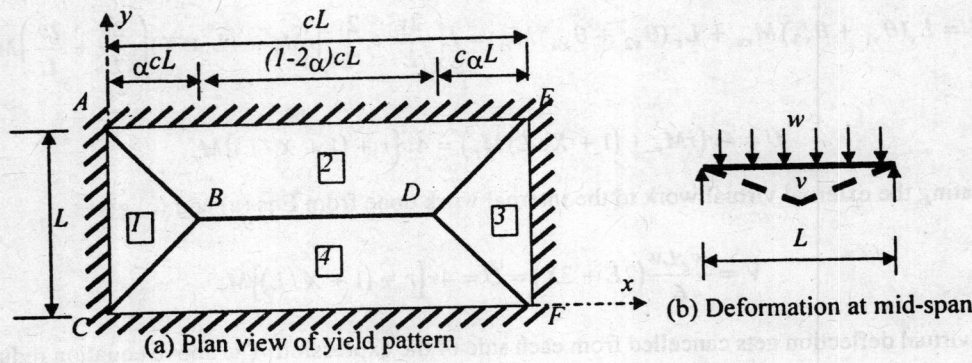

Fig. 9.13 Yield Pattern of Example 9.5.

illustrates the virtual deflection at the middle of the long span. The yield lines are *AB, BC, DE, DF, and BD*.

Let L = length of the short span that is along the y-axis,

Let cL = length of the long span that is along the x-axis,

Let $M_{ry} = M_r$ = moment capacity of the section perpendicular to the short span, (y-plane)

Let $M_{rx} = rM_r$ = moment capacity of the section perpendicular to the long span,

Let v = virtual deflection of the points B and D

Let w_c = intensity of the collapse load,

Virtual Work

The virtual work done by the loads on the segments moving through the virtual deflections are computed one by one. The virtual work done by the load on segment *ABC* (also called as segment 1) is equal to the total load on the triangle multiplied by the virtual displacement of the centroid of the load. The centroid of the load on the triangle is at the centroid of the triangle.

The corresponding displacement of the centroid of the load is = $v/3$

The work done by the load of the triangle *ABC* is = $\dfrac{1}{2} w_c (L) \alpha c L \left(\dfrac{v}{3}\right) = \dfrac{w_c \alpha c L^2 v}{6}$

The work done by the load of the triangle *ABE* is = $\dfrac{1}{2} w_c cL \left(\dfrac{L}{2}\right)\left(\dfrac{v}{3}\right) = \dfrac{w_c c L^2 v}{12}$

The work done by the load of the triangle *BDE* is = $\dfrac{1}{2} w_c cL (1 - 2\alpha)\left(\dfrac{L}{2}\right)\left(\dfrac{v}{3}\right) = \dfrac{w_c c (1 - 2\alpha) L^2 v}{6}$

There is symmetry in the yield lines and the segments, so the total virtual work done by the 6 segments is

$$V = 2w_c L^2 v \left(\frac{c\alpha}{6} + \frac{c}{12} + \frac{c(1-2\alpha)}{6}\right) = \frac{w_c L^2 v}{6}(2c\alpha + c + 2c - 4c\alpha) = \frac{w_c c L^2 v}{6}(3 - 2\alpha) \qquad \text{(a)}$$

The rotation of the segment *ABC* is = $v/\alpha cL$

Similarly the rotation of segment *ABDE* is = $2v/L$

The internal work done is computed by considering each segment and its rotation about x and y axes separately and then multiply the rotation by the corresponding moment capacity of the slab about the respective axis.

Let x_i = projection of the yield line of segment i on x-axis, similarly
                   y_i = projection of the yield line of segment i on y-axis

The internal work done can be expressed as:

$$U = \Sigma \left(x_i \theta_{xi} M_{rxi} + y_i \theta_{yi} M_{ryi} \right)$$

where θ_{xi} = rotation of segment I about the x-axis, similarly the other notations,
                   M_{rxi} = moment capacity of the segment i about x-axis

Segments 1 and 3 rotate about y-axis only and the rotation is equal to

$\theta_{y1} = \theta_{y3} = v/\alpha L$. (Only magnitude of rotation is given and not the sign)
And $\theta_{x1} = \theta_{x3} = 0$; That is no rotation of the segment about x -axis.

The projection of the yield lines of segments 1 and 3 on y-axis is equal to $L_y = L$.
Similarly, segments 2 and 4 rotate about x-axis only.

$\theta_{x2} = \theta_{x4} = 2v/L$. (Only magnitude of rotation is given and not the sign)
And $\theta_{x2} = \theta_{y4} = 0$; That is no rotation of the segment about y -axis.

The projection of the yield lines of segments 2 and 4 on x-axis is equal to L_x.
The work done at the yield line is always positive because the rotation is always opposite to the direction of the resistance. The work done by the yield line of the four segments is equal to:

$$U = L_y \left(\theta_{y1} + \theta_{y3} \right) M_{rx} + L_x \left(\theta_{x2} + \theta_{x4} \right) M_{ry} = L \left(\frac{v}{\alpha cL} + \frac{v}{\alpha cL} \right) r M_r + (cL) \left(\frac{2v}{L} + \frac{2v}{L} \right) M_r$$

$$U = 2v[r/\alpha c + (2c)] M_r = 2v[r/\alpha c + (2c)] M_r$$

$$U = 2v \left[\frac{r + 2\alpha c^2}{\alpha c} \right] M_r$$

(b)

The virtual work done and the resisting internal work is a function of a parameter a. Equating the virtual work to the internal work done gives:

$$\frac{w_c cL^2 v}{6} (3 - 2\alpha) = 2v \left[\frac{r + 2\alpha c^2}{\alpha c} \right] M_r$$

The collapse load from the above expression is:

$$w_c = \frac{12(r + 2\alpha c^2) M_r}{(3 - 2\alpha) \alpha c^2 L^2}$$

(9.19)

The above expression can be rewritten for moment computation as

$$M_r = \frac{(3-2\alpha)\alpha c^2 L^2 w_c}{12(r + 2\alpha c^2)} \qquad (9.20)$$

The minimum resisting moment required for the parameter a can be obtained by setting the variation of the moment equal to zero with respect to the parameter and it is:

$$\frac{\partial M_r}{\partial \alpha} = 0 \text{ this gives}$$

$$(3 - 4\alpha)(r + 2\alpha c^2) - (3 - 2\alpha)\alpha(2c^2) = 0$$

The value of a works out to be:

$$\alpha = \frac{-r + \sqrt{r^2 + 3rc^2}}{2c^2} \qquad (9.21)$$

Or

$$2\alpha c^2 = \sqrt{r(r + 3c^2)} - \gamma$$

After the substitution of the value of a in the moment capacity expression Eq.(9.20) and simplification of the expression gives:

$$M_r = \frac{1}{24}\left[\frac{\sqrt{r + 3c^2} - \sqrt{r}}{c}\right]^2 w_c L^2 \qquad (9.22)$$

Alternatively, the collapse load is given by:

$$w_c = \frac{24 M_r}{L^2}\left[\frac{c}{\sqrt{r + 3c^2} - \sqrt{r}}\right]^2 \qquad (9.23)$$

Special case: one-way slab. In normal problems of one-way slabs, the moment capacity of the section perpendicular to the long span is considered to be zero. The substitution of r equal to zero, the collapse load expression given above results:

$$w_c = \frac{8 M_r}{L^2}$$

This is the normally accepted collapse load in one-way slabs.

Special case: square slab. In a square slab the moment capacity in both the directions is same, that means r = 1. And c = 1. The substitution of these values in the collapse expression gives

$$w_c = \frac{24 M_r}{L^2}$$

Example 9.6: Collapse Load of Rectangular Slab of Aspect Ratio of 3

A rectangular slab of short span L having a moment capacity of M_r per unit length and one-fourth moment capacity in the long span direction is simply supported on all the four edges. Determine the collapse load.

Solution

The previous example illustrates a more general case of rectangular simply supported slab. The yield line formation distance from the short support is given in terms of a in Eq. (9.21) and it is

$$\alpha = \frac{-r + \sqrt{r^2 + 3rc^2}}{2c^2}$$

in which the values of c and r are: $c = 3$ and $r = 0.25$. The substitution of the values in the above expression gives:

$$\alpha = \frac{-r + \sqrt{r^2 + 3rc^2}}{2c^2} = \frac{-0.25 + \sqrt{0.0625 + 6.75}}{18} = 0.13$$

The intersection point of the yield lines from the smaller support is equal to $= 0.13(3L) = 0.39L$
The collapse load expression is also taken from the previous example and it is:

$$w_c = \frac{24\,M_r}{L^2}\left[\frac{c}{\sqrt{r + 3c^2} - \sqrt{r}}\right]^2 = \frac{24\,M_r}{L^2}\left[\frac{3}{\sqrt{0.25 + 27} - \sqrt{0.25}}\right]^2 = \frac{9.69\,M_r}{L^2}$$

Example 9.4 illustrates the case of a long rectangular plate with zero moment capacity in the long span direction and the collapse load in that case is $9M_r/L^2$. This value is close to the one that is calculated in this example. The difference is due to the moment capacity in the long span direction is one-fourth that of the short span.

Example 9.7: A long Rectangular Slab with Fixed Boundary Conditions

A rectangular slab has its two long edges fixed against rotation. The negative bending moment capacity at support in the short span direction is same as that of the positive bending moment capacity in the middle of the slab in that direction. The positive and negative moment capacities of the slab in the long span direction are zero. The slab is subjected to uniformly distributed load. Determine the collapse load of the slab.

Solution

The rectangular slab along with yield lines is shown in Fig. 9.14, in which the negative yield lines are marked with broken lines while the positive moment yield lines are shown with firm lines. Edges AE and CF have negative yield lines shown by dotted lines. The slab is divided into four segments similar to the one of the earlier examples. For convenience of illustration, the intersection point B is chosen at $0.5L$ distance from the short support AC. Similarly at the other end of the slab. The segments are numbered 1 to 4 and are indicated on the figure.

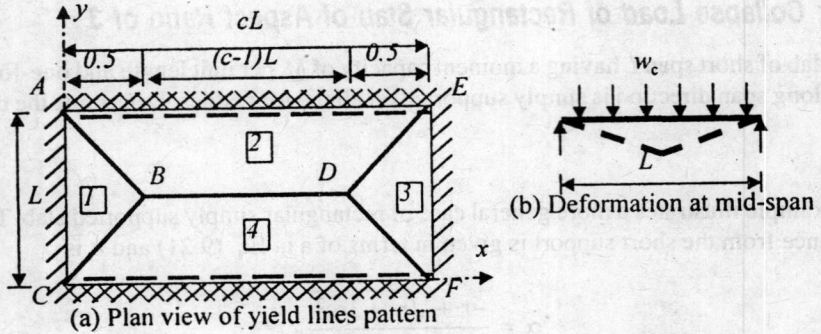

(a) Plan view of yield lines pattern

(b) Deformation at mid-span

Fig. 9.14 Yield Line Pattern of Example 9.7.

The virtual work done the segment 2 (*ABDE*) is obtained from two triangular segments *ABE* and *BDE*.

Let v = virtual displacement of the yield line *BD*,
Let M_r = positive and negative moment capacity of the short span,
Let L = length of the short span,
Let cL = length of the long span

Virtual Work

The virtual work done by the load on segment *ABC* (also called segment *1*) is equal to the total load on the triangle multiplied by the virtual displacement of the centroid of the load. The centroid of the load on the triangle is at the centroid of the triangle.

The corresponding displacement of the centroid of the load is $= v/3$

The work done by the load of the triangle *ABC* is $= \dfrac{1}{2} w_c (L) \left(\dfrac{L}{2}\right)\left(\dfrac{v}{3}\right) = \dfrac{w_c L^2 v}{12}$ (a)

The trapezium *ABDE* is divided into two triangles *ABE* and *EBD* and the work done by the loads on the two separate triangles is computed separately.

The work done by the load of the triangle *ABE* is $= \dfrac{1}{2} w_c (cL) \left(\dfrac{L}{2}\right)\left(\dfrac{v}{3}\right) = \dfrac{w_c cL^2 v}{12}$

The work done by the load of the triangle *BDE* is $= \dfrac{1}{2} w_c (c-1) L \left(\dfrac{L}{2}\right)\left(\dfrac{2v}{3}\right) = \dfrac{w_c (c-1) L^2 v}{6}$

The virtual work done by the segment 2 is $= \dfrac{w_c (3c - 2) L^2 v}{12}$ (b)

The total virtual work done by the four segments is twice the amount of the work done on the two segments for which the virtual work was already calculated. The total virtual work done is:

$$V = 2\left[\frac{w_c (1 + 3c - 2) L^2 v}{12}\right] = \frac{w_c (3c - 1) L^2 v}{6}$$ (c)

The internal work done at the yield lines is computed by considering the moment capacity in two perpendicular directions multiplied by the corresponding rotation of the segments.

Let $\quad x_i$ = projection of the yield line of segment 1 on x-axis, similarly
$\qquad\qquad y_i$ = projection of the yield line of segment 1 on y-axis

The internal work done can be expressed as

$$U = \Sigma\left(x_i\theta_{xi}M_{rxi} + y_i\theta_{yi}M_{ryi} + x_i\theta_{xi}M_{nrxi} + y_i\theta_{yi}M_{nryi}\right)$$

where

$\qquad _o\theta_{xi}$ = rotation of segment I about the x-axis, similarly the other notations,
$\qquad M_{rxi}$ = positive moment capacity of the segment i about x-axis
$\qquad M_{nrxi}$ = negative moment capacity of the segment i about x-axis

Similarly the other terms.
Segments 1 and 3 rotate about y-axis only and the rotation is equal to

$$\theta_{y1} = \theta_{y3} = 2v/L. \text{ (Only magnitude of rotation is given and not the sign)}$$
And $\quad \theta_{x1} = \theta_{x3} = 0$; That means there is no rotation of the segment about x -axis.

The projection of the yield lines of segments 1 and 3 on y-axis is equal to $L_y = L$.
Similarly segments 2 and 4 rotate about x-axis only.

$$\theta_{x2} = \theta_{x4} = 2v/L. \text{ (Only magnitude of rotation is given and not the sign)}$$
And $\quad \theta_{x2} = \theta_{y4} = 0$; no rotation of the segment about y -axis.

The projection of the yield lines of segments 2 and 4 on x-axis is equal to L_x.
Since the moment capacity in the long span direction is zero, then $M_{rx} = M_{nrx} = 0$
Similarly the positive and negative moment capacities in the short span direction are equal,

$$\text{then } M_{ry} = M_{nry} = M_r$$

The work done at the yield line is always positive because the rotation is always opposite to the direction of the resistance. The work done by the yield line of the four segments is equal to:

$$U = L_x\left(\theta_{x2} + \theta_{x4}\right)M_{ry} + L_x\left(\theta_{x2} + \theta_{x4}\right)M_{nry}$$

The rotations are equal and it is equal to $\theta = v/(L/2) = 2v/L$, the above internal work done is

$$U = (cL)4\frac{2v}{L}M_r = \frac{8cLM_rv}{L} = 8cM_rv \tag{d}$$

Equating the external virtual work to the in the internal work done gives

$$V = \frac{w_c\,(3c - 1)L^2v}{6} = U = 8cM_rv\,, \text{ This works out to be}$$

$$w_c = \frac{48cM_r}{(3c - 1)L^2} \tag{9.24}$$

Special Cases

A rectangular slab of aspect ratio 3 and fixed on the long edges, that is $c = 3$; the collapse load works out to be

$$w_c = \frac{18cM_r}{L^2}$$

Equation (9.24) can be rewritten for convenience of long or one-way slab by dividing the numerator and the denominator by c. The Eq.(9.24) reduces to

$$w_c = \frac{48M_r}{(3-1/c)L^2} \tag{9.25}$$

One-way slab

For one-way slab in which the length is very long when compared with the width of the slab, one can use $1/c$ equal to zero, so the above equation reduces to

$$w_c = \frac{16M_r}{L^2}$$

The collapse load of one-way slab with the long edges fixed is twice that of a simply supported slab.

Square slab

Incase of square slab, the value of c is equal to one, then Eq.(9.25) works out to be

$$w_c = \frac{24M_r}{L^2}$$

9.8 EQUILIBRIUM IN MECHANISM METHOD

The equilibrium of any yielded segment should be satisfied, further the bending moment at any other section of the structure doesn't exceed the moment capacity of the section. The number of equilibrium equations depends on the number of unknown variables in the system. If the choice of the yield line incorporates an unknown parameter, then more than one segment must be selected of the equilibrium. The collapse load computed from one segment need to be same as that for another segment. Consider the yield line pattern of the slab shown in Fig. 9.14 in which the point B is chosen at a distance $0.5L$. This problem takes the yield line pattern as deterministic. But on the other hand if the point B is chosen an arbitrary distance from the edge and than that distance also to be evaluated, then equilibrium of an additional segment has to be considered to determine the value. The method of determination of the collapse load is illustrated with examples.

Example 9.8: Collapse Load from Equilibrium of Collapse Mechanism of Slab with two Edges Fixed

A rectangular slab has its two long edges fixed against rotation. The negative bending moment capacity of the slab at support in the short span direction is same as that of the positive bending moment capacity at the middle of the slab in that direction. The positive and negative moment capacities of the slab in the long span direction are zero. The slab is subjected to uniformly distributed load. Determine the collapse load of the slab. (The problem is same as example 9.7)

Solution

The segments of the slab have to rotate about the support edges. As the long edges of the slab are fixed, negative moment yield lines form along the long edges. The failure yield mechanism is indicated in Fig. 9.14 in which the negative yield lines are marked with broken lines while the positive moment yield lines are shown with firm lines. Edges AE and CF have negative yield lines shown by dotted lines. The slab is divided into four segments similar to the one of the earlier example. For convenience of illustration the intersection point B of the yield lines is chosen at a distance $0.5L$ from the short support AC. Similarly at the other end of the slab. The solution is an upper bound one. The segments are numbered 1 to 4 and are indicated on the figure.

The equilibrium of segment 2 (that is $ABDE$) is considered for equilibrium.

A free-body–diagram of section of segment 2 is illustrated in Fig. 9.15. From symmetry the shear force on the line BD is zero. However there is shear force on the edges of the yield lines AB and DE. The shear and the bending moments on the line have to be included in the equilibrium. In this example, for the sake of simplicity, the shear forces on the lines AB and DE are neglected and the collapse load is computed as an approximation. The moment equilibrium of the segment is taken about support AE. The segment consists of two triangles ABE and EBD. The bending moment of the load on the segment is computed considering the two triangular segments about the support AE. The moment equilibrium of the free body diagram about the support AE is:

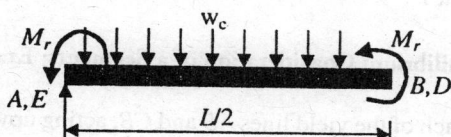

Fig. 9.15 Equilibrium Consideration of a Segment; Example 9.8.

$$M_r \left(AE + ABDE \right) = w_c \frac{cL}{2} \frac{L}{2} \frac{L}{6} + w_c \frac{(cL - L)}{2} \frac{L}{2} \frac{2L}{6} = w_c L^3 \left[\frac{c}{24} + \frac{2(c-1)}{24} \right]$$

$$M_r \left(cL + cL \right) = \frac{w_c L^3 (3c - 2)}{24} \text{ , or}$$

$$w_c = \frac{48 c M_r}{(3c - 2) L^2}$$

The collapse load from this method is slightly different from the one computed from the virtual work method. The collapse load from the virtual work method from the previous example is:

$$w_c = \frac{48 c M_r}{(3c - 1) L^2} \tag{9.26}$$

It can be seen that the collapse load computed from the equilibrium of segment 2 is lower than that from the virtual work method. The virtual work method encompasses the totality of the equilibrium of all the segments. This difference is because of the approximation of neglecting the shear forces on the diagonal yield lines. In this example if one select the segment 1, the collapse load computed from the equilibrium of the segment by neglecting the shear forces on the yield lines AB and BC, the collapse load comes out to be equal to zero. Actually, the shear force on the yield lines can be computed from the equilibrium of the triangular segment 1.

Consider the triangular segment ABC in Fig. 9.14 which is shown in Fig. 9.16(a) along with the shear forces and the load. Similarly Fig. 9.16(b) illustrates the forces on the segment 2. The forces now include the shear on the yield lines.

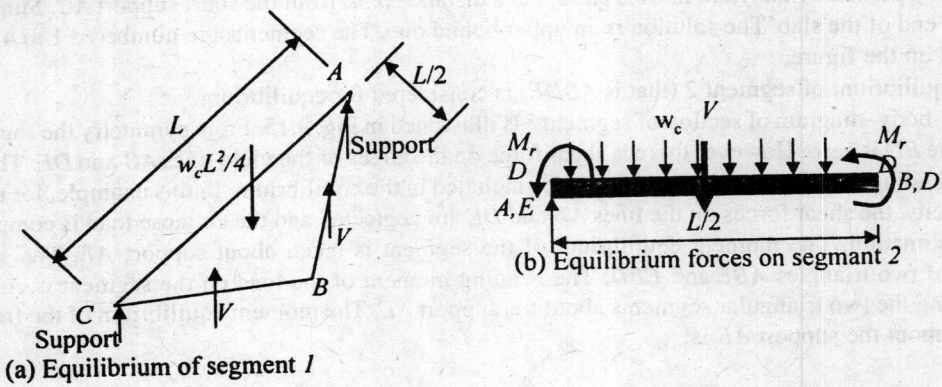

(a) Equilibrium of segment 1

(b) Equilibrium forces on segmant 2

Fig. 9.16 Equilibrium Consideration of a Segments; Example 9.8.

Let V = total shear forces on each of the yield lines AB and CB, acting upwards at the middle of the line. The shear force is assumed to be acting upward on the yield lines of the segment 1 and it is equal and opposite in segment 2. The bending moment on the x-plane is as equal to be zero so no moment is shown on the segment. There exists moment about the y-axis but the moment on the line AB is equal and opposite to that on the line BC. These moments are self balancing and hence not shown in the figure.

Taking moment equilibrium of all the forces about the support line AC, we have:

$$w_c \frac{8L}{2} \frac{L}{2} \frac{L}{6} = 2V \frac{L}{4} \text{, or the value of the shear force } V \text{ is:}$$

$$V = \frac{w_c L^2}{12} \tag{a}$$

The equilibrium of moments taken about the line AE of the segment 2 is similar to the one made earlier plus the moment caused by the shear acting at the middle of AB and DE. This gives:

$$M_r(AE + ABDE) = w_c \frac{cL}{2} \frac{L}{2} \frac{L}{6} + w_c \frac{(cL-L)}{2} \frac{L}{2} \frac{2L}{6} + 2V \frac{L}{4} = w_c L^3 \left[\frac{c}{24} + \frac{2(c-1)}{24} + \frac{1}{24} \right]$$

$$M_r\,(cL + cL) = \frac{w_c\,L^3(3c - 1)}{24}$$

$$w_c = \frac{48cM_r}{(3c - 1)\,L^2} \tag{9.27}$$

This value is exactly same as that obtained in the virtual work method.

9.9 EFFECT OF RESTRAINED CORNERS

Corners of most slabs are restrained from lifting up. In such cases Y-fork yield line pattern is generated as shown in Fig. 9.17. A negative yield line forms at the corner. Consider a simple case of a square plate of span L and having restrained corners. A typical yield line pattern is shown in Fig. 9.17(a).

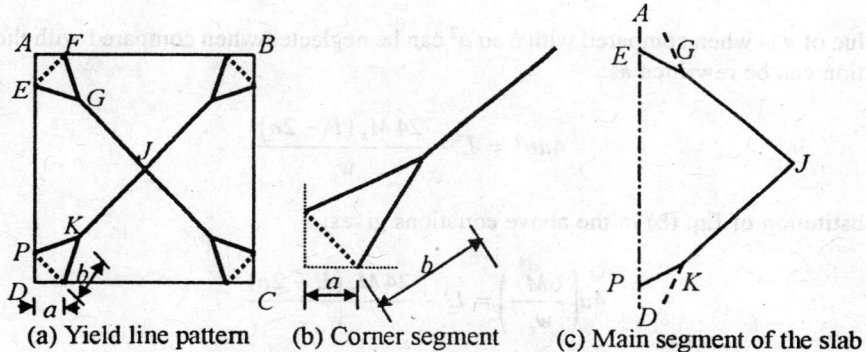

(a) Yield line pattern   (b) Corner segment   (c) Main segment of the slab

Fig. 9.17 Yield Lines Patterns of Square Slab; Example 9.9.

Example 9.9: Simply Supported Square Slab with Corner Yield Lines

A simply supported square slab is under uniform load. The corners of the slab are restrained against lifting. The slab is isotropic which means that that the moment capacity of the slab in two perpendicular directions is same. Determine the collapse load.

Solution

A typical yield line pattern of the slab is indicated in Fig. 9.17(a). The two distances, a and b are chosen arbitrarily to start with. There is a symmetry in the yield pattern because of the symmetry in the problem. The shear force at the yield lines is zero because of the symmetry. The equilibrium of the segments is used in the computation of the collapse load.

Let M_r = positive bending moment capacity of the slab, negative bending moment is zero.
Let L = span of the slab and w_c = collapse load intensity on the slab.
Let a = distance from the corner from which the forking of the triangular segment stars,
Let b = perpendicular distance from the apex to the base of the triangle *EFG*.

The slab is isotropic so the moment capacity of the slab in any direction is same.
Consider the equilibrium of the corner triangular segment *EFGJ* (Fig. 9.17(b)). The moment equilibrium of the triangle about the base *EF* is:

$$M_r(EF) = \frac{EF(b)w_c}{2}\frac{b}{3}$$

(a)

The value of $EF = \sqrt{2}\,a$, so the above equilibrium equation gives:

$$b^2 = \frac{6M_r}{w_c}$$

(b)

The equilibrium of moment about the segment $EGJKP$ taken about the support line AD is obtained by considering the triangle AJD minus the work done from the corner triangle.

$$M_r(L - 2a) = \frac{w_c L}{2}\frac{L}{2}\frac{L}{6} - \frac{\sqrt{2}\,abw_c}{2}\left(\frac{b}{3\sqrt{2}} + \frac{a}{\sqrt{2}}\right) = \left[\frac{L^3 - 4ab^2 - 12a^2b}{24}\right]w_c$$

The value of a is when compared with b so a^2 can be neglected when compared with the other values. This equation can be rewritten as:

$$4ab^2 = L^3 - \frac{24M_r(L - 2a)}{w_c}$$

(c)

The substitution of Eq. (b) in the above equations gives:

$$4a\left(\frac{6M_r}{w_c}\right) = L^3 - \frac{24M_r(L - 2a)}{w_c}$$

$$\frac{24M_r(L - 2a + a)}{w_c} = L^3 \text{ or}$$

$$w_c = \frac{24M_r(L - a)}{L^3}$$

(9.27)

This expression assume that the negative bending moment capacity at the corners of the slab over a distance of a is assume to be same as that of the positive bending moment capacity. In case the value of a is zero the solution reduces to that of a simply supported square slab.

9.10 CIRCULAR SLAB

It is easier to work with polar coordinate system in circular slabs. The moment capacities along the radial and along the circumferential directions are taken equal. The reinforcement can be placed along the radial and circumferential directions or along the x and y directions.

Figure 9.18(a) illustrates a circular slab with radial and tangential directions. It also indicates small infinitesimal element that is enlarged in Fig. 9.18(b). Piercing arrow indicates the moment along the radial or tangential direction. It is assumed that the student is familiar with the notations of circular slabs so not much explanation is indicated in the text or in the illustrations. The collapse of a circular slab occurs with radial yield lines.

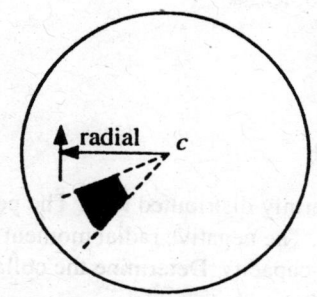
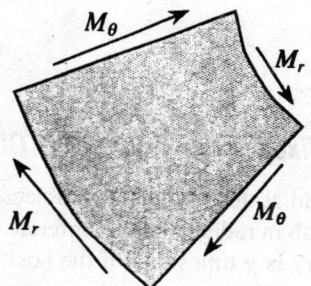

(a) Circular slab with typical element   (b) Typical small element with moments

Fig. 9.18 Circular Slab in Plan with Notations.

Example 9.10: Simply Supported Circular Slab

Determination of collapse load on a simply supported circular slab subjected to uniformly distributed load.

Solution

The slab deforms into a conical shape, typical line pattern of a simply supported slab is shown in Fig. 9.19(a). A typical segment of the yield line included in an angle of dq is shown in Fig. 9.19(b). dq is considered to be infinitely small angle that will tend to zero in the limit. The yield lines are very close, more or less continuous, as the bending moment is continuous. The collapse load can be calculated from the equilibrium of the thin segment. The segment can be considered as a triangle with included angle as dq. The base of the triangle is then $R\,dq$. The moment equilibrium of the infinitesimal triangle is taken about the base. The bending moment Mq when projected on to the base of the sector is equal to the intensity of the moment multiplied by the projected base of the two radii of the segment. The moment caused by the load on the triangular segment about the base is acting in the opposite direction. The equilibrium about the base is:

$$M_\theta \big(R(d\theta)\big) = w_c \frac{1}{2}(Rd\theta)(R)\frac{R}{3}$$

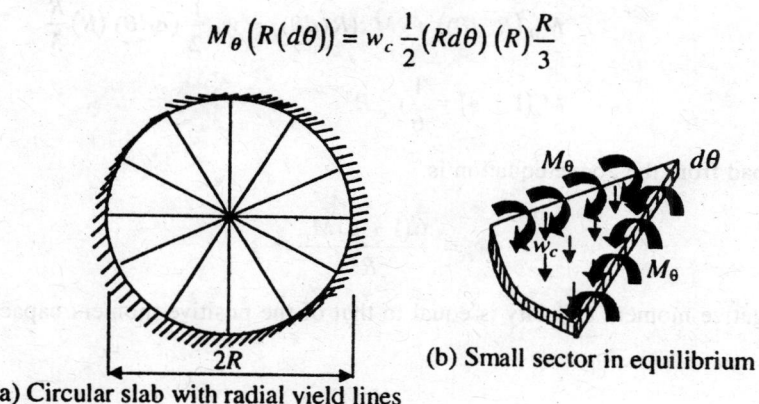

(a) Circular slab with radial yield lines

(b) Small sector in equilibrium

Fig. 9.19 Simply Supported Circular Slab with Radial Yield Line; Example 9.9.

$$w_c = \frac{6M_\theta}{R^2} \qquad (9.29)$$

Example 9.11: Fixed Edge Supported Circular Slab

A circular slab fixed at the periphery is subjected to a uniformly distributed load. The positive moment capacities of the slab in radial and circumferencial are same. The negative radial moment capacity of the slab at the periphery is g times that of the positive moment capacity. Determine the collapse load.

Solution

The problem is similar to the example 9.9 except that the edge is fixed instead of simply supported. Let the positive bending moment capacity in the radial and circumferencial direction = $M_r = M\theta$

Let the negative bending moment capacity in the radial direction at periphery = gM_r

Figure 9.20 illustrates an infinitesimal sector of the slab with positive and negative bending moments. The equilibrium of the forces acting on the sector taken about the base of the sector is:

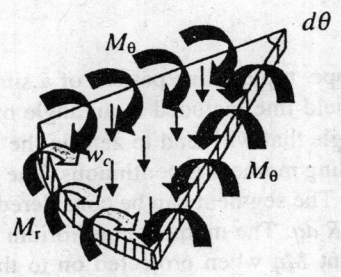

Fig. 9.20 Small Sector of Fixed Edge Circular Slab in Equilibrium; Example 9.11.

$$M_\theta \left(R(d\theta) \right) + gM_r \left(R(d\theta) \right) = w_c \frac{1}{2} (Rd\theta)(R)\frac{R}{3}$$

Or

$$M_r(1 + g) = \frac{1}{6} w_c R^2$$

The collapse load from the above equation is:

$$w_c = \frac{6(1 + g)M_r}{R^2} \qquad (9.30)$$

In case the negative moment capacity is equal to that of the positive moment capacity, $(g = 1)$, the collapse load:

$$w_c = \frac{12M_r}{R^2} \qquad (9.31)$$

The collapse load on a fixed edge circular slab is twice as much as that of a simply supported slab.

Example 9.12: Simply Supported Circular Slab Subjected to Concentrated Load at Centre

A simply supported circular slab is subjected to concentrated load at the centre of the slab. Determine the collapse load.

Solution

Figure 9.21 illustrates an infinitesimal sector of the slab along with the circumferential bending moment acting along the radius of the element. The load is concentrated load and its share on the sector is obtained by proportioning the load as per the included angle.

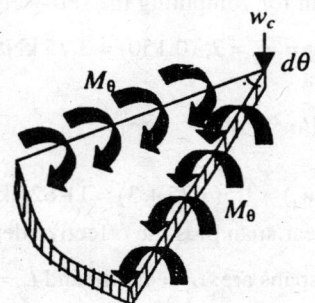

Fig. 9.21 Small Sector of Circular Slab in Equilibrium with Central Load; Example 9.12.

The corresponding load is:

$$w_c = \frac{W_c\,(d\theta)}{2\pi}$$

in which W_c = concentrated load at centre.
The equilibrium of the moments acting on the sector taken about the base is:

$$R(d\theta)M_\theta = w_c R = \frac{W_c\,(d\theta)}{2\pi}R$$

The collapse load from the above expression is:

$$W_c = 2\pi M_\theta = 2\pi M_r \qquad\qquad (9.32)$$

9.11 DESIGN OF SLABS BASED ON THE YIELD LINE THEORY

The yield line theory of slab assumes total ductility of the slab. Such ductility exists in steel structures, however the reinforced concrete slabs are not ductile. Concrete is brittle and it is possible that some portion of the concrete that is stressed reaching the limiting strain even before the redistribution of the moments takes place in the slab. The code of practice allows only a limited redistribution of moments in concrete structures. The bending moments in most slabs can be found easily by yield line theory. Some examples are presented as an exercise. It is possible that the codes will accept the design by yield line theory in due course with suitable redistribution factors.

Example 9.13: Design of Rectangular Slab by Yield Line Theory

A simply supported slab in mild environment with unrestrained corners with clear spans of 4.0 by 4.5 m is subjected to a uniformly distributed live load of 3 kN/m². Design the slab with $M25$ concrete and $HYSD$-Fe415 steel.

Solution

The given data of the example is:

Clear spans are: $L_y = 4$ m, and $L_x = 4.5$ m. The other data is $w_1 = 3$ kN/m², $f_{ck} = 25$ N/mm², $f_y = 415$ N/mm²,
Let the thickness of the slab is 150 mm for computing the self-weight:

$$\text{Self weight} = w_g = 25(0.150) = 3.75 \text{ kN/m}^2,$$

Add extra for floor finish: 1 kN/m².
Then the total dead weight is 4.75 kN/m².
The factored ultimate load is:

$$w_t = 1.5(w_d + w_1) = 1.5(4.75 + 3) = 11.625 \text{ kN/m}^2.$$

The effective spans of the slabs are clear span plus the effective depth of the slab of about 0.1m.

The effective spans are: $L_y = 4.1$ m, and $L_x = 4.6$ m.

The collapse load relation based on the yield line theory is given in Eq. 9.21 and it is:

$$M_r = \frac{1}{24}\left[\frac{\sqrt{(r + 3c^2)} - \sqrt{r}}{c}\right]^2 w_c L^2$$

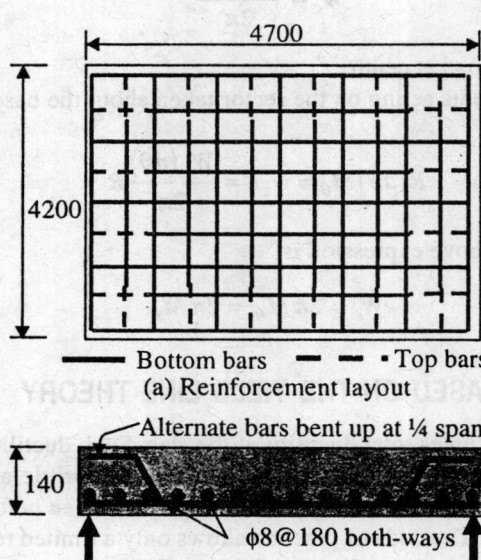

Bottom bars —— — — • Top bars
(a) Reinforcement layout

Alternate bars bent up at ¼ span

φ8 @ 180 both-ways
(b) Cross section-reinforcement

Fig. 9.22 Slab Reinforcement; Example 9.13.

Where $$c = \frac{L_x}{L_y} = \frac{4.6}{4.1} = 1.12$$

And r is the ratio of the moment capacities of the slab in the two directions. $r = M_{rx}/M_{ry}$

The aspect ratio is 1.12, further the reinforcement in the long span direction will be placed above that of the short span direction. Therefore effective depth of the reinforcement in the long span direction will be atleast 10 mm less than that of the short span direction. It is therefore assume the moment capacity of the slab in both the directions same. That is $r = 1$. The substitution of the values of c and r in the collapse load relation gives.

$$M_r = \frac{1}{24} \left[\frac{\sqrt{(r + 3c^2)} - \sqrt{r}}{c} \right]^2 w_c L^2 = \frac{1}{24} \left[\frac{\sqrt{1 + 3.7632} - 1}{1.12} \right]^2 w_c L^2 = 9.096 \, \text{kNm/m}$$

The collapse moment is equated to limit moment capacity of the section to obtain the thickness required.

$$Kbd^2 f_{ck} = M_r = 9.096(10^6)$$

In which $K = 0.138$, $b = 1000$ mm, $f_{ck} = 25$ N/mm². The substitution of these values in the above relation gives:

$$d = \sqrt{\frac{9.096(10^6)}{0.138(1000)25}} = 52 \, \text{mm}$$

The effective depth required is only 52 mm but it is too small, demands higher percentage of reinforcement and may have large deflection. It is therefore recommended to select 105 mm effective depth for the reinforcement in the long span. The effective depth for the short span will be 115 mm. Provide a cover of 20 mm, the overall depth of the slab works out to be $115 + 5 + 20 = 140$ mm. This value is quite close to the depth assumed for the purpose of the dead weight. The area of reinforcement required in the short span direction is:

$$A_{st} = \frac{1.15 M_c}{jdf_y} = \frac{1.15(9.096)10^6}{0.8(115)415} = 274 \, \text{mm}^2/\text{m}$$

Provide 8mm bars at 180 mm spacing in the short span direction, also in long span. The actual reinforcement provided is: $A_{st} = 50/0.180 = 277$ mm²/m

The percentage of the reinforcement is: $$p = \frac{277 * 100}{140 * 1000} = 0.2$$

The spacing of the reinforcement must be: less than twice the thickness of the slab. This condition is also satisfied. Figure 9.21 illustrates the layout of the reinforcement and a cross-section across the width of the slab.

Example 9.14: Design of Rectangular Slab with Fixed Edges by Yield Line Theory

A simply supported slab exposed to moderate environment with clear spans of 4.0 by 4.5 m has fixed edges along the long span. It is subjected to a uniformly distributed live load of 3 kN/m². Design the slab with M25 concrete and HYSD-Fe415 steel.

Solution

The given data of the example is:

Clear spans are: $L_y = 4$ m, and $L_x = 4.5$ m. The other data is: $w_1 = 3$ kN/m², $f_{ck} = 25$ N/mm², $f_y = 415$ N/mm², and moderate environment. Moderate environment means the cover has to be 30 mm.

Let the thickness of the slab be 150 mm for the purpose of computing the self-weight

$$\text{Self weight} = w_g = 25(0.150) = 3.75 \text{ kN/m}^2,$$

Add extra for floor finish of about 40mm thick that corresponds to 1 kN/m². Then the total dead weight is 4.75 kN/m². The factored ultimate load is:

$$w_t = 1.5(w_d + w_1) = 1.5(4.75 + 3) = 11.625 \text{ kN/m}^2.$$

The effective spans of the slabs are clear span plus the effective depth of the slab of about 0.1m. The effective spans are: $L_y = 4.1$ m, and $L_x = 4.6$ m.
The collapse load relation based on the yield line theory is given in Eq. 9.25 and it is:

$$w_c = \frac{48cM_r}{(3c - 2)L^2}$$

Where
$$c = \frac{L_x}{L_y} = \frac{4.6}{4.1} = 1.12$$

Further the ratio of the positive moment capacity at mid span is assumed equal to the negative moment capacity of the slab. The substitution of the value of c and other values in the collapse load relation gives:

$$M_r = \frac{(3c - 2)L^2 w_c}{48c} = 4.95 \text{ kNm/m}$$

The collapse moment required is equated to limit moment capacity of the section.

$$Kbd^2 f_{ck} = M_r = 4.95(10^6)$$

In which $K = 0.138$, $b = 1000$ mm, $f_{ck} = 25$ N/mm². The substitution of these values in the above relation gives:

$$d = \sqrt{\frac{4.95(10^6)}{0.138(1000)25}} = 38 \text{ mm}$$

The effective depth required is only 38 mm but it is too small. It is therefore recommended to select 105 mm effective depth for the reinforcement in the long span. The effective depth for the short span will be 115 mm. Provide a cover of 30 mm, the overall depth of the slab works out to be $115 + 5 + 30 = 150$ mm. This value is same as that assumed for the dead weight. The area of reinforcement required in the short span direction is:

$$A_{st} = \frac{1.15 M_c}{jdf_y} = \frac{1.15(4.95)10^6}{0.8(115)415} = 150 \text{ mm}^2/\text{m}$$

The minimum percentage of the reinforcement is:

$$A_{smin} = 0.12(150 * 1000)/100 = 180 \text{ mm}^2/\text{m}$$

The minimum reinforcement controls the design. Select 8 mm bars.

Provide 8 mm bars at 275 mm spacing in the short span direction. The spacing is also used in the long span direction. The actual reinforcement provided is:

$$A_{st} = 50/0.275 = 181 \text{mm}^2/\text{m}$$

The percentage of the reinforcement is:

$$p = \frac{181 * 100}{150 * 1000} = 0.12$$

Figure 9.23 illustrates the layout of the reinforcement and a cross-section across the width of the slab.

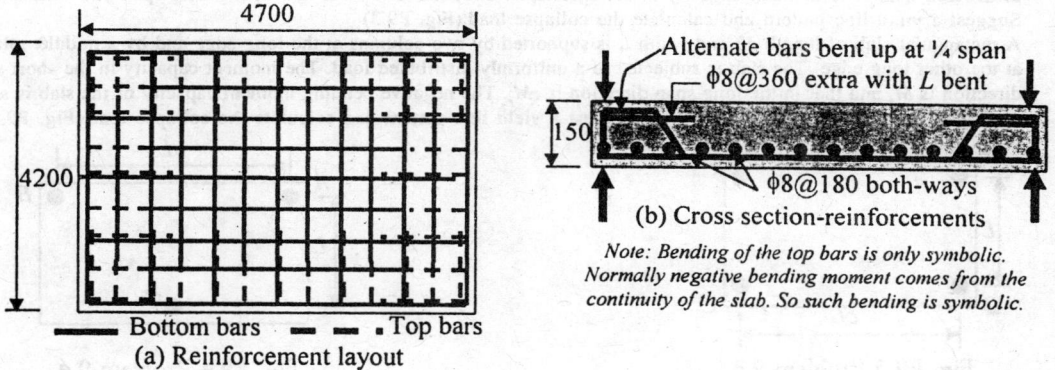

Note: Bending of the top bars is only symbolic. Normally negative bending moment comes from the continuity of the slab. So such bending is symbolic.

Fig. 9.23 Slab Reinforcement; Example 9.14.

Top Reinforcement

Reinforcement must be provided at the top face of the long edges to resists the negative bending moment. The reinforcement at top near the edge support can be same as that at the mid-span. This reinforcement is provided by bending the alternate bottom bars to the top near the end of the span say at one-fourth of the span. The balance of the reinforcement that is needed at top can be provided by extra reinforcement. 8 mm bars at 360 mm for one-third of the span. The negative reinforcement must have adequate development length at both ends. The bars must be bent down and taken in the reverse direction to meet the development length or taken in to the next slab. The development length of 8 mm bars in $M25$ concrete is 328 mm. The depth of the slab is only 150 mm of which the actual down bent length is only (150 - 2*30 – 8 =) 82 mm. The fixidity at the slab at support is normally due to the continuity of the slab. The top bars normally extend into the next span. Alternate bottom reinforcement bar can be bent up at one-fourth of the span.

PROBLEMS

9.1 A rectangular slab of length cL and width L is supported at one long edge and supported by two columns at the other edge. The slab is subjected to a uniformly distributed load over the entire area. The moment capacity in the short span direction is M_r and that in the long span direction is rM_r. Suggest a yield line pattern and calculate the collapse load.(Fig. P9.1)

9.2 A rectangular slab of length cL and width L is supported on one long edge and by a column at the middle of the other long edge. The slab is subjected to a uniformly distributed load. The moment capacity in the short span direction is M_r and that in the long span direction is rM_r. The negative bending moment capacity of the slab is same as that of the positive bending moment. Suggest a yield line pattern and calculate the collapse load.(Fig. P9.2)

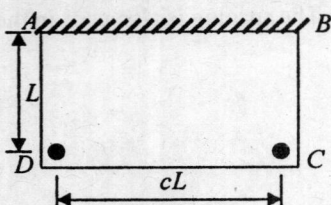

Fig. P9.1 Problem 9.1.

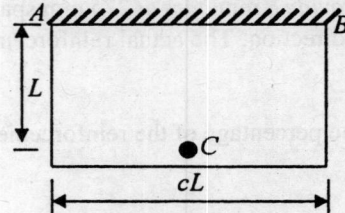

Fig. P9.2 Problem 9.2.

9.3 Four corners columns support a rectangular slab of length cL and width L. The slab is subjected to a uniformly distributed load. The moment capacity in the short span direction is M_r and that in the long span direction is rM_r. Suggest a yield line pattern and calculate the collapse load.(Fig. P9.3)

9.4 A rectangular slab of length cL and width L is supported by two columns at the long edge and by a middle column at the other long edge. The slab is subjected to a uniformly distributed load. The moment capacity in the short span direction is M_r and that in the long span direction is rM_r. The negative bending moment capacity of the slab is same as that of the positive bending moment. Suggest a yield line pattern and calculate the collapse load.(Fig. P9.4)

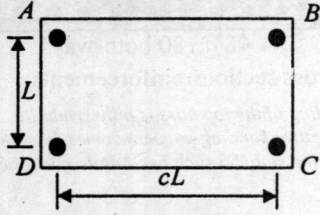

Fig. P9.3 Problem 9.3.

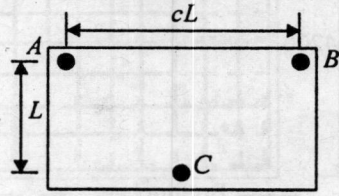

Fig. P9.4 Problem 9.4.

9.5 A rectangular slab shown in Fig. P9.5 is simply supported at one edge and the opposite edge is fixed. The other two edges are free. The slab is subjected to a uniformly distributed load. Determine the collapse load based on the yield line theory. The positive and negative moment capacity in the short span direction is M_r and that in the long span direction is one fourth of that in the short span.

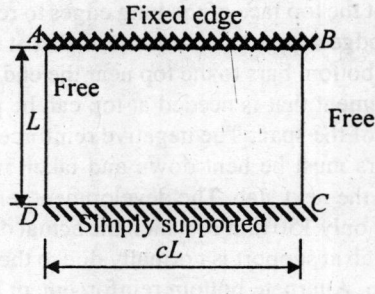

Fig. P9.5 Problem 9.5.

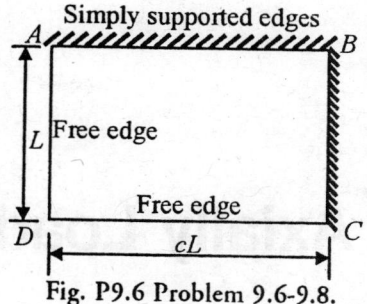

Fig. P9.6 Problem 9.6-9.8.

9.6   A rectangular slab as shown in Fig. P9.6 is simply supported on two adjacent edges and the other two adjacent edges are free. The slab is subjected to a uniformly distributed load, determine the collapse load based on the yield line theory. The moment capacity in both the directions is M_r.

9.7   A rectangular slab as shown in Fig. P9.6 is simply supported on two adjacent edges and the other two adjacent edges are free. The slab is subjected to a concentrated load of W at the free corner D. Illustrate the collapse mechanism and determine the collapse load based on the yield line theory. The positive and negative moment capacity in both the directions is M_r.

9.8   A rectangular slab as shown in Fig. P9.6 is simply supported on two adjacent edges and the other two adjacent edges are free. The slab is subjected to a line load of w per unit length along the diagonal BD and no other load is acting on the slab. Illustrate the collapse mechanism and determine the collapse load based on the yield line theory. The positive and negative moment capacity in both the directions is M_r.

9.9   A simply supported slab having six equal sides as shown in Fig. P9.7 is subjected to uniformly distributed load. The positive moment capacity in the two principal directions is Mr. Illustrate the yield line pattern of the slab and calculate the collapse load.

9.10   A circular slab of radius R and simply supported around is loaded with a central circular column having a radius equal to one-tenth of that of the slab. The load from the column including the weight of the column is W. The positive and negative moment capacity in radial and circumferencial directions is M_r. Determine the collapse load by yield line theory.

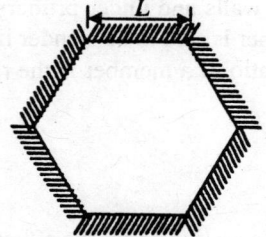

Fig. P9.7 Problem 9.9 & 9.13.

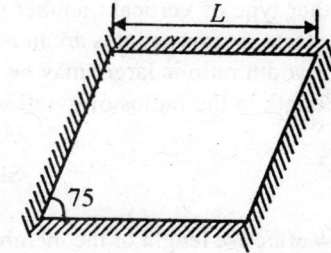

Fig. P9.8 Problem 9.11 & 9.12.

9.11   A skew slab of equal sides of span L as shown in Fig. P9.8 and simply supported on all four edges is subjected to a uniformly distributed load. The positive moment capacities in two principal directions are M_r. Illustrate the yield line pattern and calculate the collapse load of the slab.

9.12   A skew slab of equal sides of span L as shown in Fig. P9.8 and simply supported on all four edges is subjected to a line load of w along the short diagonal. The positive moment capacity in two principal directions is M_r. Illustrate the yield line pattern and calculates the collapse load of the slab.

9.13   Slab having six equal sides as shown in Fig. P9.7 is fixed on all the sides and subjected to uniformly distributed load. The positive moment capacities in the two principal directions inside the slab is M_r. The negative moment capacities at all supports is M_r. Illustrate the yield line pattern of the slab and calculate the collapse loads.

Design of Axially Loaded Columns and Struts

10.1 INTRODUCTION

Column is an upright or almost upright member subjected to compression with or without bending moments. Column transfers forces either directly or through another structural system to the foundation. A compression member oriented in an arbitrary direction and not subjected to bending moment is called *strut,* however secondary bending moment may exist. A column that is founded on a wall or a frame is referred as *floating column*. Floating column must be interconnected to the structural system to ensure safe transfer of the lateral loads to the ground. Upright members that support bridge decks are called *piers*. A very short column having height to side ratio less than four is called a *stub column*. Enlarged portion of column or short pillars base over the foundation is called *Pedestal*. Poles, pillars and posts are upright slender members and are subjected to dominant bending moment and nominal compression force. *Pilaster* is another type of vertical member interconnecting two adjacent walls and under primary bending. Columns, struts, beams and ties are mostly slender members. A member is said to be slender if its length to depth or width ratio is large, may be more than six. Slenderness ratio of a member is the ratio of the effective length to the radius of gyration of the section.

$$\text{Slenderness ratio} = k = \frac{L_e}{r} \tag{10.1}$$

Where L_e = effective length of the member, r = radius of gyration of the section about the effective length axis.

The radius of gyration of a section is equal to the under root-square of the ratio of the moment of inertia and the area of the cross section. It is expressed as:

$$\text{Radius of gyration} = r = \sqrt{\frac{I}{A}} \tag{10.2}$$

Where I = second moment of area, sometimes simply called moment of inertia, A = area of cross section.

The cross section of slender members has two axes so there are two moments of inertii. Similarly, the supports to the column can be different for the two axes.

The radius of gyration about each axis is defined separately. The higher slenderness ratio about the axes controls the design. The size of the masonry column is usually large when compared with that the metal ones. Instead of slenderness ratio, another factor called slenderness factor is used in the design of masonry and reinforced concrete columns.

10.2 BUCKLING OF COLUMNS

A member under compressive force may fail either by crushing or by buckling. Column deflects laterally if a lateral force is applied and the deflection is recovered on the removal of the lateral load with in the elastic limit. An axial load for which an undefined magnitude of deflection takes place is called *Buckling load*. The buckling load is also referred as *critical load*. Consider a column subjected to axial load *P* as shown as an inset in Fig. 10.1. As the load increased, only axial compression of the column takes place upto a point. At the buckling load, the column undergoes buckling and lateral deflection of undefined magnitude. A horizontal line on the figure indicates the lateral deflection of the column and instability of the column. In case the column is also subjected to transverse load, the column deflects as shown in the *bending problem curve* of the Fig. 10.1. The buckling of a column is initiated in the plane about which the slenderness ratio is largest. In case of reinforced concrete or masonry columns, the slenderness ratio is usually less than 60 and such columns are likely to get crushed rather than buckled. However, a reduction in the load carrying capacity of the column must be made to the slenderness effect in columns. For this purpose another factor that is called slenderness factor (sometimes improperly called as slenderness ratio) is defined. The ratio of the effective length to the depth or width of the column is called *slenderness factor*. It can be expressed as:

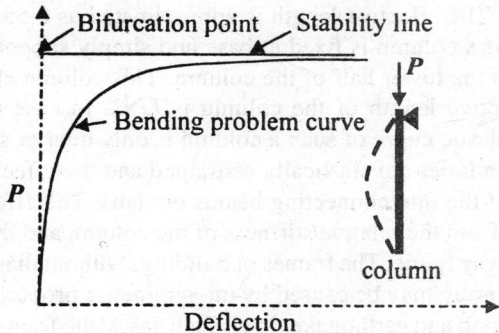

Fig. 10.1 Phenomenon of Buckling of Column.

$$\text{Slenderness factor} = g = \frac{L_e}{d} \qquad (10.3)$$

Where d = depth of the cross section of the column about which the column is likely to buckle (lateral dimension about which the buckling is likely to take place), and L_e = effective length of the column in the plane of the buckling.

The effective height (also referred as effective length) of the column plays an important role. A column hinged at both ends bends into a simple half *sine wave* curve during buckling as shown in Fig. 10.2(a). The height of a column between the two consecutive points of zero bending moments is same as the distance between the supports. Consider the column fixed at both the ends that can be called as fixed ended column as shown in Fig. 10.2(b). At the state of buckling, the column bends into a compound curve as it is restrained against sloping at one end. The top one-fourth portion of the bent curve has positive curvature and then the curvature changes to negative until one-fourth height from bottom. The balance

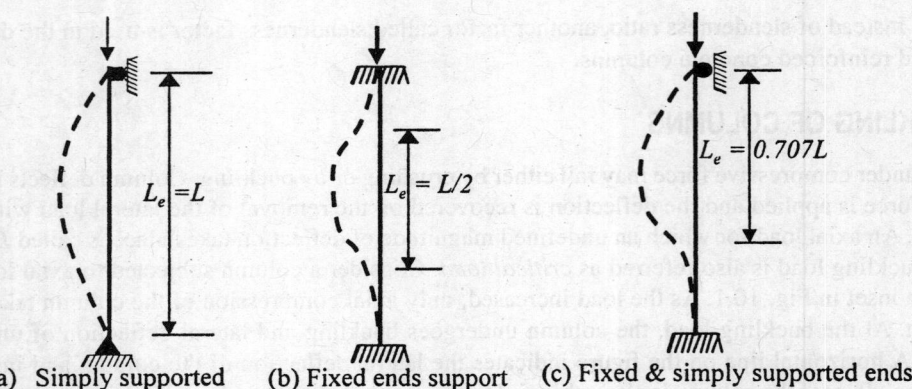

a) Simply supported   (b) Fixed ends support   (c) Fixed & simply supported ends

Fig. 10.2 Typical Buckled Shapes for Different Boundary Conditions.

of the curve has a positive curvature. The bending moments in the outer one fourth of the length of the column are negative while that at the middle half is positive. The column elastic curve has two points of contra flexure at one fourths length from the supports. The distance between the two successive points of the elastic curve is the effective height of the column. So in a fixed ended column, the theoretical effective height of the column is half of the distance between the supports. The theoretical effective height of the column is $0.5L$. The effective length is approximated as $0.65L$ to be on the safer side for practical purpose. Similarly, if a column is fixed at base and simply supported at top then one point of contra flexure is developed in the lower half of the column. This column elastic curve is shown in Fig. 10.2(c). The theoretical effective length of the column is $L/\sqrt{2}$. In case of a cantilever column, the effective height is 2L as the elastic curve of such a column is only quarter sine wave.

In most structures, the boundaries are elastically restrained and the effective length of the member is a function of the properties of the interconnecting beams or slabs. The effective length of a column in framed structures is obtained from the relative stiffness of the column and the connecting members. The frame may be a sway or non-sway frame. The frames or buildings without diagonal braces or without shear walls will sway laterally. The sway may be caused by un-symmetric properties of the frame or eccentric loads or lateral loads such as wind and earthquake. In all such cases, the frame sways if not restrained. The sway in the column increases the effective length of the column. On the other hand, the buildings with diagonal bracing or with shear walls don't sway or sway negligibly. In such frames, the effective length of the column is less than the un-supported height of the column. The designer is likely to ignore the effective length of the columns in concrete frames and consequently the design the column is not accurate. Code of practice on plain and reinforced concrete gives the effective lengths of columns as a function of the relative stiffness. Table 10.1 gives the effective lengths of some commonly encountered columns.

The radius of gyration of rectangular sectioned column is:

$$r = \sqrt{\frac{I}{A}} = \sqrt{\frac{bD^3}{12bD}} = \frac{D}{\sqrt{12}} \qquad (10.4)$$

The slenderness factor divided by the slenderness ratio for rectangular section gives:

$$\frac{g}{k} = \sqrt{12} = 3.46$$

Table 10.1 Typical Idealized Effective Length of Columns.

Boundary conditions	Effective length
1. Simply supported at both the ends	L
2. Fixed at both the ends	0.65 L
3. Fixed at one end and hinged at the other	0.8 L
4. Fixed at one end and constrained against rotation at the other end	1.2 L
5. Fixed at one end and free at the other end	2L
6. Columns in portal frames with fixed bases and having lateral sway	1.5 L
7. Columns in portal frames with hinged bases and having lateral sway	2 L
8. Interior columns in multi-storey frames above GL	0.8 L
9. Exterior columns in multi-storey frames above GL	1.2L
10. Crane carrying columns in braced buildings	L

This ratio for circular columns is 4. The concept of slenderness factor rather than the slenderness ratio gives more flexibility to architect to visualize the effective length of the columns more easily. A column having a slenderness factor less than 12 is called short column, and beyond that are long columns. The maximum slenderness factor for load carrying columns is limited to 60. The problem of amplification of the deflection due to axial load in columns is called *P-D effect*. Drift problem becomes more dominant in very long columns with lateral sway in the building.

10.3 TYPES OF COLUMNS AND OTHER LIMITS

Even though a rigid demarcation doesn't exist, the following guidelines are suggested for and nomenclature of compression members.

Table 10.2 Some Nomenclature of Compression Members.

Member type	Max. slenderness Factor
Stub column/Pedestal	4
Short column	12
Long column	60

The stub columns and pedestals are very short and fail by crushing only. Non-load bearing architectural columns may be designed with slenderness factor more than 60. Columns are classified in to four types based on the structural configuration:

(a) Tied column,
(b) Spiral column,
(c) Composite column,
(d) Infilled column.

Tied Column: A reinforced concrete column in which the main bars are interconnected by closed loops or links is called tied column. The interconnecting loops or links are called transverse reinforcement or ties. Figure 10.3 illustrates typical cross section of the tied column.

A column in which metal structural sections are embedded in reinforced concrete is called as *composite column*.

Steel tubes filled with reinforced concrete are called *infilled columns*. Infilled columns are used for architectural purpose and some times for construction convenience.

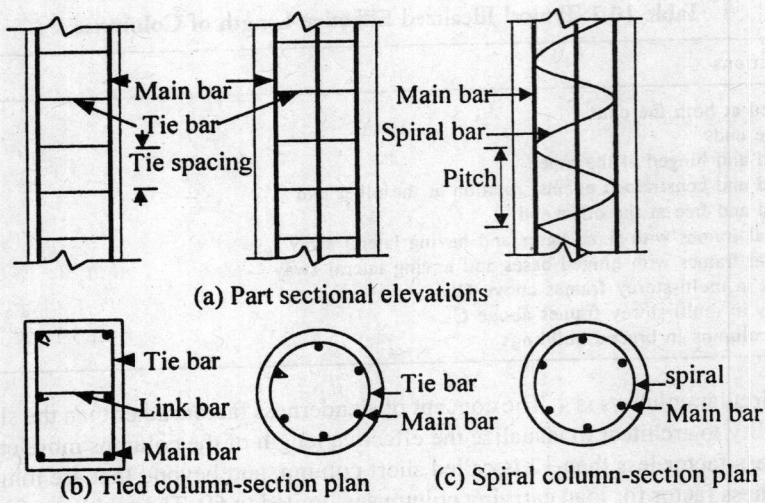

(a) Part sectional elevations

(b) Tied column-section plan   (c) Spiral column-section plan

Fig. 10.3 Tied Spiral Columns-plan and Elevation Sections.

The theory of limit state design of reinforced concrete columns is based on the following assumptions:

➤ The column is slender and the plane sections remain plane even after loading,
➤ The reinforcement is distributed to behave like a homogeneous and isotropic body,
➤ The strain compatibility between the concrete and the reinforcement is normally considered but ignored in equilibrium model,
➤ Adequate cover to the main reinforcement and transverse ties is provided to avoid local buckling of the bars,
➤ Effect of shrinkage and creep are considered negligible. Certain minimum provisions must be made in the column to ensure it satisfy the assumptions and can with stand the exposure conditions.

Maximum percentage of reinforcement is limited to avoid honeycombing. Minimum cover to the reinforcement is provided to ensure durability and avoid local buckling of the main reinforcement bars. Some of these limits are listed here for ready reference.

The limits of percentages of the main reinforcement are given by:

$$0.8 \% \leq p \leq 6 \% \qquad (10.5)$$

where p = percentage of the main reinforcement,

Minimum diameter of main reinforcement bar = 12 mm (Exception in very thin columns)
Minimum number of bars in rectangular section = 4
Minimum number of bars in circular or polygonal section = 6

The minimum requirements for the transverse reinforcement are:

The diameter of the tie should be ≥ one fourth the diameter of the main bar or
The diameter of the tie should be ≥ 5 mm, preferably 6 mm   (10.6)
The spacing of the ties should be ≤ minimum size of the cross section   (10.7)
The spacing of the ties should be ≤ 16 times the diameter of the main reinforcement bar
The spacing of the ties should be ≤ 48 times the diameter of the tie bar itself

In case of spiral columns, the transverse reinforcement is a spiral going round the all the bars on a circle. The pitch of the spiral must satisfy the following conditions.

$$Pitch \leq 75 \text{ mm } and$$
$$Pitch \leq One\text{-}sixth \text{ the diameter of the core further} \tag{10.8}$$
$$Pitch \geq 25 \text{ mm } and$$
$$Pitch \geq three \text{ times the diameter of the spiral bar} \tag{10.9}$$

The strength of the core concrete is enhanced by the confinement through the spiral reinforcement. The maximum size of the bar is normally restricted to 40 mm. Not more than one third of the bars can be lapped at a location. Providing laps to all main bars at beginning of a storey is a wrong practice. The columns in a floor are subjected to maximum bending and axial loads at the base of the floor, so avoid laps at this level.

The maximum admissible compressive strains in concrete of reinforced concrete are:.

Axial compressive strain $= 0.002$
Bending compressive strain $= 0.0035$
Maximum compressive strain in combined bending and axial force $= \varepsilon_{c1} = 0.0035 - 0.75\varepsilon_{c2}$ (10.10)

Where ε_{c2} = minimum compressive strain on the section.

10.4 AXIALLY LOADED TIED COLUMNS

The crushing strength of concrete prism when compared with the characteristic strength of 150 mm cube is:

$$f_{cc} = k_p f_{ck} = 0.67 f_{ck} \tag{10.15}$$

where f_{ck} = the characteristic strength of concrete (cube) k_p = prism strength reduction factor.

The design capacity of concrete in columns is obtained by dividing the crushing capacity of prism by a partial safety factor applied to the material and it is given by:

$$f_{cd} = k_p f_{ck}/\gamma_m = 0.67 f_{ck}/1.5 = 0.446 f_{ck}$$

where γ_m = partial safety factor applied to material = 1.5

The above factor 0.446 is approximated to 0.40 in case of axially loaded columns to account for possible eccentricity of load or other idealizations of supports.

The design capacity of reinforced concrete short column is obtained by the sum of the compression capacities of concrete and steel divided by the partial safety factor applied to the material and it is given by:

$$P_u = 0.4 A_c f_{ck} + 0.67 A_s f_y \tag{10.16}$$

where A_c = area of concrete , A_s = area of steel reinforcement, f_y = yield or proof strength of steel.
The axial load capacity of the short column can be rearranged as:

$$P_u = 0.4 f_{ck} \left(A_g - A_s \right) + 0.67 A_s f_y \tag{10.17}$$

where A_g = gross area of cross section
The above equation can be rearranged in ratio of the reinforcement as:

$$P_u = 0.4 f_{ck} \left[1 + \frac{A_s}{A_g} \left(1.67 \frac{f_y}{f_{ck}} - 1 \right) \right] \tag{10.18}$$

Let the ratio of the area of reinforcement to that of the gross area is:

$$p = \frac{A_s}{A_g} \tag{10.19}$$

and the strength ratio is $= \alpha = \dfrac{f_y}{f_{ck}}$ \hfill (10.20)

The above axial load capacity of a short column can be expressed as:

$$P_u = 0.4 A_g f_{ck} \left[1 + p(1.67\alpha - 1) \right] \tag{10.21}$$

The design criterion of short a column is that the design strength must be more than the factored load acting on the column, and it is expressed as:

$$P_u \geq P_d \tag{10.22}$$

All columns must be designed for a minimum eccentricity. The minimum eccentricity specified by the code is:

$$e_0 = \frac{\text{Lateral dim}\,ension}{30} + \frac{column\ length}{500}\ \ or\ subject\ to\ \min inimum\ of \tag{10.23}$$

Example 10.1: Design of Short Tied Column

A column in moderate exposure environment is resting on a footing and supports a beam-slab construction. The working load on the column is 690 kN. The height of the beam-slab above the footing is 4.5 m. Design the column with $M25$ concrete and $HYSD$ 415 bars.

Solution

The column is in moderate exposure condition so the minimum grade of concrete is $M25$. The characteristic strengths of concrete and steel are: $f_{ck} = 25$ MPa and $f_y = 415$ MPa.

To start with the column is assumed as short one with one percent reinforcement. Assuming the self-weight of the column as 10 kN, the design load on the column with 1.5 partial safety factor is:

$$P_d = \gamma_f (W_g + W_l) = 1.5(690 + 10) = 1050 \text{ kN}$$

The loading carrying capacity of the column is:

$$P_d = 0.4 A_g f_{ck} \left[1 + p(1.67\alpha - 1) \right]$$

in which $p = 0.01$(one percent assumed), and $\alpha = 415/25 = 16.6$. The capacity expression results:

$$P_d = 0.4(25)(1 + 0.01(1.67(16.6 - 1)))A_g = 12.6 A_g$$

Equating the capacity of the column to the factored design load, we have:

$$P_u = 1050 \text{ kN} = 1050000 \text{ N} = P_d = 12.6A_g$$

The gross area of concrete section is obtained from the above and it is:

$$A_g = 1050000/12.6 = 83333.3 \text{ mm}^2$$

The size is rounded off to 300 by 300 mm.

$$P_u = 0.4A_c f_{ck} + 0.67 A_s f_y = 0.4(300)300*(0.99)(25) + 0.67(415)A_s = 1050000$$

The area of reinforcement is:

$$A_s = \frac{1050000 - 0.4*300*300*0.99*(25)}{0.67(415)} = 540 \text{ mm}^2$$

Select four numbers of 16 mm bars, then the area of reinforcement provided is: $A_s = 4(201) = 804 \text{ mm}^2$
The percentage of reinforcement provided is $= p\% = 804(100)/(300(300)) = 0.9\%$.
The weight of the column is: $W_g = (0.3*0.3)(4.5)(25) = 10$ kN
So the assumed weight of the column is reasonable. In most cases the weight of the column is negligible when compared with that of the actual load. One to two percent of the design load can be assumed as self-weight of the column in most cases.

The column is on an independent footing so the footing end of the column can be considered as a hinge even though some constraint is provided at the base of the column from the footing. The column is connected to a beam and slab at the top so it can be considered as a partially restrained. The effective length of the column can be taken as 0.8 times the supported length.
The effective length is: $L_e = 0.8L = 0.8(4.5) = 3.6$ m

The slenderness factor of the column is $\lambda = \dfrac{L_e}{b} = \dfrac{3.6}{0.3} = 12$

The column is a short one as assumed. The minimum eccentricity for which the column has to be designed is:

$$e_{min} = \frac{L}{500} + \frac{b}{30} = 19 \text{ mm it is less than 20 mm, so use 20 mm}$$

The minimum eccentricity ratio is $= 20/300 = 0.067$. This value is more than 0.05; hence, the section should be designed for minimum eccentricity. This example is only to illustrate the design of the axial load design and so the second part is not carried out here but will be illustrated later.

Design of the Ties

The diameter of the tie can be selected about one-fourth of the diameter of the main bars subject to a minimum of 5 mm. Most commonly available bar is 6 mm, so select 6mm bar as tie.
Then the spacing of the ties should be limited to:

$s_v \leq b = 300$ mm or
$s_v \leq 16$ times the diameter of the main bar $= (16)16 = 256$ mm,
$s_v \leq 48$ times the diameter of the tie bar $= 48(6) = 288$ mm

Provide 6 mm ties at 250 mm spacing.

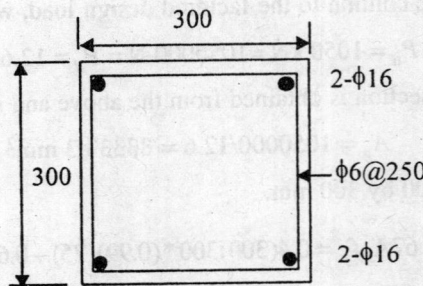

Fig. 10.4 Cross-Section of Column; Example 10.1.

Example 10.2: Design of Reinforcement for a Given the Architectural Size of Column

A working load of 690 kN acts on a column of size 330 by 300 mm. The effective height of the column is 3.6 m. Design the column with M25 concrete and HYSD 415 bars. The exposure condition is moderate.(The problem is similar to example 10.1 except that the size of the column is pre-assigned in this case.)

Solution

The characteristic strengths of concrete and steel are f_{ck} = 25 MPa and f_y = 415 MPa.

The self-weight (assuming the gross height of the column as 4.5m) is = W_g = 0.33*0.3*4.5*25 = 11kN

The factored design load at limit state of strength is:

$$P_u = \gamma_f(W_g + W_l) = 1.5(690 + 11) = 1052 \text{ kN}$$

The capacity of the column is:

$$P_d = 0.4A_c f_{ck} + 0.67A_s f_y$$

Assume about one percent of reinforcement to arrive at the net area of concrete and equate the limit load to the capacity of the section:

$$P_d = 0.99A_g(0.4f_{ck}) + 0.67A_s f_y = P_u = 1052,000 \text{ N}$$

The substitution of various quantities in the above equation and rearrangment gives:

$$A_s = \frac{1052000 - 0.99(330*300)(0.4)25}{0.67*415} = 259 \text{ mm}^2$$

The minimum percentage of reinforcement to be provided is 0.8% and it is equal to:

$$A_{smin} = 0.008*330*300 = 792 \text{ mm}^2$$

Minimum required percent of reinforcement is more than that structurally needed. So, select minimum percentage of steel. Select four numbers of 16 mm bars, then

The percentage of reinforcement provided is = $p\%$ =804(100)/330(300) = 0.81%.

The slenderness factor of the column is: $\lambda = \dfrac{L_e}{b} = \dfrac{3.6}{0.3} = 12$

The column is a short one as assumed. The minimum eccentricity for which the column has to be designed is

$$e_{min} = \frac{L}{500} + \frac{b}{30} = 19 \text{ mm } \textit{it is less than } 20 \text{ mm}, \textit{ so use } 20 \text{ mm}$$

This example is only to illustrate the design of the axial load design and so the second part is not carried out here but will be illustrated later.

Note: The area of steel required for strength is 259 mm² while that provided is 804 mm² based on the minimum requirement. The minimum percentage of steel required is computed for the minimum concrete area that will provide the strength along with the steel. The margin between the capacity and that is really required is very small and hence modification is not made in this example.

Design of the Ties

Most commonly available bar is 6 mm. The spacing of the 6 mm ties should be limited to:

$s_v \le b = 300$ mm or
$s_v \le 16$ times the diameter of the main bar = (16)16 = 256 mm, or
$s_v \le 48$ times the diameter of the tie bar = 48(6) = 288 mm
        Provide 6 mm ties at 250 mm spacing.

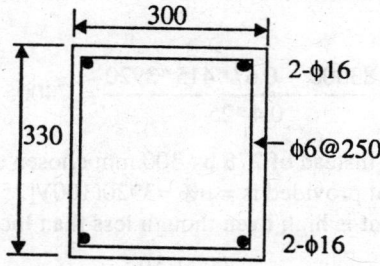

Fig. 10.5 Cross-Section of Column; Example 9.2.

Example 10.3: Column with High Percentage of Reinforcement

A column of 3.0 m effective length is subjected to an axial superimposed load of 1200 kN. The size of the column is to be desired to be as small a possible. Design the column with *M*25 concrete and *HYSD-Fe* 415 bars. The exposure condition is moderate.

Solution

The characteristic strengths of concrete and steel are: f_{ck} = 25 MPa and f_y = 415 MPa

Let the self weight of the column be = 20 kN
The factored design load at limit state of strength is:

$$P_u = \gamma_f(W_g + W_l) = 1.5(1200 + 20) = 1830 \text{ kN}$$

The reinforcement ratio is assumed as $= p = 0.05$ and the strength ratio is $= a = 415/25 = 16.6$. The column capacity expression results:

$$P_d = 0.4(25)(1+0.05(1.67(16.6-1)))A_g = 23A_g$$

Equating the capacity of column to the design load, we have:

$$P_u = 1830000 \text{ N} = P_d = 23A_g$$

The gross area of concrete section is obtained from the above and it is:

$$A_g = 1830000/23 = 79565 \text{ mm}^2$$

The column section is chosen as 275 by 300 mm. The actual area of reinforcement required from Eq. 10.16.

$$P_d = 0.4A_c f_{ck} + 0.67A_s f_y = 0.4(275)300*0.95*(25) + 0.67(415)A_s = 1830000$$

$$A_s = \frac{183000 - 0.4*275*300*0.95*(25)}{0.67(415)} = 3662 \text{ mm}^2$$

Select eight numbers of 25 mm bars, the area of reinforcement provided is: $A_s = 8(490) = 3920 \text{ mm}^2$

The actual area of the reinforcement provided is more than that is needed; use of the actual area of the reinforcement in the governing expression modifies the size of the section. The gross area of the section needed is:

$$A_s = \frac{183000 - 0.67*415*3920}{0.4*25} = 74005 \text{ mm}^2$$

Select 250 mm by 300 mm size instead of 275 by 300 mm chosen earlier.

The percentage of reinforcement provided is $= p\% = 3920(100)/[250(300)] = 5.22\%$.

5.22 percentage of reinforcement is high even though less than the maximum permissible.

The slenderness factor of the column is: $\lambda = \dfrac{L_e}{b} = \dfrac{3.0}{0.25} = 12$

The column is a short one as assumed. The minimum eccentricity for which the column has to be designed is:

$$e_{min} = \frac{L}{500} + \frac{b}{30} = 14 \text{ mm } \textit{it is less than } 20 \text{ mm}, \textit{ so use } 20 \text{ mm}$$

The section should be designed for minimum eccentricity. The design for combined bending is illustrated later.

Design of the Ties

The diameter of the tie can be selected about one-fourth of the diameter of the main bars subject to a minimum of 5mm. Select 6mm bar as tie. Then the spacing of the ties is limited to

$s_v \le b = 250$ mm or
$s_v \le 16$ times the diameter of the main bar $= (16)25 = 400$ mm,
$s_v \le 48$ times the diameter of the tie bar $= 48(6) = 288$ mm
Provide 6 mm ties at 250 mm spacing

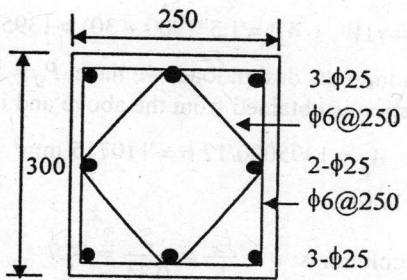

250

300

3-φ25
φ6@250
2-φ25
φ6@250
3-φ25

Fig. 10.6 Cross-Section of Column; Example 9.3.

10.5 DESIGN OF LONG TIED COLUMNS

A column is said to be long if the ratio of the effective length to side ratio is more than 12. The design of axially loaded slender column is treated as a column under bending moment, in fact it is linked with biaxial bending moment. The code has a long design procedure that is discussed in the next chapter. As a first approximation, a method similar to that recommended in the working stress design is suggested in this chapter only as initial design. The axial load carrying capacity of long column *may* be obtained from that of the short column with a correction factor. The approximate capacity may be given by:

$$P_d = C_r \left(0.4\, A_c f_{ck} + 0.67\, A_s f_y \right) \qquad (10.24)$$

where

$$C_r = 1.25 - \frac{L_e}{48b} = \text{Correction factor for long column} \qquad (10.25)$$

The L_e/b is only symbolic. It is equal to L_e/D if this value is more than L_e/b. Larger of the ratios has to be selected.

Example 10.4: Design of a Long Column

A column of effective length of 5 m in moderate environment carries an axial load of 900 kN. Design the column with M25 concrete and HYSD-Fe415 reinforcement bars. The reader is advised to read the chapter on columns under bending for actual design of slender column.

Solution

The column is exposed to moderate environment so the minimum grade of concrete is M25.

Let the column be designed with approximately one percent reinforcement. The approximate size of the column can be calculated assuming it as a short column, In which $p = 0.01$ (one percent assumed), and $a = 415/25 = 16.6$. The capacity expression results:

$$P_d = 0.4(25)(1 + 0.01(1.67(16.6 - 1)))A_g = 12.6 A_g$$

Assuming the self-weight of the column as 30 kN, the design load on the column with 1.5 partial safety factor is:

$$P_u = \gamma_f(W_g + W_l) = 1.5*(900 + 30) = 1395 \text{ kN}$$

Equating the capacity of column to the design load, we have: $P_d = 1395000 \text{ N} = P_a = 12.6 A_g$
The gross area of concrete section is obtained from the above and it is:

$$A_g = 1395000/12.6 = 110715 \text{ mm}^2$$

A 340 by 340 mm section is chosen.

The slenderness factor of the column is: $k = \dfrac{L_e}{b} = \dfrac{5}{0.34} = 14.7$

As the slenderness factor is more than 12, the column is a long column and the strength reduction factor is:

$$C_r = 1.25 - \frac{L_e}{48b} = 1.25 - \frac{5}{48*0.34} = 0.94$$

The design load for an equivalent short column is: $P_d = 1395000/0.94 = 1484043 \text{ N}$
The actual area of reinforcement required is obtained from Eq. 10.16.

$$P_d = 0.4 A_c f_{ck} + 0.67 A_s f_y = 0.4(340)340*0.99*(25) + 0.67(415) A_s = 1484043$$

The area of reinforcement is:

$$A_s = \frac{1484043 - 0.4*340*340*0.95*(25)}{0.67(415)} = 1221 \text{ mm}^2$$

Use 4 numbers of 20 mm bars, the area of reinforcement provided is: $A_s = 4(314) = 1256 \text{ mm}^2$
The percentage of reinforcement provided is $= p\% = 1256(100)/[340(340)] = 1.09\%$.
The minimum eccentricity for which the column has to be designed is:

$$e_{min} = \frac{L}{500} + \frac{b}{30} = 22 \text{ mm it is more than 20 mm}$$

The section should be designed for minimum eccentricity. The design of column under combined bending is illustrated later.

Design of the Ties

Most commonly available 6 mm bar is selected as tie. Then the spacing of the ties should be limited to:

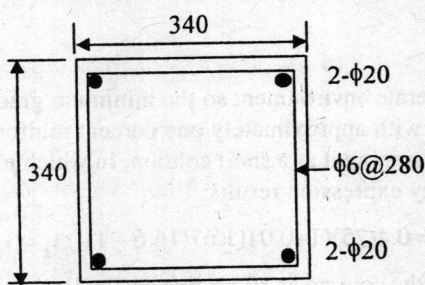

Fig. 10.7 Cross-Section of Column, Example 9.4.

$s_v \leq b = 340$ mm or
$s_v \leq (16)20 = 320$ mm, or
$s_v \leq 48(6) = 288$ mm

Provide 6 mm ties at 280 mm spacing.

10.6 DESIGN AID TABLE FOR AXIALLY LOADED COLUMNS

Table 10.3 gives the capacity of concrete section for relatively small columns for concrete grades $M20$ and $M25$. It also gives the capacities of reinforcement bars of $HYSD$-$Fe415$ grade steel. The reinforcement that is suitable for a given cross section is listed in the region of the sectional area, however, the capacities listed are independent. Recommended minimum tie reinforcement is also listed along the main reinforcement bars. The table is meant for quick and preliminary design of axially loaded columns and don't cater for columns subjected to bending moment. The use of the table illustrated through examples.

Example 10.6: Design of Axially Loaded Column with Design aid Table

A column is axially loaded with 690 kN. The effective height of the column is 3.5 m. Design the column with $M25$ concrete and $HYSD$-415 bars. The exposure condition is moderate.

Solution

To start with assume the column as short one. Assuming the self-weight of the column as 10 kN, the design load on the column with 1.5 partial safety factor is:

$$P_d = \gamma_f(W_g + W_l) = 1.5(690 + 10) = 1050 \text{ kN}$$

It is proposed to design the column with small amount of reinforcement. About 80 percent of the load is assumed to be carried by the concrete. From table 10.3, select column4 in which the capacity of $M25$ concrete is specified. Row 14 lists 850 cm^2 section and the corresponding capacity of the concrete is 842 kN.

The balance of the load to be carried by the reinforcement is: $= 1050 - 842 = 208$ kN

Column 7 of the table lists the reinforcement capacity. At row 13 and column 7, the capacity of four 16 mm bars is 223 kN, which match closely with the required capacity for the reinforcement.

Use 4 numbers of 16 mm bars.

The sectional area is 850 cm^2, so if a square section is chosen, the size of the column is 29.5 cm square. The percentage of reinforcement provided is $= p\% = 804(100)/[295(295)] = 0.92\% . > 0.8\%$.

The slenderness factor of the column is: $\lambda = \dfrac{L_e}{b} = \dfrac{3.5}{0.295} = 11.8$

The recommended ties in the table for 16 mm bars are 6 mm bar at 250 mm spacing.

10.7 SPIRAL REINFORCED CONCRETE COLUMNS

A column in which one helical tie holding the main reinforcement is provided is called a spiral reinforced column. The nomenclature may be confusing, the main reinforcement bars are straight and the transverse reinforcement is taken from bottom to top as helicoid. The helical tie provides a partial confinement to the concrete. Figure 10.3(a) illustrates the spiral reinforcement. The capacity of the column is enhanced by the circumstantial confinement action of the spiral. The ductility of the column is improved by the confinement action. The total compressive strain capacity of confined concrete can be much more than

Table 10.3 Capacity Calculation for Column Sections.

No.	Gross section area in cm²	Concrete Properties Capacity (kN) M20	M25	No.	Dia (mm.)	Longitudinal reinforcement Capacity (kN λ)	Ties Dia	Spacing (mm)
(1)	(2)	(3)	(4)	(5)	(6)	(7)	(8)	(9)
1	200	156	195	4	12	125	5	125
2	250	196	245	4	12	125	5	150
3	300	236	295	4	12	125	5	150
4	350	276	345	4	12	125	5	150
5	400	316	395	4	12	125	5	150
6	450	356	445	4	12	125	5	150
7	500	396	495	4	12	125	5	175
8	550	436	545	4	12	125	5	175
9	600	492	615	4	14	171	5	200
10	650	532	665	4	14	171	5	200
11	700	572	715	4	14	171	5	200
12	750	612	765	4	14	171	5	200
13	800	663	792	4	16	223	6	250
14	850	673	842	4	16	223	6	250
15	900	713	892	4	16	223	6	250
16	1000	793	992	4	16	223	6	250
17	1100	872	1090	4	18	283	6	275
18	1200	952	1190	4	18	283	6	275
19	1300	1030	1287	4	20	349	6	300
20	1400	1110	1387	4	20	349	6	300
21	1500	1190	1487	4	20	349	6	300
22	1600	1267	1583	8	16	446	6	250
23	1700	1347	1683	8	16	446	6	250
24	1800	1427	1783	8	16	446	6	250
25	1900	1507	1883	8	16	446	6	250
26	2000	1584	1980	8	18	566	8	275
27	2200	1744	2180	8	18	566	8	275
28	2400	1904	2380	8	18	566	8	275
29	2600	2059	2574	8	20	698	8	300
30	2800	2219	2774	8	20	698	8	300
31	3000	2379	2974	8	20	698	8	300
32	3300	2614	3268	4,4	20,25	894	8	300
33	3600	2854	3568	4,4	20,25	894	8	300
34	3900	3090	3968	4,4	20,25	894	8	300
35	4200	3329	4162	12	20	1047	8	300
36	4600	3649	4562	12	20	1047	8	300
37	5000	3963	4954	12	22	1267	8	350

that corresponding to the tied column. Load deformation behaviour of the tied and spiral columns upto a point close to the maximum load is similar. Beyond this point, there is a substantial difference in their behaviour. The concrete in tied column gets split and fall off when the load reaches the ultimate load. Whereas in the case of a spirally reinforced column, the concrete crust or the cover over the spiral falls off, lowering the resistance upto a point. There is an improved crushing strength of the concrete and ductility in the behaviour. Figure 10.8 illustrates the behaviour of tied and spiral columns.

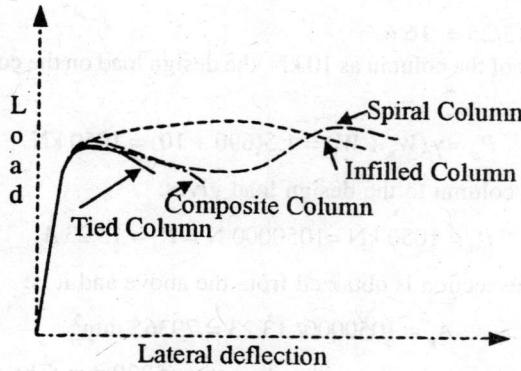

Fig. 10.8 Load Deflection Behaviour of Columns.

The load carrying capacity of a spiral column can be expressed as

$$P_a = 1.05\left(0.4 f_{ck} A_c + 0.67 f_y A_{sc}\right) \tag{10.26}$$

The ratio of the volume of the helical reinforcement to that of the core $\geq 0.36\left(\dfrac{A_g}{A_c} - 1\right)\dfrac{f_{ck}}{f_y}$ (10.27)

Minimum number of main bars is six in circular sections and limited to the number of corners of the polygon in polygonal sections. The pitch of the spiral in a spiral column is governed by:

> *Pitch ≤ 75 mm and*
> *Pitch $\leq$ One-sixth the diameter of the core further*
> *Pitch ≥ 25 mm and*
> *Pitch $\geq$ three times the diameter of the spiral bar*

The design criterion of the spiral column is rearranged for convenience of design and it is:

$$P_a = 0.42 A_g \left(1 + p(1.67\alpha - 1)\right) f_{ck} \tag{10.28}$$

in which: A_g = gross area of the section, p = ratio of the area of the main reinforcement to the gross area λ of the column and $\alpha = f_y / f_{ck}$.

Example 10.7: Spiral Reinforced Concrete Column

A column in moderate exposure is subjected to total load of 690 kN at top. The effective length of the column is 3.5 m. Design a spiral-reinforced column with $M25$ concrete and $HYSD$-415 bars.

Solution

The characteristic strengths of concrete and steel are $f_{ck} = 25$ MPa and $f_y = 415$ MPa
    The load carrying capacity of a short column is:

$$P_d = 0.42(25)[1+0.01(1.67(16.6 - 1))]A_g = 13.23 \, A_g$$

in which $p = 0.01$ and $\alpha = 415/25 = 16.6$.

Assuming the self-weight of the column as 10 kN, the design load on the column with 1.5 partial safety factor is:

$$P_d = \gamma_f(W_g + W_l) = 1.5(690 + 10) = 1050 \text{ kN}$$

Equating the capacity of column to the design load gives:

$$P_d = 1050 \text{ kN} = 1050000 \text{ N} = P_a = 13.23\, A_g$$

The gross area of concrete section is obtained from the above and it is:

$$A_g = 1050000 / 13.23 = 79365 \text{ mm}^2$$

The column section is chosen as circular with a diameter of 320 mm. The actual area of reinforcement required can be obtained from Eq. 10.26.

$$P_u = 1.05\left[0.4\, A_c f_{ck} + 0.67\, A_s f_y\right]$$

$$= 0.42 * 3.1416(320)^2 * 0.99 * (25) / 4 + 1.05 * 0.67(415) A_s = 1050000$$

The area of reinforcement is:

$$A_s = \frac{1050000 - 0.42 * 3.1416 * 320 * 320 * 0.99 * (25) / 4}{1.05 * 0.67 * (415)} = 733 \text{ mm}^2$$

Use seven numbers of 12 mm bars, the area of reinforcement is: $A_s = 7(113) = 791 \text{ mm}$

The slenderness factor of the column is: $\lambda = \dfrac{L_e}{b} = \dfrac{3.2}{0.32} = 10$

The column is a short one as assumed.

Design of Helical Reinforcement

The cover to the reinforcement is 40 mm so the core diameter of the concrete is = 320 – 2*40 = 240 mm

Most commonly available bar for spiral is 6 mm, so select 6 mm bar. Then the pitch limited to:

Pitch ≤ 75 mm and
Pitch ≤ One-sixth diameter of core = 240/6 = 40 mm, and
Pitch ≥ 25mm and
Pitch ≥ 3 times diameter of the helical bar =3*6 = 18 mm

Provide the pitch of the helical reinforcement as 40 mm. Figure 10.9 illustrates the cross section.

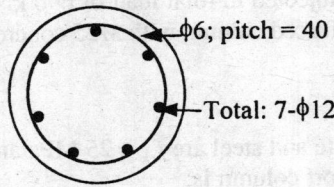

φ6; pitch = 40

Total: 7-φ12

Fig. 10.9 Cross Section of Spiral Column; Example 10.7.

Example 10.8: Spiral Reinforced Concrete Column

A column in moderate environment is subjected to total load of 690 kN at top. The effective height of the column is 3.5 m. Design a spiral reinforced column with $M25$ concrete and $HYSD\,415$ bars. The problem is same as 10.7 except that almost a maximum percentage of reinforcement is selected in this example.

Solution

The characteristic strengths of concrete and steel are $f_{ck} = 25$ MPa and $f_y = 415$ MPa
The load carrying capacity of the a short column is:

$$P_a = 0.42\,A_g f_{ck}\left(1 + p(1.67\alpha - 1)\right)$$

in which let $p = 0.05$, and the ratio of strength of material is $= \alpha = 415/25 = 16.6$.
The capacity expression results:

$$P_a = 0.42(25)[1+0.05(1.67(16.6 - 1))]A_g = 24.18\,A_g$$

Let the self-weight of the column $= 10$ kN, the design load on the column with 1.5 partial safety factor is

$$P_d = \gamma_f(W_g + W_1) = 1.5(690 + 10) = 1050 \text{ kN}$$

Equating the capacity of column to the design load, we have:

$$P_d = 1050 \text{ kN} = 1050000 \text{ N} = P_a = 24.18\,A_g$$

The gross area of concrete section is obtained from the above equation and it is:

$$A_g = 1050000/\,24.18 = 43424 \text{ mm}^2$$

The column section is chosen as circular with a diameter of 235 mm. The actual area of reinforcement required can be obtained from Eq. 10.26.

$$P_d = 0.42\,A_c f_{ck} + 1.05*0.67\,A_s f_y$$
$$= 0.42*0.95*3.1416\,(235)^2(25)/4 + 1.05*0.67(415)A_s = 1050000$$

The actual area of reinforcement is:

$$A_s = \frac{1050000 - 0.42*0.95*3.1416*235*235*(25)/4}{1.05*0.67*(415)} = 2114 \text{ mm}^2$$

Use 7 numbers of 20 mm bars, the area of reinforcement is: $A_s = 7(314) = 2198$ mm
The percentage of reinforcement provided is $= p\% = 2198 *(100)/[3.1416*235(235)/4] = 5.0\%$.

The slenderness factor of the column is: $\lambda = \dfrac{L_e}{b} = \dfrac{3.2}{0.235} = 13.6$

The slenderness factor of the column is more than 12, hence the column is treated as long one and requires a correction to the capacity of the column. The design load on the column is:

$$P_{d1} = \frac{P_d}{C_r} = \frac{1050000}{1.25 - \lambda/48} = 1086207 \; N$$

The design load is increased by about 3.4 percent, so the capacity of the column has to be increased by about the same order. The column diameter can be increase by about two percent and rounded off to 240 mm. This value decreases the slenderness factor and enhances the capacity. The load carrying capacity of the column with 240 mm diameter is:

$$P_d = 0.42 A_c f_{ck} + 1.05*0.67 A_s f_y$$

$$= 0.42*0.95*3.1416(240)^2(25)/4 + 1.05*0.67(415)A_s = 1092970 \; N$$

This capacity is adequate as the size of the column is marginally increased.

Design of Helical Reinforcement

The cover to the reinforcement is 40 mm so the core diameter of the concrete is = 240 – 2*40 = 160 mm
Most commonly available bar is 6 mm is selected. Then:
Pitch ≤ 75 mm and
Pitch ≤ One-sixth diameter of the core = 160/6 = 27 mm,
Pitch ≥ 25 mm and
Pitch ≥ three times diameter of the helical bar =3*6 = 18 mm

Provide the pitch of the helical reinforcement as 25 mm. Figure 10.10 illustrates the sectional details of the column.

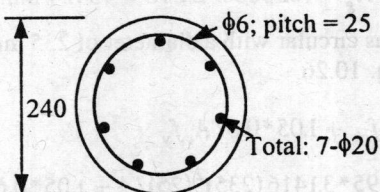

Fig. 10.10 Cross-section of Spiral Column; Example 10.8.

Example 10.9: Comparison of Tied and Spiral Column

A column of effective length 3.5 m is subjected to an axial load of 690 kN. Compare the details of the tied and spiral columns. The materials used are M25 and HYSD-Fe415 steel.

Solution

Examples 10.1, 10.7 and 10.8 illustrate the design of the column as tied column and spiral columns with different percentages of reinforcement. The design details are taken from the examples and listed in table 10.4. A simple cost comparison is made in the example considering the approximate cost of constructions (may be prevailing in year 2001 in some parts of south India) The rates are subject to variation with time, location, size of work and the nature of supervision and specifications of the contract. The rates selected that include supply and placing in position at site and they are:

Cost of $M25$ concrete	Rs. 3400 per m³
Cost of reinforcement	Rs. 2200 per kN
Formwork	Rs. 90 per m²

The cost is computed per one metre length of the column. The cost comparison gives only an idea in particular job.

It can be observed that the cost increases with increase in the percentage of the reinforcement. The strength ratio of steel to concrete is about 16 in this example. The cost of one unit volume of steel is about fifty to sixty times that of the concrete in reinforced concrete construction. So the cost of the construction increases with the increase in the percentage of reinforcement.

Table 10.4 Comparison of Different Design of Columns.

Description	Tied column with min. percent steel	Spiral column with min. percent steel	Spiral column with higher percent steel
Example	10.1	10.7	10.8
Shape	Square	Circular	Circular
Size	300*300	φ320	φ240
Main reinforcement	4-φ16; 804 mm²	7-φ12; 791 mm²	7-φ20; 2198 mm²
Transverse reinforcement	φ6@250	φ6@40	φ6@25
Cost of concrete in Rs./m	306	275	154
Cost of main steel	139	137	380
Cost of transverse steel	21	92	98
Cost of formwork	108	91	68
Total cost in Rs/m	574	595	700
Cost ratio wrt Tied Column	0.82	1.03	1.21

10.8 COMPOSITE REINFORCED CONCRETE COLUMNS

A composite reinforced concrete column consists of steel or cast iron core structural member with reinforcement in concrete. The strength of the column is based on equilibrium model. In the limit state of strength, all components are stressed to the design limit strength. Such a model is called as plastic or equilibrium model. Figure 10.11 illustrates a typical composite section.

The axial load carrying capacity of the column consists of three parts and it can be expressed as

$$P_d = 0.4 f_{ck} A_c + 0.67 f_y A_{sc} + 0.67 f_{yo} A_{so} \qquad (10.29)$$

Where

A_c = area of concrete, A_{sc} = area of reinforcement, A_{so} = area of rolled structural section

f_{sc} = yield strength of reinforcement, f_{so} = yield strength of rolled sectional material

The rolled section can be either steel or cast iron. Yield strengths of normally available rolled steel and cast iron are 250 MPa and 160 MPa respectively. Composite columns are not popular because of the problems in interconnections and constructional details. The design of the composite column is illustrated through an example. The minimum amount of the transverse reinforcement as prescribed earlier holds good for composite columns. The maximum core area is limited to 20% of the gross area of the cross

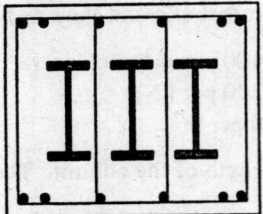

Fig.10.11 Typical Composite Column Section.

section. Sometimes the 20% of core is too large for good workability of the concrete. A minimum of 75 mm clearance between the core and the helical reinforcement, and 50 mm clearance between the core and the ties must be maintained. A reduction factor, should be applied to the long columns. Composite columns are normally provided in case of large loads and where the size restriction is severe.

The load carrying capacity of the section can be rearranged for convenience of the design as:

$$P_d = 0.4 f_{ck} A_g \left[(1 - p_1 - p_2) + 1.675 \left(p_1 \alpha_1 + p_2 \alpha_2 \right) \right] \qquad (10.30)$$

in which p_1 and p_2 are the ratios of reinforcement and the rolled sectional areas with respect to gross area, and α_1 and α_2 are the ratios of the strengths of the reinforcement and core material to that of the characteristic strength of the concrete respectively. This expression helps in selecting the section of the column. One can assume some percentages of the reinforcement and the core material to start with. Alternatively, if the size of the column is fixed ahead of the time by other considerations, the one can select the reinforcement etc.

Example 10.10: Design of Composite Column

A column in moderate exposure condition is subjected to superimposed axial load of 3000 kN. It is continuously supported by a wall and the effective length in the other direction is 6 m. Design a composite column.

Solution

The structure is exposed to moderate environment so the minimum grade of concrete is $M25$. As a preliminary exercise, select one percent reinforcement and six percent core structural steel. That is:

$$p_1 = 0.01 \text{ and } p_2 = 0.06.$$

HYSD-$Fe415$ reinforcement and mild steel rolled sections are used in the composite section. The design assumptions are: $f_y = 415$ N/mm^2, $f_{yo} = 250$ N/mm^2, this leads to

$$\alpha_1 = 415/25 = 16.6 \text{ and } \alpha_2 = 250/25 = 10.$$

The factored ultimate load on the column is: $P_u = 1.5*3000 = 4500$ kN
The substitution of the appropriate quantities in the capacity equation is:

$$P_d = 0.4 f_{ck} A_g \left[(1 - p_1 - p_2) + 1.675 (p_1 \alpha_1 + p_2 \alpha_2) \right]$$

$$= 0.4(25) A_g [1 - 0.01 - 0.06 + 1.675(0.01*16.6 + 0.06*10)] = 22.1305 A_g$$

Equating the capacity calculation to the limit load gives:

$$P_d = 22.1305 A_g = P_u = 4500,000 \text{ N}$$

Or $\quad\quad A_g = 4500000/22.1305 = 2033\ 39 \text{ mm}^2.$

Select 300 by 600 mm rectangular size section. The approximate sectional area of the core structural steel and the reinforcement are:

$$A_{so} = 0.06*203339 = 12200 \text{ mm}^2,\ A_{sc} = 0.01*203339 = 2033 \text{ mm}^2.$$

The minimum clearance required between the ties, the core is equal to 50 mm, and the cover required to the main reinforcement is 40 mm. Therefore, the largest size of the core that can fit into the selected size of 350 mm by 600 mm is:.

Maximum possible width of the core = 350 – 2*90 = 170 mm and

Maximum possible depth of the core = 600 – 2*90 = 420 mm.

One has to refer to the steel structural table to match the required area and the dimensions. One *ISMB*400 fits into the core but the area of the *ISMB*400 is 7846 mm². This area is much smaller than that required. The minimum clearances required in the composite section are generally a big constraint to select a rolled section. One of the possible choices is select two *ISMC*350 back to back and then increase the width of the section to 200 +180 = 380 mm so that the two channels can fit in. The properties of the *ISMC*350 channel section are:

$$\text{Depth} = 350 \text{ mm, width} = 100 \text{ mm, area} = 5366 \text{ mm}^2.$$

The details of the two *ISMC*350 back to back are:

$$A_{so} = 2* 5366 = 10732 \text{ mm}^2,$$

$$\text{Width} = 2*100 = 200 \text{ mm, depth} = 350 \text{ mm}$$

The area of the core steel section selected is about 20 percent less than that computed so this deficiency can be over come by increasing the size of the section and also increasing the area of the reinforcement.

$$\text{Select the section as} = A_g = 380*600 \text{ mm,}$$

Substitute the two quantities in the capacity calculation and then determine the area of reinforcement:

$$P_d = 0.4\, f_{ck}\, A_c + 0.67\, f_y\, A_{sc} + 0.67\, f_{yo}\, A_{so} = P_u = 4500,000\ N$$

From this equation we have,

$$A_{sc} = \frac{P_u - 0.4\, f_{ck}\, A_c - 0.67\, f_{yo}\, A_{so}}{0.67\, f_y} = \frac{4500000 - 0.4*25*380*600*0.93 - 0.67*250*10732}{0.67*415}$$

$$A_{sc} = 2094 \text{ mm}^2$$

Some amount of trial and correction is needed to obtain an optimal design. The closest amount of reinforcement to match that of the above is, four numbers of each 20 mm and 16 mm bars. This area is 2060 mm² that is slightly less than that is required. The following is the rationalized section:

Size = 400 mm by 600 mm
Reinforcement = A_{sc} = 2060 mm^2.
Core steel = 2-*ISMC*350: A_{so} = 10732 mm^2,

The ratios of the reinforcement and the steel are:

p_1 = 2060/(400*600) = 0.0085

p_2 = 10732/(400*600) = 0.0447

The substitution of these quantities in the capacity of the section gives:

P_a = 0.4*25*(400*600 - 2060 - 10732) + 0.67*(415*2060 +250*10732) = 4642473 N = 4642 kN > 4500

Slenderness factor of the column = 6/0.6 = 10 less than 12, hence a short column.

Design of Transverse Reinforcement

Select 6 mm ties as the diameter of the tie has to be greater than one-fourth the diameter of the main bar. The spacing of the ties is subject to the following condition:

Spacing of ties < b = 400 mm,
< 16*16 = 250 mm,
< 48*6 = 288 mm

Select 6 mm ties at 250 mm spacing. The sectional details are shown in Fig. 10.12.

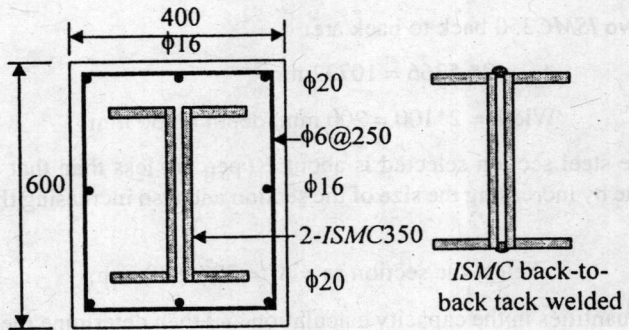

Fig. 10.12 Composite Column Section; Example 10.9.

10.9 INFILLED CONCRETE COLUMN

The outer shell of the column is usually a structural tube or sometimes an architectural material. Steel tubes or cast iron pipes are the common shells to house the concrete. The infill can be plain concrete or reinforced concrete. Reinforcement grill is first placed in the pipe and then concreted. The pipe is subjected to axial stress and nominal hoop tension depending on the axial load. If the shell is made of plastic or polythene pipe, then the pipe is not capable of resisting the axial load and it may even crack under axial load if the load transfer is not proper. Concrete when confined in a structural tube, it is subjected to tri-axial force even though the circumferential stress is only nominal that is coming from confinement. The load carrying capacity of the infilled column can be obtained from a simple plastic model with suitable correction for long column or it can also be obtained with built-in buckling effect. The infilled concrete can be treated as exposed to mild environment. The limit strength of plain concrete filled column is given by:

$$P_d = 1.1 A_c \left(0.4 f_{ck}\right) \left[1 - \frac{1}{10,000}\left(\frac{L}{a}\right)^2\right] + 0.67 f_y A_s \qquad (10.31)$$

where
  L = unsupported length of the column,
  A_c = area of cross section of the concrete core,
  A_s = area of cross section of the tube,
  a = radius of the concrete core.

The factor 1.1 in the expression reflects the effect of the confinement of concrete. In most cases of the architectural columns, the diameter of the tube is fixed and the designer may have to design the reinforcement as needed. The thickness of the structural tubes comes in three or four sizes. The tubes are classified into light, medium and heavy depending on the thickness. In case, a standard tube is not available, the tube can be rolled from standard plates.

Example 10.10: Design of Infilled Plain Concrete Column

A column of effective length 3.5 m is subjected to superimposed load of 700 kN. Design a plain infilled concrete column with M25 concrete. The tube is mild steel.

Solution

The design limit load is = P_u = 1.5(700) = 1050 kN
  Assume that the concrete carries a percentage of the load and determine the radius of the core. In the present case, select the concrete carries about seventy percent of the load, then equate the concrete capacity to the seventy percent of the load. That is:

$$0.4 f_{ck} \pi a^2 = 0.7*1050 = 735 \text{ kN} = 735000 \text{ N}$$

The value of the radius from the above expression works out to be: a = 172 mm.
  The diameter of the tube is about 340 mm and the closest available size of Indian standard tube 150 mm radius. The balance of the load has to be carried by the tube so the required thickness of the tube can be evaluated by simply equaling the 30 percent of the load to the load carrying capacity of the tube. It is given by:

$$2\pi r t (0.67 f_y) = 0.3*1050000 = 315000 \text{ N}$$

The required approximate thickness from the above equation is = 315000/(2*3.1416*150*0.67*250) = 2 mm.
  The minimum thickness of tube is also controlled by the fabrication and durability consideration. Thin plates develop corrugations during transport and fabrication. Atleast 3 mm thickness is desirable so select 3 mm thickness tube of internal radius 150 mm and check the load carrying capacity.
  The load carrying capacity of the column is computed from the expression given earlier, and it is:

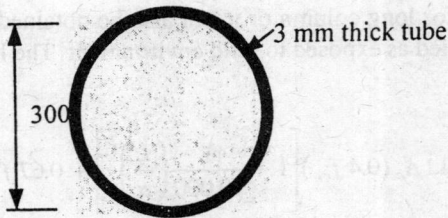

3 mm thick tube

300

Fig. 10.13 Infilled Concrete Column; Example 10.10.

$$P_d = 1.1 A_c (0.4 f_{ck}) \left[1 - \frac{1}{10,000} \left(\frac{L}{a} \right)^2 \right] + 0.67 f_y A_s$$

$$P_d = 1.1 * 3.1416 * 150^2 * 0.4 * 20 \left[1 - (3.5/0.15)^2 / 10000 \right] + 0.67 * 250 * 3.1416 * (300 + 3) * 3$$

$$= 1066503 \text{ N} = 1066 \text{ kN}$$

This capacity is just about right for the column.

PROBLEMS

(Use IS 456 - 2000)

10.1 One metre high pedestal exposed to severe environment is subjected to an axial load of 1200 kN. Design a reinforced concrete square pedestal using M35 concrete and about one percent HYSD-Fe415 reinforcement by limit state method. Sketch the reinforcement detail indicating main and transverse reinforcement.

10.2 A cantilever stub column exposed to moderate environment is 1.5 m high and subjected to an axial load of 1500 kN. Design a reinforced concrete rectangular column with width of 600 mm by limit states design using M25 concrete and about four percent HYSD-Fe415 bars. Sketch the reinforcement details.

10.3 A 3.5 m high column fixed at base and hinged at top is exposed to moderate environment. It is subjected to an axial load of 800 kN. Design a reinforced concrete rectangular column of 300 by 500 mm by limit state design using M25 concrete and HYSD-Fe415 bars. Sketch the reinforcement details of the cross section.

10.4 Column hinged at base and on roller at top in moderate exposure condition is 6 m high. It is subjected to an axial live load of 600 kN. The column size is to be limited to 250 by 300 mm. Design the reinforcement details of the column by limit state method using M25 concrete and HYSD-Fe415 bars. Sketch the reinforcement details of the cross section.

10.5 A 5 m high column, built on raft foundation and on roller support at top is subjected to an axial load of 1000 kN. Design a reinforced concrete square cross sectioned column with about two percent reinforcement by limit state method using M25 concrete and HYSD-Fe415 bars. Sketch the reinforcement details.

10.6 A 5 m high reinforcement concrete column, fixed at base and on rollers at top is subjected to an axial load of 1000 kN. Design a square cross sectioned column with five percentage of reinforcement by limit state method using M25 concrete and HYSD-Fe415 bars. Sketch the reinforcement details.

10.7 A strut of 4.5 m length is subjected to a load of 1000 kN and exposed to mild environment. Design a reinforced rectangular tied column with breadth equal to 400 mm and having about 3% of reinforcement by limit state method using M20 concrete and HYSD-Fe415 bars. Sketch the cross section.

10.8 A 2.5 m high free standing wall is subjected to an axial load of 200 kN/m. Design a reinforced concrete wall by limit state design with about one percent HYSD-Fe415 reinforcement and M25 concrete. Sketch the reinforcement details.

10.9 A column of effective height 5.0 m is subjected to an axial load of 900 kN. Design a reinforced concrete circular spiral column by limit states design using M20 concrete and HYSD-Fe415 bars. The diameter of the column is limited to 300 mm. Sketch the sectional details.

10.10 A column of effective height 4.0 m is subjected to an axial load of 900 kN. Design a reinforced concrete circular spiral column by limit states design using *M*20 concrete and *HYSD-Fe*415 bars. The percentage of the reinforcement is limited to 3 percent. Sketch the sectional details.

10.11 A column of an effective height 5.0 m is subjected to an axial load of 2200 kN. Design a rectangular composite column with *ISMB* 400 and *HYSD-Fe*415 steel bars. The area of *ISMB* 400 is 7846 mm². Yield strength of the structural steel is 250 N/mm². Design the column by limit states design using *M*25 concrete and *HYSD-Fe*415 steel bars and sketch the sectional details.

10.10 A column of effective height 7 m is subjected to an axial load of 4500 kN. The size of the column is 600 by 900 mm. Design a composite reinforced concrete column by limit states design using two rolled *ISMB*-400 steel beam sections, *M*25 concrete and *HYSD-Fe*415 bars. The area of *ISMB*-400 is 7846 mm². Sketch the sectional details.

10.11 A 3 m effective height column is subjected to an axial load 700 kN. Design an infilled concrete column by limit state design using *M*25 concrete and mild steel structural tube. Assume the thickness of the tube as 3 mm for all outer diameter of the tube varying from 250 to 400 mm.

10.12 A strut of effective length 5 m is subjected to an axial load of 1000 kN and exposed to mild environment. Design the column by limit states design using *M*25 concrete and *HYSD-Fe*415 steel bars with minimum and maximum percentages of reinforcements. Compare the cost of the two designs assuming the cost of concrete as Rs.3500 per cubic metre, cost of reinforcement as Rs.2500 per kN and cost of formwork as Rs.120 per square metre. Sketch the reinforcement details.

10.13 A column of effective length 5.5 m is subjected to superimposed load of 600 kN. Design a plain infilled concrete column with *M*25 concrete. The tube is mild steel of yield strength 250 N/mm². The tube can be made of fabricated tube with steel plate. The minimum thickness of the tube has to be 3 mm.

11

Design of Columns under Combined Bending

11.1 INTRODUCTION

Most columns in structures are subjected to axial force and bending moment. The bending moment may be due to eccentricity of the load or from the connecting beams or slabs. The eccentricity of the load may be due to boundary conditions or may be the load is applied through a bracket. Fig. 11.1 illustrates some idealized support conditions. Or columns may be subjected to lateral load caused by the wind or earthquake forces. Fig. 11.2(a) illustrates a simple frame and fig 11.2(b) illustrates bending moment on column due to eccentric load. Fig. 11.2(c) illustrates bending moment on column that is subjected to lateral load. Bending moment even if small, plays an important role in the design of columns.

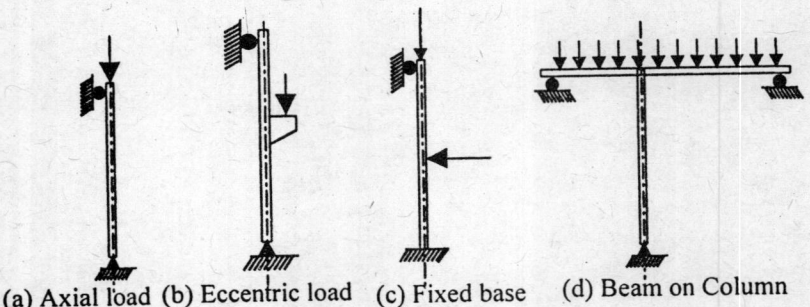

(a) Axial load   (b) Eccentric load   (c) Fixed base   (d) Beam on Column

Fig. 11.1 Columns and Load Action.

11.2 DESIGN CRITERION OF COLUMNS UNDER COMBINED BENDING AND AXIAL LOAD

For large axial load with relatively small bending moment, the entire cross section of the column can be under compression with varying degree of compressive stress. If the bending moment is the dominating force, then some part of the section can be under tension. The stress at the extreme fibre in the cross section from the elastic theory is:

$$\sigma = \frac{P}{A} \pm \frac{My}{I}$$

(11.1)

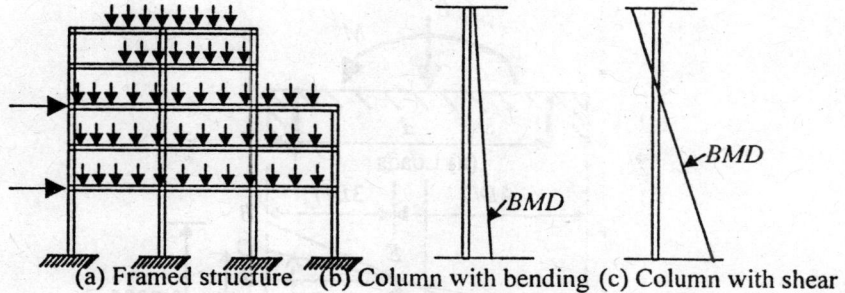

(a) Framed structure (b) Column with bending (c) Column with shear

Fig. 11.2 Framed Structure, Bending Moment on Columns in Frames.

In which

σ = stress in the section, compressive stress is considered positive.
A = area of cross section,
I = moment of inertia,
y = distance of the fibre from the centroidal axis.
P = axial load on the column,
M = bending moment on the column, then

For a rectangular section, the moment of inertia is equal to $bD^3/12$ and the maximum stress occurs at $y = D/2$. Where b = width and D = depth of the section.

The above equation can be rearranged as:

$$\sigma = \frac{P}{bD}\left[1 \pm \frac{6M}{PD}\right] = \frac{P}{bD}\left[1 \pm \frac{6e}{D}\right] \qquad (11.2)$$

where $e = M/P$ = equivalent eccentricity of the combined action.

The elastic stress distribution on uncracked section is shown in Fig. 11.3. For value of e greater than 1/6, the value within the brackets becomes negative for the negative value of the plus or minus sign. That means the stress at one extreme fibre is tension. The expression assumes homogeneous and isotropic material and the stress strain relation of the material is same both in tension and compression. This expression is not applicable to concrete if the tensile strain in the concrete exceeds the cracking strain. Even the compressive stress in concrete is not linear upto failure, so the stress distribution listed in the above expression is applicable only upto a limit. The figure illustrates only special elastic case stress distribution.

A column under combined bending and axial load can experience two possible stress distributions. They are:

Case 1: Entire cross section is under compression (Neutral axis lies outside the section)

The effective eccentricity is small and the entire cross section is under compression. Figure 11.4(a) illustrates the col-

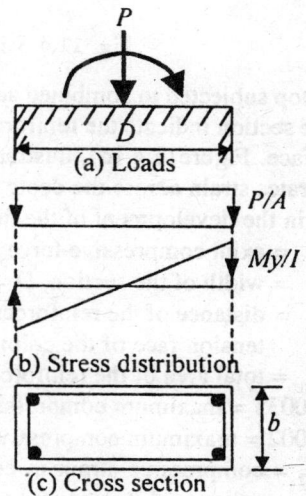

(a) Loads

P/A

My/I

(b) Stress distribution

b

(c) Cross section

Fig. 11.3 Elastic Stress Under P & M.

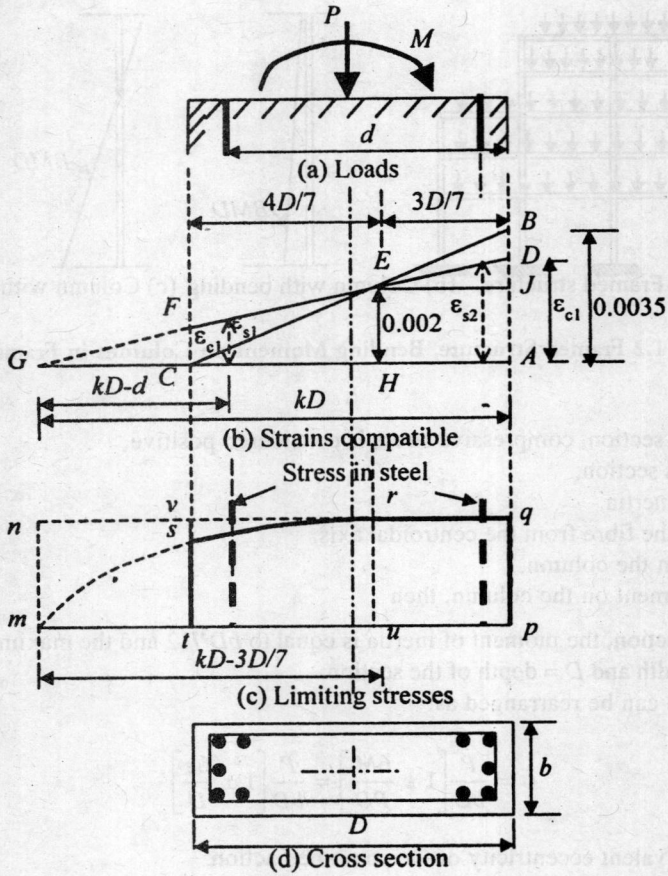

Fig. 11.4 Strain Stresses in Combined Bending & Axial Load.

umn top subjected to combined action of bending moment and axial force. Thick lines near the two faces of the section indicate the reinforcement. Each thick line represents the centroid of the reinforcement on that face. Figure 11.4 (d) illustrates the rectangular cross section under consideration. Figure 11.4 (b) illustrates strain across the depth of the section along with other notations. The following notations are used in the development of the design of the column.

P = axial compressive force on the column, M = bending moment on the column,

b = width of the section, D = overall depth of the section,

d = distance of the reinforcement that is on the tail end of the bending moment, it is also referred as tension face of the column, if tension developed,

A_s = total area of the reinforcement distributed half and half ($A_s/2$) on each face,

0.0035 = maximum compressive strain in concrete admissible under bending compression,

0.002 = maximum compressive strain concrete admissible in under axial force only,

ε_{c2} = compressive strain in concrete at the more compressed face of the column under combined action and it is equal to

= $0.0035 - 0.75\varepsilon_{c1}$, (inter- strain relation)   (11.3)

ε_{c2} = compressive strain in concrete at the least compressed face of the column,
kD = distance of the neutral axis from the extreme compression fibre,

Properties of the Compressive Force

Line CB in Fig 11.4(b) illustrates the maximum strain admissible in concrete in which point C refers to 0.0035 strain with zero strain at the other end. Line $GFED$ refers the strain variation along the depth of the section and this line has to follow the strain inter-relation. This is an assumption.

$FC = \varepsilon_{c1}$ = least compressive strain in concrete, the corresponding strain at the other extreme fibre is:
$DA = \varepsilon_{c2} = AB - 0.75*CF = AB - BD$

From the geometry of the two triangles CFE and BDE, we have $BD/CF = 0.75$ or

$$AH = 3D/7 \text{ and } CH = 4D/7; \text{ where } D = \text{overall depth of the section.}$$

Figure 11.4(c) refers to the stress distribution on the section. The rectangle $pqru$ refers to the rectangular portion of the compressive stress and the balance is a truncated parabola $rstu$. The total compressive force on the section is referred by $pqrstup$. The area of this diagram is computed by subtraction the area of part parabola rvs from the rectangle $pqvt$. The ordinate vs of the parabola can be computed from the geometry of the main parabola rsm.

The admissible compressive strength of the prism at failure is:
Prism strength $= f_{cp} = 0.67 f_{ck}/1.5 = 0.446 f_{ck}, = $ ordinate pq (11.4)
From the geometry of the parabola rsm we have:

$$vs = nm\left[\frac{tu}{mu}\right]^2 = 0.446 f_{ck}\left[\frac{4D/7}{kD - 3D/7}\right]^2 = 0.446 f_{ck}\left[\frac{4}{7k-3}\right]^2 \qquad (11.5)$$

$$\text{Let } [4/(7k-3)]^2 = g_1$$

Let this ordinate vs is given by value g and it is: $g = 0.446 f_{ck}\left(\dfrac{4}{7k-3}\right)^2 = 0.446 f_{ck} g_1$ (11.6)

The total compressive force on the section is obtained by subtracting the parabolic area rvs from the rectangle $pqvt$. That is area of $pqvt$ – area of rsv.

The total compressive force is $= F_c = 0.446 f_{ck}(bD) - \dfrac{gb}{3}\left(\dfrac{4D}{7}\right)$

This value can be expressed as:

$$F_c = C_1 bDf_{ck} \qquad (11.7)$$
Where
$$C_1 = 0.446(1 - 4g_1/21) \qquad (11.8)$$

Similarly the moment of area of the compressive force about the extreme compressive fibre can be obtained as:

$$M_c = C_2 bD^2 f_{ck} \qquad (11.9)$$
Where
$$C_2 = 0.446(0.5 - 8g_1/49) \qquad (11.10)$$

Similarly the distance of the centroid of the compressive force can be obtained from M_c/F_c and it is indicated by $c_3D = M_c/F_c$ and the value of c_3 is:

$$c_3 = \frac{C_2}{C_1} = \left[\frac{0.5 - 8g_1/49}{1 - 4g_1/21}\right] \tag{11.11}$$

Equilibrium of Forces

Let P_u = factored design limit load on the column,

M_u = factored design moment on the column,

p = percentage of the total reinforcement, equally distributed on each face of the column,

$p = 100A_s/bD$ $\tag{11.12}$

Equating the design load to the total compressive forces on the concrete and reinforcement gives:

$$P_u = C_1bDf_{ck}(1 - p/100) + \frac{A_sf_{s2}}{2*\gamma_{ms}} + \frac{A_sf_{s1}}{2*\gamma_{ms}} \tag{11.13}$$

in which

f_{s1} = compatible stress in the reinforcement on the tail end of the moment (so called tension face),

f_{s2} = compatible stress in the reinforcement on the compression face of the moment,

$A_s/2$ = reinforcement on each face of the column,

γ_{ms} = partial safety factor applied to reinforcement = 1.15

The compatible strain in the tail end of the moment is taken from Fig. 11.4(b) and it is:

$$\varepsilon_{s1} = \frac{(kD - d)*0.002}{kD - 3D/7} = \frac{(k - u)*0.002}{k - 3/7} \tag{11.14}$$

in which

$d = D - d'$ = distance of the tension face reinforcement from compression extreme face,

d' = cover to the reinforcement to the centre of tension steel,

$u = (D-d')/D$

The corresponding stress in reinforcement at the tension face of the column is:

$$f_{s1} = E_s\varepsilon_{s1} = \frac{200000*0.002(k - u)}{k - 3/7} = \frac{400(k - u)}{k - 3/7} \leq \frac{f_y}{1.15} \tag{11.15}$$

This value should be less than the factored yield stress that is $f_{s1} \leq f_y/1.15$

Similarly the compatible stress in reinforcement at the compression face is:

$$f_{s2} = E_s\varepsilon_{s2} = \frac{200000*0.002(k - d'/D)}{k - 3/7} = \frac{400(k - d'/D)}{k - 3/7} \leq \frac{f_y}{1.15} \tag{11.16}$$

Since $k - d'/D$ is much larger than $k - 3/7$, and never less than one, the stress is limited to the yield stress.

The equilibrium Eq. (11.13) can be rearranged by dividing it by bDf_{ck} and it is:

$$\frac{P_u}{bdf_{ck}} = C_1 + \frac{P}{230 f_{ck}}\left(f_y + f_{s1}\right)$$

(11.17)

Similarly the moment equilibrium about the centroid of the cross section gives:

$$M_u = C_1 bDf_{ck}\left(\frac{D}{2} - c_3 D\right) + \frac{A_{s2} f_{s2}}{1.15}\left(\frac{D}{2} - d'\right) - \frac{A_{s1} f_{s1}}{1.15}\left(\frac{D}{2} - d'\right)$$

(11.18)

Dividing the above equation by $bD^2 f_{ck}$ gives:

$$\frac{M_u}{bD^2 f_{ck}} = C_1(0.5 - c_3) + \frac{p(f_{s2} - f_{s1})(0.5 - c')}{230 f_{ck}}$$

(11.19)

in which $c' = d'/D$

Case 2 Neutral axis lies within the cross section

If the virtual eccentricity of the load is more than 0.17 times the depth if the section, tensile stress will occur on the tail end of the bending moment. This leads to a cracked section on the tension face and the tension reinforcement will be under real tensile stress. Figure 11.5 illustrates the present case. Figure 11.5(b) illustrates the compatible limiting strains on the section and Fig.11.15(c) illustrate the stresses on the section.

The compatible steel in the tension reinforcement in the cracked section is:

$$f_{s1} = \frac{0.0035(d - kD)E_s}{kD} = \frac{700(1 - k)}{k} \le \frac{f_y}{1.15}$$

(11.20)

Similarly the compressive stress in the compression reinforcement will be equal to the yield strength as the strain in the steel tends to 0.0035. All mild and high yield steels have an yield strain less than 0.0035.

Equating the total compressive force on the section to the factored design load gives:

$$P_u = 0.36kbdf_{ck} + \frac{A_s\left(f_y - f_{s1}\right)}{1.15}$$

(11.21)

Dividing the above equation by bDf_{ck} results as:

$$\frac{P_u}{bDf_{ck}} = 0.36k(1 - c') + \frac{p(f_y - f_{s1})}{230 f_{ck}}$$

(11.22)

Equilibrium of moment of the forces on the section along the generated internal resistants taken about the centre of the section gives:

$$M_u = 0.36kbdf_{ck}(0.5D - 0.42 kd) + \frac{A_{s2} f_y}{1.15}(0.5D - d') + \frac{A_{s1} f_{s1}}{1.15}(0.5D - d')$$

(11.23)

Dividing the above expression by $bD^2 f_{ck}$ results in to a non-dimensional equilibrium equation and it is:

$$\frac{M_u}{bD^2 f_{ck}} = 0.36k(1 - c')\left[0.5 - 0.42k(1 - c')\right] + \frac{p(0.5 - c')(f_y + f_{s1})}{230 f_{ck}} \qquad (11.24)$$

Design of a column subjected to combined action involves nonlinear governing equations. There is a unique location of the neutral axis for a given set of axial force and bending moment,. The determination of the neutral axis is not straightforward. First one has to assume whether the neutral axis lies within or outside the cross section. Then solve for the value of k that decides the location of the neutral axis. One can make a rough estimate of the location of the neutral axis based on the ratio virtual eccentricity. That is the ratio of the design moment and design load. A much better way of designing the column is through a set of design aid curves and it is discussed later. The trial method is illustrated by couple of examples.

Example 11.1: Design of Column under Dominant Axial Force

A column exposed to mild environment is subjected to an axial load of 900 kN and bending moment of 60 kNm. The size of the column is 300 by 500 mm.

Solution

Select $M20$ grade concrete as the column is exposed to mild environment only, and use *HYSD-Fe*415 grade reinforcement. Assume the clear cover as 40 mm and the cover to the centre of reinforcement of each face as 50 mm. The factored design loads are:

$$P_u = 1.5*900 = 1350 \text{ kN} = 1.35 \text{ MN},$$
$$M_u = 1.5*60 = 90 \text{ kNm} = 0.09 \text{ MNm},$$

The given data of the section etc. is:

$$b = 0.3 \text{ m}, D = 0.5 \text{ m}, f_{ck} = 20 \text{ MPa}, f_y = 415 \text{ MPa}, d' = 50 \text{ mm} = 0.05\text{m}$$

The cover ratio = $c' = 50/500 = 0.10$
The non-dimensional force factors are:

$$Y = \frac{P_u}{bD f_{ck}} = \frac{1.35}{0.3*0.5*20} = 0.45$$

$$X = \frac{M_u}{bD^2 f_{ck}} = \frac{0.09}{0.3*0.25*20} = 0.06$$

The virtual eccentricity ratio is:

$$\frac{M_u}{P_u D} = \frac{0.09}{1.35*0.5} = 0.133$$

This virtual eccentricity ratio is less than 0.16; therefore, the neutral axis lies outside the cross section. As a first trial let the neutral axis at the edge of the section, that means the value of $k = 1.0$. The values of C_1, C_2, c_3 etc. are calculated using the equations 11.8, 11.01, 11.15, 11.16. This involves a number of values to be computed and it is easier done by an excel sheet. However it is illustrated by long hand for

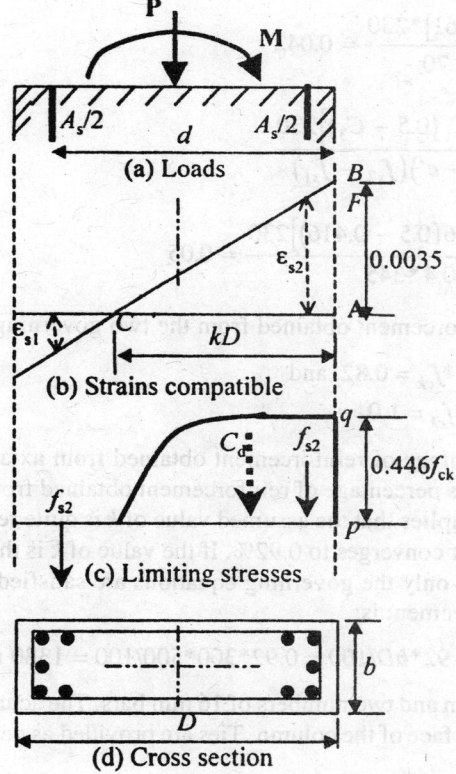

Fig. 11.5 Strains & Stresses in Combined Bending & Axial Load; Cracked Section.

better understanding of the beginners.

$$g_1 = \left(\frac{4}{7k-3}\right)^2 = (4/4)^2 = 1$$

$$C_1 = 0.446(1-4g_1/21) = 0.446(1-4/21) = 0.361$$
$$C_2 = 0.446(0.5-8g_1/49) = 0.446(0.5-8/49) = 0.15$$
$$c_3 = C_2/C_1 = 0.15/0.361 = 0.416,$$
$$u = 1-c' = 0.9,$$
$$f_{s1} = 400(k-u)/(k-3/7) = 400*0.1/(4/7) = 70 \text{ MPa},$$
$$f_{s2} = 415 \text{ MPa}$$

The percentage of reinforcement required can be obtained by rearranging the Eqs.11.17 and 11.19 and given below by assigning p_p and p_m notations for the value of p in the respective equations.

$$\frac{p_p}{f_{ck}} = \frac{[Y-C_1]230}{f_{s2}+f_{s1}}$$

$$= \frac{[0.45 - 0.361]*230}{415 + 70} = 0.042 \tag{11.25}$$

$$\frac{p_m}{f_{ck}} = \frac{[X - C_1(0.5 - C_3)]230}{(0.5 - c')(f_{s2} - f_{s1})}$$

$$= \frac{[0.06 - 0.36(0.5 - 0.416)]230}{0.4*345} = 0.05 \tag{11.26}$$

The percentages of the reinforcement obtained from the two governing conditions are:

$$p_p = 0.042*f_{ck} = 0.82, \text{ and}$$
$$p_m = 0.05*f_{ck} = 1.0$$

Where p_p and p_m are the percentage of reinforcement obtained from axial load and moment governing equations. It can be seen that the percentage of reinforcement obtained from the two equations is reasonable close to each other. This implies that the assumed value of k is quite reasonable. By another trial, the percentage of the reinforcement converges to 0.92%. If the value of k is the right one the value of p_p and p_m must be exactly same. Then only the governing equations are satisfied.

The area of the total reinforcement is:

$$A_s = 0.92*bD/100 = 0.92*300*500/100 = 1380 \text{ mm}^2.$$

Select four numbers of 20 mm and two numbers of 16 mm bars. The actual area provided results = 1660 mm². Provide three bars at each face of the column. Ties are provided as per the minimum limits specified in the earlier chapter, that is:

The diameter of the tie should be ≥ *one fourth the diameter of the main bar or*
The diameter of the tie should be ≥ *5 mm, preferably 6 mm*
The spacing of the ties should be ≤ *minimum size of the cross section*
The spacing of the ties should be ≤ *16 times the diameter of the main reinforcement bar*
The spacing of the ties should be ≤ *48 times the diameter of the tie bar itself*

Select 6 mm ties, then the spacing of the bars is controlled by:
Spacing ≤ minimum size of the column = 300 mm
Spacing ≤ 16 times the diameter of the main bar = 16*16 = 256 mm
Spacing ≤ 48 times the diameter of the tie bar = 48*6 = 288 mm
Provide 6 mm ties at 250 mm spacing

The cover to the centre of the reinforcement was assumed to be 0.1 times the depth of the column and it is 50 mm. The cover to the reinforcement is 40 mm and the centre of the 20mm bar is 10 mm, hence the cover to the centre of the reinforcement is 50 mm that checks with the value assumed. Fig. 11.6 illustrates the cross section of the column.

Example 11.2: Design of Column under Dominant Axial Force

A column exposed to mild environment is subjected to an axial load of 900 kN and bending moment of 60 kNm. The size of the column is 200 by 400 mm. This problem is similar to example 11.1 except that the size of the column is decreased to illustrate the effect of smaller size.

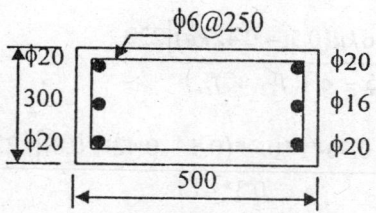

Fig. 11.6 Column Section; Example 11.1.

Solution

Select M20 grade concrete as the column is exposed to mild environment only, and use *HYSD-Fe*415 grade reinforcement. The factored design loads are:

$$P_u = 1.5*900 = 1350 \text{ kN} = 1.35 \text{ MN},$$

$$M_u = 1.5*60 = 90 \text{ kNm} = 0.09 \text{ MNm},$$

The given data of the section etc. is: $b = 0.20$ m, $D = 0.4$ m, $f_{ck} = 20$ MPa, $f_y = 415$ MPa,
The eccentricity ratio is: $e' = (M/P)/D = (60/900)/0.4 = 0.67/0.4 = 0.167$

The eccentricity ratio is about equal to the kern distance so extreme outer fibre is just subjected to tensile stress. This falls on the border case as the cover distance is still 0.2 times the depth of the section. Select the expressions of case 2 but with a value of k in the range of 1 to 1+ cover distance ratio.

Since the width is small, it is assumed that the reinforcement is arranged in two rows. Therefore the cover to the centre of reinforcement is: A cover plus another row of reinforcement + half spacing of the rows.

Cover to the centre of reinforcement = 40 + 40 = 80 mm.
The cover ratio = c' = 80/400 = 0.2; and u = 1 – c' = 0.8,
The non-dimensional force factors are:

$$Y = \frac{P_u}{bD f_{ck}} = \frac{1.35}{0.2*0.4*20} = 0.844$$

$$X = \frac{M_u}{bD^2 f_{ck}} = \frac{0.09}{0.2*0.4*0.4*20} = 0.141$$

The percentage of reinforcement required can be obtained by rearranging the Eqs.11.22 and 11.24 of cracked section. The value of k is assumed as 1, that implies that the neutral axis is at the tension reinforcement. The stress in the tension face reinforcement works out to be zero. The substitution of the trial values of $k = 1.0$, $u = 0.8$, $c' = 0.2$, $f_{s1} = 0$ etc gives:

$$\frac{p_p}{f_{ck}} = \frac{[Y - 0.36 ku]230}{f_{s2} - f_{s1}}$$

$$= \frac{[0.844 - 0.361*1.*0.8]*230}{415 + 0} = 0.172 \qquad (11.27)$$

$$\frac{p_m}{f_{ck}} = \frac{\left[X - 0.36\,ku(0.5 - 0.42\,ku)\right]230}{(0.5 - c')(f_{s2} + f_{s1})}$$

$$= \frac{[0.141 - 0.36*1*0.8*(0.5 - 0.42 * 0.8)]*230}{0.3*415} = 0.308 \qquad (11.28)$$

The percentage of reinforcement computed from the two equations differs considerably so the assumed value of the neutral axis distance is not suitable. Second trial of the neutral axis distance at $k = 1.15$ is tried. The substitution of the revised k gives the percentages of reinforcements as:

$$p_p/f_{ck} = 0.23 \text{ and}$$
$$p_m/f_{ck} = 0.24$$

It can be seen that the percentage of reinforcement obtained from the two equations is quite close to each other.

The area of the total reinforcement is:

$$A_s = 0.235*20*bD/100 = 4.7*200*400/100 = 3760 \text{ mm}^2.$$

Select eight numbers of 25 mm bars. The actual area provided is $= 8*490 = 3920 \text{ mm}^2$. Four bars in two rows on each face of the column. The percentage of reinforcement is 4.7 percent. Centre of the reinforcement from the outer face is $= 40 +25 +15 = 80$ mm.

The assumed cover distance checks with that of the actual one.

Select 8 mm ties, then the spacing of the bars is controlled by:

Spacing $\leq$ minimum size of the column = 200 mm

Spacing $\leq$ 16 times the diameter of the main bar = 25*16 = 400 mm

Spacing $\leq$ 48 times the diameter of the tie bar = 48*8 = 384 mm

Provide 8 mm ties at 200 mm spacing

The cover to the centre of the reinforcement was assumed to be 0.2 times the depth of the column and it is 80 mm. checks with the final value. Fig. 11.7 illustrates the cross section of the column.

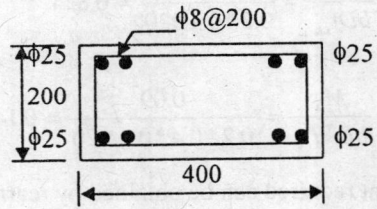

Fig. 11.7 Column Section; Example 11.2.

Example 11.3: Design of Column with the given Size is not Adequate or Percentage of Reinforcement is too High.

A column in mild environment is subjected to an axial load of 900 kN and bending moment of 60 kNm. The size of the column is 200 by 350 mm. This problem is same as example 11.2 except that the size of the column is decreased to illustrate a revision procedure.

Solution

Select M20 grade concrete as the column is exposed to mild environment only, and use HYSD-Fe415 grade reinforcement. The factored design loads are:

$$P_u = 1.5*900 = 1350 \text{ kN} = 1.35 \text{ MN},$$
$$M_u = 1.5*60 = 90 \text{ kNm} = 0.09 \text{ MNm},$$

The given data of the section is: $b = 0.20$ m, $D = 0.35$ m, $f_{ck} = 20$ MPa, $f_y = 415$ MPa,

The eccentricity ratio is: $e' = (M/P)/D = (60/900)/0.35 = 0.67/0.35 = 0.19$

The eccentricity ratio is more than the kern distance so extreme outer fibre is just subjected to tensile stress and the section is not likely to be cracked. This falls on the border of case 2. Select the expression of case 2.

Since the size of the section is small, it is assumed that the reinforcement is arranged in two rows. Therefore the cover to the centre of reinforcement is: A cover plus another row of reinforcement + half spacing of the rows.

Cover to the centre of reinforcement = 40 + 40 = 80 mm.
The cover ratio = $c' = 80/350 = 0.23$; and $u = 1 - c' = 0.77$

The non-dimensional force factors are:

$$Y = \frac{P_u}{bDf_{ck}} = \frac{1.35}{0.2*0.35*20} = 0.964$$

$$X = \frac{M_u}{bD^2 f_{ck}} = \frac{0.09}{0.2*0.4*0.35*20} = 0.184$$

The percentage of reinforcement required can be obtained from rearranging the Eqs.11.22 and 11.24 of cracked section. The substitution of the trial values of $k = 1.08$, $u = 077$, $c' = 0.23$, gives:

$$\frac{p_p}{f_{ck}} = \frac{[Y - 0.36ku]230}{f_{s2} - f_{s1}}$$

$$= \frac{[0.964 - 0.36*1.08*0.77*230]}{415 + 51} = 0.327$$

$$\frac{p_m}{f_{ck}} = \frac{[X - 0.36ku(0.5 - 0.42ku)]230}{(0.5 - c')(f_{s2} + f_{s1})}$$

$$= \frac{[0.184 - 0.36*1.08*0.77(0.5 - 0.42*1.08*0.77)]230}{0.27*415} = 0.325$$

The percentages of reinforcement computed from the two independent equations are close to each other (0.327 & 0.325). The percentage of reinforcement is: $p = 0.326*20 = 6.52\%$

The percentage of reinforcement required is more than the maximum admissible in column; therefore the size of the column suggested is not adequate. The size of the column can be taken as 200 by 375 mm. This size results in about 5.5 percent of reinforcement.

Example 11.4: Design of Column under Dominant Bending Moment

A column in moderate environment is subjected to an axial load of 900 kN and bending moment of 270 kNm. The size of the column is 300 by 550 mm. Design the column.

Solution

Select $M25$ grade concrete as the column is exposed to moderate environment, and use $HYSD$-$Fe415$ grade reinforcement. The factored design loads are:

$$P_u = 1.5*900 = 1350 \text{ kN} = 1.35 \text{ MN},$$
$$M_u = 1.5*270 = 405 \text{ kNm} = 0.405 \text{ MNm},$$

The given data of the section etc. is:

$$b = 0.30 \text{ m}, D = 0.55 \text{ m}, f_{ck} = 25 \text{ MPa}, f_y = 415 \text{ MPa},$$

The virtual eccentricity = M/P =270/900 = 0.3 m, and
The virtual eccentricity ratio with respect to depth of the section is:

$$e' = (M/P)/D = (270/900)/0.55 = 0.3/0.55 = 0.545$$

The virtual eccentricity ratio is much more than the kern distance so section is a cracked one. This falls under case 2. Select the expressions of case 2 with a value of k in the range of 0.6 plus. At the moment there is no hard and fast rule only trial and corrections is made.

Let the cover to the centre of reinforcement = 40 + 40 = 80 mm. (assumed on the higher side)
The cover ratio = $c' = 80/550 = 0.145$; use $c' = 0.15$ and $u = 1 - c' = 0.85$,
The non-dimensional force factors are:

$$Y = \frac{P_u}{bDf_{ck}} = \frac{1.35}{0.3*0.55*25} = 0.327$$

$$X = \frac{M_u}{bD^2 f_{ck}} = \frac{0.405}{0.3*0.55*0.55*25} = 0.179$$

The percentage of reinforcement required can be obtained from Eqs.11.22 and 11.24 of cracked section.

Trial 1: for k = 0.7; for c' =0.15
The value of stress in the tension reinforcement works out to be 300 MPa.

$$\frac{p_p}{f_{ck}} = \frac{[Y - 0.36ku]230}{f_{s2} - f_{s1}}$$

$$= \frac{[0.844 - 0.36*0.7*0.85]*230}{415 - 300} = 0.226$$

$$\frac{P_m}{f_{ck}} = \frac{\left[X - 0.36\,ku(0.5 - 0.42\,ku)\right]230}{(0.5 - c')(f_{s2} + f_{s1})}$$

$$= \frac{[0.141 - 0.36*0.7*0.85(0.5 - 0.42*0.7*0.85)]*230}{0.35*(415 + 300)} = 0.115$$

The percentage of reinforcement computed from the above two equations differs considerably so the assumed value of the neutral axis distance is not appropriate.

Trial 2

Another neutral axis distance at $k = 0.8$ is tried. The Substitution of $k = 0.8$ in the above two equations gives:

$$P_p/f_{ck} = 0.079 \text{ and}; \quad P_m/f_{ck} = 0.14$$

It can be seen that the nature of percentage of reinforcement obtained from the two equations for values of $k = 0.7$ and 0.8 got reversed. This implies that the value of k is between the trial values.

Trial 3: k = 0.75; with c' = 0.15

The substitution of the values result in the following percentage reinforcement ratio:

$$P_{*p}/f_{ck} = 0.123 \text{ and}; \quad P_m/f_{ck} = 0.127$$

Select p/f_{ck} equal to 0.125, then the percentage of reinforcement works out to:

$$p = 0.125*25 = 3.125\%$$

The area of the total reinforcement is: $A_s = 3.125*bD/100 = 5156 \text{ mm}^2$.

Select ten numbers of 25 mm and two numbers of 16 mm bars. The actual area provided results = $10*490 +2*201 = 5302 \text{ mm}^2$. Provide: five of 25 mm and one of 16 mm bars at each face of the column in two rows spaced at 30 mm.

The centre of the reinforcement from the outer face is = 40 +25 +15 =80 mm.

The assumed cover distance checks with that of the actual one. Tie specified are:

Select 8 mm ties, then the spacing of the bars is controlled by:

Spacing ≤ minimum size of the column = 300 mm

Spacing ≤ 16 times the diameter of main bar = 16*16 = 256 mm

Spacing ≤ 48 times the diameter of the tie bar = 48*8 = 384 mm

Provide 8 mm ties at 250 mm spacing

Fig. 11.8 illustrates the cross section of the column.

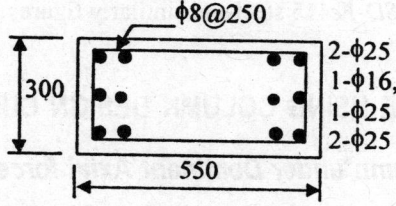

Fig. 11.8 Column Section; Example 11.4.

11.3 COLUMN DESIGN AID CURVES

The examples that were listed earlier use trial and correction method. The design of columns under combined action is normally carried by of design aid curves. Bureau of Indian Standards (*BIS*) has published design aid curves to suit the design of columns by the Indian Code of practice on limit state design. Author has developed design aid curves using the Microsoft Excel sheet and these curves are presented in this book.

The main parameters of the Column Design Cures are:

- Characteristic strength of reinforcement = f_y = 415 and 500 MPa.
- Range of Ratio of Reinforcement cover distance to depth of the section = d'/D = 0.05, 0.10, 0.15 and 0.20
- Range of Ratio of percentage of the reinforcement and strength of concrete = p/f_{ck} = 0.02 to 0.26 at intervals

The governing equations of the design of the columns were already developed and reproduced for quick reference.

Case 1: The neutral axis with in the section

$$\frac{P_u}{bdf_{ck}} = C_1 + \frac{p}{230 f_{ck}}(f_y + f_{s1}) \tag{11.17}$$

$$\frac{M_u}{bD^2 f_{ck}} = C_1(0.5 - c_3) + \frac{p(f_{s2} - f_{s1})(0.5 - c')}{230 f_{ck}} \tag{11.19}$$

in which $c' = d'/D$, and C_1, c_3, p, etc were defined earlier.

Case 2 Neutral axis lies within the cross section

$$\frac{P_u}{bDf_{ck}} = 0.36k(1 - c') + \frac{p(f_y - f_{s1})}{230 f_{ck}} \tag{11.22}$$

$$\frac{M_u}{bD^2 f_{ck}} = 0.36k(1 - c')[0.5 - 0.42k(1 - c')] + \frac{p(0.5 - c')(f_y + f_{s1})}{230 f_{ck}} \tag{11.24}$$

The right side of the above four equations is converted into curves for different values of the neutral axis factor (k) varying from $0.1D$ to $10D$ using the MS Excel sheet. The graphs are self-explanatory and the use of the curves is illustrated through a number of examples. Figures 11.9 to 11.12 list the design curves for reinforcement grade *HYSD-Fe*415 steel and similarly figures 11.13 to 11.16 list reinforcement grade *HYSD-Fe*500 steel.

11.4 ILLUSTRATIVE EXAMPLE USING COLUMN DESIGN CURVES

Example 11.5: Design of Column under Dominant Axial force by Column Design Curves

A column in mild environment is subjected to an axial load of 900 kN and bending moment of 60 kNm. The size of the column is 300 by 500 mm. Design the column.

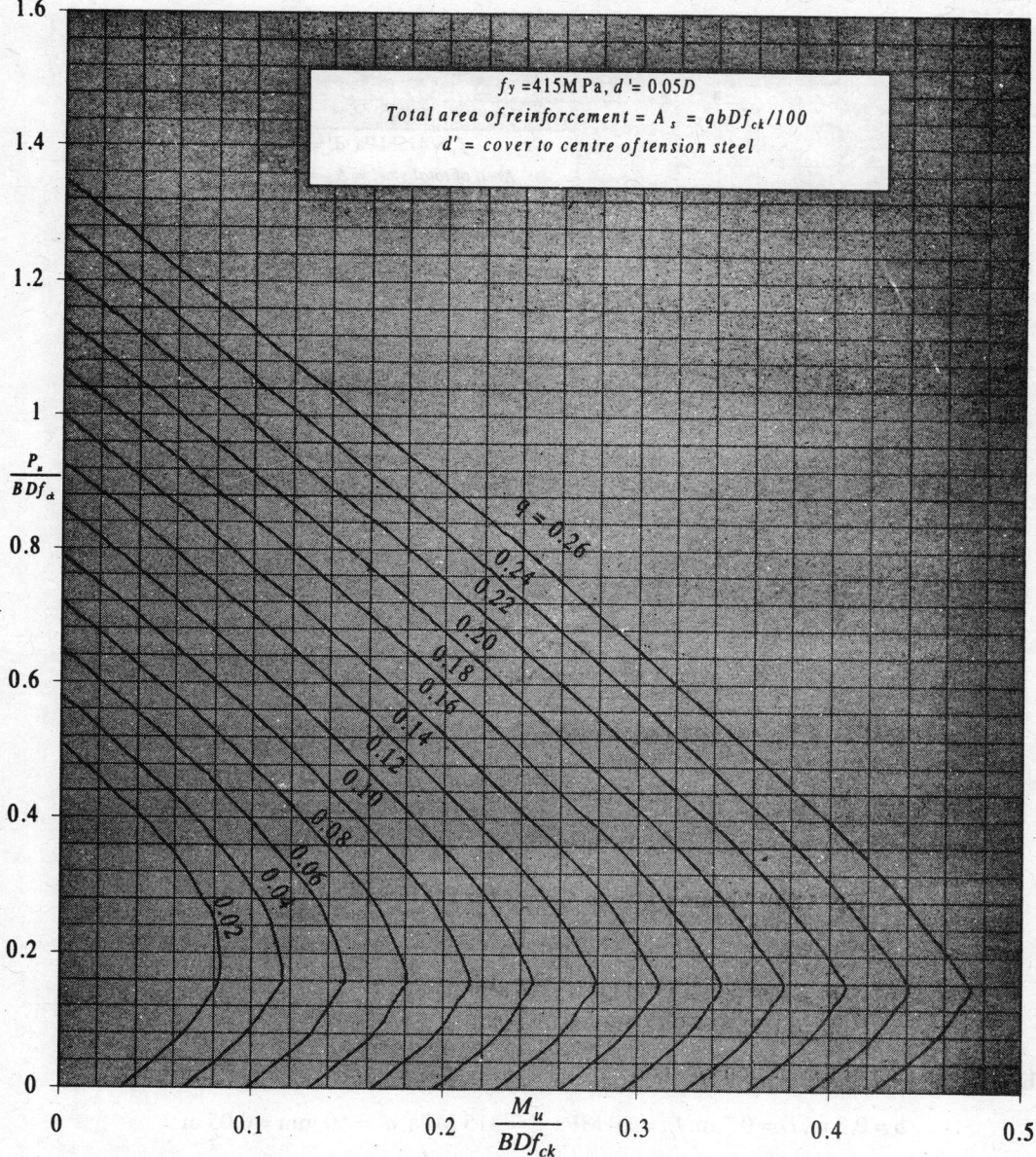

$$f_y = 415 \text{MPa}, \; d' = 0.05D$$
$$\text{Total area of reinforcement} = A_s = qbDf_{ck}/100$$
$$d' = \text{cover to centre of tension steel}$$

Fig. 11.9 Column Design Curves for Fe415 and cover of 0.05D.

Solution

Select *M*20 grade concrete and use *HYSD-Fe*415 grade reinforcement. It is assumed that the size of the column is large enough to place all the reinforcement bars in one row on two opposite faces of the section. Assume the clear cover as 40 mm and the cover to the centre of reinforcement of each face as 50 mm. The factored design loads are:

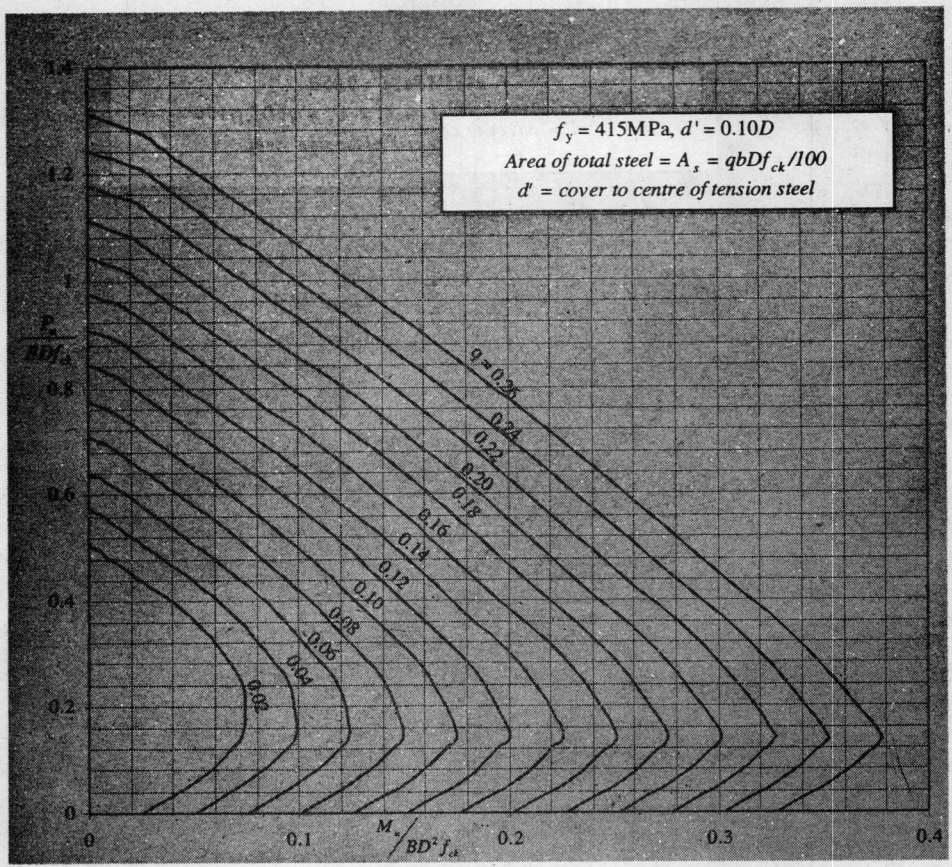

Fig. 11.10 Column Design Curves for Fe415 and cover of 0.10D.

$$P_u = 1.5*900 = 1350 \text{ kN} = 1.35 \text{ MN},$$
$$M_u = 1.5*60 = 90 \text{ kNm} = 0.09 \text{ MNm},$$

The given data of the section etc. is:

$$b = 0.3 \text{ m}, D = 0.5 \text{ m}, f_{ck} = 20 \text{ MPa}, f_y = 415 \text{ MPa}, d' = 50 \text{ mm} = 0.05 \text{ m}$$

The cover ratio = $c' = 50/500 = 0.10$
The non-dimensional force factors are:

$$Y = \frac{P_u}{bDf_{ck}} = \frac{1.35}{0.3*0.5*20} = 0.45$$

$$X = \frac{M_u}{bD^2 f_{ck}} = \frac{0.09}{0.3*0.25*20} = 0.06$$

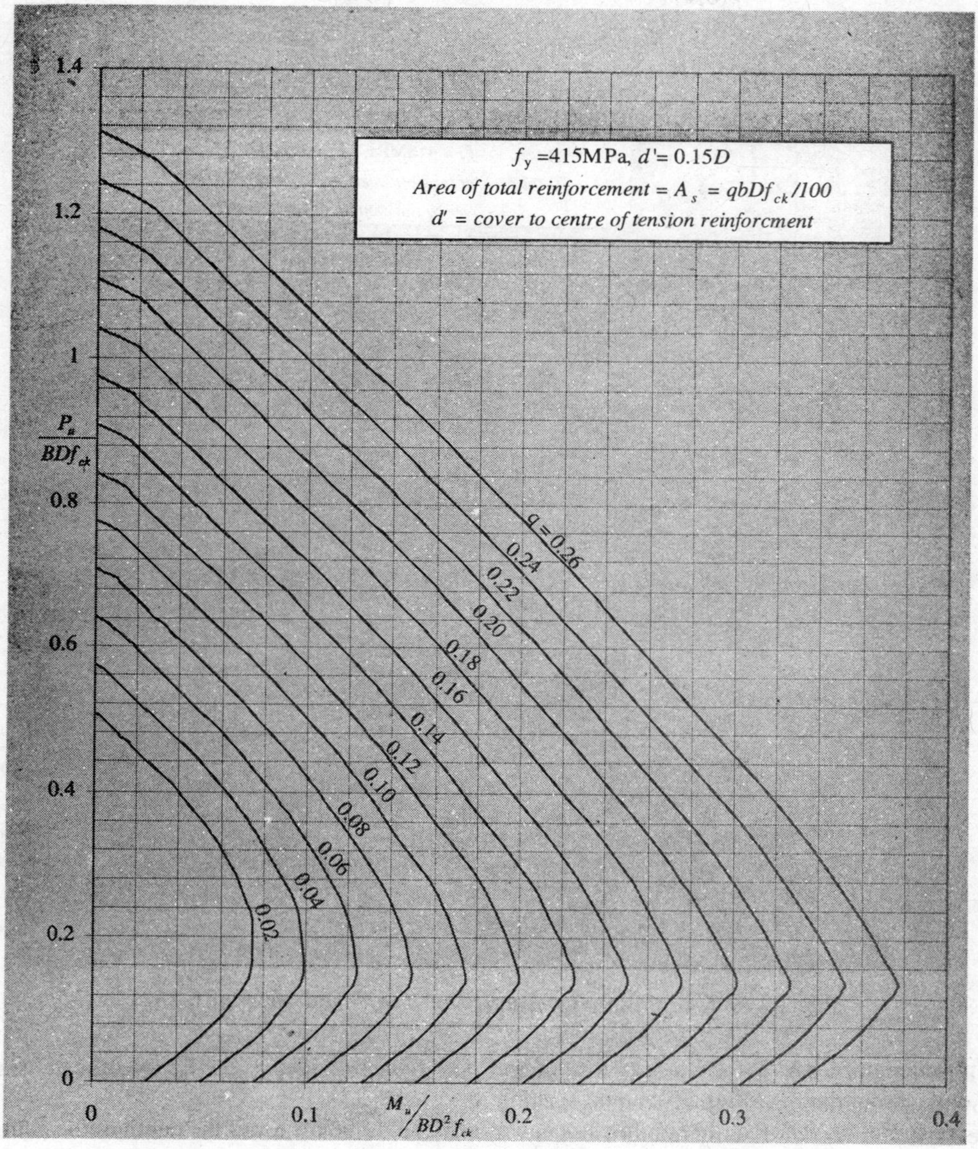

f_y =415MPa, d'= 0.15D
Area of total reinforcement = A_s = $qbDf_{ck}/100$
d' = cover to centre of tension reinforcment
1.4
1.2
1
0.8
0.6
0.4
0.2
0
$\frac{P_u}{BDf_{ck}}$
q = 0.26
0.24
0.22
0.20
0.18
0.16
0.14
0.12
0.10
0.08
0.06
0.04
0.02
0
0.1
0.2
0.3
0.4
$M_u/BD^2 f_{ck}$

Fig. 11.11 Column Design Curves for Fe415 and cover of 0.15*D*.

Figure 11.10 gives the percentage of reinforcement for reinforcement of Grade *Fe*415 and cover ratio of 0.10. Select the x - coordinate as 0.06 and y - coordinate as 0.45 from the figure 11.10 and find the intersection point of the values. Interpolation of the values of q from the curves is 0.045.

The percentage of the reinforcements obtained from the q value of the curves is: $p = 0.045*f_{ck} = 0.9$

The area of the total reinforcement is: $A_s = 0.90*bD/100 = 0.90*300*500/100 = 1350$ mm^2.

Select four numbers of 20 mm and two numbers of 16 mm bars. The actual area provided results = 1660

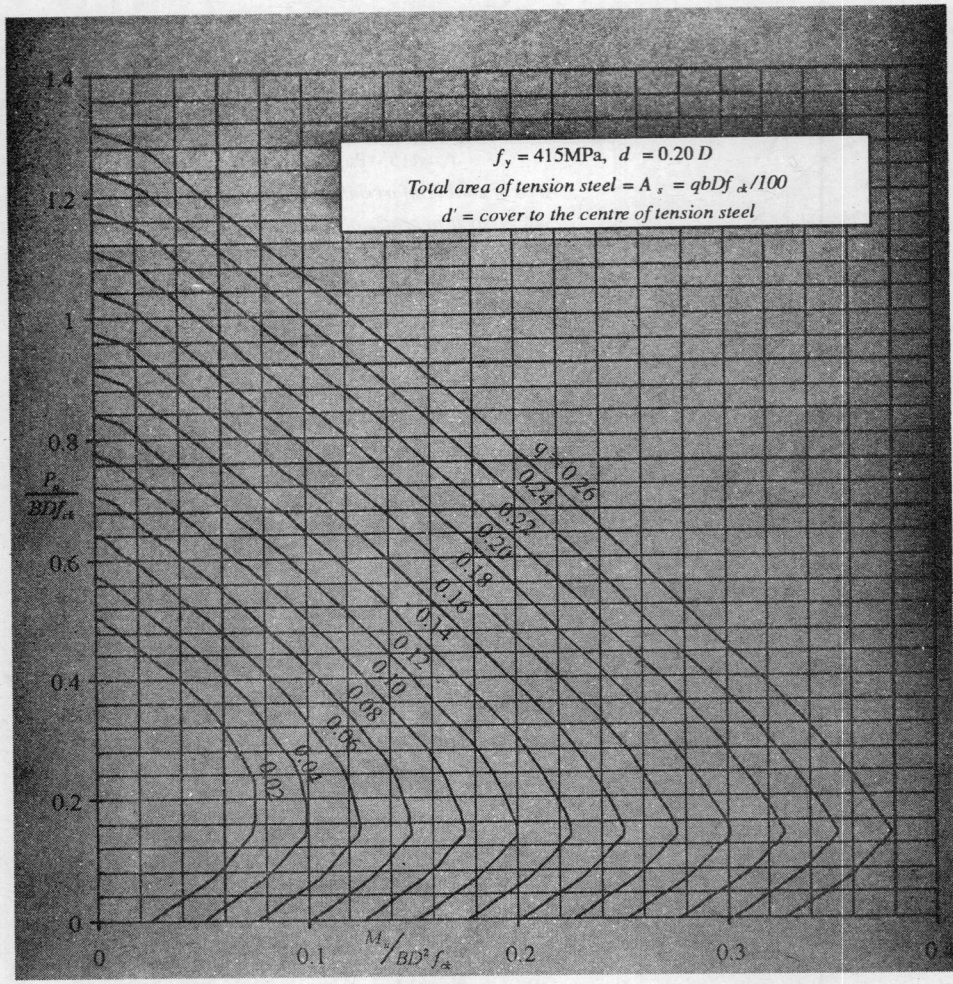

$$f_y = 415\text{MPa}, \quad d = 0.20\,D$$

Total area of tension steel $= A_s = qbDf_{ck}/100$

$d' =$ cover to the centre of tension steel

Fig. 11.12 Column Design Curves for *Fe*415 and cover of 0.20*D*.

mm². Provide three bars at each face of the column. Ties to be provided as per the specifications, that is:

Provide 6 mm diameter ties at 250 mm spacing

The cover to the centre of the reinforcement was assumed to be 0.1 times the depth of the column and it is 50 mm. The cover to the reinforcement is 40 mm and the centre of the 20 mm bar is 10 mm, hence the cover to the centre of the reinforcement is 50 mm that checks with the value assumed. This problem was solved from the governing conditions in example 11.1 and the solution is the same.

Example 11.6: Design of Column with Size that is not Adequate

A column in mild environment is subjected to an axial load of 900 kN and bending moment of 60 kNm. The size of the column is 200 by 350 mm. Design the column by design aid curves.

Solution

Select $M20$ grade concrete and use $HYSD$-$Fe415$ grade reinforcement. The factored design loads are:

$$P_u = 1.5*900 = 1350 \text{ kN} = 1.35 \text{ MN},$$
$$M_u = 1.5*60 = 90 \text{ kNm} = 0.09 \text{ MNm},$$

The data of the section is:

$$b = 0.20 \text{ m}, D = 0.35 \text{ m}, f_{ck} = 20 \text{ MPa}, f_y = 415 \text{ MPa},$$

Since the width of the section is small, it is assumed that the reinforcement is arranged in two rows on each opposite faces. Therefore the cover to the centre of reinforcement is assumed = 40 + 40 = 80 mm.

$$\text{The cover ratio} = c' = 80/350 = 0.23;$$

The non-dimensional force factors are:

$$X = \frac{M_u}{bD^2 f_{ck}} = \frac{0.09}{0.3*0.4*0.25*20} = 0.184$$

$$Y = \frac{P_u}{bD f_{ck}} = \frac{1.35}{0.2*0.35*20} = 0.964$$

Figure 11.12 gives the percentage of the reinforcement for a cover ratio of 0.2. It can be seen that the intersection of the above coordinates fall just outside the curves. This indicates that the required percentage of reinforcement is beyond six percent. Therefore the size of the column suggested is inadequate. The size need to be increased.

Example 11.7: Design of Column with Large Bending Moment

A column in moderate environment is subjected to an axial load of 900 kN and bending moment of 270 kNm. The size of the column is 300 by 550 mm. Design the column.

Solution

Select $M25$ grade concrete as the column is exposed to moderate environment, and use $HYSD$-$Fe415$ grade reinforcement. The factored design loads are:

$$P_u = 1.5*900 = 1350 \text{ kN} = 1.35 \text{ MN},$$
$$M_u = 1.5*270 = 405 \text{ kNm} = 0.405 \text{ MNm},$$

The data of the section is:

$$b = 0.30 \text{ m}, D = 0.55 \text{ m}, f_{ck} = 25 \text{ MPa}, f_y = 415 \text{ MPa},$$

Let the cover to the centre of reinforcement = 40 + 40 = 80 mm. (Reinforcement placed in two rows on each face)

$$\text{The cover ratio} = c' = 80/550 = 0.145; \text{ use } c' = 0.15,$$

The non-dimensional force factors are:

$$X = \frac{M_u}{bD^2 f_{ck}} = \frac{0.405}{0.3*0.55*0.55*25} = 0.179$$

$$Y = \frac{P_u}{bDf_{ck}} = \frac{1.35}{0.3*0.55*25} = 0.327$$

Figure 11.11 gives the data for a cover ratio of 0.15 and from it the intersection of the above coordinates indicates a value at 0.125. The percentage of reinforcement is: $p = 0.125*25 = 3.125\%$

The area of the total reinforcement is: $A_s = 3.125*bD/100 = 5156$ mm^2.

Select ten numbers of 25 mm and two numbers of 16 mm bars. The actual area provided results = $10*490+2*201 = 5302$ mm^2. Provide five of 25 mm and one of 16 mm bars at each face of the column in two rows.

The centre of the reinforcement from the outer face is = $40 +25 +15 =80$ mm.

The assumed cover distance checks with that of the actual one. Provide 8 mm ties at 250 mm spacing.

Example 11.8: Design of Column by Column Design Curves with HYSD-Fe500

A column in mild environment is subjected to an axial load of 900 kN and bending moment of 60 kNm. The size of the column is 300 by 500 mm. Design the column.

Solution

Select M20 grade concrete as the column is exposed to mild environment only, and use HYSD-Fe500 grade reinforcement. Assume the clear cover as 40 mm and the cover to the centre of reinforcement of each face as 50 mm. The factored design loads are:

$$P_u = 1.5*900 = 1350 \text{ kN} = 1.35 \text{ MN},$$
$$M_u = 1.5*60 = 90 \text{ kNm} = 0.09 \text{ MNm},$$

The data of the section is:

$$b = 0.3 \text{ m}, D = 0.5 \text{ m}, f_{ck} = 20 \text{ MPa}, f_y = 500 \text{ MPa}, d' = 50 \text{ mm} = 0.05 \text{m}$$

The cover ratio = $c' = 50/500 = 0.10$

The non-dimensional force factors are:

$$Y = \frac{P_u}{bDf_{ck}} = \frac{1.35}{0.3*0.5*20} = 0.45$$

$$X = \frac{M_u}{bD^2 f_{ck}} = \frac{0.09}{0.3*0.25*20} = 0.06$$

Figure 11.14 gives the percentage of reinforcement for reinforcement of Grade Fe500 and cover ratio of 0.10. Select the x - coordinate as 0.06 and y - coordinate as 0.45 from the figure 11.14 and find the intersection point of the values. Interpolation of the values of q from the curves is 0.045.

The percentage of the reinforcements obtained from the q value of the curves is: $p = 0.045*f_{ck} = 0.9$

The area of the total reinforcement is: $A_s = 0.90*bD/100 = 0.90*300*500/100 = 1350$ mm^2.

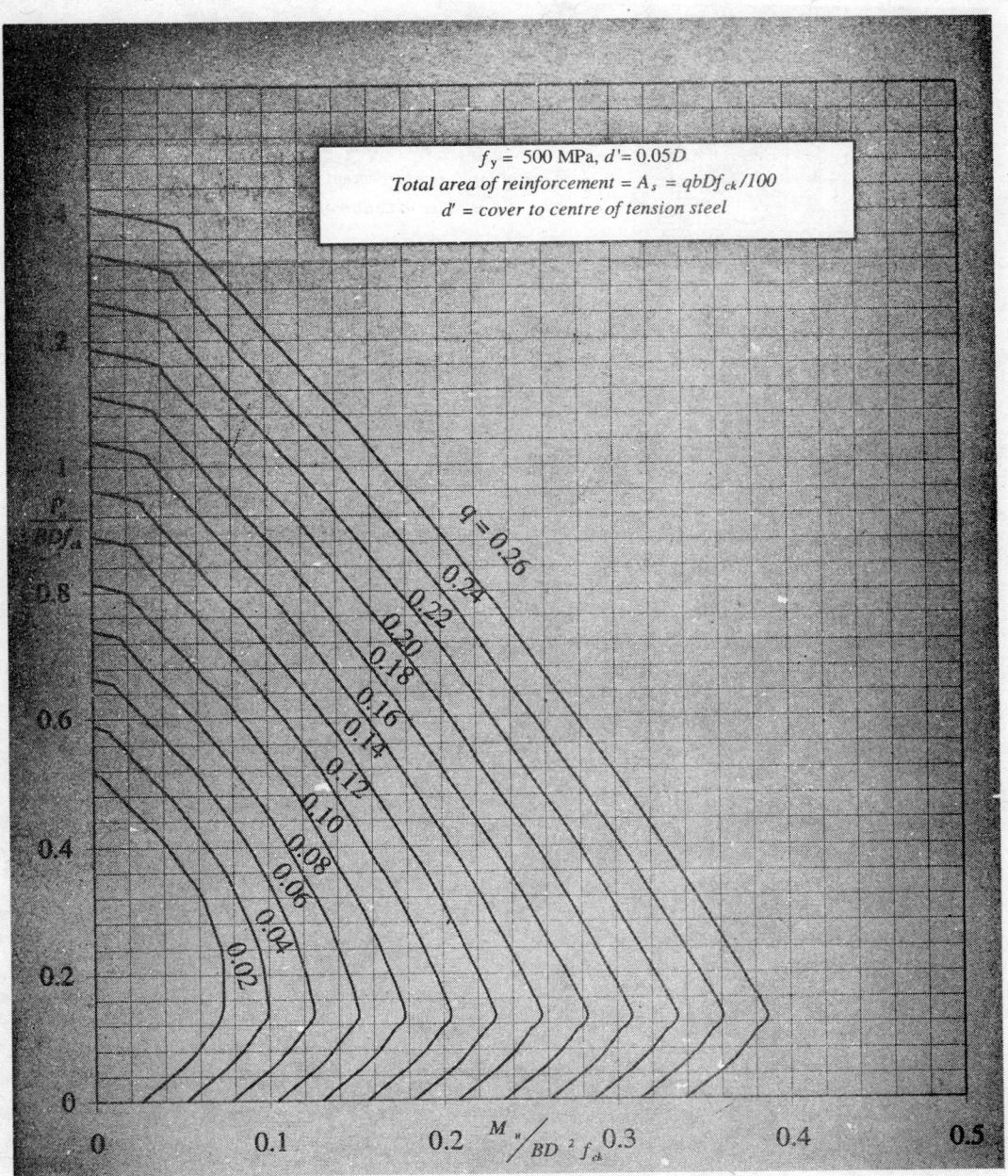

$f_y = 500$ MPa, $d' = 0.05D$
Total area of reinforcement = $A_s = qbDf_{ck}/100$
d' = cover to centre of tension steel

$\dfrac{P}{BDf_{ck}}$

$q = 0.26$
0.24
0.22
0.20
0.18
0.16
0.14
0.12
0.10
0.08
0.06
0.04
0.02

$\dfrac{M_u}{BD^2f_{ck}}$

Fig. 11.13 Column Design Curves for Fe500 and Cover of 0.05D.

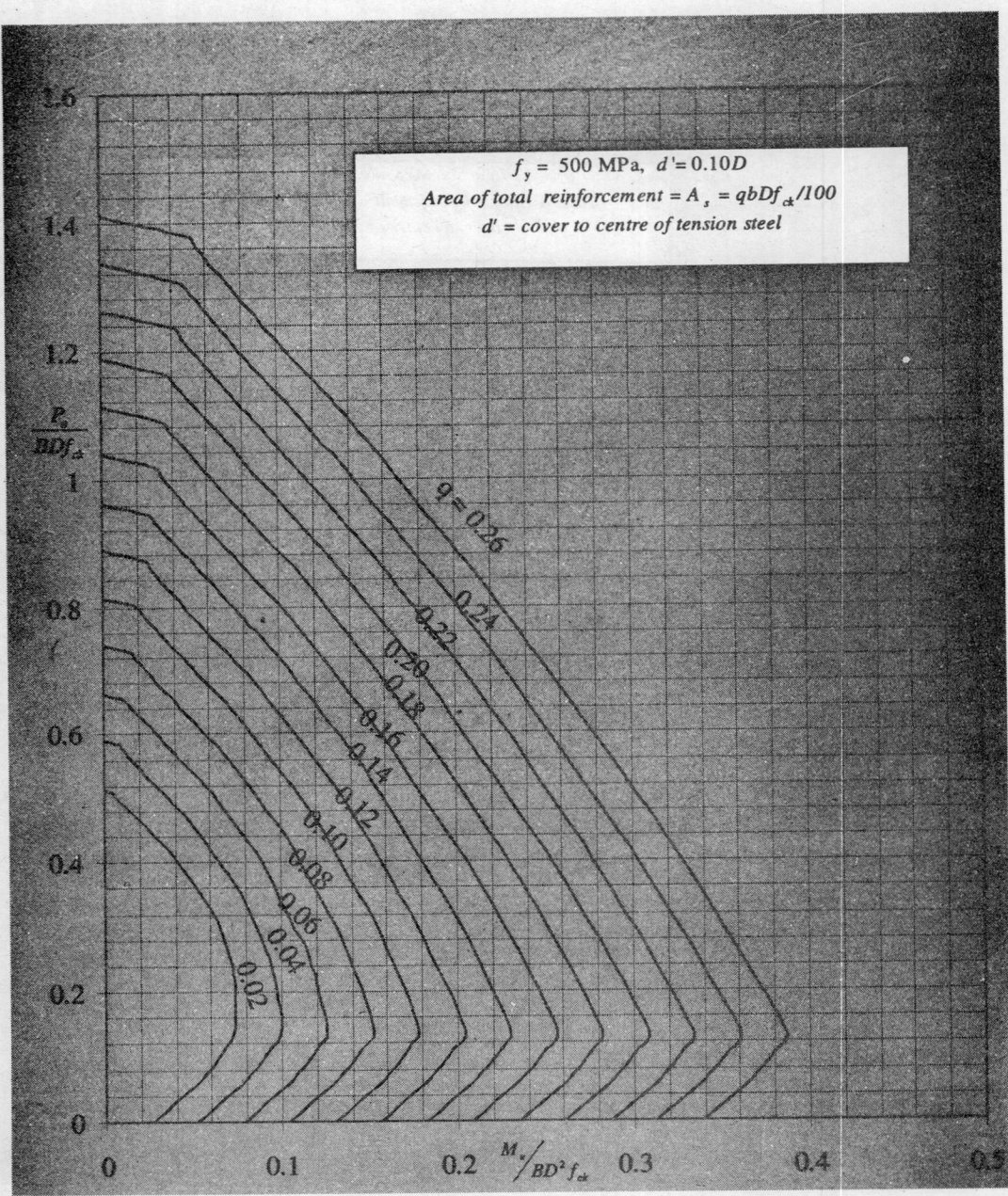

Fig. 11.14 Column Design Curves for Fe500 and Cover of 0.10*D*.

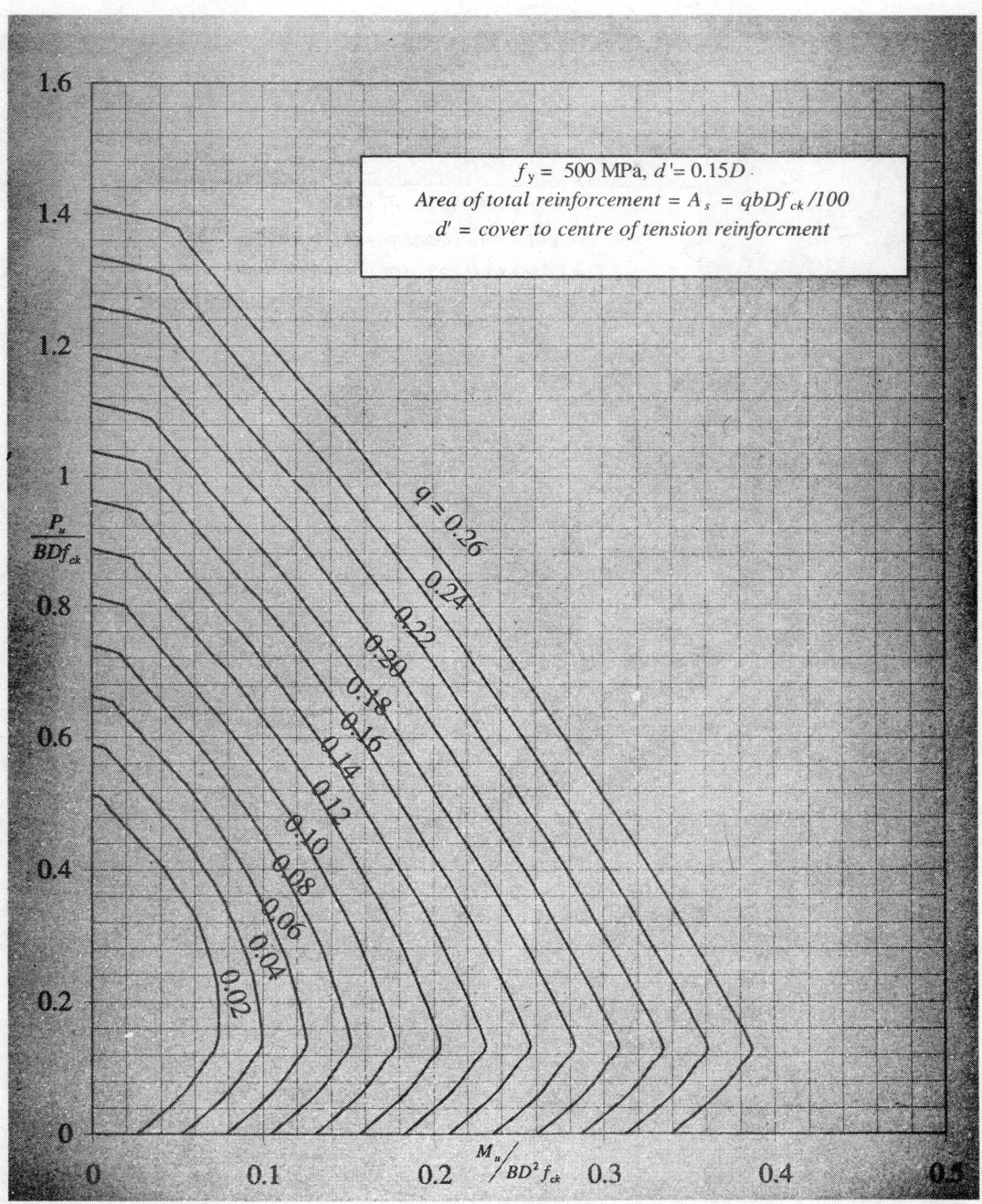

Fig. 11.15 Column Design Curves for Fe500 and Cover of 0.15D.

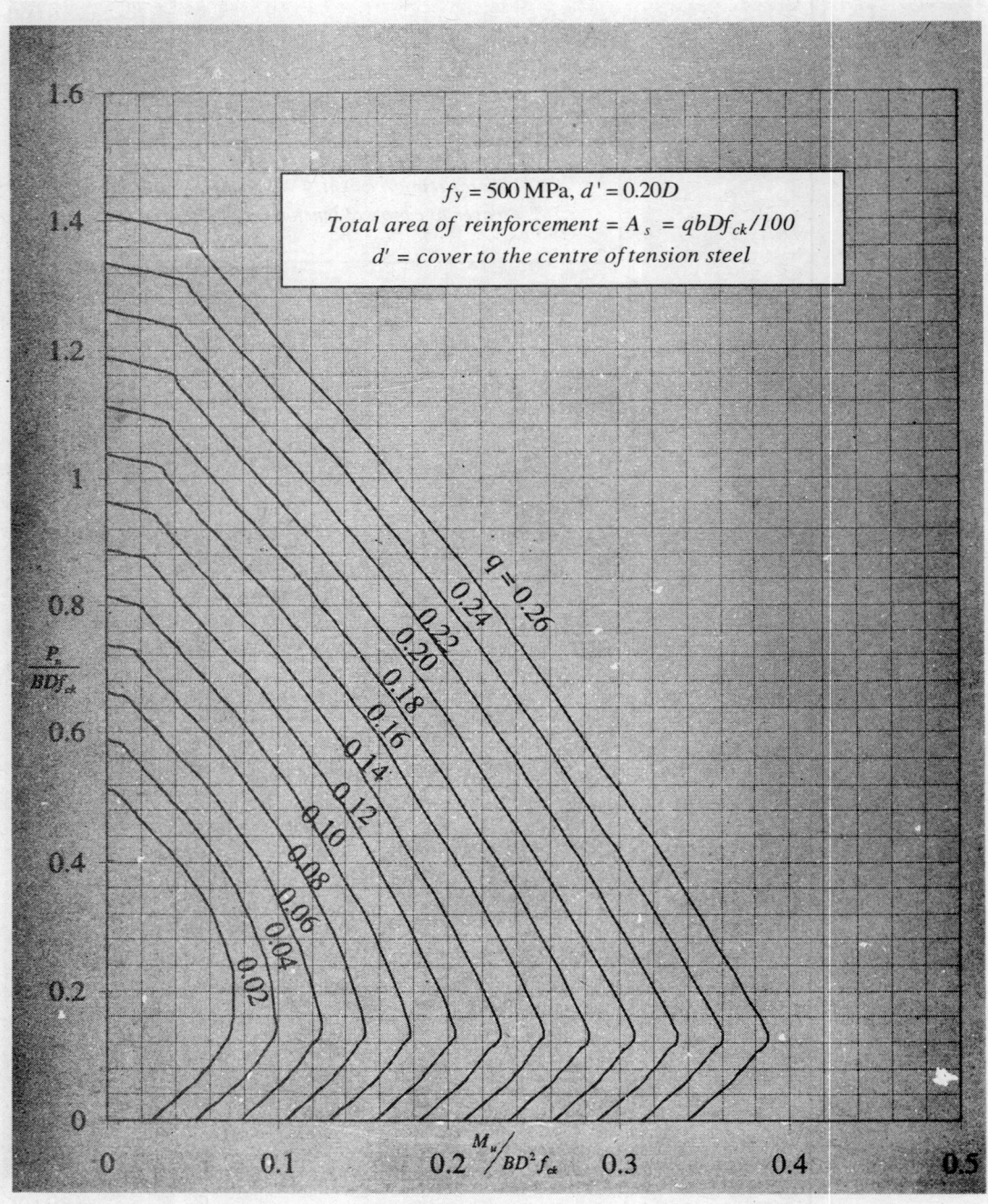

Fig. 11.16 Column Design Curves for Fe500 and Cover of 0.20*D*.

Select four numbers of 20 mm and two numbers of 16 mm bars. The actual area provided results = 1660 mm². Provide three bars at each face of the column. Tie bars to be provided as per the specifications: Provide 6 mm ties at 250 mm spacing.

The cover to the centre of the reinforcement was assumed to be 0.1 times the depth of the column and it is 50 mm. The cover to the centre of the reinforcement is 50 mm checks with the value assumed. The problem was solved in example 11.5 with $Fe415$ grade reinforcement. The solution is exactly the same even though the grade of steel used in this example is $Fe500$. The limit state design of columns limits the bending compressive stress in concrete from 0.002 to 0.0035. Consequently the compatible stress in the reinforcement is limited by the compressive strains. Therefore the compatible stress in the reinforcement may not reach the yield stress of the steel. Use of reinforcement of grade more than $Fe500$ has very little effect on the design of the reinforced concrete columns.

Example 11.9: Design of Column from the Design Curves

A column in moderate environment is subjected to an axial load of 900 kN and bending moment of 270 kNm. The size of the column is 300 by 550 mm. Design the column with $HYSD$-$Fe500$ grade reinforcement.

Solution

Select $M25$ grade concrete and use $HYSD$-$Fe500$ grade reinforcement. The factored design loads are:

$$P_u = 1.5*900 = 1350 \text{ kN} = 1.35 \text{ MN},$$
$$M_u = 1.5*270 = 405 \text{ kNm} = 0.405 \text{ MNm},$$

The data of the section is: $b = 0.30$ m, $D = 0.55$ m, $f_{ck} = 25$ MPa, $f_y = 500$ MPa,

Let the cover to the centre of reinforcement $= 40 + 40 = 80$ mm. (Assumed in two rows of bars on each face)

The cover ratio $= c' = 80/550 = 0.145$; use $c' = 0.15$,

The non-dimensional force factors are:

$$X = \frac{M_u}{bD^2 f_{ck}} = \frac{0.405}{0.3*0.55*0.55*25} = 0.179$$

$$Y = \frac{P_u}{bD f_{ck}} = \frac{1.35}{0.3*0.55*25} = 0.327$$

Figure 11.15 gives the data for a cover ratio of 0.15 and from it the intersection of the above coordinates indicates a q value at 0.125.

The percentage of reinforcement is: $p = 0.125*25 = 3.125\%$

The percentage of reinforcement obtained even in this example is same as that of example 11.6 even though higher grade of reinforcement is used here. This is because the compatible stress in the reinforcement was less than 415 N/mm².

11.5 COLUMNS UNDER BIAXIAL BENDING MOMENT

Simultaneous occurrence of bending moment about two axes of a member is called *biaxial bending*.

Many columns in three-dimensional frames are subjected to biaxial bending moment. The neutral axis in biaxial bending moment is inclined to the principal axes of the section. Analysis of an equilibrium model in biaxial bending moment is rather difficult. It is made more difficult by the stress-strain curve of concrete plus the presence of the reinforcement. The code of practice suggests a practical way of designing such columns. The governing design condition for a column subjected to biaxial bending moment is:

$$\left[\frac{M_{ux}}{M_{ux1}}\right]^{\alpha} + \left[\frac{M_{uy}}{M_{uy1}}\right]^{\alpha} \leq 1.0 \qquad (11.29)$$

in which

M_{ux}, M_{uy} = design bending moments about x and y-axes respectively,
M_{ux1}, M_{uy1} = maximum uniaxial bending moment capacities of the section for an axial load of P_u,
α is a ratio: $a = P_u/P_{uz}$,

$$P_{uz} = 0.45f_{ck}(A_c) + 0.75f_y(A_{sc}) \qquad (11.30)$$

For values of $P_u/P_{uz} = 0.2$ to 0.8, the value of a varies linearly from 1.0 to 2.0. For value less than 0.2, a is 1.0 and for value more than 0.8, it is 2.0.

This relation holds good for short and long columns, however for long columns, additional moment based on the slenderness factors is to be added to the design moments. The procedure is more of an iterative. In case of biaxial bending moment problem, the percentage of the reinforcement is estimated and checked whether the governing condition is satisfied or not. The method of design is illustrated by examples.

Example 11.10: Design of Column under Biaxial Bending Moment

A short column 300 by 500 mm in mild environment is subjected to an axial load of 900 kN and bending moments of 60 kNm and 30 kNm about major and minor axes respectively. Design the column.

Solution

Select $M20$ grade concrete and use $HYSD$-$Fe500$ grade reinforcement. Assume the clear cover as 40 mm and the cover to the centre of reinforcement of each face as 50 mm. The factored design loads are:

$$P_u = 1.5*900 = 1350 \text{ kN} = 1.35 \text{ MN},$$
$$M_{ux} = 1.5*60 = 90 \text{ kNm} = 0.09 \text{ MNm},$$
$$M_{uy} = 1.5*30 = 45 \text{ kNm} = 0.045 \text{ MNm},$$

The data of the section is: $b = 0.3$ m, $D = 0.5$ m, $f_{ck} = 20$ MPa, $f_y = 500$ MPa, $d' = 50$ mm $= 0.05$ m

The cover ratio about x-axis is = $c_x' = 50/500 = 0.10$
The cover ratio about y-axis is = $c_y' = 50/300 = 0.17$, use 0.2

One has to make an estimate of the percentage of reinforcement to compute the maximum admissible forces.

Let the percentage of the reinforcement be = 2 percent, then the value of q is: $q = 100*p/f_{ck} = 2/20 = 0.10$,

The approximate area of reinforcement for $q = 0.1$ is: $A_{sc} = 0.1*20*300*500/100 = 3000$ mm^2.

The axial limit capacity, P_{uz} is:

$$P_{uz} = 0.45f_{ck}(A_c) + 0.75f_y(A_{sc}) = 0.45*20*300*500(1-0.02) +0.75*415*3000 = 2256750 \text{ N} = 2.25\text{MN}$$

$$\frac{P_u}{P_{uz}} = \frac{1.35}{2.25} = 0.6$$

The corresponding value of α is = $1+1(0.6-0.2)/(0.8-0.2) = 1.67$

The maximum uniaxial bending moment has to be computed for the axial load. The corresponding non-dimensional force factor of the Y-axis of the column design curve is:

$$Y = \frac{P_u}{bDf_{ck}} = \frac{1.35}{0.3*0.5*20} = 0.45$$

The maximum bending moment about x-axis under uniaxial bending can be obtained from the column design curve corresponding to $c_x' = 0.05$ and $q = 0.10$ from Fig. 11.9 and it is:

$$X_x = \frac{M_{ux1}}{bD^2 f_{ck}} = 0.14$$

The actual value of M_{ux1} is: $M_{ux1} = 0.14*0.3*0.5*0.5*20 = 0.21$ MNm

Similarly the maximum bending moment about y-axis under uniaxial bending can be obtained from the column design curve corresponding to $c_x' = 0.2$ and $q = 0.10$ from Fig. 11.12 and it is:

$$X_y = \frac{M_{uy1}}{Db^2 f_{ck}} = 0.107$$

The actual value of M_{uy1} is: $M_{uy1} = 0.107*0.5*0.3*0.3*20 = 0.0963$ MNm

The biaxial bending moment governing equation 11.29 is:

$$\left[\frac{M_{ux}}{Mux1}\right]^\alpha + \left[\frac{M_{uy}}{M_{uy1}}\right]^\alpha = \left[\frac{0.09}{0.21}\right]^{1.67} + \left[\frac{0.045}{0.0963}\right]^{1.67} = 0.53$$

The value is too small when compared with one, which indicates the percentage of reinforcement selected is high or the section chosen is large. As the section was assigned by the architectural consideration, the percentage of the reinforcement is reduced to one percent and the governing condition is checked.

Let the percentage of the reinforcement be = 1 percent, then the value of q is:

$$q = 100*p/f_{ck} = 1/20 = 0.05,$$
$$A_{sc} = 0.05*20*300*500/100 = 1500 \text{ mm}^2.$$

The axial limit capacity, P_{uz} is:

$$P_{uz} = 0.45f_{ck}(A_c) + 0.75f_y(A_{sc}) = 0.45*20*300*500(1-0.02) +0.75*415*1500 = 1803375 \text{ N} = 1.8\text{MN}$$

$$\frac{P_u}{P_{uz}} = \frac{1.35}{1.8} = 0.75$$

The corresponding value of α is $= 1+1(0.75-0.2)/(0.8-0.2) = 1.91$

The maximum uniaxial bending moments have to be computed for the axial load. The corresponding non-dimensional force factor of the Y-axis of the column design curve is:

$$Y = \frac{P_u}{bDf_{ck}} = \frac{1.35}{0.3*0.5*20} = 0.45$$

The maximum bending moment about x-axis under uniaxial bending can be obtained from the column design curve corresponding to $c_x' = 0.05$ and $q = 0.05$ from Fig. 11.9 and it is:

$$X_x = \frac{M_{ux1}}{bD^2 f_{ck}} = 0.09$$

The actual value of M_{ux1} is given by:

$$M_{ux1} = 0.09*0.3*0.5*0.5*20 = 0.135 \text{ MNm}$$

Similarly the maximum bending moment about y-axis under uniaxial bending can be obtained from the column design curve corresponding to $c_x' = 0.2$ and $q = 0.05$ from Fig. 11.12 and it is:

$$X_y = \frac{M_{uy1}}{Db^2 f_{ck}} = 0.075$$

The actual value of M_{uy1} is: $M_{uy1} = 0.075*0.5*0.3*0.3*20 = 0.0675 \text{ MNm}$

The biaxial bending moment governing equation 11.29 is:

$$\left[\frac{M_{ux}}{M_{ux1}}\right]^\alpha + \left[\frac{M_{uy}}{M_{uy1}}\right]^\alpha = \left[\frac{0.09}{0.135}\right]^{1.91} + \left[\frac{0.045}{0.0675}\right]^{1.91} = 0.98 < 1.0$$

The value is 0.98 slightly less than one, therefore one percent reinforcement is the most appropriate. The area of the total reinforcement is:

$$A_s = 1.0*bD/100 = 300*500/100 = 1500 \text{ mm}^2.$$

Select 20 mm at each corner and 12 mm bar at middle of each face. The actual area provided results = 1708 mm². Provide three bars at each face of the column. Tie bars to be provided as per the minimum limits specified in the earlier chapter: Provide 6 mm ties at 190 mm spacing.

This problem was solved in example 11.5 with $Fe415$ grade reinforcement for uniaxial bending.

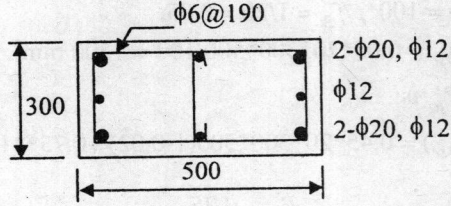

Fig. 11.8 Column Reinforcement; Example 11.10.

For quick computation of the Eq.11.29, a graph is generated by Microsoft Excel program and shown in Fig. 11.19. The use of the graph is also illustrated.

The following notation is used in the figure 11.19.

$$X = M_{ux}; \qquad X_1 = M_{ux1}$$
$$Y = M_{uy}; \qquad Y_1 = M_{uy1}$$

Example 11.11: Design of Column under Biaxial Bending Moment

A column 300 by 550 mm in moderate environment is subjected to an axial load of 900 kN and bending moments of 270 kNm about major axis and 90 kNm about minor axis. Design the column.

Solution

Select $M25$ grade concrete as the column is exposed to moderate environment, and use $HYSD$-$Fe500$ grade reinforcement. The factored design loads are:

$$P_u = 1.5*900 = 1350 \text{ kN} = 1.35 \text{ MN},$$
$$M_{ux} = 1.5*270 = 405 \text{ kNm} = 0.405 \text{ MNm},$$
$$M_{uy} = 1.5*90 = 135 \text{ kNm} = 0.135 \text{ MNm},$$

The data of the section is: $b = 0.30$ m, $D = 0.55$ m, $f_{ck} = 25$ MPa, $f_y = 500$ MPa,

Let the cover to the centre of reinforcement = $40 + 40 = 80$ mm. (Reinforcement is assumed in two rows on each face of major axis and one row in the other axis)

$$c_x' = 80/550 = 0.145$$

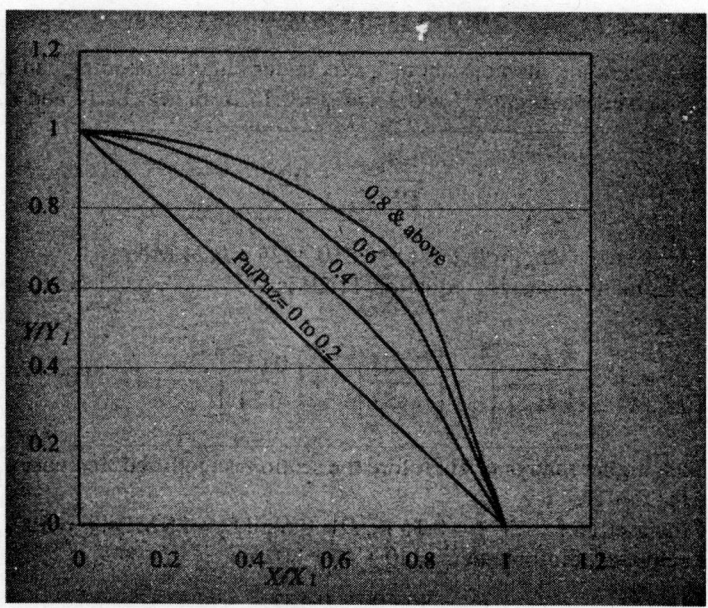

Fig. 11.19 Biaxial Bending Moment Governing Equation.

Let the percentage of the reinforcement be = 3.5 percent, then the value of q is:

$$q = 100*p/f_{ck} = 3.5/25 = 0.14,$$
$$A_{sc} = 0.14*25*300*550/100 = 5775 \text{ mm}^2.$$

The axial limit capacity, P_{uz} is:

$$P_{uz} = 0.45f_{ck}(A_c) + 0.75f_y(A_{sc}) = 0.45*25*300*550(1-0.035) +0.75*500* 5775$$

$$= 1791281 + 2165625 = 3956906 \text{ N} = 3.96 \text{ MN}$$

$$\frac{P_u}{P_{uz}} = \frac{1.35}{3.96} = 0.34$$

The corresponding value of α is = $1+1(0.34-0.2)/(0.8-0.2) = 1.2$

The maximum uniaxial bending moment has to be computed for the axial load. The corresponding non-dimensional force factor of the Y-axis of the column design curve is:

$$Y = \frac{P_u}{bDf_{ck}} = \frac{1.35}{0.3*0.5*25} = 0.33$$

The maximum bending moment about x-axis under uniaxial bending can be obtained from the column design curve corresponding to $c_x' = 0.15$ and $q = 0.14$ from Fig. 11.15 and it is:

$$X_x = \frac{M_{ux1}}{bD^2f_{ck}} = 0.2$$

The actual value of M_{ux1} is: $M_{ux1} = 0.2*0.3*0.55*0.55*25 = 0.453$ MNm

Similarly the maximum bending moment about y-axis under uniaxial bending can be obtained from the column design curve corresponding to $c_x' = 0.2$ and $q = 0.14$ from Fig. 11.16 and it is:

$$X_y = \frac{M_{uy1}}{Db^2f_{ck}} = 0.2$$

The actual value of M_{uy1} is: $M_{uy1} = 0.2*0.55*0.3*0.3*25 = 0.24$ MNm
The biaxial bending moment governing equation 11.29 is:

$$\left[\frac{M_{ux}}{M_{ux1}}\right]^\alpha + \left[\frac{M_{uy}}{M_{uy1}}\right]^\alpha = \left[\frac{0.405}{0.453}\right]^{1.2} + \left[\frac{0.135}{0.24}\right]^{1.2} = 1.43 > 1.0$$

The value is 1.43 much higher than one. Therefore the section is modified. Increase the section to 400 by 650 mm and revise the design.
The revised data of the section is: $b = 0.35$ m, $D = 0.6$ m, $f_{ck} = 25$ MPa, $f_y = 500$ MPa,
Let the cover to the centre of reinforcement = 40 + 40 = 80 mm.

$$c_x' = 80/600 = 0.133$$

Let the percentage of the reinforcement be = 3.5 percent, then the value of q is:

$$q = 100*p/f_{ck} = 3.5/25 = 0.14,$$
$$A_{sc} = 0.14*25*350*600/100 = 7350 \text{ mm}^2.$$

The axial limit capacity, P_{uz} is:

$$P_{uz} = 0.45f_{ck}(A_c) + 0.75f_y(A_{sc}) = 0.45*25*350*600(1-0.035) + 0.75*500*7350$$

$$= 2279812 \text{ N} + 2756250 = 5036062 \text{ kN} = 5.03 \text{ MN}$$

$$\frac{P_u}{P_{uz}} = \frac{1.35}{5.03} = 0.27$$

The corresponding value of α is $= 1 + 1(0.27-0.2)/(0.8-0.2) = 1.1$

The maximum uniaxial bending moment has to be computed for the axial load. The corresponding non-dimensional force factor of the Y-axis of the column design curve is:

$$Y = \frac{P_u}{bDf_{ck}} = \frac{1.35}{0.3*0.6*25} = 0.257$$

The maximum bending moment about x-axis under uniaxial bending can be obtained from the column design curve corresponding to $c_x' = 0.15$ and $q = 0.14$ from Fig. 11.15 and it is:

$$X_x = \frac{M_{ux1}}{bD^2 f_{ck}} = 0.21$$

The actual value of M_{ux1} is: $M_{ux1} = 0.21*0.35*0.6*0.6*25 = 0.6615 \text{ MNm}$

Similarly the maximum bending moment about y-axis under uniaxial bending can be obtained from the column design curve corresponding to $c_x' = 0.2$ and $q = 0.14$ from Fig. 11.16 and it is:

$$X_y = \frac{M_{uy1}}{Db^2 f_{ck}} = 0.22$$

The actual value of M_{uy1} is: $M_{uy1} = 0.22*0.6*0.35*0.35*25 = 0.404 \text{ MNm}$

The moment ratios are: $M_{ux}/M_{ux1} = 0.405/0.6615 = 0.61$
$M_{uy}/M_{uy1} = 0.135/0.404 = 0.33$

From Fig. 11.19, for $M_{ux}/M_{ux1} = 0.61$ and $P_u/P_{uz} = 1.27$, the admissible maximum value of M_{uy}/M_{uy1} is 0.41 that is greater than the actual value of 0.33. Hence the design is acceptable.

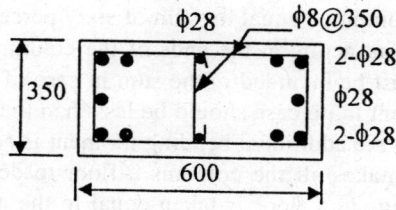

Fig. 11.20 Column Section; Example 11.12.

The area of the total reinforcement is:

$$A_s = 3.5*bD/100 = 350*600/100 = 7350 \text{ mm}^2.$$

Select 12 numbers of 28 mm bars.

The actual area provided results = 7380 mm^2. Provide five bars at each short face of the column, and one at the mid point of the longer face.

Provide 8 mm ties at 350 mm spacing. Extra 8mm link connecting the middle bars as shown in Fig. 11.20.

11.6 DESIGN OF SLENDER COLUMNS

A column is said be slender if the slenderness factor is greater than 12. The ratio of the effective length to the side of the column section is defined as the slenderness factor of the column. A slender column must be designed with bending moment coming from the *P-delta* effects or alternatively it must be designed for additional bending moments corresponding to the slenderness factors as given below:

$$M_{ax} = \frac{P_u D}{2000}\left\{\frac{l_{ex}}{D}\right\}^2 \tag{11.31}$$

$$M_{ay} = \frac{P_u b}{2000}\left\{\frac{l_{ey}}{b}\right\}^2 \tag{11.32}$$

where

P_u = axial load,

l_{ex} = effective length in respect of major axis,

l_{ey} = effective length in respect of minor axis,

D = depth of the section with respect to the major axis,

b = depth of the section with respect to the minor axis, usually called the width of the section.

Additional bending moments are added to the design moments from the structural analysis. All slender columns have to be designed for biaxial bending moment.

The code lists the following additional notes in the design of the slender columns:

1. A column is considered as braced if the lateral forces on the building are resisted by walls or bracing or buttresses or shear walls etc.
2. A column is considered unbraced if the beam-column frame resists the lateral loads on the structure. *P-delta* effects in the design of the column have to be considered in such a column.
3. In case of braced column without any lateral load the additional moment has to be added to an effective bending moment computed equal to: sum of sixty percent of the larger moment and forty percent of the smaller moment acting at the ends of the column. The negative sign if any to the smaller bending moment must be included in the sum in case of column having double curvature. The effective bending moment in no case should be less than the smaller bending moment.
4. In case of unbraced column, the additional bending moment is added to the end bending moment.
5. The stiffness of the floors makes all the columns a floor to deflect equally (more or less). The slenderness of all the columns in a floor is taken equal to the average of all the columns in that direction.

The effective length of non-sway and sway columns in a framed structure depends on the relative stiffness of the members connected at the joint. IS: 456-2000 gives two sets of graphs taken from the Journal of: *'Structural Engineer'*, *CISE, UK* that enable to determine the effective lengths of columns.

The method of design of slender columns is illustrated.

Example 11.12: Design of a Long (slender) Column

A 320 by 320 mm column of effective length of 5 m in moderate environment carries an axial load of 900 kN. Design the column with M25 concrete and HYSD-Fe415 reinforcement bars. (Same as the example of long column in chapter 10)

Solution

Assuming the self-weight of the column as 30 kN, the design load on the column with 1.5 partial safety factor is

$$P_u = \gamma_f(W_g + W_l) = 1.5*(900 + 30) = 1395 \text{ kN} = 1.395 \text{ MN}$$

The slenderness factor of the column is: $g = \dfrac{L_e}{b} = \dfrac{5}{0.32} = 15.625 > 12$.

The column is a long one and the bending moments due to the slenderness effects must be calculated.

$$M_{ax} = \frac{P_u D}{2000}\left\{\frac{l_{ex}}{D}\right\}^2 = \frac{1.395(0.32)}{2000}\left\{\frac{5}{0.32}\right\}^2 = 00544 \text{ MNm}$$

The bending moment about the y-plane is also same, and it is: $M_{ay} = 0.0544$ MNm

The minimum eccentricity for which the column has to be designed is:

$$e_{min} = \frac{L}{500} + \frac{b}{30} = 21 \text{ mm } \textit{it is more than } 20 \text{ mm}, \textit{ so use } 21 \text{ mm}$$

The minimum bending moment based on the eccentricity is:

$$M_{min} = P_u * e_{min} = 1.395*0.21 = 0.0293$$

This minimum eccentricity bending moment is smaller than that due to the slenderness factor, so use the larger of the moments. The column is designed as slender column, subjected to the biaxial bending moment caused by the slenderness factor.

The data of the section is: $b = 0.32$ m, $D = 0.32$ m, $f_{ck} = 25$ MPa, $f_y = 415$ MPa, Let $d' = 50$ mm $= 0.05$m

The cover ratio about x-axis is $= c_x' = 50/320 = 0.16$
The cover ratio about y-axis is $= c_y' = 50/320 = 0.16$,

One has to make an estimate of the percentage of reinforcement to compute the maximum admissible bending and axial capacities of the section. Let the percentage of the reinforcement be $= 2$ percent, then the value of q is:

$$q = 100*p/f_{ck} = 2/25 = 0.08,$$

Use $q = 0.10$ for the approximate area of reinforcement is: $A_{sc} = 0.1*25*320*320/100 = 2560$ mm^2.

The axial limit capacity, P_{uz} is:

$$P_{uz} = 0.45f_{ck}(A_c) + 0.75f_y(A_{sc}) = 0.45*25*320*320(1-0.02) +0.75*415* 2560 = 1925760 \text{ N} = 1.925 \text{ MN}$$

$$\frac{P_u}{P_{uz}} = \frac{1.395}{1.925} = 0.72$$

The corresponding value of α is $= 1+1(0.72-0.2)/(0.8-0.2) = 1.87$

The maximum uniaxial bending moment is computed for the axial load. The corresponding non-dimensional force factor of the Y-axis of the column is:

$$Y = \frac{P_u}{bDf_{ck}} = \frac{1.395}{0.32*0.32*25} = 0.54$$

The maximum bending moment about x-axis under uniaxial bending is obtained from the column design curve corresponding to $c_x' = 0.2$ and $q = 0.10$ from Fig. 11.12 and it is:

$$X_x = \frac{M_{ux1}}{bD^2 f_{ck}} = 0.085$$

The actual value of M_{ux1} is:

$$M_{ux1} = 0.085*0.32*0.32*0.32*25 = 0.07 \text{ MNm}$$

Similarly the maximum bending moment about y-axis under uniaxial bending can be obtained from the column design curve corresponding to $c_x' = 0.2$ and $q = 0.10$ from Fig. 11.12 is exactly same as the section is.square:

$$M_{uy1} = 0.07 \text{ MNm}$$

The biaxial bending moment governing equation 11.29 is:

$$\left[\frac{M_{ux}}{M_{ux1}}\right]^{\alpha} + \left[\frac{M_{uy}}{M_{uy1}}\right]^{\alpha} = \left[\frac{0.0544}{0.07}\right]^{1.87} + \left[\frac{0.0544}{0.07}\right]^{1.87} = 2 * 0.62 = 1.24 > 1.0$$

The governing condition is not satisfied. As the size is pre-assigned, the percentage of the reinforcement is increased to 3 percent and rechecked.

Let the percentage of the reinforcement be $= 3$ percent, then the value of q is:

$$q = 100*p/f_{ck} = 3/25 = 0.12,$$
$$A_{sc} = 0.12*25*320*320/100 = 3072 \text{ mm}^2.$$

The axial limit capacity, P_{uz} is:

$$P_{uz} = 0.45f_{ck}(A_c) + 0.75f_y(A_{sc}) = 0.45*25*320*320(1-0.03) +0.75*415* 3072 = 2073600 \text{ N} = 2.07 \text{MN}$$

$$\frac{P_u}{P_{uz}} = \frac{1.395}{2.07} = 0.67$$

The corresponding value of α is $= 1+1(0.67-0.2)/(0.8-0.2) = 1.95$

The maximum uniaxial bending moments are computed for the axial load. The corresponding non-dimensional force factor of the Y-axis of the column design curve already computed is 0.54.

The maximum bending moment about x-axis under uniaxial bending is obtained from the column design curve corresponding to $c_x' = 0.2$ and $q = 0.12$ from Fig. 11.12 is:

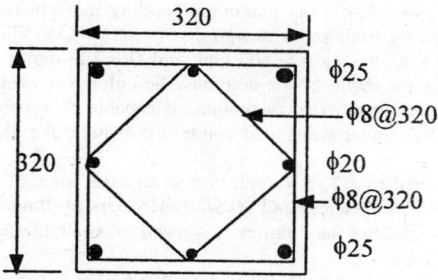

Fig. 1.22 Cross-section of Column; Example 11.12.

$$X_x = \frac{M_{ux1}}{bD^2 f_{ck}} = 0.1$$

The actual value of M_{ux1} is: $M_{ux1} = 0.1*0.32*0.32*0.32*25 = 0.082$ MNm
Similarly the maximum bending moment about y-axis is: $M_{uy1} = 0.082$ MNm
The biaxial bending moment governing equation 11.29 is:

$$\left[\frac{M_{ux}}{M_{ux1}}\right]^\alpha + \left[\frac{M_{uy}}{M_{uy1}}\right]^\alpha = \left[\frac{0.0544}{0.082}\right]^{1.95} + \left[\frac{0.0544}{0.082}\right]^{1.95} = 0.9 < 1.0$$

The value is 0.9 less than one. Therefore about three percent reinforcement is appropriate. The area of the total reinforcement is:

$$A_s = 3.0*bD/100 = 3*320*320/100 = 3072 \text{ mm}^2.$$

Select four numbers of 25 mm and four numbers of 20 mm bars. The actual area provided results = 3216 mm^2. Provide 25 mm bars at the four corners and 20 mm bars at the mid points of the four sides. Tie bars to be provided as per the specifications.

Provide 8 mm ties at 320 mm spacing. Figure 11.21 illustrates the sectional details.

Some more examples of columns under combined bending moment are solved in the chapter on design of framed building.

PROBLEMS

11.1   A column in mild environment, effective height of 4 m and size 340 by 340 mm is subjected to an axial force of 800 kN and bending moment of 100 kNm. Design a rectangular sectioned column by limit state design using *M*20 concrete and *HYSD-Fe*415 bars

11.2   A 300 by 350 mm column in moderate environment, 3.6 m effective length is subjected to an axial live load of 800 kN and bending moment of 70 kNm under wind load condition. Design the column with *M*25 concrete and *HYSD-Fe*415 bars by limit state design.

11.3 A column in moderate environment, effective length of 4.5 m is subjected to an axial live load of 1200 kN and bending moment of 280 kNm. The size of the column and reinforcement details are: $b = 400$ mm, $D = 450$, $A_s = 2512$ mm^2. Determine whether the column when made of $M25$ concrete and $HYSD$-$Fe415$ bars can withstand the above load. The reinforcement is equally divided on two opposite faces.

11.4 A column of effective height 5 m is subjected to an axial load of 1800 kN. The size and reinforcement details are: $b = 425$ mm, $D = 500$ mm; $A_s = 2512$ mm^2 equally divided on the two faces. The column is made of $M25$ concrete and $HYSD$-$Fe500$ bars. Investigate what is the maximum bending moment that the column can resist under wind load condition. Check the result by limit states design.

11.5 A column of effective height 5.8 m, having $b = 500$ mm, and $D = 700$ mm is provided with 4 Nos. of 25 mm bars at each of the opposite faces of the depth of the column. The column is made of $M25$ concrete and $HYSD$-$Fe415$ bars. Investigate by limit states design whether the column is capable of resisting 1.0 MN axial live load along with the bending moment of 200 kNm under wind load condition. Assume the clear cover to the reinforcement as 50 mm.

11.6 A column of effective height equal to 3.2 m is subjected to an axial force of 500 kN and bending moment of 200 kNm. Design the column with $M20$ concrete and $HYSD$-$Fe415$ bars by limit state design. The size of the column is 300 by 400 mm. Investigate whether the column is cracked or un-cracked section. Provide equal reinforcement at the tension and compression faces.

11.7 A long column exposed to severe environment, 6.5 m effective height is subjected to an axial load of 750 kN. The size of the column is 400 by 400 mm. Design the column by limit state design using $M30$ concrete and $HYSD$-$Fe500$ bars. Provide equal reinforcement at four faces of the column.

11.8 A column of effective height 5.5 m is subjected to an axial load of 400 kN and bending moment of 100 kNm. The size of the column is 400 by 450 mm. Design the column by limit state method using $M25$ concrete and $HYSD$-$Fe415$ bars.

11.9 A reinforced concrete column of 4.5 m effective length in both the directions is subjected to an axial force of 500 kN and biaxial bending moments of 100 and 200 kNm under live load condition. The size of the column is 350 by 350 mm. Design the column by limit state design with $M25$ concrete and $HYSD$-$Fe500$ reinforcement.

Limit State Design of Footing and Raft foundations

12.1 INTRODUCTION

Geotechnical engineering deals with mechanics of soils and design of foundations and is an important branch in civil engineering. It is presumed that the student is familiar with the basics of soil mechanics. Important characteristics of soils for foundation design are:

- Type of soil, filling if any,
- Bearing capacity of soil,
- Shear and unconfined strengths,
- Settlement and compressibility,
- Water table level,
- Permeability,
- Friction angle, angle of repose, etc.

A more detailed soil investigations need to be carried out for special foundations. Soil structure interaction becomes important in vibration problems. Foundations to structures are broadly categorized as:

- Shallow foundations, and
- Deep foundations.

A footing is a spread construction at the base of a wall or column to transfer the load to the ground. Footing, raft and mat foundations are the typical examples of shallow foundations. Deep foundation derives its strength from the depth properties through which it is taken. Pile and well foundations are the typical examples of deep foundations.

The primary design criteria in shallow foundations are:
1. Maximum pressure on the soil should not exceed the allowable pressure,
2. Settlement of the soil due to loads and environmental conditions must be less than the allowable value,
3. The minimum foundation depth should be provided so as to protect from environmental factors.
4. The soil is not capable of resisting tensile stress but only compressive and shears forces.

Some of the durability considerations are:
- Lean concrete of minimum of 75 mm is provided below the RC foundation.
- Normal Minimum cover to the reinforcement is 50 mm, and 75 mm in aggressive environment.

- The external temperature does not effect much soil beyond one metre depth when compared with those on the superstructure. A nominal reinforcement of 0.12% is provided in foundations.

12.2 SIZE OF FOOTING

The size of foundation is controlled by the allowable bearing capacity and settlement property of the soil. The design criterion of the size of a footing is that the bearing pressure under the footing must be with in the admissible limit. The bearing pressure under the footing is:

$$p = \frac{P}{A} \pm \frac{My}{I} \qquad (12.1)$$

where:

p = bearing pressure on the soil,
P = axial load on the foundation,
A = area of the foundation, M = bending moment,
y = distance of the point under consideration from the centre of gravity of the foundation, and
I = sectional moment of inertia,

The design consideration for a rectangular footing is:

$$p_1 = \frac{P}{LB} + \frac{6M}{LB^2} \le p_a \qquad (12.2)$$

$$p_1 = \frac{P}{LB} - \frac{6M}{LB^2} \ge 0 \qquad (12.3)$$

where:

B = width and L = length of the footing,
p_1 = maximum pressure on the soil,
p_2 = minimum pressure on the soil,
p_a = allowable compressive bearing pressure on the soil.

Figure 12.1 illustrates the typical pressure distribution on soil and the notations in rectangular footing.

The pressure distribution in Fig. 12.1 assumes that the total bottom face of the footing is in contact with the soil. However in some situations, the bending moment on the footing may cause uplift of the footing from the soil. Such pressure distribution in partial contact with the soil is shown in Fig. 12.2.

Equation 12.3 is not satisfied if the virtual eccentricity ratio is beyond the kern distance, which is 1/6 in case of rectangular base. The intensity of the maximum pressure on the soil from Eq. 12.2 is given by:

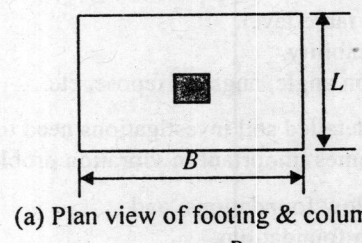

(a) Plan view of footing & column

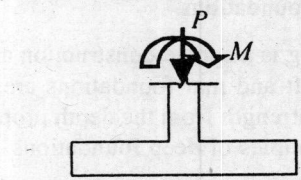

(b) Sectional elevation & loads on column

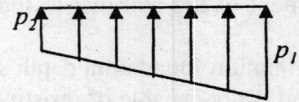

(c) Soil pressure distribution

Fig. 12.1 Footing and Pressure Distribution.

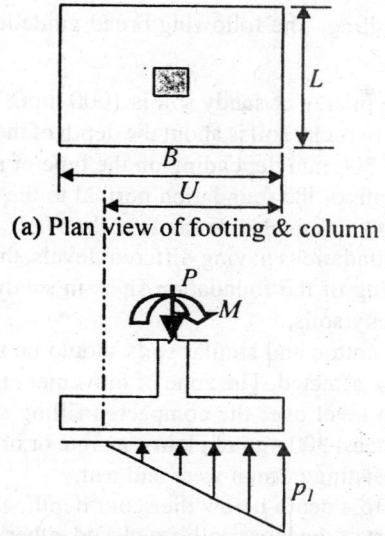

(a) Plan view of footing & column

(b) Pressure distribution in partial contact

Fig. 12.2 Pressure Distribution in Partial Contact Footing.

$$p_1 = \frac{P}{LU} + \frac{6M}{LU^2} \leq p_a$$

where U = the width of the contact of the footing and less than B.   (12.4)

Eq. 12.3 set equal to zero and rearranged as:

$$p_1 = \frac{2P}{LU} \leq p_a \qquad\qquad (12.5)$$

The bearing pressures can be computed from Eqs.12.2 and 12.3. In case the pressure comes out to be negative (tension), then the two Eqs.12.4 and 12.5, are applicable. In a design problem, the primary unknowns are the dimensions of the footing. The value of the maximum pressure (p_1) can be set equal to the allowable pressure. The lower pressure (p_2) need not be pre-assigned. Therefore, the two governing equations have three unknowns, namely, length and breadth of the footing and the minimum pressure. It is desirable to assume a reasonable width or length of the footing and then determine the other dimension.

12.3 DEPTH OF FOUNDATION

The depth measured from the ground level to the bottom face of the lean concrete is considered as the depth of foundation. A lean concrete of mix 1:5:10 or 1:4:8 or lime concrete of 1:3:6 can be used as a surface finish on the soil. The level of excavation may not be uniform and further the soil absorbs cement grout, so a lean concrete is placed over the excavated surface. The thickness of the lean concrete can be from 75 mm to 150 mm in which the lower value is used in small residential buildings or in rocky ground surface. The foundation depth should be taken to a depth to avoid any damage to the foundation concrete and to protect the soil against erosion. No foundation should be founded on any filled up soil unless special treatment is given to the filling. The following broad guidelines are recommended for depth of foundation for quick reference.

- Minimum Depth of foundation in clay or sandy soil is 1000 mm,
- Minimum Depth of foundation in rocky soil is about the depth of the weathered rock surface. It may be any where between 100 mm to 300 mm depending on the type of rock surface,
- In case of sloping soils, the depth of the foundation normal to the sloping ground should be equal to 100 mm in rocky soils and 1000 mm in clay or sandy soils,
- In case of depths of adjacent foundations having different levels, the difference between levels should not exceed half the clear spacing of the foundation slabs in sandy or clayey soils and it should not exceed the clear spacing in rocky soils,
- The foundation depth in black cotton and similar soils should be taken to a level where the swelling and shrinkage of the soil are not affected. The zone of movement is about 3.5 m in black cotton soils.
- An apron placed at the ground level over the compacted filling should protect the foundation. The width of the apron should be atleast 300 mm of plain concrete or brick or stone slabs in pointing. Such an apron must be capable of resisting normal wear and tear,
- The foundation must be taken to a depth below the scour depth with proper anchorage,
- In case there is a soil filling, part of the loose soil is replaced either by compacted sand or by gravel on which the footing can rest. Alternatively, part of the filling can also be replaced by lean concrete. The compacted sand or the lean concrete must develop the same bearing capacity as that of the base soil. Settlements must be within the limits,
- The problem of liquefaction of the soil in clayey type under earthquake load should be considered.

12.4 STRUCTURAL DESIGN OF FOOTING

Minimum structural requirements in footings are:

Minimum thickness of footing including that at free edge is = 150 mm,
Minimum thickness of pile cap even at the free edge is = 300 mm,
Minimum percentage of reinforcement in each direction = 0.12 percent,
Minimum cover to the reinforcement is 50 mm under normal conditions of exposure, it may extend to 75 mm in sever exposure conditions.

Bending Moment Consideration are

Footing must be designed to resist bending moments and shear forces. The critical bending moment locations are:

- At the face of the column in reinforced concrete column or wall,
- Half way between the centre line and the edge of the wall or column in Masonry column or wall,
- Half way between the face of the column and the edge of the gusseted base.

The bending moment is computed across the critical face extending over the total width of the footing.

The total reinforcement to resist the bending moment must be distributed as indicated below:

- In One-way reinforced footing: The total reinforcement is distributed evenly across the width,
- In Two-way square footing: The total reinforcement is evenly distributed across the width,
- In Two-way rectangular footing: The reinforcement in the long direction is distributed evenly across the width. In case of short span direction, the reinforcement is distributed in different proportion in the central band and the edge bands. The amount of the reinforcement in the central band width is given by the following:

$$\frac{Reinforcement\ in\ the\ central\ band\ width}{Total\ reinforcement\ in\ the\ short\ span} = \frac{2}{\beta + 1} \tag{12.6}$$

where β = ratio of the long side to short side of the footing.

The balance of the reinforcement is evenly distributed in the outer portions of the footing.

Shear Design Consideration are

The design for shear in footing is governed by the severe of the following two. One is transverse shear strength design across the width of the footing as a wide beam.

· The footing acts as a wide beam with critical section at a distance of effective depth from the face of the column. The student has to refer to design of beam for shear.

· A diagonal tension failure at the surface of truncated cone or pyramid around the column (concentrated load). The critical surface is at a distance of half of the effective depth from the faces of the column. The total reaction acting outside the truncated cone is considered as shear acting on the surface. In two-way footing the design is for diagonal tension shear. The critical shear plane is at a distance half the effective depth of the slab from the face of the column. The total effective width is equal to the total perimeter at a distance d/2 from the face of the column. There are three possible cases in the transverse shear strength and they are:

Case 1: The nominal shear stress at the section less than the allowable: No shear reinforcement is needed.

$$\tau_v \le \tau_{cc} = k_s \left(0.25\sqrt{f_{ck}}\right) \tag{12.7}$$

where

τ_v = nominal shear stress,

τ_{cc} = allowable shear stress,

k_s = [0.5 + *(short side/long side) of the column*], subject to a maximum limit of 1.0 $\tag{12.8}$

Case 2: For $\tau_{cc} < \tau_v < 1.5\tau_{cc}$, the shear reinforcement is provided in the form of stirrups for the shear stress in excess of $0.5\tau_{cc}$.

Case 3: For $\tau_v > 1.5\tau_{cc}$, the section has to be revised such that the nominal shear stress becomes less than 1.5 times the admissible one without the shear reinforcement.

Bond and Bearing Considerations

The design bond strength and development length in footing is same as that in the beams or slabs.

In case of masonry columns or walls, the admissible bearing pressure on the footing concrete is given by:

$$Permissible\ bearing\ pressure = 0.45 f_{ck} \left[\sqrt{\frac{A_1}{A_2}},\ but\ not\ more\ than\ 2\right] \tag{12.9}$$

where

f_b = bearing stress in direct compression,

A_1 = supporting bearing area on the footing, excluding the tapering faces with side slope more than 1 in 2

A_2 = bearing area of the column base.

The main bars or the dowel bars of column must be taken into the footing to a length equal to the development length. Sometimes the thickness of the footing is increased to accommodate the development length of the bars. Or a pedestal is provided so that the development length of the compression bars is taken care including the depth of the pedestal. Under all circumstances atleast four corner bars must have the embedment equal to the development length. In case the diameter of the dowel bars is more than that of the main bars, the development length can be suitably be adjusted to the force in the main column bars.

Example 12.1: Strip Foundation to Concrete Wall

A reinforced concrete wall 180 mm thickness is subjected to 200 kN/m and 100 kN/m of dead and live loads respectively. A bending moment of 60 kNm/m is caused by the live load at the wall base. The net bearing capacity of the soil is 150 kN/m² at 1 m below G.L. Design a foundation to the wall in mild exposure condition.

Solution

Dead load = 200 kN/m; Live load = 100 kN/m
Total superimposed load = 300 kN/m
Add extra about 7% for the weight difference between the replaced soil mass and the footing = 20 kN/m²
Design service load = P = 300 +20 =320 kN/m,
Total bending moment = M_f = 60 kNm/m
Net bearing capacity p_a = 150 kN/m²
The governing condition for bearing requirement is:

$$p_1 = \frac{P}{LB} + \frac{6M_f}{LB^2} \leq p_a$$

The footing is designed for one metre length; so $L = 1$ m, the above equation gives:

$$p_1 = \frac{320}{B} + \frac{6*60}{B^2} \leq 150$$

The above equation reduces to:

$$B^2 - 2.1B - 2.4 \geq 0$$

$$B \geq \frac{2.1 + \sqrt{2.1^2 + 4*2.4}}{2} = 2.92 \text{ m}; \text{ Use } B = 3\text{m},$$

Then $A = 3$ m², $I = 3^3/12$, $Z = 3*3/6 = 1.5$ m³,

Maximum and minimum soil pressures under the footing are:

$$p_1 = \frac{320}{3} + \frac{60}{1.5} = 106.7 + 40 = 146.7 \text{ kN}/\text{m}^2$$

$$p_2 = \frac{320}{3} - \frac{60}{1.5} = 106.7 - 40 = 66.7 \text{ kN/m}^2$$

The maximum bearing pressure is slightly less than the net bearing capacity of the soil.

Structural Design

The bearing pressure under the soil is in trapezoidal form as shown in Fig. 12.3. The weight of the footing acts downwards while the soil pressure is upwards. The net pressure that causes the bending moment on the footing is the difference between the soil pressure and the weight of the footing per unit area. Assuming the average thickness of the footing as about one fourth the cantilever span of the footing, the self-weight of the footing is about 9 kN/m². This value can be deducted from the pressure of the soil to compute the net force on the footing slab for structural design purpose. For the purpose of simplicity, no such correction is given in this example.

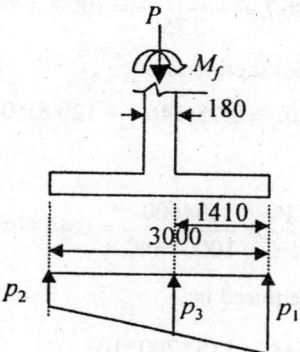

Fig. 12.3 Pressure under Footing; Example 12.1.

The critical bending moment occurs at the face of the wall. The cantilever span of the footing, an offset from the face of the wall is:

$$L_m = (3000 - 180)/2 = 1410 \text{ mm} = 1.41 \text{ m}$$

The bearing pressure (p_3) at the face of the wall is:

$$p_3 = \frac{P}{A} + \frac{My}{I} = 106.7 + \frac{60*0.09}{2.25} = 106.7 + 2.4 = 109.1 \text{ kN/m}^2$$

The factored bending moment at the face of the wall is:

$$M = 1.5\left(\frac{1}{6}\right)(2p_1 + p_3) L_m^2 = 0.25(2*146.7 + 109.1)*1.41^2 = 200 \text{ kNm/m}$$

Equating the bending moment to the moment capacity of the section gives balanced effective depth needed:

$$M = kbd^2 f_{ck} = 0.138(1)(20) d^2$$

$$d = \sqrt{\frac{0.2}{0.138(20)}} = 0.27 \text{ m}$$

Select the overall depth of the footing at the face of the wall as 0.5 m. The minimum cover to the reinforcement in mild environment is 50 mm., then the effective depth of the footing is $500 - 50 - 10 = 440$ mm.

It is better to check for the shear capacity of the section before finalizing for the bending moment.

$$\text{Shear span} = L_s = L_m - d = 1.41 - 0.44 = 0.97 \text{ m.}$$

The bearing pressure at the face of the critical shear section, that is $0.44 + 0.09 = 0.53$ m from the centre is:

$$p_4 = \frac{P}{A} + \frac{M_f y}{I} = 106.7 + \frac{60*0.53}{2.25} = 106.7 + 14.4 = 120.8 \text{ kN/m}^2$$

The factored shear force at the critical section is:

$$V = 1.5*0.5(p_1 + p_4)L_s = 0.75(146.7 + 120.8)*0.97 = 194.6 \text{ kN/m}$$

The nominal shear stress is:

$$\tau_v = \frac{V}{bd} = \frac{194600}{1000*440} = 0.44 \text{ N/mm}^2$$

The area of tension reinforcement required is:

$$A_{st} = \frac{1.15M}{jdf_{ck}} = \frac{1.15*200*10^6}{0.8*440*415} = 1575 \text{ mm}^2/\text{m}$$

Provide 16 mm bars at 125 mm spacing, then the area of reinforcement provided is = 201/0.125 = 1608 mm²/m

The percentage of the reinforcement is: $p\% = 1608*100/1000*500 = 0.32\%$

The shear strength of the concrete from table 3.4 is more than the nominal shear stress, so no shear reinforcement is needed. Provide a nominal reinforcement in the longitudinal direction (at 0.12%). 12 mm bar at 180 mm spacing is adequate. The development length of bars taken from table 3.3 is 43 times the diameter of the diameter of the bar and it is 688 mm. An adequate development length is available as the cantilever length is 1500 mm. Fig. 12.4 illustrates the reinforcement.

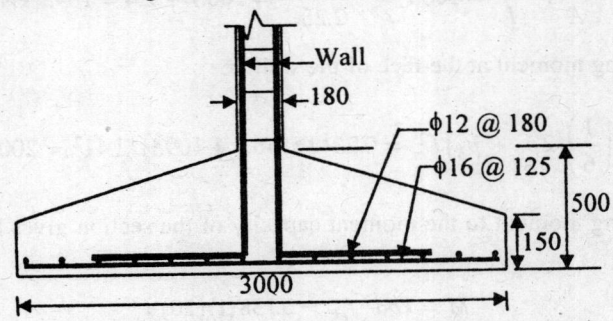

Fig. 12.4 Wall Footing Reinforcement; Example 12.1.

Example 12.2: Eccentrically Located Wall

A 180 mm thick reinforced concrete wall carries 200 kN/m and 100 kN/m a dead and live loads respectively. The wall is at a distance of 1.0 m from the boundary of the property. Net allowable bearing capacity of the soil at one metre below ground level is 150 kN/m². The soil is exposed to mild environment.

Dead load = 200 kN/m;   Live load = 100 kN/m
Total load = 300 kN/m

Add 20 kN/m extra towards the increased load due to the concrete footing replacing the soil. The design load is:

$$P = 320 \text{ kNm/m}$$

The minimum bearing width of the footing assuming no eccentricity of the wall with respect to footing is:

$$\text{Minimum symmetrical width} = P/p_a = 320/150 = 2.133 \text{ m,}$$

The width available on one side from the centre of the wall is only 1 m, whereas the minimum width required is 2.133 m, therefore the wall has to be placed eccentric with the footing. The eccentricity produces bending moment on the footing. Let the width of the footing be taken as 2.4 m, then the eccentricity if the wall load is:

$$e = 2.4/2 - 1.0 = 0.2 \text{ m}$$

The maximum bearing pressure under the footing for one metre length of the footing is:

$$p_1 = \frac{P}{A} + \frac{Pey}{I} = \frac{320}{2.4} + \frac{320*0.2*1.2}{2.4^3/12} = 133.1 + 66.7 = 199.8 \text{ kN/m}^2$$

The bearing pressure exceeds the admissible value. Increase in width of the footing, increases the eccentricity and consequently the bending moment. It's vicious circle. Select the total width of the footing as 3.2 m. The eccentricity of the footing is:

$$e = 3.2/2 - 1.0 = 0.6 \text{ m}$$

The maximum bearing pressure under the footing for one metre length of the footing is:

$$p_1 = \frac{P}{A} + \frac{Pey}{I} = \frac{320}{3.2} + \frac{320*0.6*1.6}{3.2^3/12} = 100 + 112.5 = 212.5 \text{ kN/m}^2$$

The bearing pressure at the other end is $100 - 112.5 = -12.5$, negative, so the total base is not in contact with the soil. Further the pressure is far in excess of the permissible value. The governing condition for one metre length of the footing with eccentricity is:

$$\frac{P}{A} + \frac{Pe}{Z} = \frac{P}{B} + \frac{6P(0.5B - 1)}{B^2} = \frac{320}{B} + \frac{3*320(B - 2)}{B^2} \leq p_a = 150$$

The above equation reduces to:

$$B^2 - 2.133\,B - 6.4(B - 2) = 0$$
$$B^2 - 8.533\,B + 12.8 = 0$$

The solution of the quadratic equation is:

$$B = \frac{8.533 \pm \sqrt{8.533^2 - 4*12.8}}{2} = 6.6 \text{ m } or \text{ } 1.93 \text{ m}$$

The smaller value is not acceptable. Set the width of the footing as 6.6 m and check the actual pressure distribution. The equation assumes that the total width of the footing is in contact with the soil. The bearing pressures are:

$$p_1 = \frac{320}{6.6} + \frac{6*320*(3.3 - 1.0)}{6.6^2} = 48.5 + 101.4 = 149.9 \text{ kN/m}^2 \text{ and}$$

$$p_2 = 48.5 - 101.4 = -52.9;$$

One can see that the width of the footing is unreasonable, further there is still tension on the soil. Instead it is recommended that the depth of the footing be increased such the bearing capacity of the soil is enhanced to 160 kN/m². The problem of eccentric footings can become very tricky, so one has to investigate alternate feasible foundations.

12.5 ISOLATED RECTANGULAR FOOTING

The footing slab can taper at all the four faces. Figure 12.5 illustrates the variable thickness footing and the neutral axis of the section. The balanced moment capacity of a trapezoidal section was derived in chapter 4. The overall depth of the section at the free edge is usually taken in the range of 150 mm. In case, the footing thickness is taken as uniform instead of tapering, then the effective depth can be calculated from the standard formula. The method of design is illustrated through examples. Design against shear force in tapered sections becomes more complicated even to select the location of the critical diagonal tension zone. The depth of the section up to a distance d from the face of the column is usually kept constant.

The bending moment capacity and the lever arm of the trapezoidal section are given below and the coefficients are listed in table 12.1 for quick reference.

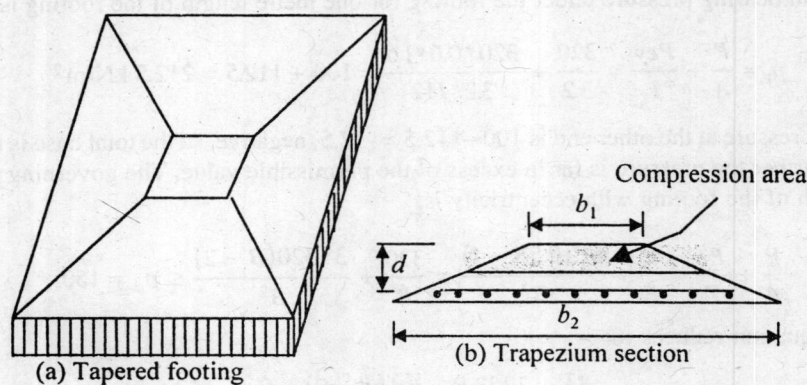

Fig. 12.5 Tapered Footing & Bending Resistant Section.

$$M_{rb} = Kb_1d^2 f_{ck} + K_2(b_2 - b_1)d^2 f_{ck} \tag{12.10}$$

$$j = \frac{(K - K_2)b_1 + K_2b_2}{0.36\,k_u b_1 + 0.204\,(b_2 - b_1)k_u^2} \tag{12.11}$$

A uniform thickness footing is more convenient in practice as laying tapering faces in not convenient.

Table 12.1 Design Moment & Neutral Axis Coefficients.

f_y (N/mm²)	K	K_2	k_u
415	0.138	0.025	0.479
500	0.133	0.023	0.456

Example 12.3: Design of Isolated Footing with Uniform Thickness under Axial Load

A 300 mm square column is transmitting 300 kN and 200 kN dead and live loads respectively to a footing in moderate exposure condition. The net allowable bearing capacity of the clayey soil is 100 kN/m². Design an isolated footing.

Solution

      The total service superimposed load from the column = 500 kN
      Add extra towards the increased load due to the concrete footing from soil at = 50 kN
      Total design service load = P = 550 kN
      Net bearing capacity at 1m below GL   = p_a = 100 kN/m²,
    Select M25 grade concrete as the footing is exposed to moderate environment.

Design of Footing Base Area

Square column under axial load, so a square footing is recommended. The bearing area required is:

$$A_f = \frac{P}{P_a} = \frac{550}{100} = 5.5 \text{ m}^2$$

Select 2.35 by 2.35 m footing, then A_f provided is = 5.5225 m²,

$$\text{Uniform bearing pressure} = \frac{550}{5.5225} = 99.6 \text{ kN/m}^2$$

Structural Design

Design for Bending Moment

The cantilever span is = (2.35 – 0.3)/2 = 1.025 m. Assume the average thickness of the footing as about one-fourth of the cantilever for the purpose of computing the self-weight.

Self-weight of footing = $(1.025/4)*25 = 6.4$ kN/m^2. Use 6.6 kN/m^2.

The net pressure on the footing for the purpose of bending moment is: $p = 99.9 - 6.6 = 93$ kN/m^2.
Cantilever span for bending moment $= L_m = 1.025 - 0.2 = 1.075$ m
Factored bending moment on the section is:

$$M_c = \frac{\gamma_f pL_m^2}{2} = \frac{1.5*93*1.025^2}{2} = 73.3 \text{ kNm/m}$$

Let the thickness of the slab is uniform, equating the moment capacity to the design moment gives:

$$Kbd^2 f_{ck} = M_c = 73.3$$

$$d = \sqrt{\frac{73.3*10^6}{0.138*1000*25}} = 145 \text{ mm}$$

Experience indicates that the shear strength controls the design, so select 300 mm as the overall thickness of the footing and the cover as 50 mm. The effective depth of the slab is equal to $d = 300 - 66 = 244$ mm. The effective depth in the other direction is $244 - 16 = 228$ mm. This assumes that 16 mm bars are used for the main reinforcement. It is suggested that an effective depth of 230 mm can be used for both the directions, as the footing is square and can be built symmetric.

The area of the tension reinforcement per metre width is:

$$A_{st} = \frac{1.15 M_c}{jdf_y} = \frac{1.15*73.3*10^6}{0.8*230*415} = 1104 \text{ mm}^2/\text{m}$$

Provide 16 mm bars at 180 mm spacing in both the directions. In a two-way square footing the reinforcement has to be distributed evenly across the width of the footing.
Percentage of reinforcement provided is $= (201/0.18)*100/300*1000 = 0.37$ %

Design for Shear

The critical section is at a distance of $(d/2)$ from the face of the column.
The length of the periphery of critical section is: $b_0 = 4(a + 2*d/2) = 4(0.3 + 0.23) = 2.12$ m
The size of the square on which the shear force is critical is: $b_s = a + 2*d/2 = 0.53$ m
The factored shear force on the periphery of the square is:

$$V_s = \gamma p(B*B - b_s*b_s) = (1.5)*93(2.35*2.35 - 0.53*0.53) = 731.3 \text{ kN}$$

The nominal shear stress at the critical section is:

$$\tau_v = \frac{V_s}{b_0 d} = \frac{731200}{2120*230} = 1.5 \text{ N/mm}^2$$

The permissible shear stress is given by:

$$\tau_{cc} = k_s \left(0.25\sqrt{f_{ck}}\right)$$

where
$k_s = [0.5 + (short\ side\ /\ long\ side)\ of\ the\ column] = 0.5 + 1 = 1.5$ but subject to a maximum limit of 1.0
Since k_s should not be greater than 1.0

$$\tau_{cc} = k_s \left(0.25\sqrt{f_{ck}}\right) = 0.25*\sqrt{25} = 1.25 \text{ N/mm}^2$$

Since the shear strength is less than the nominal shear stress, either shear reinforcement is to be provided or depth of the footing can be increased. In flat slabs and footings it is better to increase the effective depth in case the shear stress requirement is violated marginally. Let the effective depth be increased approximately proportional to the shear stress:

$$d = (1.5/1.25)*0.23 = 276, \text{ use } d = 280 \text{ mm}.$$

The peripheral distance at the critical shear zone is: $b_0 = 4(a +2*d/2) = 4(0.3 + 0.28) = 2.32$ m
The size of the square where shear force is maximum: $b_s = a + 2*d/2 = 0.58$ m
The factored shear force on the periphery of the square is:

$$V_s = \gamma p(B*B - b_s*b_s) = (1.5)*93(2.35*2.35 - 0.58*0.58) = 723.5 \text{ kN}$$

The nominal shear stress at the critical section is: $\tau_v = \dfrac{V_s}{b_0 d} = \dfrac{723500}{2320*280} = 1.11 \text{ N/mm}^2$

The permissible shear stress is 1.25 N/mm^2.

Check for Transverse Shear Stress

In addition to the design against punching shear, the footing has to be designed for transverse shear similar to the one done in beams. The critical section is at a distance d from the face of the column.

The shear span is: $L_s = (B - a)/2 - d = 2.05/2 - 0.28 = 0.745$ m

The shear force on the critical plane is:

$$V_s = \gamma p B L_s = 1.5*93*2.35*0.745 = 244.2 \text{ kN}$$

The nominal shear stress is: $\tau_v = \dfrac{V_s}{Bd} = \dfrac{244200}{2350*280} = 0.37 \text{ N/mm}^2$

The shear strength is 0.36 MPa even for 0.25% reinforcement, whereas the percentage of reinforcement computed based on the earlier depth is 0.37 and it is likely to be in the range of 0.3 percent or more. Therefore, the nominal shear stress is within the limits.
Area of steel in each direction is:

$$A_{st} = \frac{1.15M_c}{jdf_y} = \frac{1.15*73.3*10^6}{0.8*280*415} = 907 \text{ mm}^2/\text{m}$$

Provide 16 bars at 220 mm spacing in each direction.
The total depth of the footing is equal to the effective depth plus one and half times the diameter of the bar plus the cover to the reinforcement and it is equal to $= h = 280 + 16 + 8 + 50 = 354$ mm; Use 355 mm. Fig. 12.6 illustrates the reinforcement details of the footing.

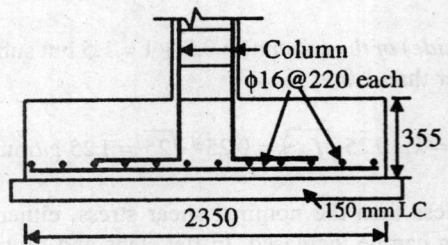

Fig. 12.6 Footing Reinforcement, Example 12.3.

Example 12.4: Design of Isolated Footing with Tapering Surface

A column of 300 by 300 mm size is transmitting 300 kN and 200 kN dead and live loads respectively to a footing in moderate exposure. The net allowable bearing capacity of the clayey soil is 100 kN/m². Design an isolated footing with tapering top face. The problem is same as that of 12.3 except for tapering top.

Solution

Vide example 12.3 for choice of the size of the footing 2.35 m square. Structural design is continued.

Structural Design

Design for Bending Moment

The cantilever span is $L_m = (2.35 - 0.3)/2 = 1.025$ m

Assume the average thickness of the footing as about one-fourth of the cantilever for the purpose of computing the self-weight of the footing.

Self-weight of footing = $(1.025/4)*25 = 6.4$ kN/m². Use 6.6 kN/m².

The net pressure on the footing for the purpose of bending moment is: $p = 99.9 - 6.6 = 93$ kN/m². Factored bending moment on the section is:

$$M_c = \frac{\gamma_f \, BpL_m^2}{2} = \frac{1.5*2.35*93*1.025^2}{2} = 172.255 \text{ kNm}$$

Let the thickness of the slab is tapering and moment capacity of such a footing is given in Eq. 12.10. Equating the moment capacity to the design moment gives:

$$M_c = 172.255 = M_{rb} = Kb_1 d^2 f_{ck} + K_2 \left(b_2 - b_1\right)d^2 f_{ck}$$

where the values are given in table 12.1 and they are: $K = 0.138$, $K_2 = 0.025$, $k_u = 0.479$. Further, the base width $b_2 = 2350$ mm, and let the top width of the footing be = $b_1 = 500$ mm.

$$M_{rb} = Kb_1 d^2 f_{ck} + K_2 \left(b_2 - b_1\right)d^2 f_{ck} = \left(0.138*0.5 + 0.025*(2.35 - 0.5)\right)25(10)^6 d^2$$

$$= 2.881(10)^6 d^2 = M_c = 172.255(1000) \text{ Nm}$$

The solution of the above equation is: $d = 0.244$ m

As the shear force controls the design, select the effective depth of the section as: $d = 320$ mm.

The lever arm of the trapezium section is given in Eq.12.12 and it is equal to:

$$j = \frac{(K - K_2)b_1 + K_2 b_2}{0.36 k_u b_1 + 0.204(b_2 - b_2)k_u^2} = \frac{(0.138 - 0.025)0.5 + 0.025*2.35}{0.36*0.479*0.5 + 0.204*(2.35 - 0.5)*0.479^2} = 0.667$$

The area of tension reinforcement is:

$$A_{st} = \frac{1.15 M_c}{jdf_y} = \frac{1.15*172.255(10^6)}{0.667*320*415} = 2236 \text{ mm}^2/\text{m}$$

The percentage of reinforcement at the deepest section is = $2236*100/(2350*400) = 0.24\%$

Punch Shear Strength

Overall depth = $320 + 16 + 8 + 50 = 396$ mm; use $h = 400$ mm and $d = 400 - 74 = 326$ mm.

Fig. 12.7 illustrates the tapering footing along with some critical sections.

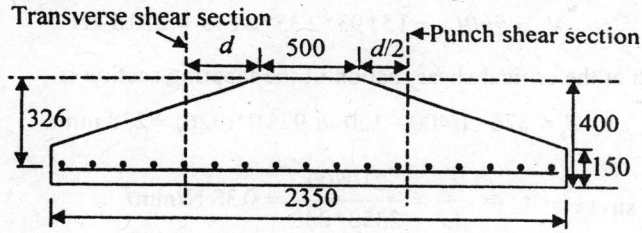

Fig. 12.7 CS of Tapering Footing with Critical Shear Sections.

The punch shear is critical at a distance $d/2$ from the edge of the column. Even though the column size is 300 by 300 mm. The top face is taken as 500 by 500 mm. The actual effective depth at a distance $d/2$ from the edge can be calculated from the geometry of the footing as shown in Fig. 12.7.

The effective depth at $d/2$ is:

$$d_s = 326 - [(400 - 150)/(925)]*(326/2) = 282 \text{ mm}.$$

The peripheral distance of the square where the punch shear is critical is:

$$b_0 = 4(0.500 + 2*d/2) = 4(0.5 + 0.282) = 3.128 \text{ m}$$

The size of the square on whose periphery the shear force is critical is:

$$b_s = 0.5 + 2*d/2 = 0.782 \text{ m}$$

The factored shear force on the periphery of the square is:

$$V_s = \gamma p(B*B - b_s*b_s) = (1.5)*93(2.35*2.35 - 0.782*0.782) = 685.1 \text{ kN}$$

The nominal shear stress at the critical section is: $\tau_v = \dfrac{V_s}{b_0 d} = \dfrac{685100}{3128*282} = 0.77 \text{ N/mm}^2$

The permissible shear stress is given by:

$$\tau_{cc} = k_s \left(0.25\sqrt{f_{ck}} \right)$$

where k_s = [0.5 + (*short side / long side*) *of the column*] = 0.5 + 1 = 1.5 but subject to a maximum of 1.0

Since k_s should not be greater than 1.0

$$\tau_{cc} = k_s \left(0.25\sqrt{f_{ck}} \right) = 0.25*\sqrt{25} = 1.25 \text{ N/mm}^2$$

The permissible shear stress is 1.25 N/mm². The section is safe in punch shear.

Check for transverse shear stress

The critical section is at a distance d from the face of the column.

The shear span is: $L_s = (B - a)/2 - d = (2.35 - 0.5)/2 - 0.282 = 0.643$ m

The shear force on the critical plane is:

$$V_s = \gamma p B L_s = 1.5*93*2.35*0.643 = 210.8 \text{ kN}$$

The effective depth at the critical shear section of the tapering section is:

$$d_s = 326 - [(400 - 150)/(925)]*(326) = 238 \text{ mm.}$$

The nominal shear stress is: $\tau_v = \dfrac{V_s}{Bd} = \dfrac{210800}{2350*238} = 0.38$ N/mm²

The shear strength is 0.36 MPa even for 0.25% reinforcement, whereas the nominal shear stress is 0.38 MPa, marginally more than the strength. Therefore, increase the depth of the section instead of providing shear reinforcement. Or alternatively increase the top face width from 500 mm to 550 mm. The second option is better as it decreases the shear span and increases the effective depth of the section at the shear zone. However the area of reinforcement is computed based on the effective depth of 326 mm instead of 320 mm.

Area of steel in each direction with effective depth of 326 mm is:

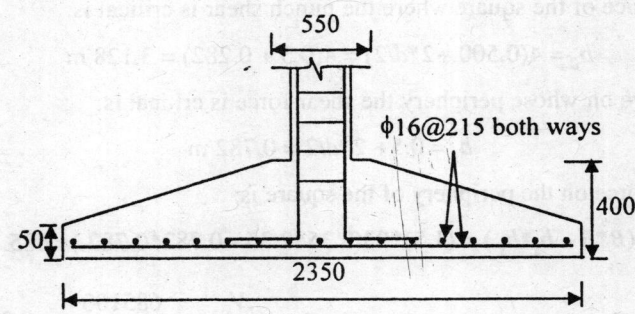

Fig. 12.8 Cross-Section of Tapering Footing; Example 12.4.

$$A_{st} = \frac{1.15*M_c}{jdf_y} = \frac{1.15*172.255(10^6)}{0.667*326*415} = 2196 \text{ mm}^2/\text{m}$$

Provide 16 bars at 215 mm spacing in each direction. The development length is adequate. Fig. 12.8 illustrates the reinforcement details of the footing.

Example 12.5: Design of Isolated Rectangular Footing under Axial Load

A column 300 by 300 mm is transmitting 300 kN and 200 kN dead and live loads respectively to a footing in moderate exposure. The size of the pedestal is 400 mm by 400 mm. The net allowable bearing capacity of the clayey soil is 100 kN/m². The clear edge distance available from the centre of the column is only one metre. Design an isolated footing.

Solution

The total service superimposed load from the column = 300 + 200 = 500 kN
Add extra towards the increased load due to the concrete footing from soil at = 50 kN
Total design service load = P = 550 kN
Net bearing capacity at 1m below GL = p_a = 100 kN/m²,
Select M25 grade concrete as the footing is exposed to moderate environment.

Design of footing base area

The bearing area required is:

$$A_f = \frac{P}{p_a} = \frac{550}{100} = 5.5 \text{ m}^2$$

A square footing is desirable, but the available space is limited to one metre in one direction, so the width of the footing is restricted to two metres. The footing be a rectangular one of 2 by 2.75 m.

Structural Design

Design for bending moment

The cantilever span on the length side is = L_m = (2.75 − 0.4)/2 = 1.175 m
Assume the average thickness of the footing as about one-fourth of the cantilever for the purpose of computing the self-weight of the footing.

Self-weight of footing = (1.175/4)*25 = 7.4 kN/m². Use 8 kN/m².

The net pressure on the footing that causes bending moment is: p = 100 − 8 = 92 kN/m².

Design of the long span

Factored bending moment on the section is:

$$M_c = \frac{\gamma_f p L_m^2}{2} = \frac{1.5*92*1.175^2}{2} = 95.3 \text{ kNm/m}$$

Let the thickness of the slab is uniform, equating the moment capacity to the design moment gives:

$$Kbd^2 f_{ck} = M_c = 95.3$$

$$d = \sqrt{\frac{95.3*10^6}{0.138*1000*25}} = 166 \text{ mm}$$

Let the effective depth of the slab be taken as $d = 330 - 50 - 10 = 270$ mm.
The area of the tension reinforcement per metre width is:

$$A_{st} = \frac{1.15 M_c}{jdf_y} = \frac{1.15*95.3*10^6}{0.8*270*415} = 1223 \text{ mm}^2/\text{m}$$

Provide 16 mm bars at 160 mm spacing along the long span. In a two-way rectangular footing the reinforcement of the length is distributed evenly across the width of the footing.
Actual % reinforcement provided is = $(201/0.160)*100/330*1000 = 0.38$ %

Design for Shear

The critical section punch shear is at a distance of $(d/2)$ from the face of the column.
The peripheral length of the critical section is: $b_0 = 4(a + 2*d/2) = 4(0.4 + 0.27) = 2.68$ m
The size of the square on whose periphery the shear is critical is: $b_s = a + 2*d/2 = 0.67$ m
The factored shear force on the periphery of the square is:

$$V_s = \gamma p(L*B - b_s*b_s) = (1.5)*92(2.75*2.0 - 0.67*0.67) = 697 \text{ kN}$$

The nominal shear stress at the critical section is: $\tau_v = \dfrac{V_s}{b_0 d} = \dfrac{697000}{2680*270} = 1.1 \text{ N/mm}^2$

The permissible shear stress is given by: $\tau_{cc} = k_s \left(0.25 \sqrt{f_{ck}} \right)$

Where $k_s = [0.5 + (short\ side\ /\ long\ side)\ of\ the\ column] = 0.5 + 1 = 1.5$, but subject to a maximum limit of 1.0

$$\tau_{cc} = k_s \left(0.25 \sqrt{f_{ck}} \right) = 0.25*\sqrt{25} = 1.25 \text{ N/mm}^2$$

The section is safe against shear force.

Check for transverse shear stress

The critical section is at a distance d from the face of the column. The shear span is:

$$L_s = (L - a)/2 - d = 2.35/2 - 0.27 = 0.905 \text{ m}$$

The shear force on the critical plane is:

$$V_s = \gamma p B L_s = 1.5*92*2.0*0.905 = 249.8 \text{ kN}$$

The nominal shear stress is: $\tau_v = \dfrac{V_s}{Bd} = \dfrac{249800}{2000*270} = 0.46 \text{ N/mm}^2$

The shear strength from table 3.4 is 0.40 MPa for 0.38% reinforcement. The nominal shear stress is more than the strength hence section has is be revised or stirrups to be provided. Stirrups provided in the long span.

Design of Stirrups

Select six legged 6 mm bar, the area of six legs of the stirrup is: 168 mm². The stirrup spacing is:

$$s_v = \frac{0.87 A_{sv} f_{yv}}{b(\tau_v - \tau_c)} = \frac{0.87(168)(415)}{2000(0.46 - 0.40)} = 505 \text{ mm}$$

The maximum spacing of the stirrups is controlled by:

$$s_{\max} \le \frac{A_{sv} f_{yv}}{0.4b} = \frac{168(415)}{0.4(2000)} = 87 \text{ mm}$$

It is suggested that six legged 8 mm stirrups be chosen, then the spacing works out to be 155 mm. The shear is not critical in the short span direction.

Provide two legged 8 mm stirrups at 155 mm spacing.

Reinforcement along the short span

The bending moment span is: $L_m = (B - a)/2 = 0.8$ m
The factored bending moment is:

$$M_c = \frac{\gamma_f p L_m^2}{2} = \frac{1.5*92*0.8^2}{2} = 44.16 \text{ kNm / m}$$

The effective depth of the reinforcement is: $d = 330 - 50 - 16 - 8 = 256$ mm
Area of steel in the short span direction is:

$$A_{st} = \frac{1.15 M_c}{jd f_y} = \frac{1.15*44.16*10^6}{0.8*256*415} = 598 \text{ mm}^2/\text{m}$$

The minimum reinforcement required is 0.12 percent and it is:

$$A_{smin} = 0.12 *330*1000/100 = 396 \text{ mm}^2/\text{m}$$

Select 10 mm bars in the short span direction. The total reinforcement that is needed for the length of 2.75 m is:

$$A_{st} = 598*2.75/1 = 1644 \text{ mm}^2$$

The total area of the reinforcement is to be distributed between the column band and the remaining. The proportion of the column band reinforcement is:

$$\frac{\text{Reinforcement in the central band width}}{\text{Total reinforcement in the short span}} = \frac{2}{1 + \text{ratio of long span to shot span}}$$

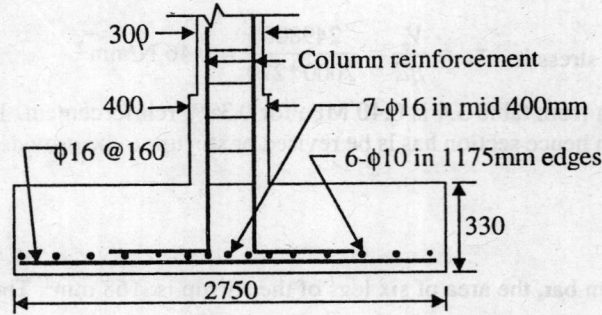

Fig. 12.9 Rectangular Footing Reinforcement; Example 12.5.

The above factor is $2/(1+2.75/2) = 0.84$

0.84 of the reinforcement is to be in the central strip, and it is $= 0.84*1644 = 1380$ mm^2.

Provide seven of 16 bars in the 400 mm width of the column strip. The balance of the reinforcement is not much but the maximum spacing of the reinforcement should preferably limit to twice the effective depth. Select 10 mm bars at 200 mm spacing in remaining width.

Example 12.6: Footing Subjected to Combined Axial and Bending Moment

A column is subjected to an axial load of 900 kN and a bending moment of 200 kNm. The net allowable bearing capacity of the soil is 110 kN/m and mild exposure condition. The size of the pedestal is 500 by 500 mm.

Solution

As exposure condition is mild, let the grade of the concrete be $M20$ and that of reinforcement is *HYSD-Fe*415.

The axial load from the column is 900 kN, The difference between the weights of the concrete and the replaced soil is about five percent of the load. Let the design axial load is $= 900 + 50 = 950$ kN. The loads are:

$P = 950$ kN, and $M_f = 200$ kNm; The virtual eccentricity is $= e = M_f/P = 200/950 = 0.21$m.

Let the length of the foundation is $= L = 3$ m.

The maximum bearing pressure caused by the combined load should be limited to the net allowable pressure:

$$\frac{P}{LB} + \frac{6M_f}{LB^2} = \frac{950}{3B} + \frac{6*200}{3*B^2} \le p_a = 110$$

The above equation reduces to:

$$B^2 - 2.878B - 3.636 = 0$$

The solution of the above quadratic is: $B = 3.83$ m or -0.95 m

The width of the foundation is atleast 3.83 m. Let $B = 3.9$ m and $L = 3$ m.

The maximum and minimum bearing pressures on the soil are:

$$p_1 = \frac{P}{LB} + \frac{6M_f}{LB^2} = \frac{950}{3B} + \frac{6*200}{3*B^2} \leq p_a = 110$$

$$p_2 = 81.2 - 26.3 = 54.9 \text{ kN/m}^2$$

Structural Design

Figure 12.10 illustrates the critical pressures.

p_3 = pressure at the face of the column, critical *BM* section;
p_4 = pressure at a distance $d/2$ from the face of the column, critical punching section;
p_5 = pressure at a distance d from the face of the column, critical transverse shear section;
Cantilever span = $L_m = (B-a)/2 = (3.9 - 0.5)/2 = 1.7$ m.

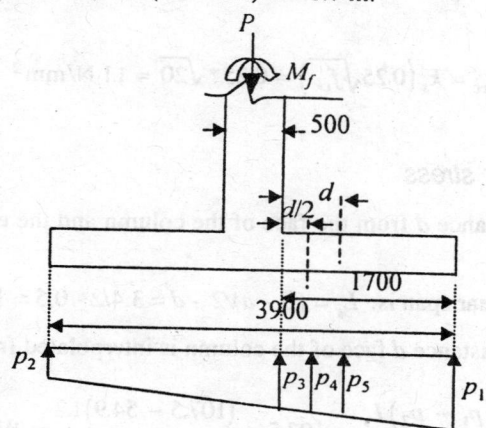

Fig. 12.10 Pressure under Footing; Example 12.6.

The bearing pressure at face of the column is interpolated from the figure and it is:

$$p_3 = p_1 - \frac{(p_1 - p_2) L_m}{B} = 107.5 - \frac{(107.5 - 54.9)1.7}{3.9} = 84.6 \text{ kN/m}^2$$

The maximum bending moment at the face of the column is:

$$M_c = \gamma L L_m^2 (2p_1 + p_3)/6 = 1.5*3*1.7^2 (2*107.5 + 84.6)/6 = 649.4 \text{ kNm}$$

Equating the bending moment capacity to the factored moment gives:

$$KLd^2 f_{ck} = M_c = 649.4 \text{ kNm; or}$$

$$d = 0.29 \text{ m}$$

Since the shear stress usually controls the thickness, assume the overall thickness of the footing as = $h = 0.56$ m.

The effective depth of the section assuming 20 mm diameter bars as: d = 560 − 50 − 10 = 500 mm.

Design for Shear

The critical section is at a distance of $(d/2)$ from the face of the column.

The length of the periphery of the section is: $b_0 = 4(a + 2*d/2) = 4(0.5 + 0.5) = 4$ m

The size of the square on which the shear force is critical is: $b_s = a + 2*d/2 = 1.0$ m

The factored shear force on the periphery of the square is:

$$V_s = \gamma(P/A_f)(L*B - b_s*b_s) = (1.5)*((950)/(3*3.9))(3*3.9 - 1*1) = 1303 \text{ kN}$$

The nominal shear stress at the critical section is: $\tau_v = \dfrac{V_s}{b_0 d} = \dfrac{1303000}{4000*500} = 0.65 \text{ N/mm}^2$

The permissible shear stress is given by: $\tau_{cc} = k_s \left(0.25\sqrt{f_{ck}}\right)$

where $k_s = [0.5 + (short\ side\ /\ long\ side)\ of\ the\ column] = 0.5 + 1 = 1.5$ but subject to a maximum limit of 1.0

$$\tau_{cc} = k_s \left(0.25\sqrt{f_{ck}}\right) = 0.25*\sqrt{20} = 1.1 \text{ N/mm}^2$$

Check for transverse shear stress

The critical section is at a distance d from the face of the column and the entire width of the footing is accounted in resistance.

The shear span is: $L_s = (B - a)/2 - d = 3.4/2 - 0.5 = 1.2$ m

The bearing pressure at a distance d face of the column is interpolated from the figure and it is:

$$p_s = p_1 - \frac{(p_1 - p_2)L_s}{B} = 107.5 - \frac{(107.5 - 54.9)1.2}{3.9} = 91.3 \text{ kN/m}^2$$

The factored shear force on the critical plane is:

$$V_s = \gamma LL_s (p_1 + p_s)/2 = 1.5*3*1.2*(107.5 + 91.3)/2 = 536.76 \text{ kNm}$$

The nominal shear stress is: $\tau_v = \dfrac{V_s}{Bd} = \dfrac{536760}{3000*500} = 0.36 \text{ N/mm}^2$

The shear strength is 0.36 MPa for 0.25% reinforcement, it is hoped that the percentage of the reinforcement is more than 0.25 percent.

Design for bending in the width direction

Area of steel in width (3.9m) direction is:

$$A_{st} = \frac{1.15M_c}{jdf_y} = \frac{1.15*649.4*10^6}{0.8*500*415} = 4499 \text{ mm}^2/\text{m}$$

Provide 15 numbers of 20 mm bars along the 3.9m length, the spacing is 3000/15 = 200 mm.
Percentage of the reinforcement is = (15*314*100)/(3000*560) = 0.28 %
The transverse shear strength is adequate.

Design of reinforcement in the length direction (Reinforcement along the short span)
The bending moment span is: $L_m = (L - a)/2 = (3 - 0.5)/2 = 1.25$ m
The average net bearing pressure that causes bending moment is = $p = 900/3*3.9 = 77$ kN/m^2,
The factored bending moment over the width of 3.9 m is:

$$M_c = \frac{\gamma_f BpL_m^2}{2} = \frac{1.5*3.9*77*1.25^2}{2} = 352 \text{ kNm/m}$$

The effective depth of the reinforcement is: $d = 560 - 50 - 20 - 10 = 480$ mm

Area of steel in the short span direction is: $A_{st} = \frac{1.15 M_c}{jdf_y} = \frac{1.15*352*10^6}{0.8*480*415} = 2540 \text{ mm}^2/\text{m}$

The minimum reinforcement required is 0.12 percent and it is:

$$A_{smin} = 0.12 *560*3900/100 = 2621 \text{ mm}^2$$

The total area of the reinforcement is to be distributed between the column band and the remaining. The proportion of the column band reinforcement is:

$$\frac{\text{Reinforcement in the central band width}}{\text{Total reinforcement in the short span}} = \frac{2}{1 + \text{ratio of long span to shot span}}$$

The above factor is 2/(1+3.9/3) = 0.87
87 % of the reinforcement has to be in the column or pedestal (central) strip, and it is = 0.87*2540 = 2210 mm^2.

Provide seven of 20 bars in the 500 mm width of the column strip. The balance of the reinforcement is not much but the maximum spacing of the reinforcement should be limit to twice the effective depth and the minimum percentage of reinforcement as 0.12 %.

Provide six numbers of 16 mm bars in each segment of 1.7 m on either side of the footing projected beyond the column. Figure 12.11 illustrates the reinforcement details of the footing.

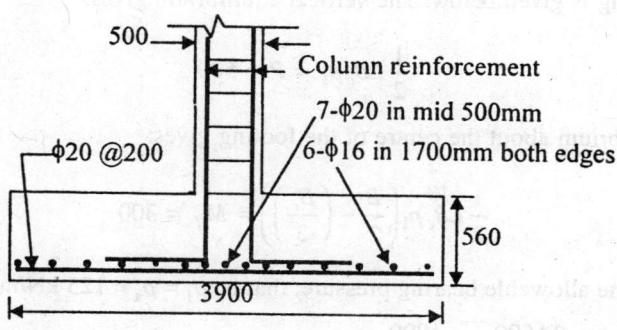

500
Column reinforcement
7-φ20 in mid 500mm
6-φ16 in 1700mm both edges
φ20 @200
560
3900

Fig. 12.11 Rectangular Footing Reinforcement; Example 12.6.

Example 12.7: Rectangular Footing under Combined Bending and Partial Contact

A square column of 500 mm is transferring 300 kN and 170 kN dead and live loads respectively and a bending moment of 300 kNm to a rectangular foundation. Net allowable bearing pressure of the soil of 125 kN/m² and the foundation is exposed to moderate environment. The length of the footing is limited to 2.5 m.

Solution

Foundation size

Select $M25$ grade of concrete along with $HYSD$-$Fe415$ reinforcement. The length of the footing is $= L = 2.5$ m. The total superimposed axial load is 470 kN. Add extra six percent account to the weight difference of soil and the concrete in the foundation. Select the design load as 500 kN. The maximum bearing pressure under combined bending moment is:

$$p_1 = \frac{P}{LB} + \frac{6M_f}{LB^2} = \frac{500}{2.5*B} + \frac{6*300}{2.5*B^2} \le p_a = 125$$

The above governing condition results into a quadratic equation: $B^2 - 1.6B - 5.76 = 0$
The solution of the quadratic is:

$$B = \frac{1.6 \pm \sqrt{2.56 + 23.04}}{2} = 3.33 \text{ m or } -1.73 \text{ m}$$

The width of the foundation has to be atleast 3.33 m. Let $B = 3.4$ m.
The maximum and minimum bearing pressures on the soil are:

$$p_1 = \frac{P}{LB} + \frac{6M_f}{LB^2} = \frac{500}{2.5*3.4} + \frac{6*300}{2.5*3.4^2} = 58.8 + 62.3 = 121.1 \le p_a = 125$$

$$p_2 = 58.8 - 62.3 = -3.5 \text{ kN/m}^2$$

Figure 12.12 (a) illustrate the footing section and (b) illustrates the pressure distribution in the first trial in which a tension on the soil is indicated. Only a part of the footing is in contact with the soil and the corresponding pressure distribution is shown in Fig. 12.12 (c). The equilibrium of forces taken about the centre of the footing is given below. The vertical equilibrium gives:

$$\frac{1}{2} LB_c p_1 = P = 500 \tag{a}$$

The moment equilibrium about the centre of the footing gives:

$$\frac{1}{2} LB_c p_1 \left(\frac{B}{2} - \left(\frac{B_c}{3} \right) \right) = M_f = 300 \tag{b}$$

Let p_1 be equal to the allowable bearing pressure, that is: $p_1 = p_a = 125$ kN/m².

From Eq.(a): $$B_c = \frac{2*500}{Lp_a} = \frac{1000}{2.5*125} = 3.2 \text{ m}$$

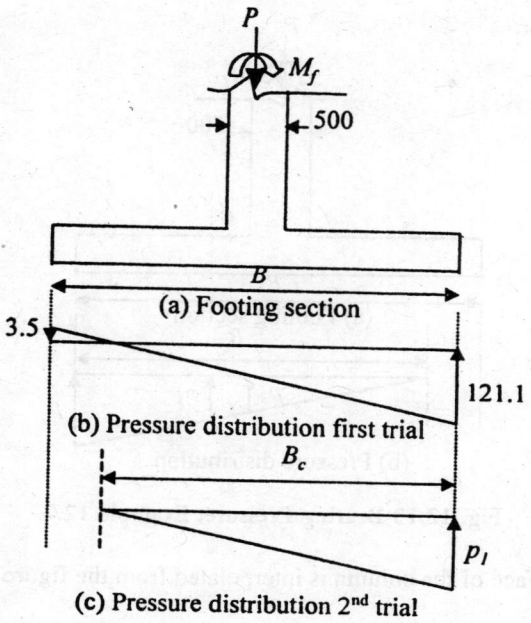

(a) Footing section

(b) Pressure distribution first trial

(c) Pressure distribution 2ⁿᵈ trial

Fig. 12.12 Pressure under Footing; Example 12.7.

From Eq.(b): $B = 2\left[\dfrac{2M_f}{LB_cP_a} + \dfrac{B_c}{3}\right] = 2\left[\dfrac{600}{2.5*3.2*125} + \dfrac{3.2}{3}\right] = 3.333$ m

Select the width of the footing as $B = 3.4$ m, then rework out the actual pressure for a triangular distribution.

Dividing Eq.(b) by Eq.(a) gives:

$$\frac{B}{2} - \frac{B_c}{3} = \frac{M_f}{P} = \frac{3}{5}$$ (c)

Substitution of the value of B in the above equation gives:

$$B_c = 3\left[\frac{B}{2} - \frac{3}{5}\right] = 3.3 \text{ m}$$

From Eq.(a): $P_1 = \dfrac{2P}{LB_c} = 121.2 < 125$

Structural Design

P_3 = pressure at the face of the column, critical BM section;
P_4 = pressure at a distance $d/2$ from the face of the column, critical punching section;
P_5 = pressure at a distance d from the face of the column, critical transverse shear section;
The cantilever span $= L_m = (B - a)/2 = (3.4 - 0.5)/2 = 1.45$ m.

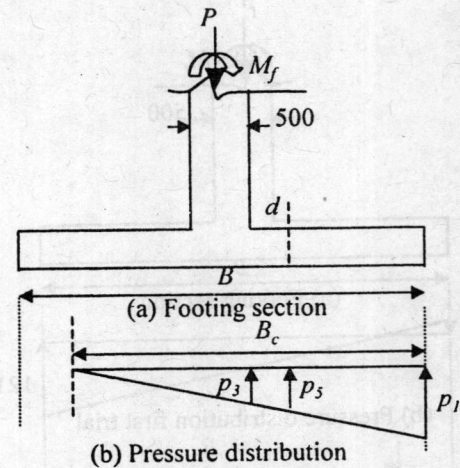

(a) Footing section

(b) Pressure distribution

Fig. 12.13 Bearing Pressure; Example 12.7.

The bearing pressure at face of the column is interpolated from the figure and it is:

$$p_3 = p_1 - \frac{(p_1) L_m}{B_c} = 121.2 - \frac{(121.2)1.45}{3.3} = 68 \text{ kN/m}^2$$

The factored bending moment at the face of the column is:

$$M_c = \gamma L L_m^2 (2p_1 + p_3)/6 = 1.5*2.5*1.45^2 (2*121.2 + 68)/6 = 407.9 \text{ kNm}$$

The thickness of the footing is assumed to be uniform thick. Equating the bending moment capacity to the factored moment gives:

$$KLd^2 f_{ck} = M_c = 407.9 \text{ kNm; or } d = 0.217 \text{ m}$$

Since the shear stress usually controls the thickness, assume the overall thickness of the footing as = $h = 0.5$ m.

The effective depth of the section assuming 20 mm diameter bars as: d = 500 – 50 –10 = 440 mm.

Design for shear

Punching type of shear stress is not critical in footing having the triangular distribution of bearing stress. The transverse shear stress on the critical section at a distance of d from the face of the column on the high intensity reaction face is critical.

Check for transverse shear stress

The critical section is at a distance d from the face of the column. The shear span is:

$$L_s = (B - a)/2 - d = 2.9/2 - 0.44 = 1.01 \text{ m}$$

The bearing pressure at a distance d face of the column is interpolated from the figure and it is:

$$P_5 = P_1 - \frac{(p_1)L_s}{B_c} = 121.2 - \frac{(121.2)1.01}{3.3} = 84.1 \text{ kN/m}^2$$

The factored shear force on the critical plane is:

$$V_s = \gamma LL_s (p_1 + p_5)/2 = 1.5*2.5*1.01*(121.2 + 84.1)/2 = 388.8 \text{ kN}$$

The nominal shear stress is: $\tau_v = \dfrac{V_s}{Ld} = \dfrac{388800}{2500*440} = 0.35 \text{ N/mm}^2$

The shear strength is 0.36 MPa for 0.25% reinforcement.
Design for bending in the width direction
Area of steel in width (3.9m) direction is:

$$A_{st} = \frac{1.15M_c}{jdf_y} = \frac{1.15*407.9*10^6}{0.8*440*415} = 3211 \text{ mm}^2/2.5 \text{ m}$$

Provide 11 numbers of 20 mm bars along the 3.4 m length, the spacing is 2500/11 = 227 mm.
Percentage of the reinforcement is = (11*314*100)/(2500*500) = 0.28 %
The transverse shear strength is adequate.

Design of reinforcement in the length direction (Reinforcement along the short span)
The bending moment span is: $L_m = (L - a)/2 = (2.5 - 0.5)/2 = 1. \text{ m}$
The average net bearing pressure that causes bending moment is = $p = 500/2.5*3.4 = 59 \text{ kN/m}^2$,
The factored bending moment over the width of 3.9 m is:

$$M_c = \frac{\gamma_f pBL_m^2}{2} = \frac{1.5*3.4*59*1.2}{2} = 150.45 \text{ kNm/m}$$

The effective depth of the reinforcement is: $d = 500 - 50 - 20 - 10 = 420 \text{ mm}$
Area of steel in the short span direction is:

$$A_{st} = \frac{1.15M_c}{jdf_y} = \frac{1.15*150.45*10^6}{0.8*420*415} = 1241 \text{ mm}^2/2.5 \text{ m}$$

The minimum reinforcement required is 0.12 percent and it is:

$$A_{smin} = 0.12 *500*3400/100 = 2040 \text{ mm}^2$$

The total area of the reinforcement is to be distributed between the column band and the remaining. The proportion of the column band reinforcement is:

$$\frac{Reinforcement \ in \ the \ central \ band \ width}{Total \ reinforcement \ in \ the \ short \ span} = \frac{2}{1 + ratio \ of \ long \ span \ to \ shot \ span}$$

The above factor is 2/(1+3.4/2.5) = 0.85
85 of the reinforcement has to be in the column or pedestal (central) strip, and it is = 0.85*2040 = 1734 mm².

Provide six of 20 bars in the 500 mm width of the column strip. The balance of the reinforcement is not much but the maximum spacing of the reinforcement should preferably limit to twice the effective depth and the minimum percentage of reinforcement as 0.12 %.

Provide six numbers of 16 mm bars in each segment of 1450 mm on either side of the footing projected beyond the column. Figure 12.14 illustrates the reinforcement details of the footing.

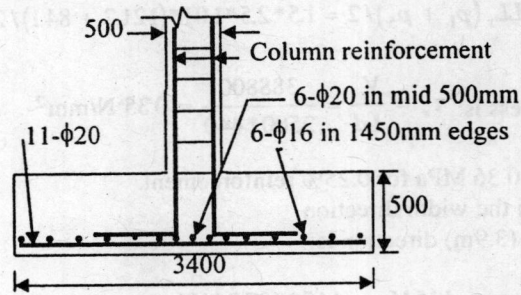

Fig. 12.14 Rectangular Footing Reinforcement; Example 12.7.

12.6 CIRCULAR FOOTING

Circular footings are used in case of circular sectioned columns, well foundations, circular overhead water tanks etc. The loads and bending moment are normally represented in the polar coordinates. The reinforcement is provided in the radial and circumferential directions or in rectangular grid. Three common types of footings are:

- Circular slab with central column,
- Circular slab with ring beam or shaft at the periphery of the slab and
- Circular slab with a ring beam or shaft inside the slab.

The critical bending moments in the three types of circular slab are given for quick reference.

Circular slab with central column load

Maximum bending moments at the periphery of the central column:

$$M_R = \frac{W}{4\pi}\left(\ln\frac{R}{a} - 1 + \left(\frac{a}{R}\right)^2\right) \qquad (12.13)$$

$$M_\theta = \frac{W}{4\pi}\left(\ln\frac{R}{a}\right) \qquad (12.14)$$

Where

M_R = maximum radial bending moment at the periphery of the column,
W = superimposed load on the footing through central column,
ln = natural logarithm,
a = radius of the circular column, and
R = radius of the circular footing.

M_θ = circumferential bending moment at the periphery of the column..

Simply supported slab with uniformly distributed load

$$M_R = M_\theta = \frac{3wR^2}{16}$$

(12.15)

where

M_R and M_θ = maximum radial and circumferential bending moments that occur at the centre of the slab,

w = intensity of the uniformly distributed load.

Circular footing subjected to concentric ring load at a radius
   Maximum bending moments at the centre of the slab:

$$M_R = M_\theta = \frac{W}{16\pi}\left[2\left(1 + 2\ln\frac{R}{a} - \left(\frac{c}{R}\right)^2\right) - 3\right]$$

(12.16)

where c = radius of the circular line load from shaft
   The maximum bending moments at the face of the line load

$$M_R = \frac{W}{16\pi}\left[2\left(1 + 2\ln\frac{R}{a} - \left(\frac{c}{R}\right)^2\right) - 3\left(1 - \left(\frac{c}{R}\right)^2\right)\right]$$

(12.17)

$$M_\theta = \frac{W}{16\pi}\left[2\left(1 + 2\ln\frac{R}{a} - \left(\frac{c}{R}\right)^2\right) - \left(3 - \left(\frac{c}{R}\right)^2\right)\right]$$

(12.18)

The design and detailing of the circular footing is illustrated with examples.

Example 12.8: Circular Footing with Central Column Load

A 500 mm diameter column is transferring a load of 900 kN to a circular footing mild environment. Net safe bearing capacity of the soil is 90 kN/m² at one metre below the ground level. Design the footing.

Solution

The exposed environment is mild so *M*20 concrete and *HYSD-Fe*415 reinforcement are selected. Bearing area of the footing is obtained for a superimposed load of 960 kN, in which the extra 60 kN accounts for the extra weight difference between the footing concrete and the soil. The bearing area required is:

$$\text{Bearing area required} = \frac{W_s}{P_a} = \frac{960}{90} = 10.67$$

Select the radius of the footing as = R = 1.9 m. The net soil reaction for bending and shear force calculation is:

$$p = \frac{W}{\pi R^2} = \frac{900}{3.1416 * 1.9^2} = 79.4 \text{ kN/m}^2 \text{ ; use } p = 80 \text{ kN/m}^2.$$

The maximum factored bending moments on the slab are:

$$M_R = \gamma \frac{W}{4\pi} \left(\ln \frac{R}{a} - 1 + \left(\frac{a}{R} \right)^2 \right) = \frac{1.5 * 900}{4 * 3.1416} \left[\ln \frac{1.9}{0.25} - 1 + \left(\frac{0.25}{1.9} \right)^2 \right] = 112.4 \text{ kNm/m}$$

$$M_\theta = \gamma \frac{W}{4\pi} \left(\ln \frac{R}{a} \right) = 217.9 \text{ kNm/m}$$

The circumferential bending moment is about twice of the radial bending moment. Equating the moment capacity to the circumferential bending moment gives:

$$d = \sqrt{\frac{M_\theta}{kbf_{ck}}} = \sqrt{\frac{217.9(10)H^6}{0.138 * 1000 * 20}} = 281 \text{ mm}$$

The radial cantilever span is about 1.7 m so let the overall depth of the footing at the column face be 600 mm. The effective depth of the section is $600 - 50 - 10 = 540$ mm. The area of the circumferential reinforcement is:

$$A_{st} = \frac{1.15 M_\theta}{jdf_y} = \frac{1.15 * 217.9(10^6)}{0.8 * 540 * 415} = 1400 \text{ mm}^2 / 2.5 \text{ m}$$

Provide 20 mm bars at 220 mm spacing.
Similarly the area of the radial reinforcement is:

$$A_{st} = \frac{1.15 M_R}{jdf_y} = \frac{1.15 * 112.4(10^6)}{0.8 * 520 * 415} = 749 \text{ mm}^2 / 2.5 \text{ m}$$

Provide 16 mm bars at 260 mm spacing.
The minimum percentage of the reinforcement required is:

$$A_{s\,min} = 0.12 * 1000 * 600 / 100 = 720 \text{ mm}^2 / \text{m}$$

So the radial reinforcement when curtailed, the spacing of the 16 mm bars must be atleast 275 mm.

Design for Shear Force

The punching shear is critical at distance $d/2$ from the column face. The diameter of the punch circles is:

$$u = a + d = 500 + 540 = 1040 \text{ mm}$$

The perimeter of the punch circle is: $b = \pi u = 3.267$ m.
Area on which the punching force acts is on the annular circle with internal diameter as u, and it is:

$$A = \frac{\pi (D^2 - u^2)}{4} = \frac{3.1416 (3.8^2 - 1.04^2)}{4} = 10.49 \text{ m}^2$$

The net punching shear force is: $V = pA = 80*10.49 = 839$ kN

Punching nominal shear stress is: $\tau_v = \dfrac{V}{bd} = \dfrac{839000}{3267*540} = 0.48$ N/mm^2

The punching shear strength is: $\tau_{cc} = k_s\left(0.25\sqrt{f_{ck}}\right) = 0.25*\sqrt{20} = 1.1$ N/mm^2

The footing is safe against punching shear.

The transverse shear stress normally controls the design of the footings as seen in the previous examples. The transverse shear strength is also important. Figure 12.15 illustrates a transverse section *afb* that

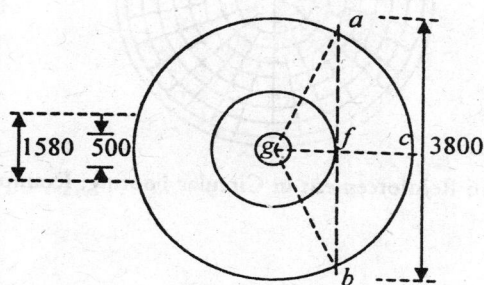

Fig. 12.15 Transverse Shear Section in Example 12.8.

is at a distance *d* from the face of the column.

Let the semi-subtended angle *agf* (Fig. 12.15) be equal to a, and it is:

$$\cos\alpha = (250+540)/1900 = 0.416 \text{ and } \alpha = 65.4°, \text{ or } 1.9 \text{ radians}$$

The length of the section *afb* = 2R sin α = 3.8*0.909 = 3.46 m.
The area of the segment *acb* is:

$$A = R^2(\alpha)/\pi - 0.5*3.46*0.79 = 3.8^2(1.9)/3.1416 - 1.367 = 7.366 \text{ m}^2$$

The transverse shear force is: $V = A*p = 7.366*80 = 589.28$ kN
The nominal shear stress is:

$$\tau_v = \frac{V}{bd} = \frac{589280}{3460*540} = 0.31 \text{ N/mm}^2$$

The shear strength is 0.36 MPa for 0.25% reinforcement, it is hoped that the percentage of the reinforcement will be more than0.25 percent. The reinforcement details are shown in Fig. 12.16.

Example 12.9: Circular Footing Supporting a Shaft

A circular shaft of 3.5 m external radius is transferring a load of 11,000 kN to a circular footing. The foundation is exposed to mild environment and the net safe bearing capacity of the soil is 120 kN/m^2 at one metre below the ground level.

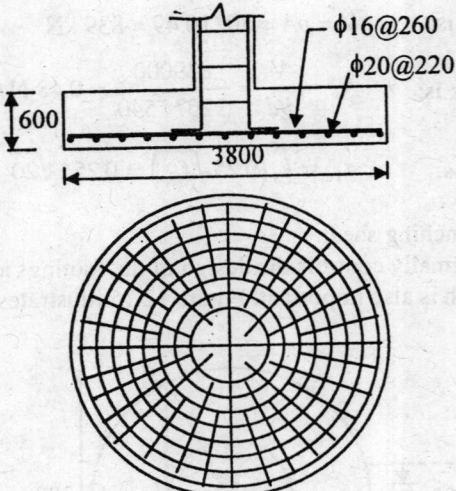

Fig. 12.16 Reinforcement in Circular Footing; Example 12.8.

Solution

The exposed environment is mild so *M*20 concrete and *HYSD-Fe*415 reinforcement is used. The foundation is placed at one metre below the ground. Bearing area of the footing is obtained for a superimposed load of 11,700 kN, in which the extra load of 700 kN accounts for the extra weight difference between the footing concrete and the soil. The bearing area required is:

$$\text{Bearing area required} = \frac{W_s}{p_a} = \frac{11700}{120} = 97.5 \text{ m}^2$$

Select the radius of the footing as = $R = 5.7$ m.
The net soil reaction for bending and shear force calculation is:

$$p = \frac{W}{\pi R^2} = \frac{11000}{3.1416 * 5.7^2} = 107.8 \text{ kN}/\text{m}^2 \text{ ; use } p = 108 \text{ kN/m}^2.$$

Maximum bending moments at the centre of the slab:

$$M_R = M_\theta = \frac{W}{16\pi}\left[2\left(1 + 2\ln\frac{R}{a} - \left(\frac{c}{R}\right)^2\right) - 3\right]$$

where c = radius of the circular line load from shaft = 3.5 m and $R = 5.7$ m

The maximum bending moment occurs at the centre of the slab with in the concentric ring of 3.5 m.

$$M_R = M_\theta = \frac{W}{16\pi}\left[2\left(1 + 2\ln\frac{R}{a} - \left(\frac{c}{R}\right)^2\right) - 3\right]$$

$$= 218.84 \left[2 \left(1 + 2 \ln\left(5.7/3.5\right) - 0.377\right) - 0.377 \right] = 44 \text{ kNm}/\text{m}$$

The maximum bending moments at the face of the line load is:

$$M_R = \frac{W}{16\pi} \left[2 \left(1 + 2 \ln\frac{R}{a} - \left(\frac{c}{R}\right)^2 \right) - 3 \left(1 - \left(\frac{c}{R}\right)^2 \right) \right]$$

$$= 218.84 \left[2 \left(1 + 2 \ln 1.63 - 0.377 \right) - 3 \left(1 - 0.377 \right) \right] = 291.3 \text{ kNm}/\text{m}$$

$$M_\theta = \frac{W}{16\pi} \left[2 \left(1 + 2 \ln\frac{R}{a} - \left(\frac{c}{R}\right)^2 \right) - \left(3 - \left(\frac{c}{R}\right)^2 \right) \right] = 126.34 \text{ kNm}/\text{m}$$

The radial bending moment at the face of the shaft is the largest and it controls the thickness of the footing. Equating the moment capacity to the bending moment gives:

$$d = \sqrt{\frac{M_R}{kbf_{ck}}} = \sqrt{\frac{291.3(10)^6}{0.138*1000*20}} = 325 \text{ mm}$$

The radial cantilever span is about 2.2 m. Let the overall depth of the footing at the column face be 600 mm. The effective depth of the section would then be 600 −50 −10 = 540 mm. The area of the radial reinforcement is:

$$A_{st} = \frac{1.15 M_R}{jdf_y} = \frac{1.15*291.3(10)^6}{0.8*540*415} = 1869 \text{ mm}^2/\text{m}$$

Provide 20 mm bars at 165 mm spacing.
Similarly the area of the circumferential reinforcement is:

$$A_{st} = \frac{1.15 M_R}{jdf_y} = \frac{1.15*126.34(10)^6}{0.8*520*415} = 842 \text{ mm}^2/\text{m}$$

Provide 16 mm bars at 235 mm spacing.
The minimum percentage of the reinforcement required is

$$A_{s\,min} = 0.12*1000*600/100 = 720 \text{ mm}^2/\text{m}$$

Spacing of the radial reinforcement of 16 mm bars must be atleast 275 mm. The bending moment in the middle portion of the footing is small so only nominal reinforcement is needed. The thickness of the footing at the mid-point is reduced to 300 mm. The calculations are very similar. The shear stress is not critical. Figure 12.17 illustrates the details in brief. The curtailment of the radial reinforcement is not shown here.

12.7 COMBINED FOOTINGS

Two or more columns are placed on one footing is called combined footing. Such a need may arise if the bearing capacity of the soil is low or the columns are placed close to each other or independent footings overlap. A footing for a row of columns is called a combined *strip footing* and designed as a continuous

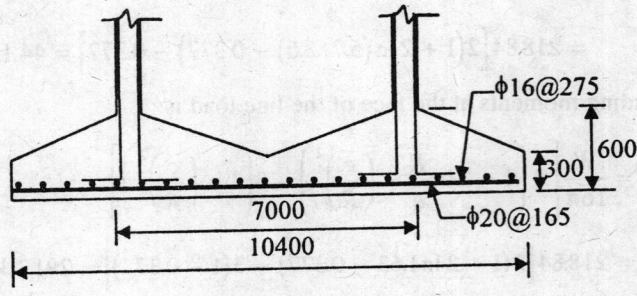

$\phi 16@275$

600

300

7000

$\phi 20@165$

10400

Fig. 12.17 Details of Example 12.9.

beam. The centroid of the foundation coincides with that of the loads on the footing is likely to be the smallest the size. Typical combined footings are illustrated in Fig. 12.18. The centroid of the loads may be placed eccentric with the footing in case of unfavorable site conditions. Critical shear force occurs at a distance equal to the effective depth of the foundation from the face of the column.

Example 12.10: Design of Combined Footing

Two columns of each 500 by 500 mm size are spaced at 4 m apart. The loads from the columns are 500 kN and 600 kN. The soil is in moderate exposure condition and with a net safe bearing capacity of 90 kN/m² at 1.2 m below the ground level. The width of the foundation is not to exceed 2 m. Design a foundation.

Solution

Select the minimum grade of the concrete is $M25$, as the exposure condition is moderate. Also select $HYSD$-$Fe415$ reinforcement. The loads from the columns are:

$$P_1 = 500 \text{ kN}, \quad P_2 = 600 \text{ kN}.$$

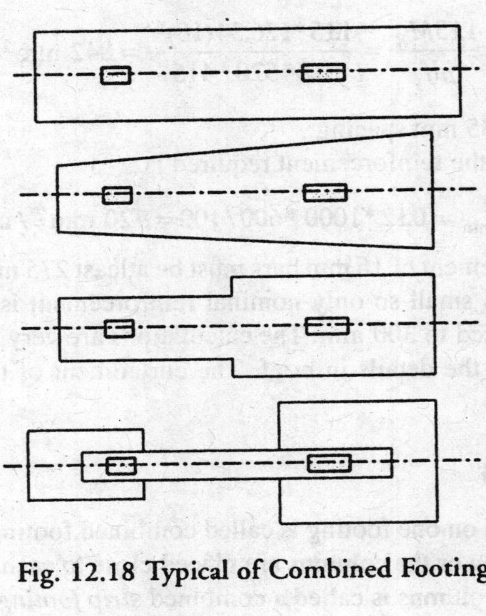

Fig. 12.18 Typical of Combined Footings.

Add extra of about six percent of the load to account for the increased load due to concrete from the replaced soil. The total superimposed load is taken as equal to 500 + 600 +70 = 1170 kN.

Minimum foundation area required is:

$$A_f = 1170/90 = 13 \text{ m}^2.$$

The width of the foundation is chosen equal to = B = 2 m.
The minimum length required is = 13/2 = 6.5 m.
Select the foundation size as: B = 2 m, L = 6.5 m.
The net pressure on the footing slab for the purpose of structural design of footing is:

$$p = \frac{P_1 + P_2}{BL} = \frac{1100}{2*6.5} = 84.6 \text{ kN/m}^2$$

Use the net reaction pressure on the footing as: p = 85 kN/m².

The pressure under the footing is uniform as the centroid of the loads coincides with the centroid of the footing. Fig. 12.19 (a) illustrates the notations.

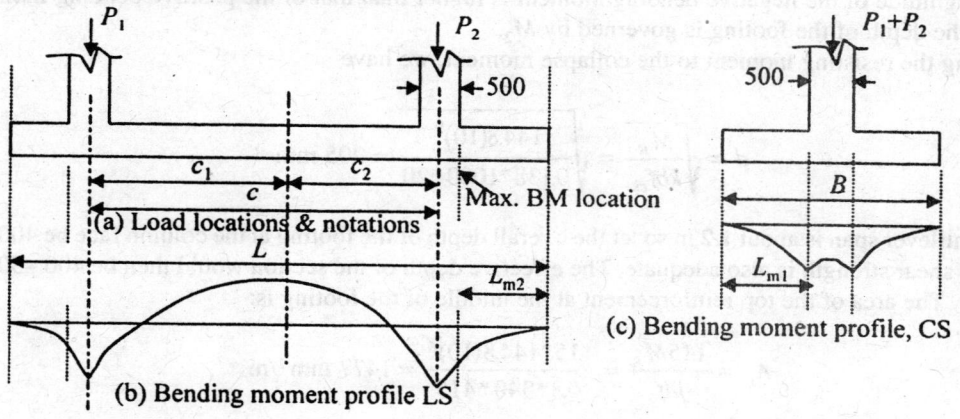

Fig. 12.19. Combined Footing Notations.

Let c = distance between the column loads,
  c_1 and c_2 are the distances of the loads P_1 and P_2 from the centroid of the loads.

The distance of the centroid of the loads from the P_1 is obtained by the taking moments of the two loads about the line of action of P_1 and it is:

$$c_1 = \frac{P_2 c}{P_1 + P_2} = \frac{600*4}{500 + 600} = 2.182 \text{ m}$$

$$c_2 = c - c_1 = 4.0 - 2.182 = 1.818 \text{ m}$$

Design of the section

The cantilever span that is beyond the face of the columns in the width is:

$$L_{m1} = (B - a)/2 = (2.0 - 0.5)/2 = 0.75 \text{ m}$$

The cantilever span that is beyond the face of the column on the right side is:

$$L_{m2} = L/2 - c_2 - a/2 = 3.25 - 1.818 - 0.25 = 1.182 \text{ m}$$

Since L_{m2} is larger than the L_{m1} span L_{m2} is critical. Maximum factored positive bending moment (causing compression on top fibre) on the footing is:

$$M_1 = \frac{\gamma \, p L_{m2}^2}{2} = \frac{1.5 * 85 * 1.182^2}{2} = 89.1 \text{ kNm/m}$$

The maximum negative bending (tension at top fibre) occurs at the centroid of the footing and it is:

$$M_2 = \gamma \left(\frac{P_2 c_2}{B} - \frac{pL^2}{8} \right) = 1.5 \left(\frac{600 * 1.818}{2} - \frac{85 * 6.5 * 6.5}{8} \right) = 144.8 \text{ kNm/m}$$

The magnitude of the negative bending moment is higher than that of the positive bending moment; therefore the depth of the footing is governed by M_2.

Equating the resisting moment to the collapse moment, we have

$$d = \sqrt{\frac{M_R}{kbf_{ck}}} = \sqrt{\frac{144.8(10)^6}{0.138 * 1000 * 20}} = 205 \text{ mm}$$

The cantilever span is about 1.2 m so let the overall depth of the footing at the column face be 400 mm so that the shear strength is also adequate. The effective depth of the section would then be $400 - 50 - 10 = 340$ mm. The area of the top reinforcement at the middle of the footing is:

$$A_{st} = \frac{1.15 M_R}{jdf_y} = \frac{1.15 * 144.8(10)^6}{0.8 * 340 * 415} = 1477 \text{ mm}^2/\text{m}$$

Provide 20 mm bars at 210 mm spacing at top face in the middle zone.

Similarly the area of the cantilever reinforcement is:

$$A_{st} = \frac{1.15 M_R}{jdf_y} = \frac{1.15 * 89.1(10)^6}{0.8 * 340 * 415} = 908 \text{ mm}^2/\text{m}$$

Provide 16 mm bars at 220 mm spacing.

The percentage of the reinforcement provided is:

$$percentage \ of \ reinforcement = \frac{201 * 100}{0.22 * 340 * 1000} = 0.26$$

Design for the transverse shear

The critical section is at a distance d from the face of the column and the shear span is:

$$L_s = L_m - d = 1.182 - 0.34 = 0.842 \text{ m}$$

The factored transverse shear force is:

$$V = 1.5pBL_s = 1.5*85*2*0.842 = 214.71 \text{ kN}$$

The nominal shear stress is

$$\tau_v = \frac{V}{Bd} = \frac{214710}{2000*340} = 0.34 \text{ N/mm}^2$$

The value is just less than 0.36 MPa, which is the shear capacity for the 0.25 % of reinforcement. The critical punching diagonal tension is not a problem in this case. The section is safe in shear strength. The minimum percentage of the reinforcement is: $A_{smin} = 0.12*400*1000/100 = 480 \text{ mm}^2/\text{m}$ Provide 12 mm bars at 230 mm spacing as the nominal reinforcement.

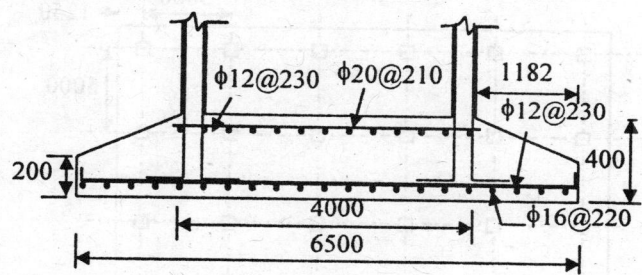

Fig. 12.20 Details of Example 12.10.

12.8 DESIGN OF RAFT FOUNDATION

A shallow single foundation unit that supports all columns and walls of a structure or part of a structure may be called a raft foundation. A raft foundation is also called as mat foundation. They are usually provided for multistory buildings, overhead water tanks, chimneys, etc. A raft foundation becomes unavoidable in submerged structures, in some multistory structures with basement and in retaining walls, etc. The raft foundation is designed as a flat slab. The reader is advised to read the section on flat slabs.

Example 12.11: Design of Rectangular Raft Foundation

A multistory building is provided with columns spaced at 5 m apart in two perpendicular directions. There are four rows of columns widthwise and six rows lengthwise. The size of the columns is 500 mm square and transmitting 2000 kN to the foundation. The height of the column above the foundation is 5 m. The soil in moderate environment is silty clay with safe net bearing capacity of 110 kN/m². Design the foundation.

Solution

M25 concrete and HYSD-Fe415 reinforcement bars are selected. A raft foundation is foundation adopted.
   The total number of columns = 4*6 = 24
   Total load transferred by the columns is = 24*2000 = 48,000 kN
Add extra of 6% for additional load from the concrete foundation.

$$\text{The foundation area required for raft is} = \frac{1.06*48000}{110} = 462.5 \text{ m}^2$$

The centre-to-centre area covered by the columns is = 3*5 by 5*5 = 15 m by 25 m. It is desirable to have a cantilever projection of the raft beyond the outer column line to reduce the bending moment in the outer most bay.

Provide a raft foundation area as 18 m by 28 m. This provides an area of 504 m², with projection of 1.5 m beyond the centre line of the outer columns. The actual projection beyond the column face is 1.25 m. The proposed layout of the raft foundation along with the column locations is shown in Fig. 12.21.

The safe effective pressure for structural design can be calculated by considering the actual load transmitted by the columns to the foundation soil.

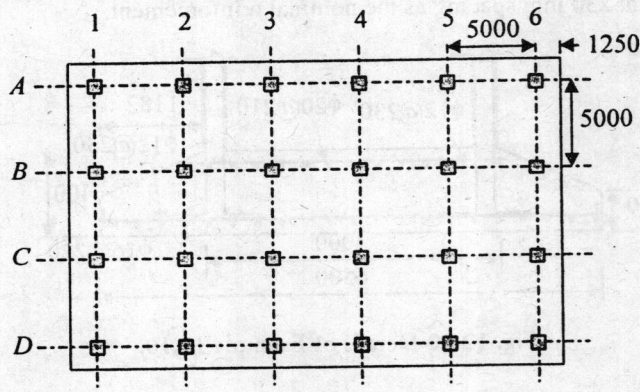

Fig. 12.21 Column Layout on Raft; Example 12.11.

$$\text{The effective bearing pressure is} = p = \frac{24*2000}{504} = 95.3 \text{ kN/m}^2$$

Design of the raft foundation

The panel size in both directions is same and it is $= L = 5$ m.
The clear span is $= L_o = L - a = 5 - 0.5 = 4.5$ m

The factored load from the soil pressure on one panel is:

$$W = \gamma p \, L_o L = 1.5*95.3(4.5)(5) = 3216.375 \text{ kN/one panel}$$

Sum of the magnitudes of positive and negative bending moments in a panel is:

$$M_0 = \frac{WL_0}{8} = \frac{3216.375*4.5}{8} = 1809 \text{ kNm / panel}$$

There is a correction due to cantilever bending moment in the edge. This is not considered to be on the safer side.

The relative stiffness of the columns and the slab determine the distribution of the negative and positive bending moments in the exterior panel. The building is not restrained against lateral sway so the effective column height can be taken equal to 1.2 times the clear height of the column. Hence the effective height of the column is:

$$L = H - t = 5 - 0.25 = 4.75m$$

The relative stiffness of the column is:

$$K_{cl} = \frac{I_c}{L_e} = \frac{a^4}{12L_e} = \frac{0.5^4}{12*4.75} = 0.0011$$

Let the thickness of the raft slab be 500 mm for the purpose of computing the relative thickness. The relative stiffness of the slab panel is:

$$K_s = \frac{I_s}{L_{se}} = \frac{bt^3}{12L_{se}} = \frac{5*0.5^3}{12*5} = 0.01$$

The exterior columns must be designed to have a minimum relative stiffness so as to withstand the panel bending moment. Such a desired minimum relative stiffness is given in the table 8.9 for different aspect ratios and live load to dead load ratios. From table 8.9, it can be seen that the minimum required relative stiffness ratio is zero for the live load ratio of 0.5 and the aspect ratio of 1.0. The relative stiffness ratio in the present case is:

$$\alpha_c = \frac{K_{cl}}{K_s} = \frac{0.0011}{0.01} = 0.11$$

The *Alfa* factor that decides the relative distribution of bending moment between the negative and the positive bending moments is:

$$\alpha = 1 + 1/\alpha_c = 1 + 1/0.11 = 10$$

Table 8.7 gives the relative negative and the positive bending moment coefficients in the exterior panel:

2. End span

Exterior negative bending moment coeff. $= c_{n0} = 0.65/\alpha = 0.065$
Interior negative bending moment coeff. $= c_{ni} = 0.75 - 0.1/\alpha = 0.74$
Positive bending moment coeff. $= c_{p0} = 0.63 - 0.28/\alpha = 0.6$

The corresponding end panel bending moments are:

$$M_{no} = c_{n0}M_0 = 118 \text{ kNm/panel}$$
$$M_{ni} = c_{ni}M_0 = 0.74*1809 = 1339 \text{ kNm/panel}$$
$$M_{po} = c_{p0}M_0 = 0.6*1809 = 1085 \text{ kNm/panel}$$

The negative bending moment corresponding to the interior panel support is:

$$M_{nii} = 0.65M_0 = 0.65*1809 = 1176 \text{ kNm/panel}$$

In which the subscript n refers to the negative bending moment, p refers to positive bending moment, and the subscript i refers interior support or interior panel and o refers to the outer panel. The negative bending moment is the one that causes tension on the face of the soil.

The bending moment is distributed between the column and middle strips as per table 8.8. The subscript c refers to column strip and m refers to the middle strip. The corresponding bending moments in the outer (exterior) panel strip are:

$$M_{cno} = (1)\, M_{no} = 118$$
$$M_{cni} = 0.75 M_{ni} = 0.75*1339 = 1004 \text{ kNm}/2.5\text{m}$$
$$M_{cpo} = 0.65 M_{po} = 0.65*1085 = 705 \text{ kNm}/2.5\text{m}$$

The negative bending moment corresponding to the interior panel column strip is:

$$M_{nii} = 0.75 M_{cnii} = 0.75*1176 = 882 \text{ kNm}/2.5\text{m}$$

The remaining bending moment is assigned to the middle strip. First determine the reinforcement in the column strip and the reinforcement spacing in the middle strip can be adjusted proportionally. The most critical bending moment occurs at the interior support of the outer panel column strip and it is:

$$M_{cni} = 1004 \text{ kNm}/2.5\text{m}$$

The effective balanced depth required is:

$$d = \sqrt{\frac{M_c}{Kbf_{ck}}} = \sqrt{\frac{1004\left(10^6\right)}{0.138(2500)25}} = 342 \text{ mm}$$

Let the overall thickness of the raft slab can be chosen as 800 mm. The effective depth in one direction for the lower layer of the reinforcement is = 800 – 50 – 10 = 740 mm. The diameter of the reinforcement is assumed as 20 mm. Since the panel is a square panel, it is desirable to have spacing of the reinforcement in both the directions. The effective depth in the other direction is 740 – 20 = 720 mm. Use effective depth of the slab that can be used is 720 mm. Area of the reinforcement in column strip at the support is:

$$A_{st} = \frac{1.15 M_c}{jdf_y} = \frac{1.15*1004(10^6)}{0.8*720*415} = 4830 \text{ mm}^2/2.5 \text{ m}$$

Let 25 mm bars are selected, and then the spacing of the bars is 250 mm.

Provide 25 mm bars at 250 mm spacing at bottom face of the slab in the column strip. Alternate bars are continued from end to end and the other are curtailed at one-third span on either side of the column line.

The actual reinforcement at different locations of the column strip can be obtained by proportioning with the bending moments that were obtained earlier. The positive bending moment reinforcement in the column strip placed at the top of the slab.

The spacing of the bars is proportioned with respect to the bending moment and it is = 1004*250/705 = 356 mm.

Place 25 mm bars at 350 mm spacing at top in the column strip.

The spacing of the reinforcement is rationalized later after the reinforcement in the middle strip is computed.

The bending moments in middle strip are the balance of the panel moments and the column strip, and they are:

$$M_{mno} = 0.$$
$$M_{mni} = 0.25 M_{ni} = 0.25*1339 = 335 \text{ kNm}/2.5\text{m}$$
$$M_{mpo} = 0.35 M_{po} = 0.35*1085 = 320 \text{ kNm}/2.5\text{m}$$

The effective depth of the section in the middle strip is same as that of the column strip, therefore the reinforcement can be obtained by prorating with respect to the bending moment. However the minimum percentage and maximum spacing of the reinforcement (twice the depth of the slab) must be maintained.

The minimum reinforcement in the slab is:

$$A_{smin} = 0.12*800*2500/100 = 2400 \text{ mm}^2 \text{ per 2.5 m together on both faces}$$

The minimum reinforcement is 20 mm bars at 330 mm spacing. Select 16 mm bars in the middle strip and 20 mm bars in the column strip.

Provide 20 mm bars at 330 mm at the bottom face and at 330 mm on the top face of the middle strip. These bars are continued from end to end.

Design of Section for Shear

The critical shear plane is the peripheral plane at a distance $0.5d$ from the face of the column. The length of the critical section is: $b = 4(a + d) = 4(0.5 + 0.72) = 4.8$ m

The shear force on the plane is:

$$V_u = \gamma p\, (L*L - (a + d)(a + d)\,) = 1.5*95.3(5*5 - 1.2*1.2) = 3368 \text{ kN}$$

The nominal shear stress is:

$$\tau_v = \frac{V_u}{bd} = \frac{3368000}{4800*720} = 1.0 \text{ N/mm}^2$$

The shear strength of the concrete without shear reinforcement is given by:

$$\tau_c = 0.25\left(0.5 + \beta_c\right)\sqrt{f_{ck}}$$

The aspect ratio of the column is one, so the maximum admissible value of β is only 0.5. Therefore the shear strength of the *M*25 concrete is:

$$\tau_c = 0.25(5) = 1.25 \text{ N/mm}^2.$$

The nominal shear stress is less than the shear capacity of the *M*25 concrete; therefore no need for stirrups. However some nominal stirrups in the column strip near the column are provided to distribute the column load.

Select eight legged 12 mm stirrups over a width of 2.5 m column strip, the area of the stirrup steel is: Select eight legged 12 mm stirrups.

$$A_{sv} = 8*113 = 904 \text{ mm}^2$$

The maximum spacing of the stirrups is controlled by:

$$s_{max} \leq \frac{A_{sv}f_{yv}}{0.4b} = \frac{904(415)}{0.4(2500)} = 375 \text{ mm} \text{ , further the maximum spacing is limited to:}$$

Provide eight legged 12 mm stirrups at 375 mm spacing.

Figure 12.22 illustrates the reinforcement details of the column and middle strips of the raft slab. The thickness of the raft slab is kept almost to the minimum. Therefore stirrups were required. Further the depth should be such that the development length of the column reinforcement must be accommodated

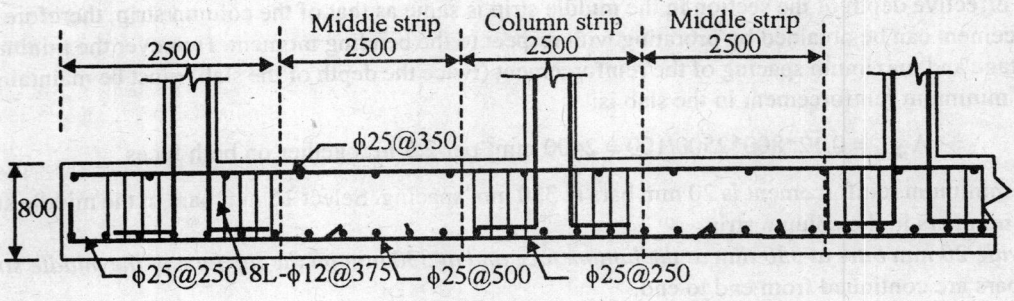

(a) Column strip reinforcement details

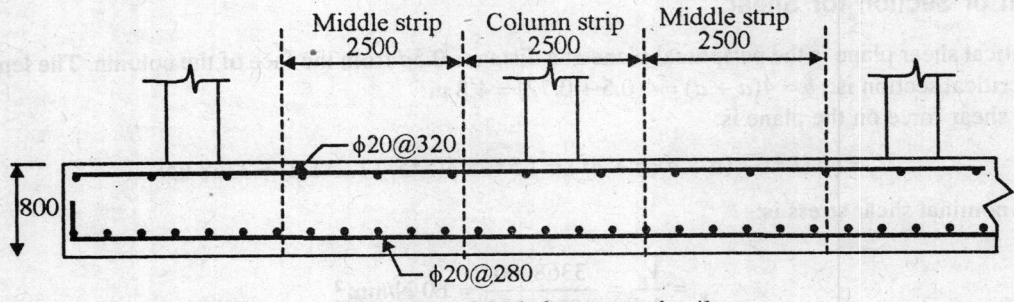

(b) Middle strip reinforcement details

Fig. 12.22 Reinforcement Details; Example 12.11.

with in the depth of the raft. This example is given to illustrate the design of raft including the design of the stirrups.

PROBLEMS

Use IS 456-2000 is permitted. Sketches of reinforcement details must illustrate all solutions.

12.1 A 345 mm thick brick masonry wall is transmitting a load of 160 kN/m to a soil. Net allowable bearing capacity of the soil is 100 kN/m² and in mild exposure condition. Design a reinforced concrete strip foundation using $M20$ concrete and $HYSD\text{-}Fe$ 415 reinforcement bars.

12.2 A reinforced concrete wall of 250 mm thickness is transmitting an axial load of 300 kN/m to a soil having a net allowable bearing capacity of 100 kN/m². The property limit is 1.2 m on one side of the central line of the wall. Design a reinforced concrete foundation using $M25$ concrete and $HYSD\text{-}Fe415$ bars.

12.3 A 250 mm thick reinforced concrete wall of is transmitting an axial load of 400 kN/m and bending moment of 50 kNm/m. The net allowable bearing capacity of the soil is 100 kN/m². Design a reinforced concrete foundation using $M25$ concrete and $HYSD\text{-}Fe415$ bars.

12.4 A square column of size 500 mm is transmitting a load of 800 kN to a soil having a net allowable bearing capacity of 110 kN/m². The exposure condition is severe. Design a square reinforced concrete footing using $M30$ concrete and $HYSD\text{-}Fe415$ reinforcement. The thickness of the footing is kept constant.

12.5 A 500 mm square column is transmitting a load of 800 kN to a soil having a net allowable bearing capacity of 110 kN/m². Design a square reinforced concrete footing using $M30$ concrete and $HYSD\text{-}Fe415$ reinforcement. The footing thickness tapers 150 mm at the free edge. The exposure condition is severe.

12.6 A 300 mm square column of is transmitting 600 kN axial load to a soil having a net allowable bearing capacity of 100 kN/m². The width of the footing is 1.8 m. Design an isolated footing with $M25$ concrete and $HYSD\text{-}Fe415$ reinforcement.

12.7 A 400 mm square column of is transmitting an axial load of 800 kN and bending moment of 100 kNm to a soil having a net allowable bearing capacity of 100 kN/m². Design a reinforced concrete square footing using M30 concrete and HYSD-Fe415 reinforcement bars.

12.8 A square column of 400 mm size is transmitting an axial load of 800 kN and bending moment of 100 kNm in seismic load condition to a soil having a net allowable bearing capacity of 100 kN/m². Design a reinforced concrete footing using M25 concrete and HYSD-Fe415 reinforcement bars.

15.9 A 350 mm diameter circular column is transmitting 700 kN to a soil having a net allowable bearing capacity of 90 kN/m². Design a circular footing using M20 concrete and HYSD-Fe415 reinforcement bars. Maximum bending moments at the periphery of the central column are:

$$M_R = \frac{W}{4\pi}\left(\ln\frac{R}{a} - 1 + \left(\frac{a}{R}\right)^2\right); \quad M_\theta = \frac{W}{4\pi}\left(\ln\frac{R}{a}\right)$$

Where: M_R = maximum radial bending moment at the periphery of the column,
W = superimposed load on the footing through central column,
$\ln$ = natural logarithm, a = radius of the circular column, and
R = radius of the circular footing.
M_θ = circumferential bending moment at the periphery of the column.

12.10 A circular cylindrical shaft of radius 3.5 m and thickness of 200 mm is transmitting an axial load of 10,000 kN to a soil having a net allowable bearing capacity of 100 kN/m². Design a reinforced concrete circular raft foundation using M25 concrete and HYSD-Fe415 reinforcement bars. The bending moments for the circular raft are:

$$M_R = M_\theta \frac{W}{16\pi}\left[2\left(1 + 2\ln\frac{R}{c} - \left(\frac{a}{R}\right)^2\right) - 3\right]$$

where: c = radius of the circular line load from shaft, R = radius of the raft slab.
The maximum bending moments at the face of the line load

$$M_R = \frac{W}{16\pi}\left[2\left(1 + 2\ln\frac{R}{c} - \left(\frac{c}{R}\right)^2\right) - 3\left(1 - \left(\frac{c}{R}\right)^2\right)\right]$$

$$M_\theta = \frac{W}{16\pi}\left[2\left(1 + 2\ln\frac{R}{c} - \left(\frac{c}{R}\right)^2\right) - 3\left(1 - \left(\frac{c}{R}\right)^2\right)\right]$$

12.11 Two 350 mm square columns spaced at 3 m apart transmitting 500 kN each to the foundation. The net allowable bearing capacity of the soil is 90 kN/m² at one metre below GL. Design a combined reinforced concrete footing using M25 concrete and HYSD-Fe415 bars. The width of the footing is 2 m.

12.12 Two columns spaced at 4 m apart transmit 600 and 800 kN to the foundation. The net allowable bearing capacity of the soil is 80 kN/m² at one metre below GL. Design a combined footing having a width of 3 m using M25 concrete and HYSD-Fe415 reinforcement bars. The size of each column is 350 mm square.

12.13 Two 500 mm square columns spaced at 5 m apart transmit 600 kN each to a soil having a net allowable bearing capacity of 65 kN/m². Design a combined beam and slab footing with a width of slab as 2.5 m and 500 mm beam width. Use M25 concrete and HYSD-Fe415 reinforcement bars.

12.14 A raft foundation is provided to a multi-storey building with columns spaced at 5 m apart in both directions. There are 4 rows of columns and four columns in each row. The net allowable bearing capacity of the soil is 100 kN/m². Each column transmits a load of 2400 kN. The size of the column is 500 mm square. Design a raft foundation using M25 concrete and HYSD-Fe415 reinforcement bars.

12.15 A multi-storey building has columns spaced at 4.5 m apart in two perpendicular directions. The building has 6 rows of columns with 5 columns in each row. Each column is transmits 1500 kN axial load and a bending moment of 200 kNm in each column under seismic. Design a raft foundation using M25 concrete and HYSD-Fe415 reinforcement bars.

13

Pile Foundations

13.1 INTRODUCTION

A foundation that derives primary strength from the depth of the soil is called *deep foundation*. The design of a deep foundation has two distinct parts. The first is the determination of the depth of the foundation and second is the structural design. The deep foundations are classified into two main groups:

1. Pile foundation, and
2. Well foundations.

The pile foundations are further classified into three sub-groups:

(a) Bearing piles,
(b) Friction piles, and
(c) Compaction piles.

Based on the construction method, the piles are classified as:

● Precast driven pile,
● Bored cast-in-situ pile,
● Driven cast-in-situ pile; encased or not encased ones,
● Under-reamed pile.

The cast-in-situ pile is either compacted or non-compacted. Normally the piles are upright but in some situations, piles are placed in an inclined manner and such piles are called *batter or Rake piles*.

Reinforced or prestressed concrete piles of square or circular or polygonal in cross section are precast in desired length. Such piles are driven in to the ground by drop hammer. A cushion of hard wood is placed at the top of the pile to receive the blows of the hammer. This cushion is called *dolly*. A weight is lifted up by a winch and allowed to fall by gravity on dolly. This weight is called *drop hammer* or ram or monkey. Hammer can also be operated by compressed air. A movable pile frame or Pile rig is used to position, align and drive the pile.

Driven cast-in-situ pile is the one in which a casing is driven first into the soil and then concrete is cast along with reinforcement. The casing can be either withdrawn or left in the ground. Pile with casing left in position is called encased pile. The casing is usually a steel tube that withstands the driving force. The diameter of the bored pile varies from 300 mm to 2500 mm. *Bored pile* is the one in which a bore is drilled into the ground and concrete is cast with reinforcement in the bore. If concrete is placed first in the bore and then the reinforcement cage is driven through the wet concrete, then it is called *bored compaction pile*. *Under-reamed pile* is the one in which one or more bulbs are formed in the soil during drilling and concrete is cast with the reinforcement. Under-reamed pile is used in cohesive soil, especially in black cotton soils.

A number of piles of about the same capacity are grouped into group. A group is integrated by a pile cap. The top of a pile group is provided with a pile cap on which the column or wall or super-structures stands. The cut-off-level of the pile is the level at which the pile is cut to accommodate the pile cap. The top of a pile should have anchor into the pile cap. Similarly the reinforcement of the pile should have adequate development length into the pile cap.

The capacity of a pile even though computed theoretically, but established by tests because of many uncertainties of the soil. One or more trial piles can be cast and load tested for the capacity. A trial pile is subjected to a test load and the total and regained displacements after unloading are measured. Certain percentage of piles must be tested for the capacity and such piles are called test piles. The test load is one-fifty percent of the working load of the pile. The ultimate load of a pile is the load at which the pile fails either structurally or by the collapse of the supporting soil. Safe load on the pile is obtained by dividing the load carrying capacity of the pile by a factor of safety which is normally equal to 2.5.

13.2 BEARING PILES

A pile that transmits the load to the soil through bearing at the tip of the pile is called a *bearing pile*. Hard or rocky strata should be available at reasonable depth below the ground level for sound bearing. The rate of settlement of bearing piles is nominal; consequently, friction between the soil and the pile along its depth may not be mobilized. The ultimate capacity of the bearing pile is given by:

$$Q_u = A_b P_u \qquad (13.1)$$

Where A_b = bearing area of the at tip of the pile, P_u = ultimate bearing capacity of the strata.

The minimum spacing of the piles must be twice the diameter of the pile or 750 mm so as to derive the full benefit of the bearing capacity of the soil strata. The capacity of a group of piles can be taken equal to the direct sum of the capacities of the piles in the group. Piles resting on gravel or stiff clay have a tendency to sink into the soil thus mobilizing the frictional resistance along the length of the pile. The load carrying capacity of such piles is combination of bearing friction, and discussed later.

13.3 FRICTION PILES

A pile that transmits the load to the soil primarily by skin friction between the pile and the soil is called a *friction pile*. The load carrying capacity of the pile can be determined either by static formula or dynamic formula. Ultimate load capacities of piles as per Indian Code of practice static formula is listed for ready reference.

Pile capacity in granular soil

$$Q_u = \sum_{i=1}^{n} (SKq_d \tan\delta)_i + A_b\left(\frac{\gamma}{2} DN_r + q_{dn}N_q\right) \qquad (13.2)$$

Where
Q_u = ultimate capacity of friction pile,
S_i = surface area of the pile in the ith soil layer,
K_i = coefficient of lateral earth pressure at ith soil layer, it varies from 1 to 3 in loose to medium sands,
q_{di} = effective over burden pressure at the ith layer of soil,
δ_i = angle of wall friction between pile and soil in the ith layer, it may be taken equal to ϕ_i = angle of internal friction of the soil,

A_b = area of cross section at bearing tip,
D = size of the pile at toe,
γ = unit weight of soil at toe,
N_r and N_q = bearing capacity factors of the soil at toe,
n = number of different layers through which the pile passes through, nth layer is at the toe of the pile.

The first part of the expression is from the friction and the second part is that due to bearing at the tip of the pile. The capacity of the pile can be approximated for a uniform type of soil as:

$$Q_u = S\left(q_0 + \frac{\gamma}{2}H\right)K\tan\phi + \gamma A_b\left(\frac{DN_r}{2} + HN_q\right) \tag{13.3}$$

where
S = surface area of the pile,
q_0 = permanent surcharge,
H = embedded depth of the pile in the natural soil,

The values of N_r and N_q are given in Table 13.1

Table 13.1 Values of Soil N Factors.

ϕ	N_r	Nq	Nc
10	—	—	9
20	5	8	18
25	10	15	26
30	20	24	37
35	50	40	55
40	120	70	—
44	260	—	—

Pile capacity in cohesive soils is:

$$Q_u = A_b N_c C_b + \alpha CS \tag{13.4}$$

where
N_c = bearing capacity factor and it is taken as 9 for clays or soft soils,
C_b = cohesion of soil at the tip of the pile,
C = average cohesion of the soil,
α = reduction coefficient,

Approximate values of α and cohesion values are given in Table 13.2

Table 13.2 Reduction Coefficients and Cohesive in Soils.

Soil Type	N_c-value Bored	α Driven	C (kN/m²)
Very soft to Soft	4 or less	0.7	1.0 0 to 35
Medium	4 to 8	0.5	1.0 to 0.535 to 70
Stiff	8 to 15	0.4	0.7 to 0.470 to 150
Hard	15 or +	0.4 to 0.3	150

Dynamic Formula

Dynamic formula is based on the laws governing the dynamic impact of the elastic bodies. The energy of the hammer blow is equated to the work done by the pile in penetration with allowance for the loss of energy due to elastic compression of the pile. Indian code of practice has adopted the Hiley formula.

The ultimate bearing capacity of a pile based on dynamic formula of Indian Code practice is:

$$Q_u = \frac{Wh\eta}{S + C/2} \qquad (13.5)$$

where

Q_u = ultimate driving resistance in tonnes, the safe load is obtained by dividing it by a factor of safety. The factor of safety is equal to 2.5.

W = mass of the hammer or ram in tonnes,

h = height of free fall of the hammer in cm taken at its full value for trigger operated drop hammers, 80 percent of the fall for normally proportioned winch operated drop hammers, and 90 percent of the stroke of the single acting hammers.

η = efficiency of the blow, ratio of the energy after impact to the striking energy of ram,

S = final set or penetration per blow in cm,

C = sum of the temporary elastic compressions in cm of the pile, dolly, packing and ground.

The reader is advised to refer to the Indian code of practice for detailed explanation of the formula. In any case the capacity of a pile is determined by trial test and then from the test pile.

Uplift Capacity

Piles are subjected to downward load along with shear and bending moment. But there are situations in which a pile is also subjected to upward load called *uplift*. The uplift may arise out of the bending moment acting a group of piles.

The uplift capacity of a pile is given by:

$$Q_f = W_g + S(C + p \tan\phi) \qquad (13.6)$$

where

W_g = weight of the pile plus that portion of the pile cap and overburden,

S = surface area of the pile,

C = cohesion of the soil,

p = normal pressure on the pile,

ϕ = angle of internal friction.

In non-cohesive soils, the weight of the pile limits the uplift capacity. Trial pile is also to be tested for the uplift.

13.4 STRUCTURAL DESIGN OF PILE

Pile is designed as short column except in filled up soil. Bending moment on the pile cap causes additional axial force and secondary bending moment on the piles. The lateral force on the pile cap causes shear and bending moment on the piles. The lateral load carrying capacity of a pile depends on the structural strength and the sub-grade modulus. The top portion of the pile cantilevers as the lateral deflection takes place. The cantilever span of the pile is a function of the horizontal sub-grade property. As per the suggestion of the Indian Code of Practice, the equivalent cantilever is obtained with the help of graphs. First estimate the value of horizontal sub-grade reaction k_s or the modulus of reaction k_c from Tables 13.3 or 13.4 respectively. The equivalent cantilever span for a set of values of k_s and k_c are given in Table 13.5. Figure 13.1 illustrates a simple pile group and the free body diagram of a typical pile. The equivalent cantilever of the pile is also indicated in the figure.

Table 13.3 Values of Horizontal Sub-grade Reaction k_s (kg/cm³).

	Dry	Submerged
Loose sand	0.26	0.15
Medium sand	0.78	0.53
Dense sand	2.07	1.24

Table 13.4 Sub-grade Constant for Clays k_c (kg/cm²).

Unconfined compressive Strength (kg/cm³)	Range	Approximate
0.2 to 0.4	7 to 42	7.7
1 to 2	32 to 65	49
2 to 4	65 to 130	100

The bending moment caused by the lateral force that is the shear force on the pile as:

$$M = LF_h \tag{13.7}$$

where F_h = lateral force on the pile, shear force on the group distributed,
L = cantilever span taken from Table 13.5

The head of the pile is assumed fixed against rotation if connected by a pile cap. The equivalent cantilever is obtained in terms of the size of the pile. The design of the pile by limit states design is made using the procedure applied to columns. The design is illustrated by examples.

Table 13.5 Equivalent Cantilever of Piles.

	Sand				Clay		
k_s(N/mm³)	L_c /D			k_s(N/mm³)	L_c /D		
	Fixed head	Free head			Fixed head	Free head	
0.005	8	11		0.7	6.5	7.5	
0.01	6	10		2.0	5.5	6.5	
0.05	5	7		5.0	4.0	5.2	
0.10	3.5	5.5		10	3.4	5.0	
0.20	3.0	5.0		20	3.0	4.0	

L_c = cantilever span, D = size of pile; k_s = sub-grade reaction

13.5 PRECAST REINFORCED CONCRETE PILE

Precast Concrete Piles of square, hexagonal, octagonal cross section are precast in $M20$ to $M45$ concrete with adequate reinforcement. These piles are designed to withstand handling and impact in addition to the working loads.

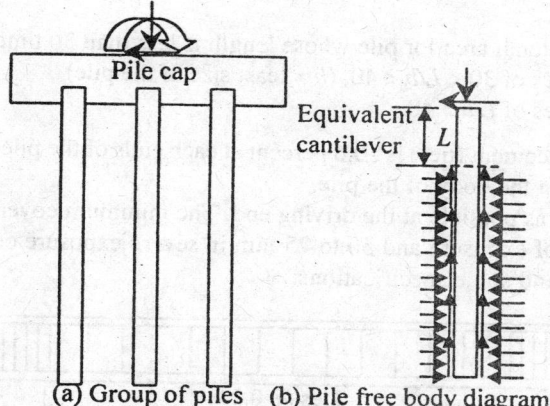

(a) Group of piles   (b) Pile free body diagram

Fig. 13.1 Pile Group and Forces on a Pile.

Handling condition: The precast concrete pile is cast and transported to the actual site. Handling and transportation forms an important design condition. Two or three pick-up points are chosen to lift the pile either to transport or to erect in position. The location of the pick-up points for minimum bending moments and the design moments are listed in Table 13.6

Table 13.6 Pick-up Points and Bending Moments in Precast Piles.

No. of points	Location of pick-up points	Design BM
Three	0.145L from each end and at mid span	0.0215WL
Two	0.027L from each end	0.011 WL
One	0.0293L from head	0.043 WL

(L = length of the pile, W = weight of the pile)

The impact stress during driving is given by the code of practice and it is:

$$\sigma = \frac{F\left(2/\sqrt{\eta} - 1\right)}{A_t} \qquad (13.8)$$

Where

F = resistance against driving,
A_t = transformed area of the cross section,
η = efficiency of the blow,
$A_t = A_c + 1.5m A_s$,
A_c = area of the concrete section,
A_s = area of the reinforcement,
m = modular ratio

Minimum reinforcement in precast concrete pile

Minimum reinforcement is provided to withstand the impact load. The minimum reinforcement recommended by the Indian Code of Practice is:

(a) Main reinforcement:
   1.25   percent of sectional area for pile whose length is less than 30 times the least size,
   1.50   percent for piles of $30 < L/b < 40$, (b = least size of the pile)
   2.00   percent for piles of $L/b > 40$

(b) Minimum lateral reinforcement (ties) is : 0.6 percent at each ends of the pile for a distance of 3 times
   depth and 0.2 per cent in the body of the pile.

The ties must be as close as possible at the driving end. The minimum cover to the ties should be 40 mm in ordinary conditions of exposure and 50 to 75 mm in severe exposure condition. Fig. 13.2 illustrates a typical precast pile and some specifications.

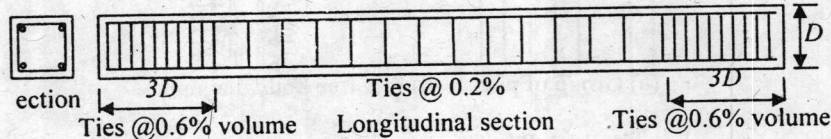

ection   3D   Ties @ 0.2%   3D
       Ties @0.6% volume   Longitudinal section   Ties @0.6% volume

Fig. 13.2 Typical Precast Pile Profile.

The minimum spacing of the bearing piles is twice the size of the piles and that of the friction piles is 2.5 times the size of the piles. The capacity of the group of piles is equal or less than the sum of the individual piles, may be taken about ninety percent of the sum of the individual piles.

Example 13.1: Determination of Length of Pile and Design of RC Pile

A 300 mm square precast pile carries an axial load of 400 kN, placed in loose to medium sandy soil having friction angle of 25 degrees. The density of the sand is 18 kN/m³ and the surcharge on the soil is 20 kN/m². Estimate the length of a concrete pile and suggest reinforcement details.

Solution

The total length of the pile from the static formula Eq. 13.3 is:

$$Q_u = S\left(q_0 + \frac{\gamma}{2}H\right)K\tan\phi + \gamma A_b\left(\frac{DN_r}{2} + HN_q\right)$$

Properties and data associated with the problem are:
   Surcharge = q_o = 20 kN/m², Density of soil = γ = 18 kN/m³,
   Angle of internal friction = ϕ = 25°, $\tan\phi$ = 0.466,

The N values of the soil that are estimated from the sandy soil are: N_r = 10, and N_q = 15,
The coefficient of lateral pressure for the sandy soil is estimated as 1.75 = K.
The size (D) of the pile is 0.3 m so the surface area of the pile in square metres is:

$$S = 4DH = 1.2H,$$

where: H = the length of the pile.
   The area of the cross section is = A_b = 0.3*0.3 = 0.09 m²,
   The ultimate load on the pile is obtained by multiplying the service load by a factor of safety of 2.5 and it is:

$$Q_u = 2.5*P = 1000 \text{ kN},$$

Substitution of the various quantities in the static capacity equation gives:

$$Q_u = S\left(q_0 + \frac{\gamma}{2}H\right)K\tan\phi + \gamma A_b\left(\frac{DN_r}{2} + HN_q\right)$$

$$1000 = 1.2H(20 + 9H)1.75(0.466) + 18*0.09(0.3*10/2 + 15H)$$

The rearrangement of the above expression gives:

$$H^2 + 4.98H - 113.3 = 0$$

The solution of the above quadratic gives the value of the depth of the pile.

$$H = 8.44 \text{ m. Select the length of the pile as 8.5 m.}$$

Structural Design

The factored limit load is : 1.5*400 = 600 kN
The ratio of length to side of the pile is = 8.5/0.3 = 28.33.

The minimum percentage of the main reinforcement for a slenderness of 28.3 is taken from that listed in this book is: 1.25%. The axial load capacity of the pile is checked for the minimum percentage of the reinforcement.

$$A_{smin} = 1.25*300*300/100 = 1125 \text{ mm}^2$$

Select four numbers of 12 mm and 16 mm bars, then the area of the reinforcement is = A_s = 1256 mm², The axial load capacity of short reinforced concrete column is:

$$P_u = 0.4A_c*f_{ck} + 0.67A_s*f_y = 0.4*0.09*20 + 0.67*0.001256*415 = 1.069 \text{ MN} = 1069 \text{ kN.}$$

The capacity of the pile is adequate.
Assume 8 mm ties with a 45 mm cover to the center of the tie bar.
The total length of one tie bar is = s = 4*(300 – 2*44) = 848 mm.
The volume of one tie bar is = v = 848*50 = 42400 mm³
The minimum volume of the ties in the end zone is = 0.6*300*300*900/100 = 486 000 mm³.
The number of ties in the end 900 mm length = 486000/42400 = 15.
The spacing of the ties = 900/15 = 60 mm.
Spacing of ties in the middle zone *three* times that at the end = 180 mm.

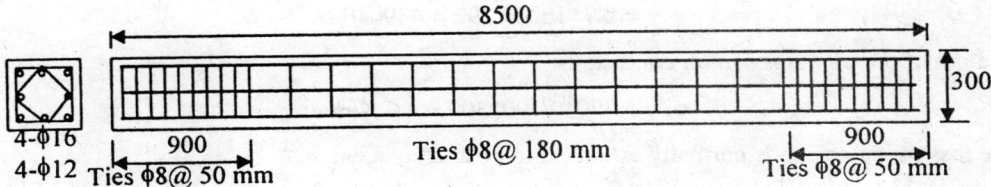

8500

300

4-φ16
4-φ12 Ties φ8@ 50 mm   Ties φ8@ 180 mm   Ties φ8@ 50 mm
         900   900

Fig. 13.3 Precast RC Pile, Example 13.1.

13.6 PRECAST PRESTRESSED CONCRETE PILE

High strength concrete of grade $M40$ and above is used along with high tensile steel wires or cables in prestressed concrete pile. Such piles can be transported and driven with ease without causing cracks. The piles are suitable as raking piles. The minimum pre-compression is of the order of 2 to 6 N/mm^2 depending on the relative weight of the hammer. Similarly the maximum axial compression in the pile is limited to twenty-five percent of the characteristic strength of the concrete. 0.12 percent nominal reinforcement is provided. Ties spacing is same as that of reinforced concrete piles. The recommended maximum spacing of the ties at the end zone is 50 mm.

Example 13.2: Design of Precast Prestressed Concrete Pile

A 300 mm square precast prestressed concrete pile is provided in medium sandy soil having an angle of friction of 25 degrees to carry an axial load of 400 kN. The density of the sand is 18 kN/m^3 and the surcharge on the soil is 20 kN/m^2. Design prestress and the reinforcement details of the pile.

Solution

The required length of the pile is 8.5 m as per the example 13.1, so only structural design is illustrated.

Structural Design

The minimum grade of concrete for prestressed precast piles is $M40$. The same grade is selected. The choice of the prestressed concrete pile here is mainly for handling, transportation.

The length to side ratio of the pile is = 8.5/0.3 = 28.33.

The desired initial prestress on to the concrete is taken as about 5 N/mm^2. The prestress force required is:

$$P_i = 5*300*300 = 450000 = 450 \text{ kN}.$$

The area of the pre-tension wires of 5 mm wires with proof strength of 1600 MPa is

$$= P_i/0.7f_p = 450000/0.7*1600 = 402 \text{ mm}^2.$$

The number of 5 mm wires required = 402/19.6 = 21.
20 wires of 5 mm, four rows of five wires in each row are recommended to maintain symmetry.
The actual area of tensioned steel and the initial pretension force are:

$$A_p = 20*19.6 = 392 \text{ mm}^2.$$

$$P_i = 0.7*1600* 392 = 439040 \text{ N}$$

The initial pre-compression on concrete is:

$$\sigma_{ci} = 439040/300*300 = 4.9 \text{ N/mm}^2.$$

The loss of pretension is normally estimated to be at 25 percent.

The effective prestress at working load is $= 0.75*\sigma_{ci} = 3.6$ N/mm^2.

The axial compression caused by the load on the pile is = 400 000/90000 = 0.44 N/mm².

The net stress on concrete at working load = 3.6 + 0.44 = 4.04 N/mm².

This stress is much less than the allowable value of 0.25 times the characteristic strength, that is = 10 N/mm².

The Pile must be provided with nominal non-tensioned reinforcement of 0.12% and ties.

Area of non-tensioned reinforcement = 0.12*90000/100 = 108 mm².

Provide four numbers of 10 mm bars as the nominal longitudinal reinforcement. The ties confirm to the 0.6% of the volume at the end zones and 0.2% in the middle zone. The calculation made in the previous example holds good here also. The details of the pile are shown in Fig. 13.4.

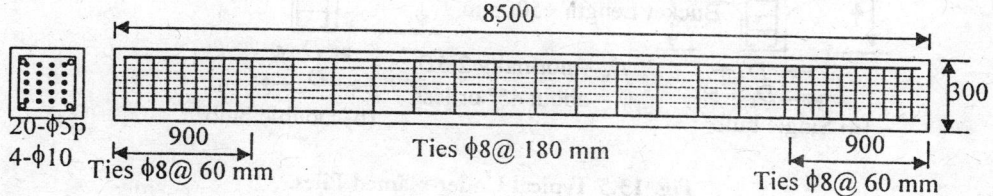

Fig. 13.4 Precast Prestressed Concrete Pile; Example 13.2.

For convenience of driving, a pile can be fitted with a suitably designed shoe.

13.7 BORED PILES

A hole is driven into the ground by an auger for a desired size and depth. Reinforcement cage is inserted into the borehole, and then concrete is placed. No casing is provided while boring in stiff soil. Mud slurry (Bentonite) filling is done in unstable soils to stabilize borehole surface. High slum concrete is placed through a chute in the borehole. The mud slurry will rise to the surface as the concrete fills from the bottom of the hole. In some cases a temporary steel tube casing is driven. The casing can be withdrawn while laying the concrete. In some marine or underwater situations, a steel tube is driven and left as it is even after placing the concrete. The cohesion between the soil and the pile in bored piles is less when compared with that in the driven piles. The reduction factors for cohesion are given in Table 13.2. Bored piles should be placed at a reasonable spacing so as to derive full benefit of the friction. Minimum spacing of the piles is 2.5 times the diameter and the minimum percentage of the main reinforcement in piles is 0.4. The nominal reinforcement is not accounted towards the strength of the pile. The diameter of the bored pile ranges from 300 mm to 2500 mm. Large diameter piles are provided for bridge piers with depth going to more than 30 m.

13.8 UNDER-REAMED PILES

Under-reamed pile is a bored cast-in-situ pile with enlarged bulbs formed along the depth of the pile. The number of bulbs may vary from one to three and the size of the bulb is as per the specifications of the code. The enlarged bulb provides additional bearing area thus increasing the load carrying capacity of the pile. Typical under-reamed piles are shown in Fig. 13.5. Expansive soils swell with change in the moisture in the soil. The bulb acts as an anchor in such cases of swelling and provides better stability to the pile.

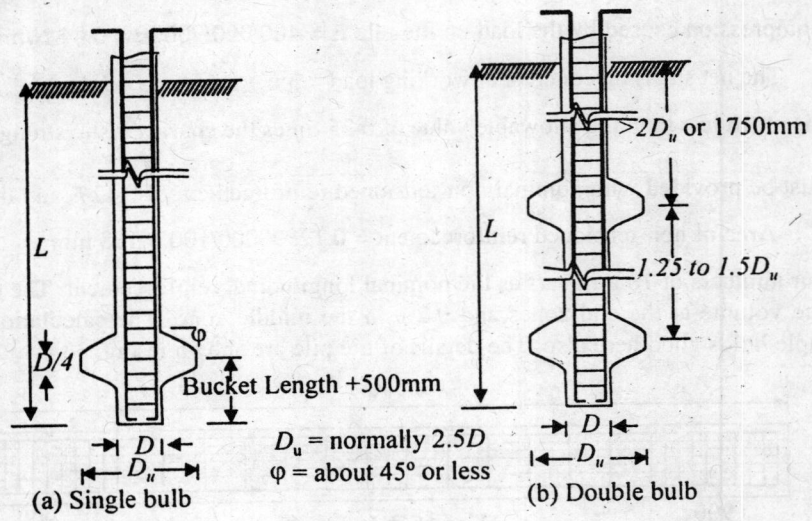

Fig. 13.5 Typical Under-reamed Piles.

Under-reamed piles are most suitable in expansive soils such as black cotton soil. Some minimum requirements specified by the code are:

- Minimum depth of pile is 3.5 m and minimum stem diameter is 250 mm.
- The diameter of the bulb is normally 2.5 times the diameter of the stem, and two times the diameter of the stem for bored compaction piles.
- The minimum spacing of the piles is 1.5 times the diameter of the bulb in case of 300 mm diameter piles. A reduction of 10 per cent is allowed if spaced slightly closer. Maximum spacing of the pile is 3 metres.
- The spacing of the bulbs along the length is 1.25 times the diameter of the bulb for 300 mm diameter pile.
- Top most bulbs should be beyond a distance twice the diameter of the bulb and with a minimum of 1750 mm.
- Minimum distance between the under side of the pile cap and the bulb is 1.5 times the diameter of the bulb.
- Location of bottom most bulbs from toe is equal to bucket length plus 500 mm.
- Minimum percentage of *HYSD* reinforcement is 0.3 percent of the sectional area. The pile is designed as a short column. Four number of 10 mm bars should be the absolute minimum reinforcement.
- Minimum diameter ties is 6 mm, and the spacing is equal to the diameter of the pile or subject to a 300 mm.
- The minimum cover to the reinforcement is 40 mm in mild environment, and the bottom cover is 50 mm.
- The minimum cover in other exposure condition conditions should be as per the IS:456-2000.

The safe load carrying capacities of the under-reamed piles in sandy, clayey and black cotton soils are suggested by Indian code of practice. Sample safe loads are listed in table 13.7 for quick reference. The reader is advised to refer to the code for detailed information. The capacities indicated are indicative for design but the actual safe load has to be obtained by actual tests.

Table 13.7 Safe Load Capacities of under-reamed Piles in Sandy, Clayey and Black Cotton Soils. As per IS:1911-1980; Part 3.(Dimensions in mm and load in tonnes)

Stem Diameter (mm)		250	300	375	400	450	500
Bulb diameter (mm)		625	750	940	1000	1125	1250
Length (mm)	Single bulb	3500	3500	3500	3500	3500	3500
	Double bulb	3500	3500	3750	4000	4500	5000
Safe Load (t)	Single bulb	12	16	24	28	35	42
	Double bulb	18	24	36	42	52.5	63
Increase Load for 300 mm	Single bulb	1.15	1.4	1.8	1.9	2.15	2.4
	Double bulb	0.9	1.1	1.4	1.5	1.7	1.9
Reinforcement	Main bars	4-ϕ10	4-ϕ12	5-ϕ12	6-ϕ12	7-ϕ12	9-ϕ12
	6 mm ties at	220	250	300	300	300	300
Uplift Load	Single bulb	6	8	12	14	17.5	21
	Double bulb	9	12	18	21	25.7	31.5
Lateral Load	Single bulb	1.5	2	3	3.4	4	4.5
	Double bulb	1.8	2.4	3.6	4	4.8	5.4

Example 13.3: Design of under-reamed Pile

Design an under-reamed pile to carry an axial load of 400 kN in silty-clay soil under mild exposure condition. The top 1.5 m soil is filled up.

Solution

400 kN can be taken equal to 40 tonnes load. Select M20 concrete, as the exposure is only mild. From table 13.7, select 375 mm diameter double under-ream bulb pile. The safe load of 3750 mm length of the pile is 360 kN. The rate of increase of the safe load is 14 kN per 300 mm. Increase the length of the pile by 900 mm which gives an additional capacity of 3*14 = 42 kN. The effective length of the pile is 3750 + 900 = 4650 mm. As there is 1500 mm filled up soil, total length of the pile is equal to 4650 + 1500 = 6150 mm. Let 100 mm pile projects embedded in the pile cap. Atleast an additional length of 300 mm be cast at the top that can be dismantled before casting the pile cap.

The total as cast length of the pile is = 6150 +100 + 300 = 6550 mm.

The minimum reinforcement recommended by the code be five numbers of 12 mm main bars with 6 mm ties spaced at 300 mm. The size of the bulb is 940 mm.

13.9 DESIGN OF PILE CAP

A concrete slab that interconnects a group of piles and transmits the load from superstructure to the piles is called a *pile cap*. The pile cap is normally made of reinforced or prestressed concrete. A group of piles that support a column may consist of two or more piles. A cap that supports more than one column is called a combined pile cap. A raft slab that supports all the foundation piles can also be designed. The dispersion of load from the column through the pile is at 45°. The pile cap is usually treated as rigid and

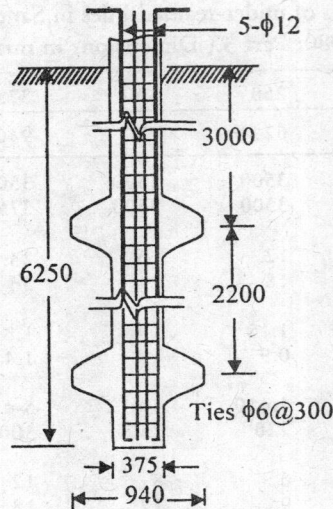

5-ϕ12

3000

6250

2200

Ties ϕ6@300

375

940

Fig. 13.6 Under-reamed Pile; Example 13.3.

the load distribution from the column to the piles is taken on a rigid support basis. In other words, if the centroid of the load coincides with that of the piles, the load is equally distributed to the supporting piles. The following are the minimum requirements to be met while designing a pile cap.

(a) Minimum thickness of the pile cap is governed by:
   - Minimum anchorage length of the main reinforcement of the column must be provided in the thickness of the cap. Similarly adequate anchorage for the pile reinforcement is available in the pile cap.
   - Pile cap be rigid enough to distribute the load uniformly to the piles.
   - Minimum 300 mm thickness at the free edges and 500 mm thickness in the body of the cap.

(b) Clear edge overhang of the pile cap beyond the outermost pile should be 150 mm.
(c) A leveling course of 75 mm to 100 mm thick lean concrete is provided.
(d) Clear cover to the main reinforcement shall be 60 mm or more depending on the exposure condition.
(e) Pile should project atleast 50 mm into the pile cap.
(f) It is desirable to arrange the piles and the load from the columns such that the centroid of the load is same as the centroid of the piles.

Typical arrangements of pile with respect to the pile cape are indicated in Fig. 13.7. Design of pile, pile group and pile cap are inter linked. The thickness of the pile cap influences the axial force on the piles. The shear force the column introduces bending moment at the level of the top of the piles. This bending moment as a couple in turn causes downward force in some piles and uplift in some other piles. Therefore there is an active interaction of forces and the size of the pile cap.

Example 13.4: Force Transfer from Pile Cap to Pile Heads

A 400 mm square column transfers 1000 kN vertical load, 100 kNm bending moment and shear force of 50 kN on to a pile cap covering four piles. Compute load on each of the four piles. The pile size is 300 mm square.

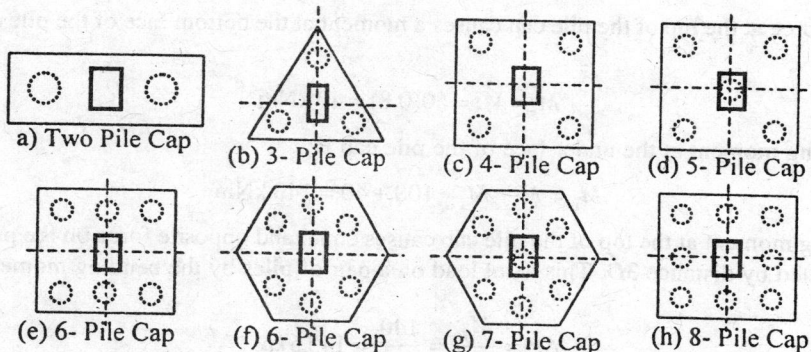

Fig. 13.7 Typical Pile Cap Layouts.

Solution

Assume the thickness (t) of the pile cap as 800 mm for the purpose of estimating self-weight. Higher thickness of the pile cap is assumed to be on the safer side. The actual thickness may be in the range of 500 mm.

Weight of the pile cap = $(1.5)(1.5)(0.8)(25) = 45$ kN

Let the spacing of the piles = $s = 3D = 900$ mm

Also assume the size of the pile cap about five times the pile diameter = $5D = 5*300 = 1500$ mm

The loads from the column to the pile cap are:

Axial load = $P = 1000$ kN,

Bending moment = $M = 100$ kNm,

Shear force = $Q = 50$ kN

Total vertical load on four piles = $1000 + 45 = 1045$ kN,

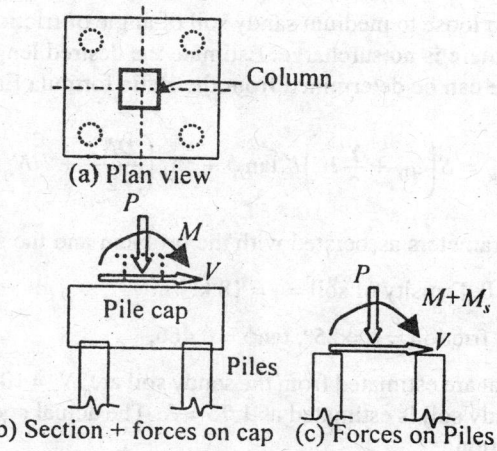

(a) Plan view

b) Section + forces on cap   (c) Forces on Piles

Fig. 13.8 Pile Group with Cap; Example 13.4.

The shear force at the top of the pile cap causes a moment at the bottom face of the pile cap and this is equal to:

$$M_s = V t = 50(0.8) = 40 \text{ kNm}$$

Total bending moment at the under face of the pile cap is:

$$M_p = M + M_s = 100 + 40 = 140 \text{ kNm}$$

The bending moment at the top of the pile cap causes equal and opposite force on the pair of the piles that are separated by distance $3D$. This axial load on a pair of piles by the bending moment M_p is:

$$\Delta P = \frac{M_p}{3D} = \frac{140}{0.9} = 145.6 \, kN$$

Maximum working load on each pile at the forward end is:

$$W = P/4 + \Delta P/2 = 261.25 + 145.6/2 = 334.05 \text{ kN, use } W = 350 \text{ kN and}$$

Minimum working load on each pile at the rear end is:

$$W = P/4 - \Delta P/2 = 261.25 - 145.6/2 = 188.45 \text{ kN}$$

There is no uplift on the pile. The pile can be designed for maximum axial force, shear and bending moment.

$$\text{Shear force} = V = Q/4 = 12.5 \text{ kN}$$

Applying a factor of safety of 2.5 to the piles, the ultimate loads acting at the head of the pile are:

$$P_u = 2.5 P = 2.5(350) = 875 \text{ kN}$$

$$V_u = 2.5 V = 2.5(12.5) = 31.25 \text{ kN}$$

Determination of the Length of the Pile

The precast pile is provided in loose to medium sandy soil of angle of friction as 25 degrees. The density of the sand is 18 kN/m³ and there is no surcharge. Estimate the desired length of a concrete pile.

The total length of the pile can be determined from the static formula Eq. 13.3 and it is:

$$Q_u = S\left(q_0 + \frac{\gamma}{2}H\right)K \tan\phi + \gamma A_b\left(\frac{DN_r}{2} + HN_q\right)$$

Various properties and parameters associated with the problem and the soil are:

Surcharge = $q_0 = 0$, Density of soil = $\gamma = 18$ kN/m³,

Angle of internal friction = $\phi = 25°$, $\tan\phi = 0.466$,

The N values of the soil that are estimated from the sandy soil are: $N_r = 10$, and $N_q = 15$. The coefficient of lateral pressure for the sandy soil is estimated as $1.75 = K$. The actual coefficients of the soil are to be determined by soil investigation.

The size (D) of the pile is 0.3 m so the surface area of the pile in square metres is:

$$S = 4DH = 1.2H,$$

The ultimate load on the pile is:

$$P_u = = 875 \text{ kN},$$

Substitution of the various quantities in the static capacity equation gives:

$$Q_u = S\left(q_0 + \frac{\gamma}{2}H\right)K\tan\phi + \gamma A_b\left(\frac{DN_r}{2} + HN_q\right)$$

$$875 = 1.2H(9H)1.75(0.466) + 18*0.09(0.3*10/2 + 15H)$$

where: H = the length of the pile.

$$875 = 8.8H^2 + 2.43 + 24.3H$$

The rearrangement of the above expression gives:

$$H^2 + 2.76H - 99.5 = 0$$

The solution of the above quadratic gives the value of the depth of the pile and it is:

$$H = 8.7 \text{ m. Select the length of the pile as 9 m.}$$

Example 13.6: Calculation of Cantilever Length of Pile and Structural Design

The pile group mentioned in the previous example is in a soil whose sub-grades reaction is 0.1 N/mm³. Compute the bending moment on the pile and design the pile.

Solution

The length of the pile based on the soil capacity from the previous example is 9.0 m.

The sub-grade reaction of the soil is given as: $k_s = 0.1$. The equivalent cantilever span (taken from Table 13.5) is:

$$L_c/D = 6 \text{ or } L_c = 6D = 6(0.3) = 1.8 \text{ m}$$

Design bending moment on the pile is:

$$M = VL_c = 31.25(1.8) = 25 \text{ kNm}$$

The design has to be done for an axial load of 875 kN and a bending moment of 25 kNm.

$$e = M/P = 25/875 = 0.029,$$

$$e/D = 29/300 = 0.1$$

The design of the pile has to be made using the column design curves.

Since the percentage of reinforcement is small, only one row of reinforcement is assumed; therefore the cover to the centre of reinforcement can be taken as 50 + 6 mm = 56 mm, say 60 mm.

The cover ratio = $c' = 60/300 = 0.2$. The grade of the concrete is assumed to be $M25$.

The non-dimensional force factors are:

$$Y = \frac{P_u}{bDf_{ck}} = \frac{0.875}{0.3*0.3*25} = 0.39$$

$$X = \frac{M_u}{bD^2 f_{ck}} = \frac{0.025}{0.3*0.3*0.3*25} = 0.037$$

From Fig. 11.13, the column interaction curve, the percentage of reinforcement for the above is:

$$p/f_{ck} = 0.02$$

The area of the total reinforcement is:

$$A_s = 0.02*f_{ck}*bD/100 = 0.02*25*300*300/100 = 4.5 \text{ mm}^2.$$

The slenderness factor for the pile is = $L/D = 9/0.3 = 30$.

The minimum percentage of reinforcement in a precast concrete pile is 1.25%. The area of main reinforcement is:

$$A_s = 1.25*300*300/100 = 1125 \text{ mm}^2.$$

Select four numbers of 20mm. The actual area provided results = $4*314 = 1256 \text{ mm}^2$.

The assumed cover distance checks with that of the actual one.

Provide 8 mm stirrups at 60 mm spacing for a length of 900 mm at each end, and 8 mm stirrup at 180 mm spacing in the middle zone.

Example 13.7: Design of Pile Cap

A square column of 400 mm transfers 1000 kN vertical load, 100 kNm bending moment and shear force of 50 kN on to a pile cap. The ultimate capacity of the available 300 mm square precast concrete pile is 900 kN. Design a suitable pile cap. The pile cap is in severe environment.

Solution

Let the spacing of the piles be = $s = 3D = 900$ mm

Let the size of the pile cap about = $5D = 5*300 = 1500$ mm

This provides 150 mm clear edge beyond the pile face.

The length of the pile cap is = $900 + 300 + 2*150 = 1500$ mm.

The loads from the column to the pile cap are:

Axial load = $P = 1000$ kN,

Bending moment = $M = 100$ kNm,

Shear force = $Q = 50$ kN

Add extra of 5 percent for the weight of the pile cap, then:

Total vertical load on four piles = $1000 + 50 = 1050$ kN,

The shear force at the top of the pile cap causes an overturning moment at the top of the piles and it is equal to:

$$M_s = V_t = 50 *0.5 = 25 \text{ kN}.$$

Total bending moment at the under face of the pile cap is:

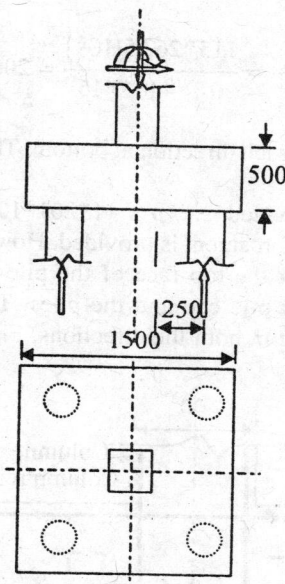

Fig. 13.9 Layout of Pile & Cap; Example. 13.7.

$$M_p = M + M_s = 100 + 25 = 125 \text{ kNm}$$

The bending moment at the top of the pile cap causes equal and opposite force on the pair of the piles that are separated by distance 3D. This axial load on a pair of piles by the bending moment M_p is:

$$\Delta P = \frac{M_p}{3D} = \frac{125}{0.9} = 139 \text{ kN}$$

Maximum working load on each pile at the forward end is:

$$W = P/4 + \Delta P/2 = 256.25 + 139/2 = 325.75 \text{ kN}$$

Use $W = 350$ kN and

The effective cantilever span from the face of the column to the centre of the pile is $= 900/2 - 400/2 = 250$ mm.

The bending moment caused by the pair of the pile under maximum reaction at the face of the column is:

$$M_w = 2*350(0.25) = 175 \text{ kNm per } 1.5 \text{ m width}$$

The limit bending moment with a partial safety of 1.5 is:

$$M_u = 1.5*175 = 262.5 \text{ kNm per } 1.5 \text{ m.}$$

The depth of the pile cap is selected as 500 mm so that adequate development length of the main bars is available. The pile cap is in the severe environment so select $M30$ concrete grade with cover to the reinforcement as 50 mm.

The tension reinforcement that is needed at the bottom face of the pile cap is:

$$A_s = \frac{1.15 M_u}{jdf_y} = \frac{1.15*262.5(10^6)}{0.8*440*415} = 2067 \text{ mm}^2$$

Provide 11 number of 16 mm bars in each direction at bottom. The spacing of the bars works out to be 150 mm.

The percentage of reinforcement provided is: $= p = 11*201*100/500*1500 = 0.3\%$

The minimum percentage of the reinforcement is provided. However, the author is of the opinion that a nominal reinforcement is provided at the top face of the pile cap. Such a reinforcement prevents microcracks and integrates the column, pile cap and the piles. This can be as low as 0.075 percent. Provide 12 mm bars at 300 mm spacing in both the directions. Fig. 13.10 illustrates the reinforcement details.

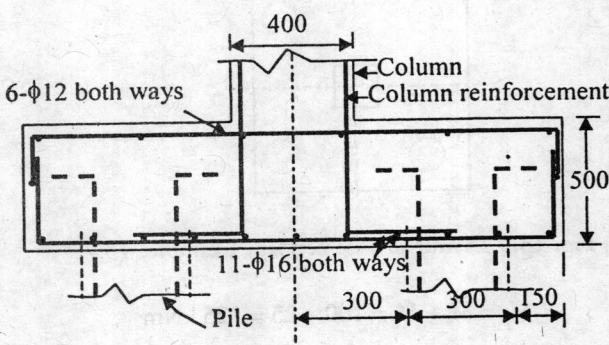

Fig. 13.10 Details in Pile Cap; Example 13.7.

Example 13.8: Design of a Group of Under-reamed Piles

A column is transferring a vertical load of 1500 kN, bending moment of 200 kNm and shear force of 60 kN. Design a pile foundation that is exposed to moderate environment. The diameter of the column is 500 mm and the soil is clayey sand.

Solution

The minimum grade of concrete that is acceptable is $M25$ in the moderate exposure condition. Use $M25$ concrete with $HYSD$-$Fe415$ bars. On a trial basis, 400 mm diameter double under-ream pile is selected. The details of the pile taken from the table 13.7 are:

Stem diameter = 400 mm, Bulb diameter = 1000 mm, Basic minimum length of pile = 4000 mm, The compression safe load is 42 t = 420 kN, Uplift safe load = 210 kN, Shear safe load = 40 kN Increasable load per 300 mm is = 15 kN, Main reinforcement = 6-ϕ12, Ties of 6 mm at 300 mm. Number of piles = 4

Let a group of four piles is selected to carry the column. The minimum spacing of the pile is 1.5 times the diameter of the bulb. Let the pile be spaced at 1800 mm, and the clear edge distance as 200 mm. This leads to a total length of the pile cap as = 1800 + 1000 + 2*200 = 3200 mm. Select a square pile cap of 3200 mm side. Let the thickness of the pile cap (t) be 600 mm. Add extra of 5% to account for the weight of the pile cap, and another 5% for superimposed filling. Then,

Total vertical load on four piles = 1500 (1.1) = 1650 kN,
The shear force at the top of the pile cap causes an overturning moment equal to:

$$M_s = V_t = 60 * 0.6 = 36 \text{ kN.}$$

Total bending moment at the under face of the pile cap is:

$$M_p = M + M_s = 200 + 36 = 236 \text{ kNm}$$

The bending moment at the top of the pile cap causes equal and opposite force on the pair of the piles that are separated by distance 1.8 m. This axial load on a pair of piles caused by the bending moment M_p is:

$$\Delta P = \frac{M_p}{1.8} = \frac{236}{1.8} = 132 \text{ kN}$$

Maximum working load on each pile at the forward end is:

$$W = P/4 + \Delta P/2 = 1650/4 + 132/2 = 478.5 \text{ kN, Use } W = 490 \text{ kN.}$$

The cantilever span of the pile cap from the centre of pair of pile to the face of the column is: = 1800/2 − 500/2 = 650 mm
The bending moment caused by the pair of the pile under maximum reaction at the face of the column is:

$$M_w = 2*490(0.65) = 637 \text{ kNm per 3.2 m width}$$

The limit bending moment with a partial safety of 1.5 is: $M_u = 1.5 * 637 = 955.5$ kNm
The depth of the pile cap is selected as 600 mm so that adequate development length of the main bars from the columns is available in the pile cap.
The tension reinforcement that is needed at the bottom face of the pile cap is:

$$A_s = \frac{1.15 M_u}{j d f_y} = \frac{1.15 * 955.5 (10^6)}{0.8 * 530 * 415} = 6245 \text{ mm}^2$$

Provide 20 number of 20 mm bars in each direction at the bottom. The spacing of the bars is 160 mm.
The percentage of reinforcement provided is:

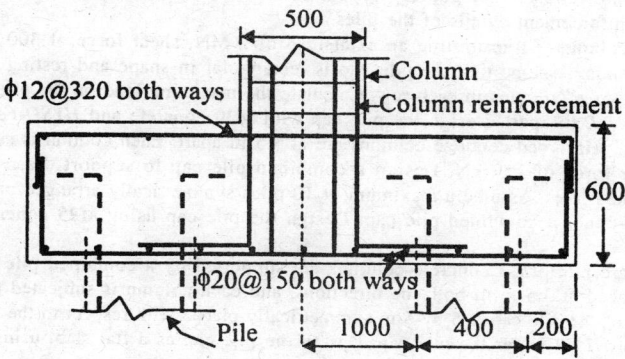

Fig. 13.11 Details in Pile Cap; Example 13.8.

$$p = 20*314*100/600*3200 = 0.33\%$$

The author is of the opinion that a nominal reinforcement is provided at the top face of the pile cap. Provide 16 mm bars at 320 mm spacing in both the directions. Fig. 13.11 illustrates the details of the pile cap.

PROBLEMS

13.1 A 300 mm square precast reinforced concrete pile is in medium dense sandy soil. The angle of repose of the soil is 30 degrees. The soil parameters are: $N_q = 24$, and $K = 2$. Determine the depth of the pile to carry an axial load of 200 kN. Assume any other suitable data.

13.2 A 300 mm square precast reinforced concrete pile is subjected to an axial load of 300 kN, bending moment of 50 kNm and shear force of 30 kN. The pile is driven in clayey soil up to 7 m depth as a friction pile. Design reinforcement details of the pile for two pick-up points. The soil is exposed to sever environment so use M30 concrete.

13.3 A 350 mm diameter circular reinforced concrete pile made of M35 grade concrete is subjected to an axial load of 300 kN and shear force of 20 kN. The length of the pile is 7 m in sandy soil. Design HYSD-Fe415 reinforcement for the pile.

13.4 A 300 mm square precast reinforced concrete is in loose sandy soil having 20 degrees angle of friction, and carries an axial load of 250 kN. The density of the sand is 18 kN/m³ and the surcharge on the soil is 25 kN/m². Estimate the desired length of the pile. Suggest the reinforcement details.

13.5 A 250 mm square precast prestressed concrete bearing pile carries a service load of 300 kN and rests on granite bedrock at 15 m depth. The grade of the concrete is M40 and pre-compression on the pile is 4 N/mm². Determine the ultimate load carrying capacity of the pile.

13.6 Design a double under-reamed pile to carry an axial load of 350 kN in silty clay soil. The stem diameter of the pile is 350 mm. Use the design aid table provided by in table 13.7 or IS: 1911.

13.7 A 500 mm diameter circular column is transferring a compressive load of 1200 kN, a bending moment of 200 kNm and shear force of 80 kN on to a 500 mm thick pile cap. The pile cap is 2000 mm square and supported by four piles of 350 mm diameter. Compute load on each of the four piles.

13.8 A 250 mm thick reinforced concrete wall of is transferring an axial load of 500 kN/m to a strip pile foundation. Two rows of piles supports the 2.1 m wide pile cap. The pile rows are at 800 mm on either side of the centre line of the wall. Each pile can carry a safe load of 250 kN. The depth of the pile based on the soil characteristics is 7 m. Design precast concrete piles and pile cap using M20 grade concrete and HYSD-Fe415 reinforcement bars. Sketch the reinforcement details.

13.9 A 600 mm square column is subjected to an axial load of 1.5 MN. An equilateral triangle pile cap of 2 m side and thickness 700 mm supports the column. Three piles are placed near three corners of the pile cap. Design bored compaction reinforcement concrete pile using M20 concrete and HYSD-Fe415 reinforcement. Depth of the pile based on the soil characteristics is 8 m. The minimum spacing between the piles should be 2.5 times diameter of the pile. The minimum projection of the pile cap beyond the face of the pile should be 200 mm. Sketch the layout of the piles and the reinforcement details of the piles.

13.10 A 550 mm square column is transmitting an axial load of 2 MN, shear force of 300 kN and 600 kNm bending moment about one axis. Assume that the pile cap is rectangular in shape and resting on six piles. The piles are placed in two rows, three columns in each row. Calculate the maximum load on the piles assuming a rigid cap and the piles are spaced 2.0 m apart. Design the pile cap with M30 concrete and HYSD-Fe415 bars.

13.11 Two 400 mm square reinforced concrete columns are at 3.5 m apart. Each column is subjected to an axial load of 1800 kN and shear force of 150 kN. Design a combined pile cap to support the two columns using 350 mm diameter under-reamed piles. Assume a maximum of 10 piles symmetrically arranged under the pile cap and locate them appropriately under a combined pile cap. Design the pile cap using M25 concrete and HYSD-Fe415 reinforcement bars.

13.12 Four, 300 mm square reinforced concrete columns are supported by a combined pile cap. The four columns are in a square grid spaced at 3.5 m in both the directions. and each column is subjected to an axial load of 600 kN. The maximum size of the pile cap is 5 by 5 m symmetrically placed with respect to the 4 columns. Select a suitable pile. 16 piles are provided under the pile cap. Design the pile cap as a flat slab, using M30 concrete and HYSD-ssssFe415 reinforcement bars.

Design of Industrial Frames

14.1 INTRODUCTION

Reinforced and prestressed concrete construction of industrial frames is gaining importance. Concrete members are becoming slender with the use of high strength and high performance concretes. Improved mechanical handling and transportation facilities in construction are helping the use of concrete in industrial buildings. Further, concretes with strengths upto $M120$ are now in practice and concretes of grades as much as $M150$ are in offering in the world. Prestressed concrete is necessary in long span constructions. Even multistory constructions are being adopted in light industry. Two typical types of reinforced concrete industrial frames shown in Figures 14.1 and 14.2 are discussed in this chapter.

Fig. 14.1 illustrates flat roof portal frames. Fig. 14.2 illustrates gabled frames that are more common in industry for economic and functional needs. High ceiling requirements, natural ventilation and lighting are important in industrial buildings. Artificial ventilation and light are becoming popular with the increased power supplies. High columns and light claddings are the needs in industrial buildings. The effective length of a column is controlled by the relative stiffness of the member with respect to the other members connected at the joint. Indian code of practice IS:456-2000 in its appendix gives a set of effective length factors for braced and unbraced frames. Wind loads on industrial structures control the designs in many cases. The design of the frames is illustrated with couple of examples.

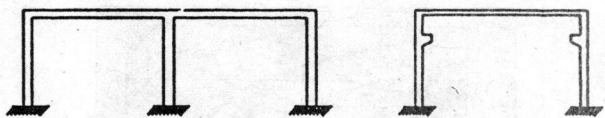

Fig. 14.1 Typical Industrial Portal Frames.

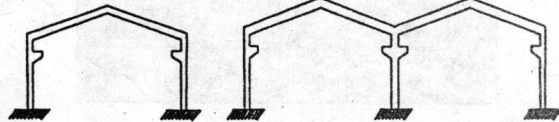

Fig. 14.2 Typical Gabled Frames.

14.2 EFFECTIVE LENGTHS OF COLUMNS

Framed buildings are classified into braced and unbraced for the purpose of computing the effective lengths of the columns. In braced frames, the effective length is less than the distance between the joints, while in the unbraced frames; the effective length is more than that of the distance between the joints. The effective length is connected with the relative stiffness of the member with respect to the total stiffness of all the members connected at the joint. This is called a b factor and it is equal to:

$$\beta = \frac{K_c}{\Sigma K_c + \Sigma K_b} \qquad (14.1)$$

where

K_c = stiffness of the column,
ΣK_c = sum of the stiffness of the columns that are connected at the joint,
ΣK_b = sum of the stiffness of the beams that are connected at the joint.

The effective length ratios of the columns in braced frames is shown in Fig. 14.3 and similarly Fig. 14.4 gives the effective ratios of the columns in unbraced buildings.

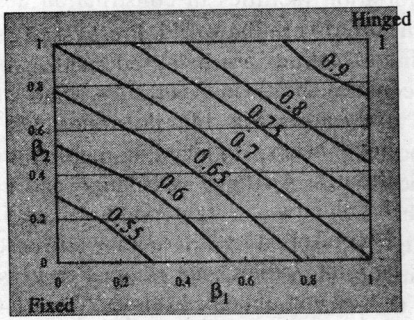

Fig. 14.3 Effective Length Ratios for Braced Frames.

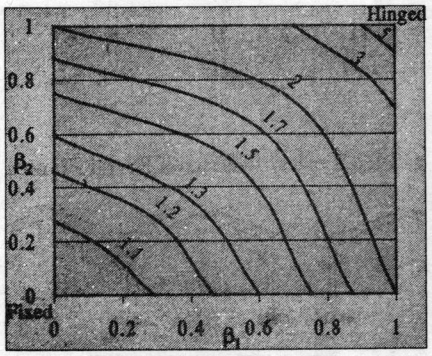

Fig. 14.4 Effective Length Ratios for Unbraced Frames.

14.3 ILLUSTRATIVE EXAMPLES

Example 14.1: Two Bay Single Storey Portal Frame

A two bay portal frame of 6 m each span houses a light industry. The ceiling height of the building is 4.5 m with columns resting on independent footings. The depth of the foundation is 1.2 m below the plinth. The exposure condition of the building is mild. The lateral basic wind pressure is 1.1 kN per square metre, the building is in earthquake zone three. The frames are spaced at 4.5 m and the roof is made of 120 mm reinforced concrete slab with waterproof treatment over. Independent walls are provided as cladding to the building. Figure 14.5 illustrates the frame. Design the frame.

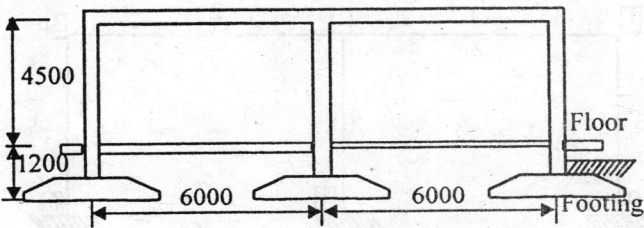

Fig. 14.5 Frame of Example 14.1.

Solution

A number of computer-aided programs to analyse and design frames and tall buildings are available. The author has developed a well-tested and applied program called *HiRise* for computer-aided design of tall Reinforced concrete buildings. Similarly another program that analyses frames. The preprocessor computes the loads including that of the walls, slabs, beams and columns, cross walls etc and distributes as per the standard format. The best way to design the frame is to apply such programs. This example illustrates few steps for the benefit of the student.

Water-proof treatment load	$= 2.5 \text{ kN/m}^2$
Weight of the slab = 0.12*25	$= 3.0 \text{ kN/m}^2$
Total dead load	$= 5.5 \text{ kN/m}^2$
Live load on the roof	$= 1.5 \text{ kN/m}^2$ (flat roof with access provided)

The walls of the building are built independently so wind load on the walls is not transferred to the frame. The basic seismic coefficient for the zone is taken as 0.04.

The sizes of the beams and columns are chosen based on the architectural and economic considerations. The span of the frame is 6 m so the size of the beams is selected as 250 by 450 mm and columns are taken as 300 by 400 mm. The basic loads are: dead, live and earthquake. The design loads are obtained by multiplying the basic load by appropriate partial safety factors. The most critical combination of design loads are:

Load combination 1 = 1.5 * Dead load + 1.5 * Live load = 1.5(*DL*) +1.5(*LL*)
Load combination 2 = 1.5 * Dead load + 1.5 * Earthquake load = 1.5(*DL*) +1.5(*EQL*)
Load combination 3 = 1.5 * Dead load - 1.5 * Earthquake load = 1.5(*DL*) -1.5(*EQL*)
Load combination 4 = 1.2 * Dead load + 1.2 * Live load + 1.2 * Earthquake load =1.2(*DL* +*LL* + *EQL*)

The program lists member forces, joint displacement, and supports reactions for each load combination. Further it also gives the design details. The deflection and the bending moment diagrams are also printed for each load combination. Since this is a simple example, to give a feel of the design procedure to the designer, some details of the software are not given here. The notations for members and joints used in the computer program are listed in Fig. 14.6. The bending moment diagram of the frame for the dead plus live load combination is shown in Fig. 14.7. Table 14.1 gives the values of the axial and bending moments on the members. For convenience of understanding, the beam notations of bending moments are used. That is the bending moment causing compression on the (top) load face is considered positive. The bending moment is indicated the tension face in the figure.

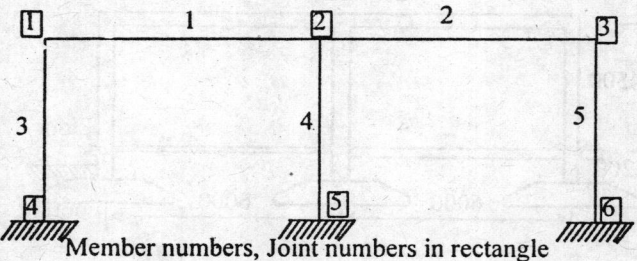

Fig. 14.6 Numbering of Members & Joints; Example 14.1.

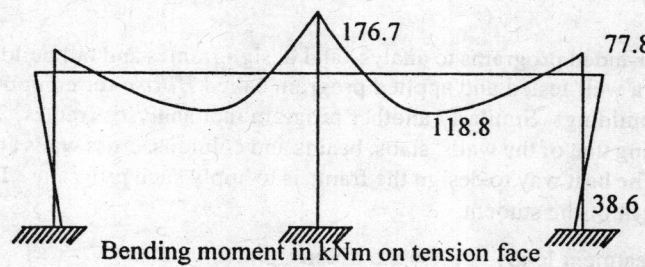

Fig. 14.7 BMD for One Dead Plus Live Loads; Example 14.1.

The load combination of $1.2(DL + LL + EQL)$ is not critical in this case. The dead and live load combination controls the design of the beams. The design of the outer columns is controlled by the load combination 2 because of the bending moment in the columns. Because of the symmetry, the middle column experiences little bending moment and dominant axial load.

Design of Outer Column

The effective length of the column is decided by the relative stiffness of the column with respect to the other members at a joint. The base end of a column is considered as fixed. The student is advised to refer to the two sets of graphs given in the appendix of the code IS:456-2000. A compact form of the effective length ratios of the columns is given Figs. 14.3 and 14.4.

The relative stiffness of a rectangular sectioned member can be expressed as:

$$K = \frac{bD^3}{L}$$

$$(14.2)$$

Table 14.1 Member Forces for Different Load Combinations Forces in kN or KNm.

Load combination	Member	Compression	Shear	BM (L)*	BM(R)	BM (M)
	1	25.9	155.1	-77.8	-176.7	118.8
	2	25.9	155.1	-176.7	-77.8	118.8
1.5DL +1.5LL	3	190	25.9	77.8	-38.6	0
	4	435	0	0	0	0
	5	190	26	-77.8	38.6	0
	1	11	70	-53.5	-72.6	57.1
	2	10.9	79.3	-97.2	-21.3	57.1
1.5DL +1.5EQL	3	105.4	21.3	53.5	-42.1	5.7
	4	222.8	11.7	24.7	-27.9	-1.6
	5	95.9	3.6	-21.3	-5	-13.2

*BM(L) = bending moment at left; BM(R) = bending moment at right, BM(M) = BM at middle.

where b = width of the member, D = depth of the member, L = length

The base of the column is fixed by the heavy flooring along with the footing so one end beta factor of the column is zero ($\beta_1 = 0$) and the other end factor is computed below using the relative stiffness factors of the beam and column at the end.

$$\beta_2 = \frac{(300)(400)^3}{300(400)^3 + 150(450)^3} = 0.46$$

Figure 14.4 gives the effective ratios of columns for an unbraced frame and the effective ratio of the column from the figure for beta factors of 0 and 0.46 is 1.2. So the effective length of the column is:

$$L_e = 1.2(\text{distance between the joints of the column}) = 1.2*4.5 = 5.4 \text{ m.}$$

The column is continuously supported laterally by infilled wall in the longitudinal direction of the building. So the slenderness factor of the column out of plane of the frame is zero. The slenderness factor of the column in the plane if the frame is:

$$g = \frac{L_e}{D} = \frac{5.4}{0.4} = 13.5$$

The design axial and bending moments in the earthquake load condition from the table 14.1 are:

$$P_u = 190 \text{ kN}$$

$$M_{xu} = 77.8 \text{ kNm}$$

The bending moment due to the slenderness effect of the column is:

$$M_{ax} = \frac{P_u D}{2000} \left\{ \frac{l_{ex}}{D} \right\}^2 = \frac{0.190(0.4)}{2000} \left\{ \frac{5.4}{0.4} \right\}^2 = 0.0069 \text{ MNm}$$

The minimum eccentricity for which the column has to be designed is:

$$e_{min} = \frac{L}{500} + \frac{b}{30} = 19 \ mm \ it \ is \ less \ than \ 20 \ mm,$$

The column is designed as slender column, the data is:

$$b = 0.3 \ m, D = 0.4 \ m, f_{ck} = 20 \ \text{MPa}, f_y = 415 \ \text{MPa}, \text{Let} \ d' = 50 \ mm = 0.05 \ m$$

The cover ratio about x-axis is = $c_x' = 50/400 = 0.125$
The total bending moment for which the column to be designed is:

$$M_u = M_{ux} + M_{ax} = 0.077.8 + 0.0069 = 0.0847 \ \text{MNm}$$

$$P_u = 190 \ \text{kN} = 0.190 \ \text{MN}$$

The non-dimensional force factors are:

$$Y = \frac{P_u}{bDf_{ck}} = \frac{0.190}{0.3*0.4*20} = 0.079$$

$$X = \frac{M_u}{bD^2 f_{ck}} = \frac{0.025}{0.3*0.16*20} = 0.088$$

From the column design aid curves, Fig. 11.11, the percentage of the reinforcement for the above parameters is:

$$p/f_{ck} = 0.046 \ \text{or}$$

$$p = 0.046*20 = 0.92\%$$

The area of reinforcement is:

$$A_s = 0.92*bD/100 = 0.92*300*400/100 = 1104 \ mm^2.$$

Select four numbers of 20 mm bars. The actual area provided results = 4*314 = 1256 mm².
The centre of the tension reinforcement from the outer face is = 40 +10 = 50 mm.
The assumed cover distance checks with that of the actual one.
Tie bars limits are:
Select 6 mm ties, then the spacing of the bars is controlled by:

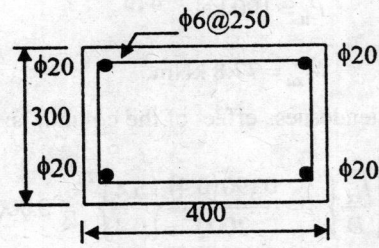

Fig. 14.8 Column Section, Example 14.1.

Spacing ≤ minimum size of the column = 300 mm
Spacing ≤ 16 times the diameter of the main bar = 16*20 = 320 mm
Spacing ≤ 48 times the diameter of the tie bar = 48*6 = 288 mm
Provide 6 mm ties at 250 mm spacing.

The column size can be reduced to 250 by 350 mm, however no revision is done in this example. The design details of the central column are chosen same as the outer column, because only the minimum reinforcement was provided.

Design of Beam

The beam size given in the problem is 250 by 450 mm. The maximum positive and negative bending moments and the shear forces on the beam are:

Maximum positive BM = M_p = 118.8 kNm.
Maximum negative BM = M_n = 176.7 kNm.
Shear force is = V = 155.1 kN

The effective depth of the section can be taken as = d = 450 – 20 –10 = 420 mm.
The balanced moment capacity of the section is:

$$M_b = Kbd^2 f_{ck} = 0.138*0.25*0.42*0.42*20 = 0.1217 \text{ MNm} = 121.7 \text{ kNm}$$

The beam is designed as singly reinforced for positive bending moment at mid-span. The negative bending moment is more than the balanced moment capacity of the section so the section is designed as doubly reinforced one. The reinforcement at the bottom face of the beam is:

$$A_{stb} = \frac{1.15*M_p}{jdf_y} = \frac{1.15*118.8(10^6)}{0.8*420*415} = 980 \text{ mm}^2$$

Provide Two numbers of 20 mm bars from end to end and two number of 16 mm bars curtailed at 1500 mm from both the ends (one-fourth span). The total area provided is 1030 mm².

Design for Negative Bending Moment

The area of reinforcement required for the balanced capacity of the section is:

$$A_{stb} = \frac{1.15*M_b}{jdf_y} = \frac{1.15*121.7(10^6)}{0.8*420*415} = 1003 \text{ mm}^2$$

The area of reinforcement required for the bending moment beyond the balanced bending moment is:

$$A_{st2} = \frac{1.15*(M_n - M_b)}{(d - d_c)f_y} = \frac{1.15*(176.7 - 121.7)(10^6)}{(420 - 30)*415} = 391 \text{ mm}^2$$

The two 20 mm bars extended at the bottom has area more than that required. The reinforcement that is needed at the top face of the beam at the corners is:

$$A_{st} = A_{stb} + A_{st2} = 1003 + 391 = 1394 \text{ mm}^2$$

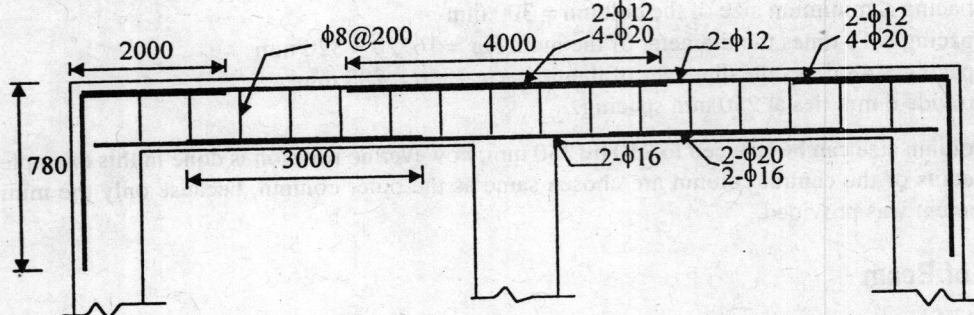

Fig. 14.9 Reinforcement Details in Beam; Example 14.1.

Provide four numbers of 20 mm bars at the top face of the beam at corners, and two 12 mm bars extend end to end of the beam to serve also as hanger bars. The bars must extend beyond the point of contraflexure. The point of contraflexure changes with the load condition so it is recommended that the bars should go for a distance of one third of the span that is 2000 mm. Figure 14.9 illustrate the reinforcement details in the beams

Design for Shear Force

The maximum shear force in the beam is 155.1 kN. The actual shear force at the face of the column is less than that listed above. In most cases it can be estimated as 90% of the one listed. Using this assumption, the shear stress is:

$$\tau_v = \frac{V}{bd} = \frac{0.9*155100}{300*420} = 1.1 \text{ N/mm}^2$$

The percentage of tension reinforcement at the critical zone is:

$$p_s = \frac{2*201*100}{250*450} = 0.36$$

The allowable shear strength for the above percentage of reinforcement is 0.42 N/mm². Select two legged 8 mm for stirrups, the area of two legs of the stirrup is: 100 mm². The spacing of the stirrup reinforcement is:

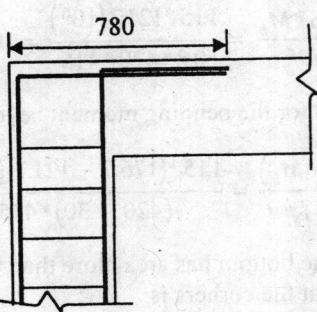

Fig. 14.10 Column Bars Extended into Beam.

$$s_v = \frac{0.87\, A_{sv}\, f_{yv}}{b(\tau_v - \tau_c)} = \frac{0.87(100)(415)}{250(1.1 - 0.42)} = 212 \text{ mm}$$

The maximum spacing of the stirrups is controlled by:

$$s_{max} \le \frac{A_{sv}\, f_{yv}}{0.4b} = \frac{100(415)}{0.4(250)} = 415 \text{ mm}$$

Or $s_{max} = d = 420$ mm or 300 mm. Least of all.

Provide two legged 8 mm stirrups at 200 mm spacing.

The development length is 43 times the diameter of the bar. This length must be provided for all the bars. The bars from the beam extend into the column with an *L* bent. Similarly the column bars must extend into the beam.

Example 14.2: Two Bay Gabled Frame

A two span gable frame of 8 m span each, column height of 4.5 m is resting on independent footings. The rise of the gable is 2 m. The foundation depth is 1.2 m below the ground level. The exposure condition of the building is mild. The basic wind pressure computed as per the code is 1.2 kN per square metre. The frames are spaced at five meters. Independent walls are provided as cladding to the building. Figure 14.11 illustrates the frame. Design the frame.

Solution

Frame is analysed by Software, *FRAME* developed by the author.

    The spacing of the frames is   = 5 m.

    Weight of the RC roof slab + *WPT* = 5.0 kN/m² = 25 kN/m, on rafter

Live load on the roof   = 1.kN/m²

Net Wind load on the windward side for the sloped roof = 4 kN/m on rafter

Net Wind load on the leeward side for the sloped roof = 5.4 kN/m on rafter

The walls of the building are built independently so wind load is not transferred to the frame.

    Sizes of the rafters = 250 by 400 mm; Size of the columns = 300 by 400 mm

The critical design load combinations are:

    Load combination 1 = 1.5 * (Dead load + Live load)

    Load combination 2 = 1.5 * (Dead load + wind load)

    Load combination 3 = 1.2 * (Dead load + Live load + wind load)

The program gives member forces, joint displacement, and supports reactions for each load combina-

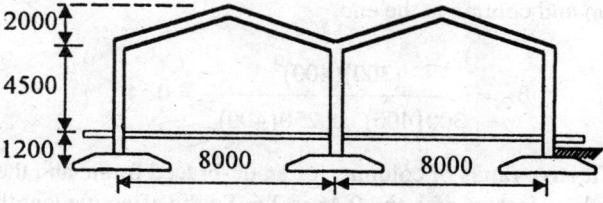

Fig. 14.11 Frame of Example 14.2.

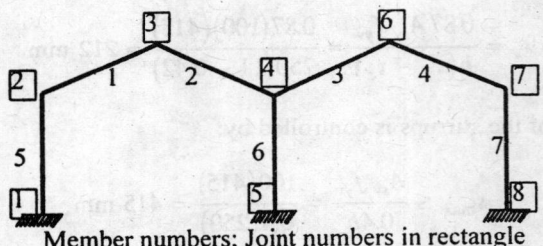

Fig. 14.12 Numbering of Members & Joints; Example 14.2.

tion. Further it also gives the design details. Many details of the software are not given here. The critical bending moments on the frame are shown in Fig. 14.13. The bending moment causing compression on the (top) load face is considered positive. The bending moment is marked on the tension face. The dead and wind load combination controls the design.

Design of Outer Column

The effective length of the column is decided by the relative stiffness of the column with respect to the other members at a joint. The student is advised to refer to the two sets of graphs given in the appendix of the code IS:456-2000 or Figs. 14.3 and 14.4.

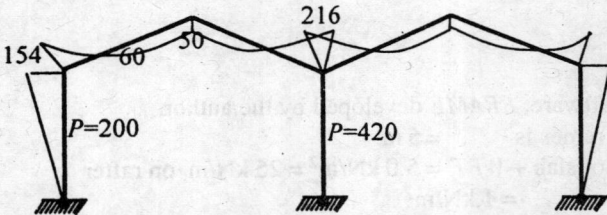

Fig. 14.13 BM in Frame; Marked on the Tension Face; Example 14.2.

The relative stiffness of a rectangular sectioned member can be expressed as:

$$K = \frac{bD^3}{L}$$

The base of the column is considered as hinged since an isolated footing supports it. Beta factor for one end of the column is one ($\beta_1 = 1$) and the other end factor is computed below using the relative stiffness factors of the beam and column at the end.

$$\beta_2 = \frac{(300)(400)^3}{300(400)^3 + 250(400)^3} = 0.54$$

Figure 14.4 gives the effective ratios of columns for an un-braced frame and the effective ratio of the column from the figure for beta factors of 1 and 0.46 is 2.6. So the effective length of the column is:

$$L_e = 2.6(\text{distance between the joints of the column}) = 2.6*4.5 = 11.7 \text{ m}.$$

The column is continuously supported by infilled wall along the longitudinal direction of the building. So the slenderness factor of the column out of plane of the frame is zero. The slenderness factor of the column is:

$$g = \frac{L_e}{D} = \frac{11.7}{0.4} = 29.2$$

The factored design axial and bending moments in the end column are:

$$P_u = 200 \text{ kN}$$

$$M_{xu} = 154 \text{ kNm}$$

The bending moment due to the slenderness effects is:

$$M_{ax} = \frac{P_u D}{2000}\left\{\frac{l_{ex}}{D}\right\}^2 = \frac{0.200(0.4)}{2000}\{29.2\}^2 = 0.034 \text{ MNm}$$

The minimum eccentricity for which the column has to be designed is:

$$e_{min} = \frac{L}{500} + \frac{b}{30} = 19 \text{ mm } \textit{it is less than } 20\,\text{mm}$$

The column is designed as slender one and subjected to the bending moment and axial force. The given data of the section is:

$$b = 0.3 \text{ m}, D = 0.4 \text{ m}, f_{ck} = 20 \text{ MPa}, f_y = 415 \text{ MPa, Let } d' = 50 \text{ mm} = 0.05 \text{ m}$$

The cover ratio about x-axis is $= c_x' = 50/400 = 0.125$
The total bending moment for which the column to be designed is:

$$M_u = M_{ux} + M_{ax} = 0.154 + 0.034 = 0.188 \text{ MNm}$$

$$P_u = 200 \text{ kN} = 0.20 \text{ MN}$$

The non-dimensional force factors are:

$$Y = \frac{P_u}{bDf_{ck}} = \frac{0.20}{0.3*0.4*20} = 0.083$$

$$X = \frac{M_u}{bD^2 f_{ck}} = \frac{0.188}{0.3*0.16*20} = 0.196$$

From the Fig. 11.11, the column design aid curves, the percentage of the reinforcement for the above parameters and cover ratio of 0.15 is:

$$p/f_{ck} = 0.13 \text{ or}$$

$$p = 0.13*20 = 2.6\%$$

The area of reinforcement is:

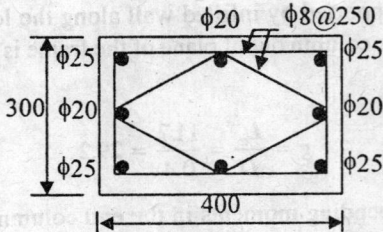

Fig. 14.14 Column Section; Example 14.2.

$$A_s = 2.6*bD/100 = 2.6*300*400/100 = 3120 \text{ mm}^2.$$

Select four numbers of 20 mm and 25 mm bars. The actual area provided is = 4*(314 + 490) = 3216 mm².

Provide 8 mm ties at 250 mm spacing.

Design of Rafter

The forces on the inner and outer rafters differ marginally. Both the rafters are designed as columns. The maximum positive and negative bending moments and the shear forces are:

> Maximum positive BM = M_p = 60 kNm.
> Maximum negative BM = M_n = 216 kNm.
> Axial force = P = 149 kN;   Shear force = V =15 kN

The virtual eccentricity ration is = e = 216/149 = 1.45 m.
Since the eccentricity ratio is high, the rafter should be designed as beam column.

$$b = 0.25 \text{ m}, D = 0.4 \text{ m}, f_{ck} = 20 \text{ MPa}, f_y = 415 \text{ MPa, Let } d' = 50 \text{ mm} = 0.05 \text{ m}$$

The cover ratio about x-axis is = c_x' = 50/400= 0.125
The roof slab continuously supports the rafter; the effective length is computed neglecting the T-beam action of the rafter and it is:

$$L_e = \sqrt{4^2 + 2^2} = 4.472 \text{ m}$$

Slenderness factor is = g = 4.472/.25 = 17.9 m
The non-dimensional force factors are:

$$Y = \frac{P_u}{bDf_{ck}} = \frac{0.149}{0.2*0.4*20} = 0.093$$

$$X = \frac{M_u}{bD^2 f_{ck}} = \frac{0.216}{0.25*0.16*20} = 0.215$$

From the Fig. 11.11, the column design aid curves, the percentage of the reinforcement is:

$$p/f_{ck} = 0.14 \text{ or}$$

$$p = 0.14*20 = 2.8\%$$

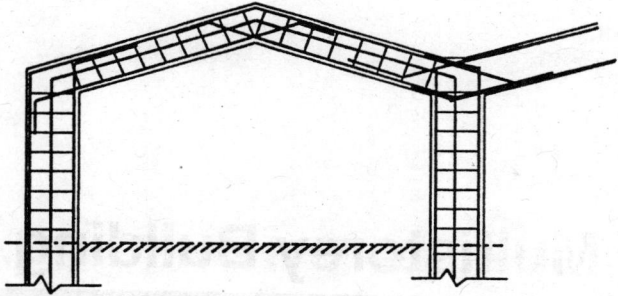

Fig. 14.15 Reinforcement Details in One Portal; Example 14.2.

The area of reinforcement is:

$$A_s = 2.8*bD/100 = 2.8*250*400/100 = 2800 \text{ mm}^2.$$

Select four numbers of 20 mm and 25 mm bars. The actual area provided results $= 4*(314 + 490) = 3216$ mm². This is same as that of the column reinforcement. Provide 8 mm ties at 250 mm spacing.

Figure 14.15 illustrates the reinforcement details of one bay of the frame. Bending and curtailment of the bars need to be done keeping in view the development lengths and reentrant corner detail.

PROBLEMS

14.1 One bay portal frame of 6 m span has a ceiling height of 5 m. The columns are founded on raft at 1.2 m below the ground level and the plinth height is 0.4 m. The exposure condition of the building is moderate. The basic wind pressure is 1.5 kN per square metre, and the building is in earthquake zone three. The frames are spaced at 4.0 m and covered by 110 mm thick reinforced concrete slab with waterproof treatment. Independent walls are provided as cladding to the building. The size of the column is 250 by 350 mm and that of the beam is 250 by 400 mm. Design the frame.

14.2 Two bay portal frame of 5 m each span has a ceiling height of 6 m. The columns are founded on isolated footings at 1.2 m below the plinth level. The exposure condition of the building is severe. The basic wind pressure is 1.5 kN per square metre, and the building is in earthquake zone two. The frames are spaced at 4.5 m and covered by 120 mm thick reinforced concrete slab with waterproof treatment. 250 mm thick walls are supported on plinth beams. The size of the column is 300 by 350 mm and that of the beam is 250 by 350 mm. Design the frame and sketch the reinforcement details.

14.3 One bay gable frame of 7 m span, column height of 5.0 m is resting on independent footings. The rise of the gable is 1.5 m. The depth of foundation is 1.5 m below the plinth level. The exposure condition of the building is mild. The basic wind pressure computed as per the code is 1.2 kN per square metre. The frames are spaced at 4.5 m. Independent walls are provided as cladding to the building. The size of the column and the rafter is 300 by 400 mm. Design the portal and sketch the reinforcement details.

14.4 A Two bay gable frame of 8 m each span has a ceiling height of 6 m. The columns are founded on isolated footings at 1.2 m below the plinth level. The exposure condition of the building is moderate. The basic wind pressure is 1.5 kN per square metre, and the building is in earthquake zone four. The frames are spaced at 4 m and covered by 100 mm thick reinforced concrete slab with waterproof treatment. 250 mm thick walls are supported on plinth beams. The size of the column is 300 by 350 mm and that of the beam is 250 by 450 mm. Design the frame and sketch the reinforcement details.

Design of Multistorey Building Frames

15.1 INTRODUCTION

Tall buildings are built for office, commercial and residential purposes. Reinforced concrete buildings are popular in large cities for apartments and office complexes. The character of tall building introduces new dimensions to the architectural and functional design. Structural design of tall buildings brings in dynamic behavior of the structure under wind and seismic loads. Wind loads are critical in tall towers, and industrial frames. Mass of concrete floor slabs, partition walls, beams and columns in reinforced concrete construction is high when compared with steel structures. Because of high mass of reinforced concrete buildings, the earthquake induces large bending moments in columns and beams. Design of tall building is also controlled by several other parameters other than structural design. Fire protection, security, ventilation, vertical transportation by lifts or by staircases, Plumbing materials and the arrangements, water supply and sewage disposal, optimal space arrangement for movement of people and goods, maintenance and repair arrangements, city bylaws, etc. In other words, a tall building is to be planed as a mini city depending on the size of the building. Reinforced concrete buildings as tall as 120 storeys are built and managed with large population in the building. According to building bylaws in most cities in India, Four storey buildings with basement need not be provided with lifts for vertical transportation, so such size of reinforced concrete buildings have become common in all the cities of India. High Yield Steel Deformed bars of 415 N/mm^2 yield strength is the most commonly used reinforcement. Yield strength HYSD bars of 500 and 550 N/mm^2 are also used to a limited extent. Cement of grades 43 and 53 are extensively used in concrete production. The same grades of cements are also used in the plastering and floor finishes even though not essential. Such extensive use is because of the availability and marketing strategies of the cement companies. Blended cements are becoming popular in all the advanced countries and it is hoped such cements will be used increasingly in building construction.

15.2 DESIGN METHODOLOGY

The limit state design is the main method of structural design of reinforced concrete multistorey frames. Working stress design is also acceptable as per the present code of practice even though not encouraged. The structural analysis is linear-elastic in both the design methods. There was a time when a sub-structure analysis technique was accepted for multistorey frames. It is a very crude model with unreliable accuracy so not recommended for the last four decades. Today, a number of computer software available for analysis and design of tall buildings. Many software packages deal with the structural analysis, as the analysis is more of a universal application. Some takes care of the structural design based on the code of practice. Even though there is no exclusive code for design of multistorey buildings, the Indian code of practice for plain and reinforced concrete structures is applicable. The basic loads on the structure are: Dead Load, Live Load, Wind Load, Earthquake Load or any other special load in special purpose structures. The main load combinations are:

1. Design Load = 1.5(Dead Load) + 1.5(live load)
2. Design Load = 1.5(Dead Load) + 1.5(Wind load)
3. Design Load = 1.5(Dead Load) + 1.5(Earthquake load)
4. Design Load = 1.2(Dead Load) + 1.2(Live load) + 1.2(Wind load or Earthquake)

The member sizes are selected based on the architectural, structural, and economic considerations. In most software, the loads are computed manually and placed at the appropriate nodes or locations. The method of analysis is finite element with linear and plate elements. Dynamic analysis is carried out in case of earthquake or wind load design in some software. Spectral analysis based on the IS:1893-2002 is commonly used in important projects. The building can be modeled in three-dimensional skeleton, loads placed on the joints or members and analysed. Modeling is the most important aspect. The software gives results of the skeletal model quite accurately but how good is the model that represents the actual building depends on the experience and engineering concepts of the engineer. Modeling of a standard building is not difficult but of buildings with discontinuities, auditoriums, shell roofs etc need good structural engineering concepts. Similarly load computations and the corresponding locations are not easy unless good experience is acquired. The reinforcement detailing, especially at beam-column joints is important. Ensuring the development lengths in the starting and closing of members in the frame is necessary. The design of a multistory building is a project by itself, so an attempt is made in this chapter with one simple example. Design of beams and columns explained earlier chapters form a part of the multistorey frames. However rationalization of the reinforcement details along the length of the multi-columns and similarly for the continuous beams considering the economy, speed of construction and minimizing the errors of confusion at site need sound engineering judgment.

15.3 ILLUSTRATIVE EXAMPLES

Example 15.1: Four Plus Basement framed Multistorey Frame

A four-storey building with basement having three bays of each 5 m in one direction and four bays of each 4.5 m in the other direction is built in reinforced concrete. The height of each floor except the basement is 3.75 m and the effective height of the basement is 2.2 m. Partition walls with suitable door openings are placed on all the beams. All the walls except of the basement walls are 150 mm thick. The building is designed for office purpose in moderate exposure condition. The thickness of the RC slabs is 140 mm plus the floor finish is 40 mm. The building is located in seismic zone three and wind basic speed is 39 m/s. Figure 15.1 illustrates the profile of the cross frame and Fig. 15.2 illustrates lay out of the columns. Design a typical RC frame.

Solution

The problem is analysed and designed by using software *HiRise* developed by the author. The software has four components. The program is based on the pseudo three-dimensional model. The floors are rigid in their planes and the joint displacements in each floor are compatible. The preprocessor generates the frame configuration, numbers the members and joints. The module generates loads on joints and beams

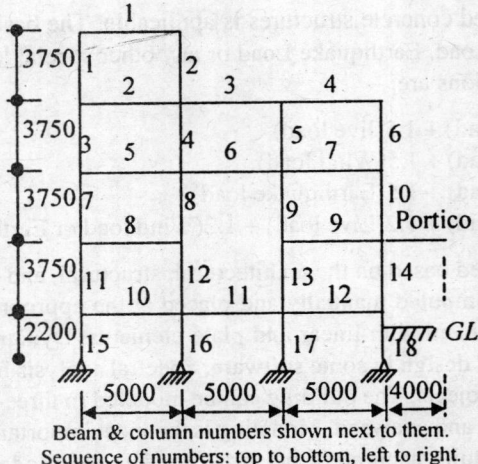

Fig. 15.1 Cross Frame Profile; Example 15.1.

from the floor slabs due to live and dead loads. The loads of the walls on main and cross beams are also generated depending on data of the location and thickness of the walls. The main program takes care of the analysis and design. The program generates the earthquake loads from the spectrum analysis. The reinforcement design and details is done by design module. The program gives very detailed information in six files, from which some extracts about the properties of the members along with forces and displacements are given in one file. The files list floor wise joint coordinates, loads and displacement. Similarly, member sizes, forces, and equivalent uniformly distributed load on beams are listed in another table. One file is of the analysis, and another design. The design file gives the areas of reinforcement at different locations of beams, and columns, and percentage of reinforcement in beams and columns. The same file gives the recommended number and diameter of bars at different locations, top and bottom, stirrups in beams and ties in columns. The postprocessor generates deflected profiles of the frame for different load combinations. Similarly, bending moment diagrams for different load combinations are generated in another file. Some extracts of the analysis and design files are given in tables 15.1 to 15.4. The seismic coefficients of the floors from top to bottom are: 0.051, 0.50, 0.43, 0.34, 0.019, and 0.004. The post processor files and diagrams are not shown here. A limited number of reinforcement details of typical beams are given in Figs. 15.2 to 15.4. Cross sections of column reinforcement are shown in Fig. 15.5.

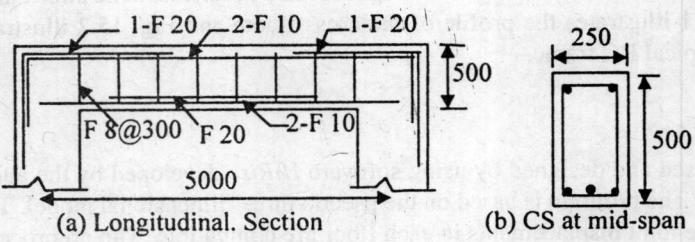

Fig. 15.2 Beam Reinforcement in 4th Floor (Top) Beams.

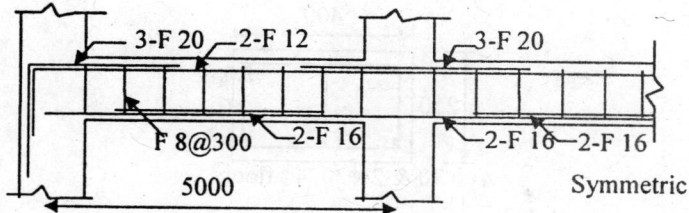

Fig. 15.3 Beam Reinforcement in 3rd & 2nd Floor Beams.

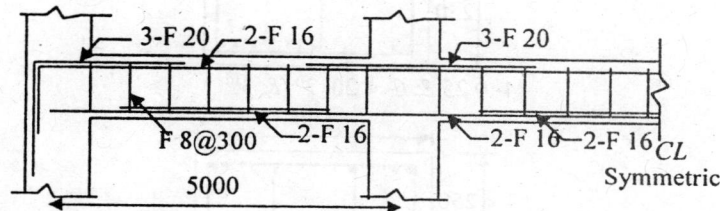

Fig. 15.4 Beam Reinforcement in Basement Floor Beams.

Table 15.1 Small Extract from Analysis File. Only Limited Dated is Shown. (Units: metres, kN, kNm).

Design Load = 1.50(DL)+ 1.50(LL) + .00(EQ or WL)

	Member	Properties			Forces & Moments: (Moment causing Comp top +)					
Num	Floor no.	L	B	D	Axial	Shear	L-BM	R-BM	EUDL	Mid-BM
					BEAMS					
1	5	5.0	0.25	0.50	21.4	90.2	-46.2	-48.9	35.9	64.5
2	4	5.0	0.25	0.50	16.4	154.8	-107.7	-134.5	59.8	106.7
3	4	5.0	0.25	0.50	44.3	155.1	-114.0	-142.2	59.8	106.7
4	4	5.0	0.25	0.50	45.1	159.7	-142.4	-91.0	59.8	106.7
8	2	5.0	0.25	0.50	-0.3	152.6	-104.3	-119.9	59.8	106.7
9	2	5.0	0.25	0.60	0.4	212.8	-3.5	-308.6	60.7	108.4
10	1	5.0	0.25	0.50	13.9	158.8	-100.6	-147.6	59.8	106.7
11	1	5.0	0.25	0.50	44.3	154.3	-117.3	-141.4	59.8	106.7
12	1	5.0	0.25	0.50	53.1	156.5	-137.9	-102.5	59.8	106.7
					COLUMNS					
1	5	3.75	0.25	0.40	153.1	21.4	46.2	-34.2	.0	6.0
2	5	3.75	0.25	0.40	154.2	21.4	-48.9	31.5	.0	-8.7
3	4	3.75	0.25	0.50	528.1	37.8	73.5	-68.4	.0	2.5
4	4	3.75	0.25	0.50	709.0	6.5	11.0	-13.3	.0	-1.1
5	4	3.75	0.25	0.50	564.0	0.8	0.3	-2.8	.0	-1.3
6	4	3.75	0.25	0.50	363.1	45.1	-91.0	78.2	.0	-6.4
15	1	2.20	0.25	0.50	1652.9	38.6	60.2	-24.8	.0	17.7
16	1	2.20	0.25	0.50	2175.9	4.1	8.9	.0	.0	4.4
17	1	2.20	0.25	0.50	1991.7	5.7	9.5	-3.2	.0	3.2
18	1	2.20	0.25	0.50	1694.5	48.4	-72.3	34.3	.0	-19.0

L-BM = left end bending moment, *R-BM* = right end bending moment, *Mid-BM* = mid-span BM

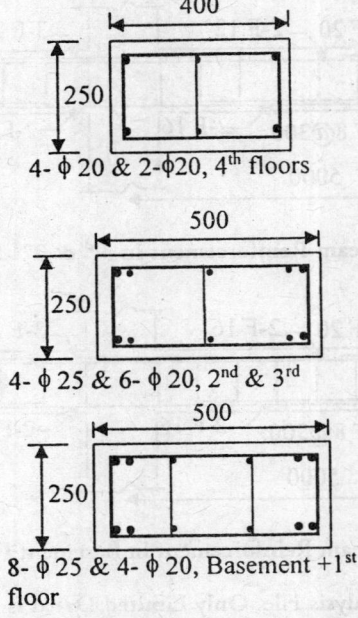

400

250

4- φ 20 & 2-φ20, 4th floors

500

250

4- φ 25 & 6- φ 20, 2nd & 3rd

500

250

8- φ 25 & 4- φ 20, Basement +1st
floor

Fig. 15.5 Column Reinforcement; Example 15.1.

Table 15.2 Support Reactions.

Joint	F_x	F_y	M_x	Joint	F_x	F_y	M_x	Joint	F_x	F_y	M_x	Joint	F_x	F_y	M_x
19	39.	1658.	-25.	20	4.2	2181.	0.	21	6.	1997	3.	22	-48.	1700	34.

Total Support Reactions are: $R_x = .0$; $R_y = 7535.6$ kN ; $M_z = 6.3$ kNm

Detailed forces and displacements for all load combinations are given by the program but not listed here to maintain compactness of the example. The reinforcement areas are computed for the most critical forces. A minimum of 0.18 percentage of reinforcement is assigned at top and bottom of the beam section in the present frame for practical considerations. In tall building construction, the sequence of construction, striking of the formwork, construction loads etc. are not well defined in many cases. The analysis program takes the total integrity of the frames but may not include the part constructed frame analysis. The author suggests a higher minimum reinforcement in tall buildings to maintain integrity and ensure safety in the construction.

The maximum column design load on the soil is 2181 kN with combined partial load factor of 1.5. The foundation size for a net bearing pressure of 120 kN/m^2 is 3.5 by 3.5 m. Each column is provided with isolated footing and the details of the footing are shown in Fig. 15.7.

Table 15.3 Reinforcement details of Beams Example 15.1 (Dimensions in "mm").

| N | Flr | L | B | D | F_{ck} | Areas of steels at | | | | | | % of steel | | | |
| | | | | | | Top | | | Bottom | | | Top | | Bottom | |
						L	M	R	L	M	R	L	M	L	M	M
1	5	5000	250	500	20	337	225	357	225	441	225	0.27	0.18	0.35
2	4	5000	250	500	20	899	225	1033	225	730	225	0.72	0.18	0.58
3	4	5000	250	500	20	950	225	1069	225	730	225	0.76	0.18	0.58
4	4	5000	250	500	20	1041	225	777	225	730	225	0.83	0.18	0.58
5	3	5000	250	500	20	1199	225	1254	225	730	225	0.96	0.18	0.58
6	3	5000	250	500	20	1149	225	1256	225	730	225	0.92	0.18	0.58
7	3	5000	250	500	20	1254	225	1198	225	730	225	1.0	0.18	0.58
8	2	5000	250	500	20	1451	225	1497	375	730	421	1.16	0.18	0.58
9	2	5000	250	500	20	1151	270	2080	785	618	790	0.77	0.18	0.41
10	1	5000	250	500	20	1174	225	1331	225	730	255	0.94	0.18	0.58
11	1	5000	250	500	20	1178	225	1278	225	730	225	0.94	0.18	0.58
12	1	5000	250	500	20	1335	225	1218	259	730	225	1.07	0.18	0.58

B, D & L = width, depth and length of members. F_{ck} = grade of concrete, Flr = floor No.
M = middle, L = left face, R = right face

Table 15.4 Reinforcement details of Columns (in "mm").

N	Flr	L	B	D	F_{ck}	steel	% steel		Ties
1	5	3750	250	400	25	1551	1.55	8	200
2	5	3750	250	400	25	1651	1.65	8	200
3	4	3750	250	500	25	2204	1.76	8	200
4	4	3750	250	500	25	1694	1.36	8	200
5	4	3750	250	500	25	1673	1.34	8	200
6	4	3750	250	500	25	2550	2.04	8	200
7	3	3750	250	500	25	2814	2.25	8	200
8	3	3750	250	500	25	3555	2.84	8	200
9	3	3750	250	500	25	2995	2.40	8	200
10	3	3750	250	500	25	2390	1.91	8	200
11	2	3749	250	500	25	3983	3.19	8	200
12	2	3749	250	500	25	4569	3.66	8	200
13	2	3749	250	500	25	4105	3.28	8	200
14	2	3749	250	500	25	4098	3.28	8	200
15	1	2200	250	500	30	4567	3.65	8	200
16	1	2200	250	500	30	5147	4.12	8	200
17	1	2200	250	500	30	4931	3.94	8	200
18	1	2200	250	500	30	4649	3.72	8	200

The reinforcement details of slabs, lintels, staircases, chejjas etc. are not listed here. Fig. 15.7 illustrates the foundation layout.

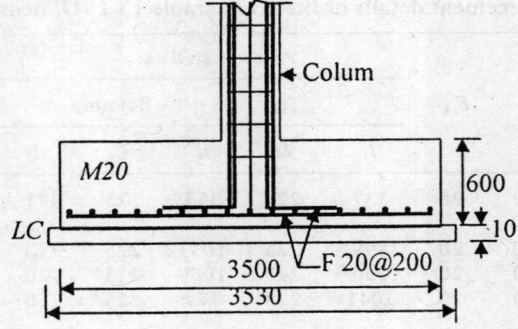

Fig. 15.7 Details of Typical Footing; Example 15.1.

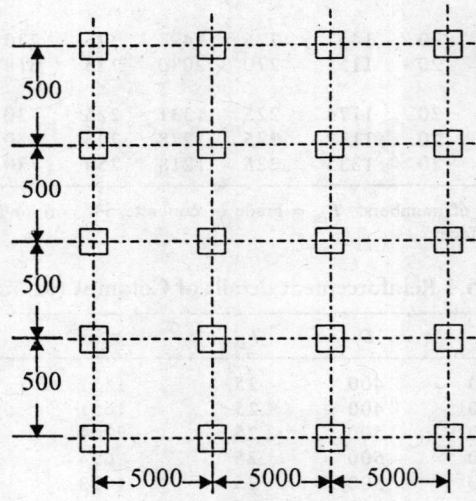

Fig. 15.8 Foundation Layout; Example 15.1.

PROBLEMS

15.1 A three-storey building having three bays of each 4.5 m in one direction and three bays of each 5 m in the other direction is built in reinforced concrete. The height of each floor is 3.25 m Partition walls of 125 mm thick are placed on all the beams. The building is designed for residential purpose in mild exposure condition. The thickness of the RC slabs is 140 mm plus the floor finish is 40 mm. The building is located in seismic zone three and wind basic speed is 44 m/s. Design a typical RC frame.

15.2 A four-storey building with basement having three bays of each 4 m in one direction and four bays of each 4.5 m in the other direction is built in reinforced concrete. The height of each floor except the basement is 3.5 m and the effective height of the basement is 2.5 m. All the walls except of the basement walls are 150 mm thick. The building is designed for commercial purpose in moderate exposure condition. The thickness of the RC slabs is 150 mm plus the floor finish is 50 mm. The building is located in seismic zone three and wind basic speed is 33 m/s. Design a typical RC frame.

16

Reinforced Concrete Retaining Walls

16.1 INTRODUCTION

Soils have a tendency to slide down and repose in a particular inclination unless retained by other means. Different types of structures are used to retain soil in embankments and cuttings. A broad classification of retaining walls is:

(i) Gravity, (ii) Cantilever, (iii) Counterfort, (iv) Buttressed and (v) Guyed

Figure 16.1 illustrates a gravity retaining wall. The gravity retaining walls are normally made of masonry such as stone, Brick and concrete.

The stability and safety considerations of the structure are:

(a) The stabilizing moment must be more than the overturning moment. A factor of safety of 1.5 to 2 against overturning is desirable.
(b) The horizontal resistance at the interface at the soil at base and the wall must be more than the net horizontal earth pressure.
(c) The bearing pressure on the soil must be less than the allowable pressure of the soil.
(d) The stresses produced in the masonry must be within the permissible limits.

Figure 16.2 illustrates a typical *cantilever retaining* wall. The *stem* retains the earth and the base (or footing) of the wall supports the stem. The base is further sub-divided into *heel* and the *toe* as shown in the figure. The base of the stem should be designed such that the structure is safe against overturning,

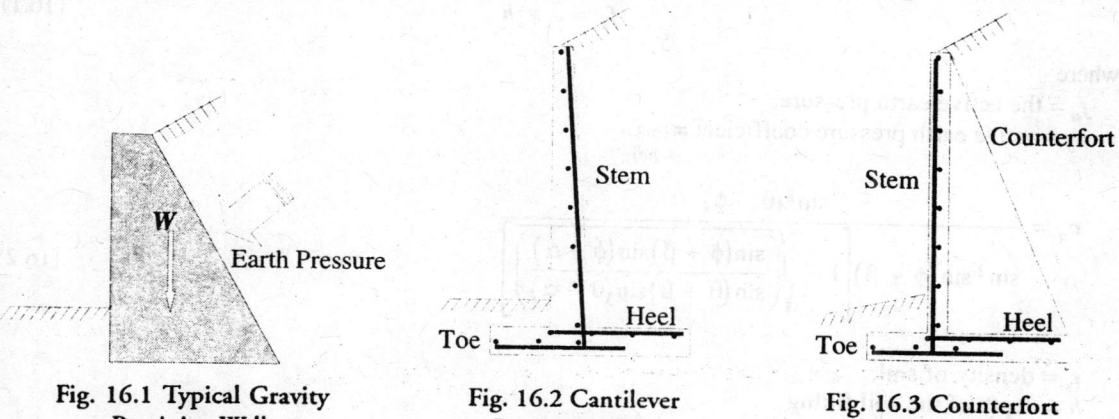

Fig. 16.1 Typical Gravity
Retaining Wall.

Fig. 16.2 Cantilever
Retaining Wall.

Fig. 16.3 Counterfort
Retaining Wall.

sliding and bearing pressure. The stem, heel and the toe are cantilever elements. The cantilever type is usually used up to 5 m height.

Counterfort and Buttress retaining walls: As the height of the retaining wall increases, the moment due to the earth pressure increases more rapidly on the cantilever stem and it becomes too heavy. Therefore, vertical walls spaced at intervals are placed on the footing supporting the stem. Such walls are called counterforts. The counterforts are built on the earth filling side. Fig 16.3 illustrates a typical counterfort retaining wall. In case the support to the stem is on other side of the filling, then it is called a *buttress retaining* wall. Fig. 16.4 illustrates a typical buttress retaining wall.

Anchor Retaining wall. The cantilever stem can also be propped by guy wires instead of counterfort. Fig. 16.5 illustrates anchored retaining wall.

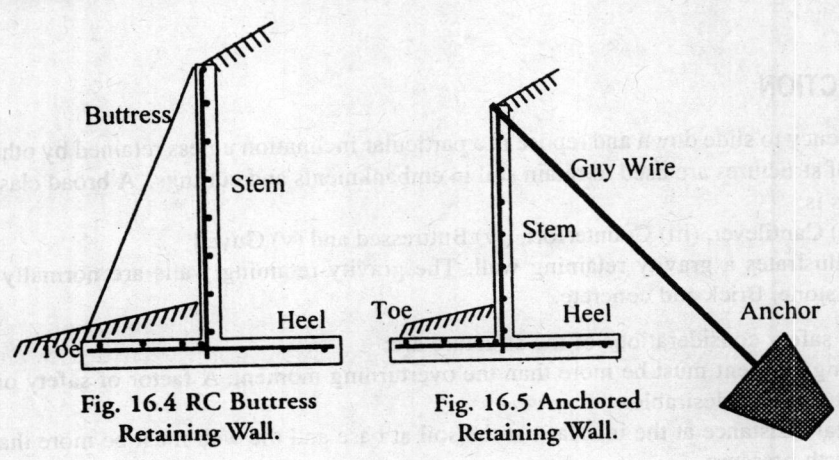

Fig. 16.4 RC Buttress Retaining Wall.   **Fig. 16.5 Anchored Retaining Wall.**

16.2 SIMPLE THEORY OF EARTH PRESSURE

Fig. 16.6 illustrates typical forces on the retaining wall. The active earth pressure intensity on the retaining wall is:

$$f_a = c_a \gamma_s h \tag{16.1}$$

where
f_a = the active earth pressure,
c_a = active earth pressure coefficient =

$$c_a = \frac{\sin^2(\theta - \phi)}{\sin^2 \sin(\phi + \beta)\left[1 + \sqrt{\left(\dfrac{\sin(\phi + \beta)\sin(\phi - \alpha)}{\sin(\theta + \beta)\sin(\theta - \alpha)}\right)}\right]} \tag{16.2}$$

γ_s = density of soil,
h = depth of the soil filling
α = slope of the embankment of or backfill

ϕ = angle of internal friction of the soil
β = friction angle between the soil and the wall surface where $\tan \beta = \mu$
μ = coefficient of friction between the soil and the surface of the wall
θ = inclination of the wall face

Retaining walls with upright on back fill face (for $\theta = 90$, and zero back fill slope ($\alpha = 0$) the active earth pressure coefficient from Eq. (16.2) is:

$$c_a = \frac{1 - \sin\phi}{1 + \sin\phi}$$ (16.3)

The earth offers a resistance if the retaining wall tends to move against the earth, and such resistance is called passive earth pressure. The passive earth pressure intensity is given by:

$$f_p = c_p \gamma_s h$$ (16.4)

where

c_p = the passive earth pressure coefficient, for the simplest case of upright retaining wall it is:

$$c_a = \frac{1 + \sin\phi}{1 - \sin\phi}$$ (16.5)

If the top of the backfill is loaded, this load is experienced as uniform along the depth of the wall of intensity equal to the intensity of the load multiplied by the coefficient of the active earth pressure.

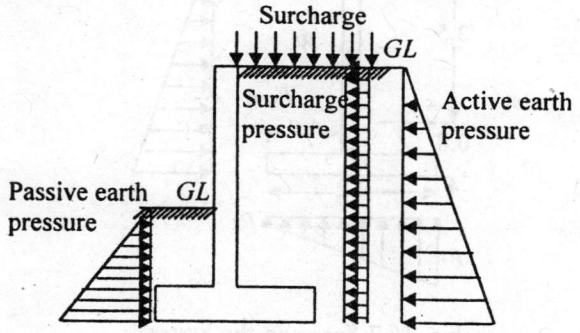

Fig. 16.6 Retaining Wall, Active & Passive Earth Pressures.

16.3 CANTILEVER RETAINING WALL

Figure 16.6 illustrate the active and passive earth pressures on a simple upright retaining wall with surcharge.

As a rough guideline, the footing length may be selected in the range of $0.3h$ to $0.4h$; and the length of the heel as about forty five to fifty percent of the length of the footing. Smaller the angle of internal friction more is the intensity of the pressure. Higher range of the lengths of the footing is chosen in clayey and silty soils.

Example 16.1: Design of Cantilever Retaining Wall

A retaining wall to retain a sandy soil of 3.5 m high above the ground level is to be built in soil of safe bearing pressure of 100 kN/m². The unit weight of the soil is 18 kN/m³ and the angle of internal friction is 30°. The exposure condition is moderate. Design a reinforced concrete retaining wall.

Solution

Select $M25$ grade concrete and $HYSD$-$Fe415$ reinforcement. A cantilever type is chosen as the height of the retaining wall is 3.5 m above GL. The minimum depth of foundation is:

$$\text{Minimum depth of foundation} = \frac{P_a}{\gamma_s}\left[\frac{1-\sin\phi}{1+\sin\phi}\right]^2 = \frac{100}{18}\left[\frac{1-\sin 30}{1+\sin 30}\right]^2 = 0.617 \text{ m}$$

Select the depth of foundation = 0.8 m

The overall depth of the retaining wall = $h = 3.5 + 0.8 = 4.3$ m

The bearing pressure can be obtained with the known values of the weights of the elements of the retaining wall from the equilibrium consideration. Fig. 16.7 illustrates various forces acting in the system.

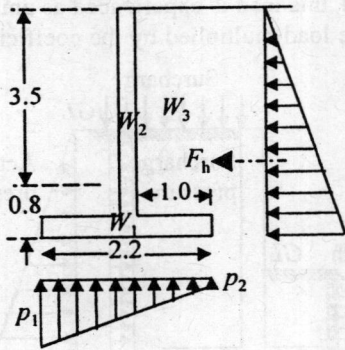

Fig. 16.7 Forces in the System.

Select the following preliminary sizes to estimate the self-weights.
    Width of footing = $B = 0.5h = 0.5*4.3 = 2.15$, use $B = 2.2$ m
    Width of heel = $0.45B = 0.96$, use 1 m
    Average thickness of the wall = (about $h/15$) = 0.3 m
    Average thickness of the footing = (about $h/20$) = 0.25 m

Determination of Pressure the Soil

The pressure on the soil is obtained from the equilibrium of the forces acting on the wall. Table 16.1 illustrates the sum of the forces and the moments. This type of tabulation of forces systematizes the data and minimizes the possible errors.

Table 16.1 Forces and Moments of One Metre Length of the Retaining Wall, Example 16.1.

No.	Notation	Item	Force (kN)	Distance From toe tip (m)	Moment about tip of toe (kNm)
1	W_1	Footing	0.25*2.2*25 = 13.75	2.2/2 =1.1	15.125
2	W_2	Wall	0.3*4.05*25 = 30.375	1.3 − 0.3/2=1.15	34.93
3	W_3	Soil on heel	1.0*4.05*18 = 72.9	1.2 +1/2 =1.7	123.93
		Sum	117.025		173.985 (cw)
4	F_h	Soil Pressure	$\dfrac{c_a \gamma_s h^2}{2}$ = 55.46	4.3/3	78.499 (acw)
5	p	Soil reaction	2.2*$(p_1 + p_2)$/2	$B^2(p_1 +2\,p_2)/6 = 0.806(p_1 +2\,p_2)$ (acw)	

cw = clockwise (stabilizing moment), acw = anticlockwise (overturning moment).

The active earth pressure on the down side of the wall is neglected. The equilibrium of the vertical forces from the quantities in the column 4 of table 16.1 give:

$$2.2*(p_1 + p_2)/2 = 117.025, \text{ or}$$

$$(p_1 + p_2) = 106.37 \tag{a}$$

The equilibrium of the moments about the toe of the footing from the quantities of table 16.1 give:

$$0.806(p_1 +2\,p_2) + 78.499 = 173.985, \text{ or}$$

$$p_1 +2\,p_2 = 118.47 \tag{b}$$

From Eqs.(a) and (b), the values of p_1 and p_2 are:

$$p_2 = 12.1, p_1 = 94.27 \text{ kN/m}^2. < 100 \text{ kN/m}^2$$

The maximum pressure is less than the permissible value, therefore the width of the footing is right. For the purpose of design the rounded values are used; $p_2 = 12$, $p_1 = 95$ kN/m^2

Check for Stability

The factor of safety against overturning = (stabilizing moment/overturning moment)
= 173.985/78.499 = 2.21
Frictional resistance between the base and the soil is

$$F = \mu W = 0.55(117.025) = 64.36 \text{ kN}$$

Horizontal active soil pressure = F_h = 55.46 kN

Factor of safety against sliding = Frictional resistance/Horizontal thrust = 64.36/55.46 = 1.16
Normally desired factor of safety against sliding is 1.5. To improve the factor of safety against sliding, either a key is provided, the depth of the foundation is increased, or the footing size is increased. In the present case, a key is provided. The passive earth pressure on the key plus the frictional resistance must provide the required factor of safety.
Let the depth of the key below the ground level be h_o.
A simple stability safety is: Passive earth pressure + frictional resistance >1.5 times the active earth pressure.

$$\frac{C_p \gamma_s (h_0)^2}{2} + F \geq 1.5 \frac{c_a \gamma_s (h_1 + h_0)^2}{2}$$

$$\frac{3(18)(h_0)^2}{2} + 64.36 \geq 1.5 \frac{0.33(18)(3.5 + h_0)^2}{2}$$

$$(27 - 4.455)h_0^2 - 31.185 h_0 + (64.36 - 54.57) = 0$$

The solution of the above quadratic expression gives $h_o = 0.93$ m.

Provide a key of 0.15 m depth below the footing. The same thickness of the wall at base can be used as the thickness of the footing with the reinforcement of the wall extending into the key.

Design of the Footing (Toe and Heel)

The thickness of the wall at base is assumed equal to 0.3 m for the purpose of design spans of the toe and the heel. The footing consists of toe and the heel. The toe is designed as a cantilever subjected to the soil reaction neglecting the weight of the soil over the toe. The forces acting on the toe and heel slabs are

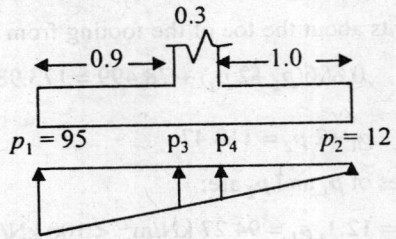

Fig. 16.8 Bearing Pressures on Footing.

taken from Fig. 16.8 The weight of the filling over the heel slab must be considered.

Cantilever span of the toe = $L_t = 2.2 - 1.0 - 0.3 = 0.9$ m; average thickness is assumed as 250 mm.

The bearing pressure at the base of the toe is:

$$p_3 = p_1 - (p_1 - p_2)L/B = 95 - (95 - 12)(0.9)/2.2 = 61 \text{ kN/m}^2$$

Self-weight = $0.25(25) = 6.25$ kN/m

The factored bending moment at base of the toe is:

$$M = 1.5\left(\frac{(2p_1 + p_2)L_t^2}{6} - \frac{w_g L_t^2}{2}\right) = 47 \text{ kNm/m}$$

The effective depth of the toe is obtained by equating the factored moment to the moment capacity.

$$d = \sqrt{\frac{M}{Kbf_{ck}}} = \sqrt{\frac{47(1000)}{0.138(1)25}} = 116 \text{ mm}$$

Let the effective depth at the base of the toe is taken as 240 mm and the overall depth as 300 mm. The thickness of the footing at free end is 150 mm. The area of the tension reinforcement is:

$$A_{st} = \frac{1.15 * M}{jdf_y} = 678 \text{ mm}^2/\text{m}$$

Provide 16 mm bars at 290 mm spacing. Percentage of reinforcement is = 0.23 %

The distribution reinforcement is to be provided at 0.15%. 10 mm bars at 200 mm spacing are provided.

The factored shear force at a distance of effective depth from stem is:

$$V = 1.5 * (p_1 + p_3)(L - d)/2 = 77 \text{ kN/m}$$

The nominal shear stress on the section is:

$$\tau_v = \frac{V}{bd} = \frac{77000}{1000 * 240} = 0.32 \text{ N/mm}^2$$

The shear strength is more than the nominal shear stress.

The development length is = 47ϕ = 752 mm

The tension bars provided at the bottom must extend at least 760 mm beyond the base of the toe.

Design of the Heel

Cantilever span of the heel = L_h = 1 m. The average thickness of the hell is assumed as 250 mm.

Soil pressure at the base of the heel is:

$$P_4 = P_1 - (P_1 - P_2)(L_t + 0.3)/B = 95 - (95 - 12)(1.2)/2.2 = 50 \text{ kN/m}^2$$

Weight of the soil over the heel is:

$$W_3 = 18(4.3 - 0.25)(1) = 72.9 \text{ kN/m}$$

Self-weight of the heel is = 0.25(25) = 6.25 kN/m

The factored bending moment at the base of the heel is: (tension causing on the top face)

$$M = 1.5 \left[\frac{W_3 L_h}{2} + \frac{wL_h^2}{2} - (2p_2 + p_4)\frac{L_h^2}{6} \right] = 41.2 \text{ kNm/m}$$

The effective depth of the heel required at the base is:

$$d = \sqrt{\frac{M}{Kbf_{ck}}} = \sqrt{\frac{41.2(1000)}{0.138(1)25}} = 110 \text{ mm}$$

Let the effective depth at the base of the toe is taken as 240 mm and the overall depth as 300 mm. The thickness of the footing at free end is 150 mm. The area of the tension reinforcement is:

$$\cdot A_{st} = \frac{1.15 * M}{jdf_y} = 595 \text{ mm}^2/\text{m}$$

Provide 16 mm bars at 330 mm spacing.

The distribution reinforcement is to be provided at 0.15%. 10 mm bars at 200 mm spacing are provided.

The shear strength is found to be more than the nominal shear stress.

Design of Stem (Vertical Wall)

$$\text{Height of stem} = h_s = 4.3 - 0.3 = 4 \text{ m.}$$

The maximum bending on the stem occurs at the base of the stem is:

$$M = 1.5 \frac{\gamma_s h_s^3}{6} = 288 \text{ kNm/m}$$

The effective depth required is:

$$d = \sqrt{\frac{M}{Kbf_{ck}}} = \sqrt{\frac{288000000}{0.138(1000)25}} = 289 \text{ mm}$$

Use the effective depth at base as 340 mm and the overall depth as 400 mm. Let the thickness at top be 150 mm.

Area of the tension reinforcement at the base of the stem is:

$$A_{st} = \frac{1.15 * M}{jdf_y} = 2934 \text{ mm}^2/\text{m}$$

Provide 20 mm bars at 100 mm spacing. Percentage of reinforcement is = 0.75 %

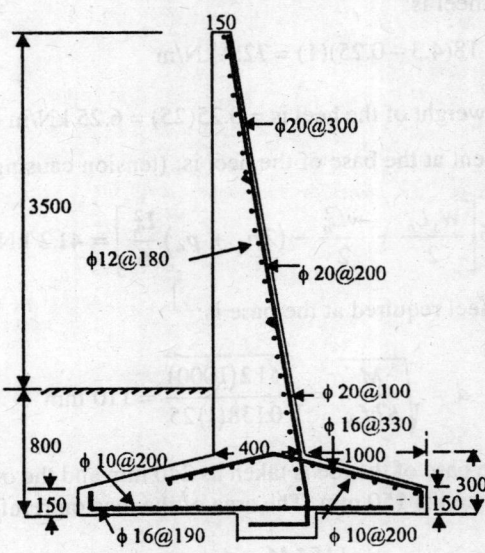

Fig. 16.9 Details of Retaining Wall; Example 16.1.

The distribution reinforcement is to be provided at 0.15%. 12 mm bars at 180 mm spacing are provided.

The factored shear force at a distance of effective depth from the base of the stem is:

$$V = 1.5 \frac{\gamma_s h^2}{2} = 1.5 \frac{18*(4 - 0.34)^2}{2} = 180 \text{ kN/m}$$

The nominal shear stress is:

$$\tau_v = \frac{V}{bd} = \frac{180000}{1000*340} = 0.53 \text{ N/mm}^2$$

The shear strength is 0.58 as against the nominal stress of 0.53 N/mm².
The bars must extend at least 752 – 8*20 = 592 mm into toe.

Curtailment of Reinforcement in the Vertical Wall

The bending moment decreases proportion to cube of the depth of filling while the thickness decreases proportional to the depth. One-third of the vertical bars can be curtailed at 1.5 m from base and another one-third bars at 3 m from the base. Fig. 16.9 illustrates the reinforcement details.

16.4 COUNTERFORT RETAINING WALL

The bending moment due to the horizontal thrust of the soil increases rapidly as the height of the retaining wall increases,. The bending moment on the vertical cantilever slab is proportional to the cube of the height of the wall. It is desirable to change the structural action of the vertical wall in such a way that the wall does not act as a cantilever but acts as a slab with supports. If a series of upright beams that support the wall are provided along the length of the wall, then the vertical wall acts as a continuous slab in one direction. The upright beams when provided at the rear of the vertical wall, they are called counterforts. If the beams are provided in the front of the wall, then they are called buttresses. The counterforts that rest on the heel, make the heel to act as a continuous slab. Each panel can be taken as a slab fixed at three edges and free at one edge. The toe acts as a cantilever subjected to the soil reaction. The thickness of the cantilever slab tends to be too thick, therefore it is desirable to have buttresses so that the toe also acts like a slab. Fig. 16.10 illustrates the Counterfort retaining wall. The slab portions within the counterforts acts as slab with fixed boundaries at the counterforts.

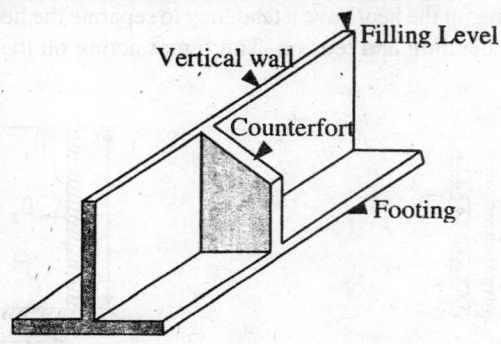

Fig. 16.10 Counterfort Retaining Wall.

Design Criteria for Footing and Vertical Slab

Bending moment coefficients of slabs subjected to uniformly distributed load and triangularly distributed load are available from plate theory. Critical moment coefficients of slabs for design are given in Table 16.2. Magnitudes of the bending moments are maximum at the middle of the fixed edges. The critical bending moment locations are indicated in Fig. 16.11. The loads acting on the toe or heel can be divided into uniformly and triangularly distributed loads. The bending moments at the critical locations on the slab can be expressed as:

$$M = \alpha_i w L^2 \tag{16.6}$$

where

α_1 = bending moment coefficient at ith location
w = uniformly distributed load,
L = clear span between the fixed edges.
i = 1,2,3 and 4 are the locations and the directions as marked in Fig. 16.11.

The bending moments at the critical locations on the slab caused by a triangular load can be expressed as:

$$M = \beta_i w L^2 \tag{16.7}$$

where

β_i = bending moment coefficient at ith location
w = load intensity at the base of the triangular load
L = clear span between the fixed edges.
i = 1,2,3 and 4 are the locations and the directions as marked in Fig. 16.11

Fig. 16.11 indicates the notations and critical location for bending moments in slabs.
The values of a and b are function of h/L and are given in Table 16.2

16.5 CRITERIA OF DESIGN OF COUNTERFORTS

In a vertical slab, the bending moment is proportional to the height of the wall and square of the spacing of the counterforts. Therefore, the spacing of the counterforts be less than the height of the slab. The spacing may be taken equal to 0.3h to 0.8h. The soil pressure induces bending moment and axial force on the Counterfort. The forces acting on the heel have a tendency to separate the heel slab from the counterfort. Reinforcement is provided for bending and tension. The forces acting on the front counterfort are earth

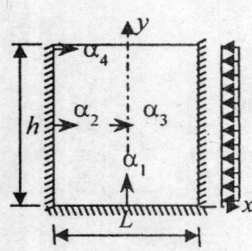

Fig. 16.11 Moment Coefficients.

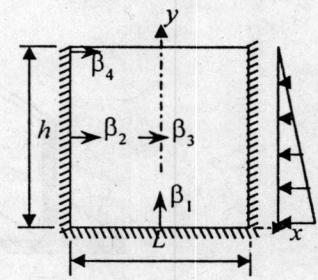

Fig. 16.12 Moment Coefficients.

Table 16.2 Bending Moment Coefficients in Plates with Fixed 3 Edges.

(a) Uniformly distributed load intensity

h/L	$y = 0$ $x = 0$	$y = h/2$ $x = \pm L/2$	$y = h/2$ $x = 0$	$y = h$ $x = \pm L/2$
	α_1	α_2	α_3	α_4
0.6	-0.055	-0.036	0.017	-0.074
0.7	-0.054	-0.044	0.021	-0.078
0.8	-0.053	-0.051	0.025	-0.081
0.9	-0.052	-0.056	0.029	-0.083
1.0	-0.051	-0.061	0.032	-0.083
1.25	-0.047	-0.071	0.037	-0.083
1.50	-0.042	-0.075	0.040	-0.083
2.0	-0.040	-0.083	0.041	-0.083

(b) Triangularly distributed load with maximum intensity at base

h/L	$y = 0$ $x = 0$	$y = h/2$ $x = \pm L/2$	$y = h/2$ $x = 0$	$y = h$ $x = \pm L/2$
	β_1	β_2	β_3	β_4
0.6	-0.024	-0.013	0.006	0.0
0.7	-0.026	-0.017	0.008	0.0
0.8	-0.02	-0.021	0.010	0.0
0.9	-0.029	-0.024	0.012	0.0
1.0	-0.030	-0.027	0.013	0.0
1.25	-0.031	-0.033	0.017	0.0
1.50	-0.029	-0.034	0.019	0.0
2.0	-0.029	-0.040	0.021	0.0

pressure through the wall and the soil reaction from the toe slab. Both the front and rear counterforts can be designed as beams with variable depth. The width of the base and that of the heel can be similar to that of the cantilever wall.

Example 16.2: Design of Counterfort Retaining Wall

Design a retaining wall to retains a filling of 7.0 m above the ground level. The unit weight and the net allowable bearing capacity of the sandy soil are 18 kN/m³ and 150 kN/m² respectively. The angle of internal friction of the soil and coefficient of friction are 30 and 0.65 respectively. The exposure condition is moderate.

Solution

The exposure condition is moderate so select M25 grade concrete with HYSD-Fe415 reinforcement. A counterfort wall is chosen, as the height is 7m. Both rear and front counterforts are provided. The design consists of width of footing for stability and bearing pressure. The soil is sandy. The footing, wall and the counterforts are designed for structural safety.

Design of the footing

Minimum depth of foundation is = 1 m.
Overall height = 7.0 + 1 = 8.0 m.
Let the width of the footing = $B = 0.5H = 4$ m.
Provide a heel length = $L_h = 2$ m.

The weights of various elements of the retaining wall are needed for the purpose of calculation of the pressure on the soil. So approximate average thickness of various elements are assumed.

Vertical wall (about $H/40$) = 0.2 m;
Base slab (about $H/30$) = 0.3 m;
Clear spacing of counterforts = $L = 3$ m;
Thickness of counterfort $(L/10) = 0.3$ m
Let the height of the front counterfort = 3 m;
Centre to centre spacing of counterforts = 3.3 m.

The first trial dimensions are indicated in Fig. 16.13. These may be modified later if needed. The reaction of the soil is obtained from the equilibrium criterion. The forces and the moments on the retaining wall for one counterfort spacing are computed in Table 16.3 in trial 1: with footing width = 4 m. The stability of the retaining wall is also computed from the same table. The equilibrium of the forces obtained from the sub-totals of columns (4) and (5) of Table 16.3:

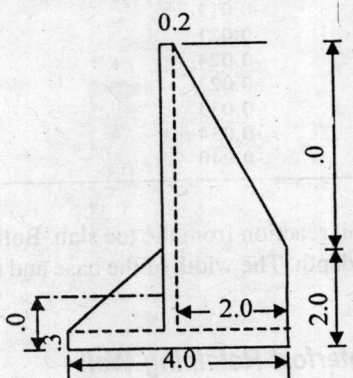

Fig. 16.13 Initial Dimensions; Example 16.2.

$$4*3.3(p_1 + p_2)/2 = 1135 \text{ kN}$$

$$4*3.3*(4/6)(p_1 + 2p_2) = 3114 - 1690 \text{ kNm}$$

The above equations reduce to:

$$p_1 + p_2 = 172; \text{ and } p_1 + 2p_2 = 162$$

$$p_2 = -10 \text{ kN/m}^2; \text{ and } p_1 = 182 \text{ kN/m}^2$$

The value of p_2 works out to be negative indicating tension on the soil at the tip of the heel. The soil cannot resist tension, so separation takes place between the soil and the tip of the heel thus resulting into only a partial contact. The width of the footing is revised to 4.2 m and the heel is kept at 2 m. The width of the toe is 4.2 –2.0 – 0.2 = 2.0 m. Table 16.3 gives the computation under trial 2. Therefore, assume a pressure distribution as:

$$4.2*3.3(p_1 + p_2)/2 = 1140 \text{ kN}$$

$$4.2*3.3*(4/6)(p_1 + 2p_2) = 3340 - 1690 \text{ kNm}$$

The above equations reduce to:

$$p_1 + p_2 = 164; \text{ and } p_1 + 2p_2 = 179$$

$$p_1 = 149 \text{ kN/m}^2; \text{ and } p_2 = 15 \text{ kN/m}^2$$

The bearing pressure is with in the limits.

Table 16.3 Forces Acting on the Retaining Wall for One Counterfort Spacing.

No.	Element	Force(kN) Trial 1: Footing width = 4 m		x_i	M	Force(kN) Trial 2: Footing = 4.2 m	x_i	M
1	Footing	0.3(4.0)(3.3)(25)	= 99	2.0	198	104	2.1	218
2	Wall	0.2(7.7)(3.3)(25)	= 127	1.9	241	127	2.1	267
3	Rear CF	0.3(7.7)(2.0)(25)/2	= 58	2.67	155	58	2.87	166
4	Front CF	0.3(3)(1.8)(25)/2	= 20	1.33	27	20	1.53	30
5	Rear Soil	(7.7)(2)(3)(18)	= 831	3.0	2493	831	3.2	2659
6.	Sub totals		1135		3114	1140		3340
7	Active pressure 18(8)(8)(3.3)/6		= 634	-8/3	-1690	634	-8/3	-1690

x_i = distance of the tip of the toe from CG of the member, M = Bending moment about the tip of the toe.

Check for the stability of the retaining wall. The factor of safety against overturning and sliding can be computed by using the forces listed in Table 16.3.

Factor of safety against overturning is: = 3340/1690 = 2
Factor of safety against sliding is = 0.65(1140)/634 = 1.17

As the factor of safety against sliding is only 1.17, which is less than 1.5, therefore a key has to be provided.

Design of Toe

Fig. 16.14 illustrates the soil reaction on the footing with the pressures. The soil pressures at the bases of

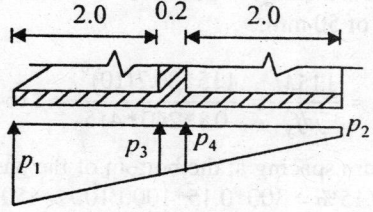

Fig. 16.14 Pressure Distribution.

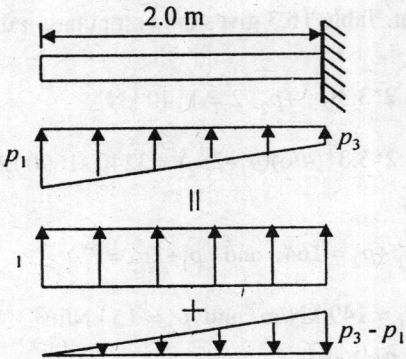

Fig. 16.15 Toe slab with Soil Pressure.

the toe and the heel are computed and the elements are designed. The earth pressure under the toe slab is shown in Fig. 16.15. It is divided into combination of uniformly and triangularly distributed loads. The triangular distribution is selected in such a way that the base of the triangle is at the base of the toe slab. The equivalent distributions along with the net distribution of the pressure under the toe are given in Fig. 16.15. The soil pressure at the base of the toe is:

$$p_3 = p_1 - (p_1 - p_2)(2/4.2) = 85 \text{ kN/m}^2.$$

The weight of the toe slab is taken equal to that of 300 mm thick slab.
The net uniform pressure = 149 − 7.5 = 141.5 kN/m²
The intensity of the triangular load at base is = 149 − 85 = 64 kN/m².

$$\text{Aspect ratio of the toe slab} = r = 2.0/3 = 0.67$$

The bending moment coefficients for the aspect ratio interpolated from table 16.2 are listed here.

$\alpha_1 = -0.054;$ $\alpha_2 = -0.042;$ $\alpha_3 = 0.02;$ $\alpha_4 = -0.077$
$\beta_1 = -0.026;$ $\beta_2 = -0.06;$ $\beta_3 = 0.008;$ $\beta_4 = 0$

The factored absolute maximum bending moment occurs at the middle of the base as seen from the coefficients.

$$M = (1.5)\left[(0.054)(141.5) - (0.026)64\right]3^2 = 80.7 \text{ kNm/m}$$

The effective depth required is obtained by equating the factored *BM* to the moment capacity expression. It works out to be 151 mm. Select the overall thickness of the toe as 300 mm and the effective depth works out to be 240 mm for a cover of 50 mm.

$$\text{The area of reinforcement is: } A_{st} = \frac{1.15M}{jdf_y} = \frac{1.15*80.7(10)^6}{0.8*240*415} = 1164 \text{ mm}^2/\text{m}$$

Provide 16 mm bars at 175 mm spacing at the bottom of the base of the toe slab.
Minimum reinforcement = 0.15% = 300*0.15*1000/100 = 450 mm²/m
The factored negative bending moment at middle distance at the counterfort support line is:

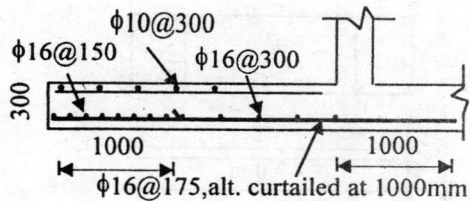

Fig. 16.16 Toe Section Near Counterfort.

$$M = 1.5[0.042*141.5 - 0.06*64]9 = 28.4 \text{ kNm/m}$$

The factored negative bending moment at the tip of the toe at the counterfort support is:

$$M = 1.5[0.077*141.5 - 0.0*64]9 = 147 \text{ kNm/m}$$

This is the bending moment at the free edge but it is at middle of the length (at 1 m from the free edge) is only 28.4 kNm/m. It is zero at the base. That is 147 kNm/m at free edge, 28.4 at the middle and zero at the base. Take an average of the two first values to compute the area of tension reinforcement. However, provide closer spacing of the reinforcement at the free edge. Use $M = (147 + 28.4)/2 = 87.7$ kNm/m

The area of reinforcement is: $A_{st} = \dfrac{1.15M}{jdf_y} = \dfrac{1.15*87.7(10)^6}{0.8*240*415} = 1266 \text{ mm}^2/\text{m}$

Provide 16 mm bars at 150 spacing at the free edge for one metre length and at 300 mm spacing in the balance one metre. Fig. 16.16 illustrates the reinforcement details in the toe slab at the counterfort support section. The positive bending moment at mid span taken from the moment coefficients is about 30 kNm/m. The reinforcement required is only nominal. 16 mm bars at 300 mm are provided at the top. A distribution reinforcement of 10 mm at 300 mm is provided.

Design of Heel Slab

The earth pressure under the heel slab is shown in Fig. 16.17. It is divided into combination of uniformly and triangularly distributed loads. The triangular distribution is selected in such a way that the base of the triangle is at the base of the heel slab. The equivalent distributions along with the net distribution of the pressure under the heel are given in Fig. 16.17. The soil pressure at the base of the heel is:

$$p_4 = p_1 - (p_1 - p_2)(2.2/4.2) = 78 \text{ kN/m}^2.$$

The average weight of the heel slab is taken that of 300 mm slab.
Weight of the soil above heel = $w_s = g_s h = 18*7 = 126$ kN/m^2.
The net uniform pressure = $w_s + w_g - p_2 = 118.5$ kN/m^2
The intensity of the triangular load at base is = $p_4 - p_2 = 63$ kN/m^2.

Aspect ratio of the toe slab = $r = 2.0/3 = 0.67$

The bending moment coefficients for the aspect ratio interpolated from table 16.2 are listed here.

$\alpha_1 = -0.054;$ $\alpha_2 = -0.042;$ $\alpha_3 = 0.02;$ $\alpha_4 = -0.077$
$\beta_1 = -0.026;$ $\beta_2 = -0.06;$ $\beta_3 = 0.008;$ $\beta_4 = 0$

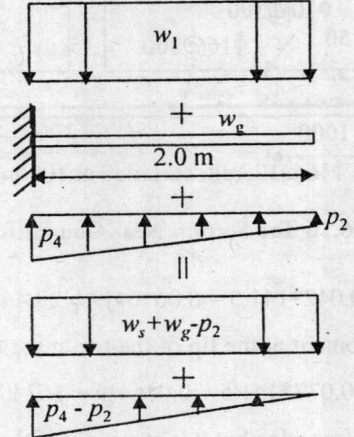

Fig. 16.17 Heel with Soil Pressure.

The factored absolute maximum bending moment occurs at the middle of the base as seen from the coefficients.

$$M = (1.5)\big[(0.054)(118.5) - (0.026)63\big]3^2 = 64.3 \text{ kNm/m}$$

The effective depth required is obtained by equating the factored *BM* to the moment capacity expression. It works out to be 138 mm. Select the overall thickness of the toe as 300 mm and the effective depth works out to be 240 mm for a cover of 50 mm.

The area of reinforcement is:

$$A_{st} = \frac{1.15M}{jdf_y} = \frac{1.15*64.3(10)^6}{0.8*240*415} = 929 \text{ mm}^2/\text{m}$$

Provide 16 mm bars at 210 mm spacing at the top of the base of the heel slab.

The factored negative bending moment at middle distance at the counterfort support line is:

$$M = 1.5[0.042*118 - 0.06*63]9 = 15.9 \text{ kNm/m}$$

The factored negative bending moment at the tip of the toe at the counterfort support is:

$$M = 1.5[0.077*118 - 0.0*63]9 = 122.7 \text{ kNm/m}$$

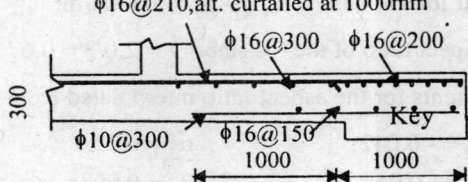

Fig. 16.18 Heel Section Near Counterfort.

Bending moment of 122.7 kNm/m is at free edge and it is 15.9 at the middle and zero at the base. Take an average of the two first values to compute the area of tension reinforcement. However provide closer spacing of the reinforcement at the free edge. Use $M = (118 + 15.9)/2 = 67$ kNm/m

The area of reinforcement is:

$$A_{st} = \frac{1.15M}{jdf_y} = \frac{1.15*67(10)^6}{0.8*240*415} = 967 \text{ mm}^2/\text{m}$$

Provide 16 mm bars at 200 spacing at the free edge for one metre length and at 300 mm spacing in the balance one metre. Fig. 16.18 illustrates the reinforcement details in the heel slab at the counterfort support section. The positive bending moment at mid span taken from the moment coefficients is small. The reinforcement required is only nominal. 16 mm bars at 300 mm are provided at the top. A distribution reinforcement of 10 mm at 300 mm is provided.

Design of Vertical Slab: Stem

The counterforts supports the vertical wall with clear span of 3 m and cantilever height = $8 - 0.3 = 7.7$ m.

$$\text{Aspect ratio of the toe slab} = r = 7.7/3 = 2.6$$

The bending moment coefficients for the aspect ratio taken from table 16.2 are:

$$\beta_1 = -0.029; \qquad \beta_2 = -0.04; \qquad \beta_3 = 0.021; \qquad \beta_4 = 0$$

The active earth pressure at the base of the wall is:

$$p = 0.33*18*7.7^2/2 = = 177.8 \text{ kN/m}^2.$$

The factored absolute maximum bending moment occurs at the counterfort support at middle height, and it is:

$$M = (1.5)\left[(0.054)(177.8)\right]3^2 = 96 \text{ kNm}/\text{m}$$

The effective depth required is obtained by equating the factored BM to the moment capacity expression. It works out to be 167 mm. Select the overall thickness as 300 mm and the effective depth works out to be 240 mm for a cover of 50 mm. The thickness of the stem at top can be reduced gradually to 200 mm from 300 mm at the height of 4.3 m from the base.

The area of reinforcement is: $A_{st} = \dfrac{1.15M}{jdf_y} = \dfrac{1.15*96(10)^6}{0.8*240*415} = 1388 \text{ mm}^2/\text{m}$

Provide 16 mm bars at 145 mm spacing at the middle height at the support on earth pressure side. Bars extend 1000 mm on either side of the counterfort.

The factored positive bending moment at middle height is:

$$M = 1.5[0.021*177.8]9 = 50.4 \text{ kNm/m}$$

The area of reinforcement is: $A_{st} = \dfrac{1.15M}{jdf_y} = \dfrac{1.15*50.4(10)^6}{0.8*240*415} = 728 \text{ mm}^2/\text{m}$

Provide 16 mm bars at 275 spacing, alternate bars can be curtailed at 750 mm from the supports. The factored negative bending moment at the base is:

$$M = 1.5[0.029*177.8]9 = 69.6 \text{ kNm/m}$$

The area of reinforcement is: $A_{st} = \dfrac{1.15M}{jdf_y} = \dfrac{1.15*69.6(10)^6}{0.8*240*415} = 1006 \text{ mm}^2/\text{m}$

Provide 16 mm bars at 200 spacing and curtail at 1500 mm from the base. Fig. 16.19 illustrates the reinforcement details in the stem.

Design of Counterfort

Height of the front counterfort = 7.7 m.

The counterfort acts as a beam supporting the stem and the footing. Fig. 16.20 illustrates the counterfort. The force acting on the counterfort is the earth pressure.

The factored bending moment on one counterfort is:

$$M = \frac{1.5c_a\gamma_s Lh^3}{6} = \frac{1.5*0.33*18*3.3*7.7^3}{6} = 2238 \text{ kNm}$$

The compression face of the beam is not uptight but at an inclination of 1.7 in 2. That is 31.7 degrees with the vertical. The moment capacity of such a section is:

$$M_r = Kbd^2 f_{ck} \cos\theta = 0.138(300)(0.85)(25)d^2 = 880d^2 \text{ Nmm}$$

The effective depth required is: $d = \sqrt{\dfrac{2238000000}{880}} = 1594 \text{ mm}$

The actual overall depth provided is 4200 mm. The effective depth can be taken as: $d = 4200 - 150 = 4050$ mm.

The area of tension reinforcement in the counterfort is:

$$A_{st} = \frac{1.15*M}{jd\cos\theta*f_y} = \frac{1.15*2238(10)^6}{0.8*4050*415} = 1914 \text{ mm}^2$$

Provide seven numbers of 20 mm bars. Actually provided is = 2198 mm². The percentage of reinforcement is = 0.18%

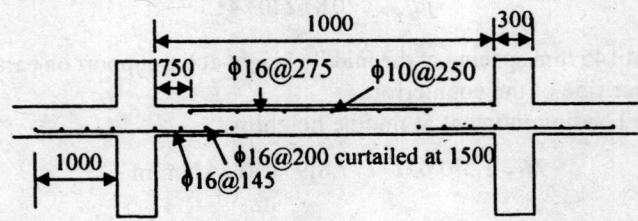

Fig. 16.19 Reinforcement in Stem at Base (Plan).

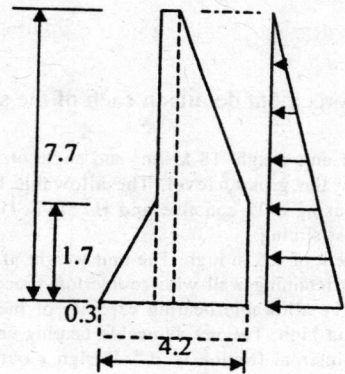

Fig. 16.20 Counterfort; Example 16.2.

The shear force is critical at a depth of $7.7 - 4.05 = 3.65$ m. and it is:

$$V = \frac{1.5 c_a \gamma_s L h^2}{2} = \frac{1.5 * 0.33 * 18 * 3.3 * 3.65^2}{2} = 196 \text{ kN}$$

The effective depth of the counterfort at this section is:

$$d = 0.3 + (2/6)3.65 = 1.51 \text{ m}$$

The nominal shear stress is:

$$\tau_v = \frac{196000}{300 * 1510} = 0.43 \text{ N/mm}^2$$

Provide 8 mm stirrups at 300 mm spacing.

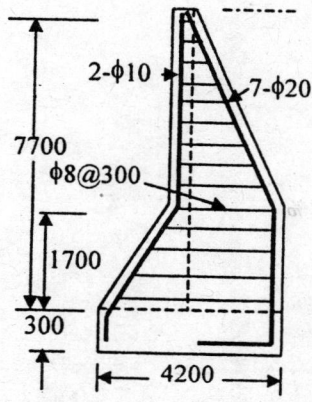

Fig. 16.21 Reinforcement in Counterfort; Example 16.2.

PROBLEMS

[Use IS 456 - 2000 and sketch reinforcement details in each of the solutions)

16.1 A retaining wall retains a 3 m soil of unit weight 18 kN/m^3 and angle of internal friction is 26^0. The footing of the retaining wall is one metre below the ground level. The allowable bearing capacity is 90 kN/m^2. Design a reinforced concrete retaining wall using $M25$ concrete and HYSD-Fe415 reinforcement. Factor of safety against overturning is 2 and it is 1.5 against sliding.

16.2 A retaining wall holds an embankment of 5.5 m high. The unit weight of the soil is 16 kN/m^3 and angle of internal friction is 20^0. Design a counterfort retaining wall with counterfort spaced at 2.5 m apart using $M25$ concrete and $HYSD$-Fe415 reinforcement bars. The allowable bearing capacity of the soil is 120 kN/m^2.

16.3 A wall holds an embankment of 5 m high. The net allowable bearing capacity is 100 kN/m^2. The unit weight of the soil is 16 kN/m^3 and angle of internal friction is 20^0. Design a buttressed reinforced concrete retaining wall using $M20$ concrete and $HYSD$-Fe415 reinforcement bars.

Index